T0310308

OFDM FOR UNDERWATER ACOUSTIC COMMUNICATIONS

OFDM FOR UNDERWATER ACOUSTIC COMMUNICATIONS

Shengli Zhou

University of Connecticut, USA

Zhaohui Wang

Michigan Technological University, USA

WILEY

This edition first published 2014
© 2014 John Wiley & Sons, Ltd

Registered office
John Wiley & Sons Ltd, The Atrium, Southern Gate, Chichester, West Sussex, PO19 8SQ, United Kingdom

For details of our global editorial offices, for customer services and for information about how to apply for permission to reuse the copyright material in this book please see our website at www.wiley.com.

The right of the author to be identified as the author of this work has been asserted in accordance with the Copyright, Designs and Patents Act 1988.

All rights reserved. No part of this publication may be reproduced, stored in a retrieval system, or transmitted, in any form or by any means, electronic, mechanical, photocopying, recording or otherwise, except as permitted by the UK Copyright, Designs and Patents Act 1988, without the prior permission of the publisher.

Wiley also publishes its books in a variety of electronic formats. Some content that appears in print may not be available in electronic books.

Designations used by companies to distinguish their products are often claimed as trademarks. All brand names and product names used in this book are trade names, service marks, trademarks or registered trademarks of their respective owners. The publisher is not associated with any product or vendor mentioned in this book.

Limit of Liability/Disclaimer of Warranty: While the publisher and author have used their best efforts in preparing this book, they make no representations or warranties with respect to the accuracy or completeness of the contents of this book and specifically disclaim any implied warranties of merchantability or fitness for a particular purpose. It is sold on the understanding that the publisher is not engaged in rendering professional services and neither the publisher nor the author shall be liable for damages arising herefrom. If professional advice or other expert assistance is required, the services of a competent professional should be sought.

Library of Congress Cataloging-in-Publication Data applied for.

ISBN: 9781118458860

Set in 10/12pt TimesLTStd by Laserwords Private Limited, Chennai, India
Printed and bound in Malaysia by Vivar Printing Sdn Bhd

1 2014

To Juanjuan, Daniel, Joyce, and my parents Heting and Caiyun
S. Z.

To my parents Yongcheng and Jiuqin
Z.-H. W.

Contents

Preface

Underwater acoustic (UWA) channels have been regarded significantly different from wireless radio channels, due to their unique characteristics, such as large temporal variations, abundance of transmission paths, and wideband property in nature. Although there are a plethora of digital and wireless communication textbooks, most of them are tailored towards wireless radio channels, where simplified channel models are usually adopted to streamline presentation. Following standard receiver designs in textbooks, a practitioner might often be frustrated by the receiver performance in real underwater acoustic environments. This book is written to unfold and to address the challenges in UWA communications particularly for the multicarrier modulation in the form of orthogonal frequency-division multiplexing (OFDM).

The last decade has witnessed the tremendous development and revolutionary impact of OFDM on high data-rate radio communications. It is the workhorse of many wireless communication standards, such as WiFi (IEEE 802.11 a/g/n), WiMAX (IEEE 802.16), digital audio and video broadcasting (DAB/DVB), and the fourth generation (4G) cellular systems. The popularity of OFDM stems from its capability to convert a long multipath channel in the time domain into multiple parallel single-tap channels in the frequency domain, thus considerably simplifying receiver design. Such a feature makes OFDM an attractive choice for UWA channels. However, the feasibility of underwater acoustic OFDM had not been validated with experimental data sets until the mid 2000s, although OFDM has been tested in UWA environments since the 1990s. Considerable progress for OFDM has been observed in the UWA community since the late 2000s.

This book is dedicated to the techniques for OFDM in UWA channels, and different chapters are focused on addressing different challenges. Readers are expected to have certain signal processing and communication background. For readers within the UWA community, this book could deepen their understanding in the design aspects specific to underwater systems. For readers outside the UWA community, this book will help them to appreciate the distinctions of system design in different domains.

The technical content of this book mainly originates from the research performed within the UnderWater Sensor Network (UWSN) lab at the University of Connecticut (UCONN), which is co-directed by Dr. Jun-Hong Cui and the first author Dr. Shengli Zhou. The past and existing members who have contributed to the content of the book include: postdoctoral researchers: Drs. Jie Huang, Hao Zhou, and Xiaoka Xu; past Ph.D. students: Drs. Baosheng Li, Christian Berger, Jianzhong Huang; current Ph.D. students: Patrick Carroll, Lei Wan, Yi Huang; past M.S. students: Sean Mason, Weian Chen, Wei Zhou; and visiting scholars: Yougan Chen,

Haixin Sun, Yuzhi Zhang, Xiaomei Xu. The authors have benefited tremendously from collaborations with faculty members affiliated with UWSN, in particular, Drs. Peter Willett, Jun-Hong Cui, Zhijie Shi, James O'Donnell, and Thomas Torgersen. The sincere gratitude of the authors also goes to the colleagues in the Systems Group at UCONN, especially Drs. Yaakov Bar-Shalom, Peter Luh, Krishna Pattipati, and Peter Willett, for promoting an atmosphere for academic excellence.

The authors would like to thank Mr. Lee Freitag, Dr. James Preisig, and their teams from the Woods Hole Oceanographic Institute (WHOI), and Dr. Josko Catipovic and his team from the Navy Undersea Warfare Center (NUWC) for providing multiple experimental opportunities. The data sets from those experiments, especially from the SPACE08 experiment, the MACE10 experiment and the AUTEC network, are instrumental to our receiver development and validation. The experimental opportunities offered by Dr. T. C. Yang have also been very helpful for our research development. We would also like to acknowledge Dr. Milica Stojanovic for stimulating discussions at the early stage of research and Dr. Zhengdao Wang for his valuable comments through regular discussions.

The feedback from the reviewers have helped to improve the presentation of this book. We acknowledge Drs. Christian Berger, Tolga Duman, Dennis Goeckel, Georgios Giannakis, Geert Leus, Aijun Song, Milica Stojanovic, Zhengdao Wang, Peter Willett, Chengshan Xiao, and Ms. Xiaoyi Hu for reviewing different chapters with a short notice. Mr. Mark Hammond, Ms. Liz Wingett, and Ms.Sandra Grayson from the publisher have been very patient and supportive during this project.

The work in this book has been supported by the Office of Naval Research (ONR) and National Science Foundation (NSF). We would like to thank the program managers: Dr. Robert Headrick from ONR who has managed the YIP and PECASE projects, and Dr. Scott Midkiff, Dr. David Du, Dr. Zygmunt Haas, and Dr. Zhi Tian from different programs at NSF. Dr. Keith Davidson from ONR has provided a lot of encouragement during annual ONR PI meetings. The University of Connecticut has provided matching funds to our NSF projects at various occasions. The first author acknowledges the support of the United Technologies Corporation (UTC) Associate Professorship in Engineering Innovation (2008–2011), and the Charles H. Knapp Associate Professorship in Electrical Engineering (2012–2013).

The training from our advisors has laid foundation for the authors to pursue this project. Dr. Shengli Zhou would like to thank his Ph.D. advisor Dr. Georgios B. Giannakis and his MSc. advisor Dr. Jinkang Zhu, and Dr. Zhaohui Wang would like to thank her MSc. advisor Dr. Huizhi Cai, for their mentoring during the graduate studies.

Last but not least, we are grateful to our family members for their continuous support and encouragement throughout the project.

Shengli Zhou
University of Connecticut

Zhaohui Wang
Michigan Technological University

Acronyms

AF	Amplify and Forward
ANC	Analogy Network Coding
AoA	Angle of Arrival
ARQ	Automatic Repeat Request
AUTEC	Atlantic Undersea Test and Evaluation Center
AUV	Autonomous Underwater Vehicle
BCJR	The Bahl-Cocke-Jelinek-Raviv Algorithm
BICM	Bit Interleaved Coded Modulation
BP	Basis Pursuit
BPSK	Binary Phase-Shift Keying
BER	Bit Error Rate
BLER	Block Error Rate
CC	Convolutional Code
CCDF	Complementary Cumulative Distribution Function
CCI	Cochannel Interference
CDF	Cumulative Distribution Function
CDMA	Coded-Division Multiple Access
CF	Compress and Forward
CFO	Carrier Frequency Offset
CP	Cyclic Prefix
CRLB	Cramer-Rao Lower Bound
CS	Compressive Sensing
CSI	Channel State Information
CZT	Chirp Z-Transform
DBC	Dynamic Block-Cycling
DCC	Dynamic Coded Cooperation
DF	Decode and Forward
DFE	Decision-Feedback Equalization
DFT	Discrete Fourier Transform
DSSS	Direct Sequence Spread Spectrum
FDM	Frequency Division Multiplexing
FFT	Fast Fourier Transform
FH	Frequency Hopping
FG	Factor Graph

FSK	Frequency Shift Keying
GIB	GPS Intelligent Buoy
GLRT	Generalized Log-Likelihood Test
GMP	Gaussian Message Passing
GPS	Globe Positioning System
HFM	Hyperbolic-Frequency Modulation
IBI	Interblock Interference
ICI	Intercarrier Interference
i.i.d.	Independent and Identically Distributed
IMM	Interacting Multiple Model
ISI	Intersymbol Interference
LASSO	Least Absolute Shrinkage and Selection Operator
LBL	Long Baseline
LDPC	Low Density Parity Check Code
LFM	Linear-Frequency Modulation
LLR	Log-Likelihood Ratio
LLRV	Log-Likelihood Ratio Vector
LMMSE	Linear Minimum Mean-Square Error
LPF	Low Bandpass Filtering
LPM	Linear-Period Modulation
LS	Least Squares
MAC	Medium-Access Control
MACE10	Mobile Acoustic Communication Experiment in 2010
MAP	Maximum *A Posteriori* Probability
MCMC	Markov Chain Monte Carlo
MIMO	Multi-Input Multi-Output
ML	Maximum Likelihood
MP	Matching Pursuit
MSE	Mean Square Error
MMSE	Minimum Mean Square Error
MRC	Maximum Ratio Combining
MUD	Multiuser Detection
MUI	Multiuser Interference
NC	Network Coding
NCM	Nonbinary Coded Modulation
NLNC	Network-Layer Network Coding
OFDM	Orthogonal Frequency-Division Multiplexing
OMP	Orthogonal Matching Pursuit
PAM	Pulse Amplitude Modulation
PAPR	Peak-To-Average-Power Ratio
PDA	Probabilistic Data Association
PER	Packet Error Rate
PLNC	Physical-Layer Network Coding
PSNR	Pilot Signal-To-Noise Ratio
QAM	Quadrature Amplitude Modulation
QC	Quasi-Cyclic

QMF	Quantize, Map and Forward
QPSK	Quadrature Phase Shift Keying
RIP	Restricted Isometry Property
RMSE	Root Mean-Squared Error
S2C	Sweep-Spread Carrier
SBL	Short Baseline
SDA	Sphere Decoding Algorithm
SIMO	Single-Input Multi-Output
SINR	Signal-to-Interference-and-Noise Ratio
SIR	Signal-to-Interference Ratio
SISO	Single-Input Single-Output
SNR	Signal-to-Noise Ratio
SOFAR	Sound Fixing and Ranging
SONAR	Sound Navigation and Ranging
SPA	Sum Product Algorithm
SPACE08	Surface Processes and Acoustic Communication Experiment in 2008
SPRT	Sequential Probability Ratio Test
SUD	Single-User Detection
TCM	Trellis Coded Modulation
TDOA	Time Difference of Arrival
TVR	Transmitter Voltage Response
USBL	Ultra-Short Baseline
UUV	Unmanned Underwater Vehicle
UWA	Underwater Acoustic
VA	Viterbi Algorithm
ZF	Zero Forcing
ZP	Zero Padding

Notation

Scalars

K	Number of subcarriers in one OFDM symbol
B	Frequency bandwidth of one OFDM symbol
Δf	Subcarrier spacing in one OFDM symbol, $:= B/K$
T	Time-duration of one OFDM symbol, $:= 1/\Delta f$
T_{g}	Time-duration of guard interval for one OFDM symbol
T_{bl}	Time-duration of one OFDM block, $:= T + T_{\mathrm{g}}$
f_{c}	Center frequency of communication system
f_k	Frequency of the kth subcarrier, $:= f_{\mathrm{c}} + k/T$
$\mathcal{S}_{\mathrm{N}}$	The set of null subcarriers in one OFDM symbol
$\mathcal{S}_{\mathrm{P}}$	The set of pilot subcarriers in one OFDM symbol
$\mathcal{S}_{\mathrm{D}}$	The set of data subcarriers in one OFDM symbol
$\mathcal{S}_{\mathrm{A}}$	The set of active subcarriers in one OFDM symbol $:= \mathcal{S}_{\mathrm{P}} \bigcup \mathcal{S}_{\mathrm{D}}$
$h(t;\tau)$	Time-varying channel impulse response
$A_p(t)$	Time-varying amplitude of the pth path
A_p	Time-invariant amplitude of the pth path
$\tau_p(t)$	Time-varying delay of the pth path
τ_p	Initial delay of the pth path
a_p	Doppler rate of the pth path
N_{pa}	Number of paths in the channel
a	The main Doppler scaling factor in the UWA channel
ϵ	The residual Doppler shift after removing the main Doppler effect
ξ_p	The equivalent amplitude of the pth path in the baseband
$\overline{\tau}_p$	The equivalent scaled delay of the pth path in the baseband
b_p	The equivalent residual Doppler rate of the pth path in the baseband
D	ICI depth
$\mathcal{N}(\mu,\sigma^2)$	Real Gaussian distribution with mean μ and variance σ^2
$\mathcal{CN}(0,\sigma^2)$	Circularly symmetric complex Gaussian distribution with zero mean and variance σ^2
$\tilde{x}(t)$	The waveform in passband
$x(t)$	The waveform in baseband; Conversion between $\tilde{x}(t)$ and $x(t)$:

$$\tilde{x}(t) = 2\Re\{x(t)e^{j2\pi f_c t}\}$$
$$x(t) = \text{LPF}[\tilde{x}(t)e^{-j2\pi f_c t}]$$

Vectors and Matrices

$\mathbf{z}$	Measurement vector formed by frequency samples at all the OFDM subcarriers
$\mathbf{s}$	Transmitted symbol vector formed by symbols at all the OFDM subcarriers
$\mathbf{w}$	Ambient noise vector formed by the ambient noise at all the OFDM subcarriers
η	Equivalent noise vector formed by the equivalent noise at all the OFDM subcarriers
$\mathbf{H}$	Channel mixing matrix
$\mathcal{CN}(0, \mathbf{\Sigma})$	Circularly symmetric complex Gaussian random vector with zero mean and covariance matrix $\mathbf{\Sigma}$

Operations

$\propto$	Equality of functions up to a scaling factor
$\|S\|$	Cardinality of set S
$[\mathbf{a}]_m$	The mth entry of vector $\mathbf{a}$
$[\mathbf{A}]_{m,k}$	The (m, k)th entry of matrix $\mathbf{A}$
$\{\mathbf{a}\}_{\ell=i}^{j}$	A set formed by elements $\{[\mathbf{a}]_i, \quad [\mathbf{a}]_{i+1}, \cdots, [\mathbf{a}]_j\}$
$\hat{a}$	The estimate of scale a
$\hat{\mathbf{a}}$	The estimate of vector $\mathbf{a}$
$\hat{\mathbf{A}}$	The estimate of matrix $\mathbf{A}$
$\mathbf{A}^{\mathrm{T}}$	The transpose of matrix $\mathbf{A}$
$\mathbf{A}^{\mathrm{H}}$	The complex conjugate transpose of matrix $\mathbf{A}$
$\mathbf{A}^{\dagger}$	The pseudo-inverse of matrix $\mathbf{A}$
$\mathrm{tr}(\mathbf{A})$	Trace of matrix $\mathbf{A}$
$\Pr\{A\}$	Probability of an event A
$\mathbb{E}(X)$	Expectation of random variable X
$\mathbb{E}(\mathbf{x})$	Expectation of random vector $\mathbf{x}$
$\mathrm{Cov}(X, Y)$	Covariance of two random variables
$\mathrm{Cov}(\mathbf{x}, \mathbf{y})$	Covariance matrix of two random vectors
$\Re\{x\}$	Real part of a complex number x
$\Im\{x\}$	Imaginary part of a complex number x

1

Introduction

1.1 Background and Context

1.1.1 Early Exploration of Underwater Acoustics

The Earth is a water planet, with two-thirds of the surface covered by water. Exploration of the mysterious underwater world has never ceased in human history. As early as 400 BC, Aristotle had noted that sound could be heard in water as well as in air. In AD 1490, Leonardo da Vinci wrote: "If you cause your ship to stop and place the head of a long tube in the water and place the other extremity to your ear, you will hear ships at great distances" [268]. In 1826, Charles Sturm and Daniel Colladon made the first accurate measurement of sound speed in water at Lake Geneva, Switzerland. The first practical application of underwater sound appeared in the 1900s: the underwater bells equipped on lightships were simultaneously sounded with a fog horn to measure the offshore distance based on the difference of the airborne and waterborne arrivals, and meanwhile the stereo headphones were also used for directions [397]. With the sinking of Titanic in 1912, L. F. Richardson successively filed a patent of echo ranging with sound in air and a patent application of echo ranging in water.

Along with the application of submarine and underwater mines in World War I (1914–1918), considerable progress has been made in underwater acoustics, especially on the underwater echo ranging for submarine and mine detection. In 1914, Constantin and Chilowski conceived the idea of submarine detection by underwater echo ranging. Based on the discovery of the piezoelectric effect by Jacques Curie and Pierre Curie in 1880, Paul Langevin in 1918 used quartz (piezoelectric) transducers as source and receiver to extend one-way sound transmission to 8 km, and for the first time observed clear echoes from a submarine at distances as large as 1500 m. Between World War I and World War II, scientists started to understand some fundamental concepts of sound in water, such as sound refraction due to changes of water temperature, salinity and pressure. Development of underwater sound applications during this period can be found in echo ranging for commercial use, underwater tomography and fisheries acoustics. The research effort on underwater acoustics during World War II (1941–1945) was mainly focused on improving echo ranging systems which were later coined as "sonar" (for SOund Navigation And Ranging). During this period, topics relative to sonar system performance were extensively investigated, including the high-frequency acoustics, low-frequency sound propagation, ambient noise, etc. By the end of World War II, the underwater sound had

OFDM for Underwater Acoustic Communications, First Edition. Shengli Zhou and Zhaohui Wang.
© 2014 John Wiley & Sons, Ltd. Published 2014 by John Wiley & Sons, Ltd.

been primarily used for navigation and threat-finding. In 1945, an underwater telephone, which was developed by the Navy Underwater Sound Laboratory in the United States for the purpose of communication with submerged submarines, was the first application of underwater sound for communications [321]. Since then, development on underwater acoustic communications has been made in various underwater acoustic applications.

1.1.2 Underwater Communication Media

To establish communications among underwater assets and systems floating on the surface, four different communication media have been used.

- Cables. There have been many cabled observatories established over the years. Cables provide robust communication performance; however, the deployment and maintenance cost is very high. This motivates the use of wireless data transmission.
- Acoustic waves. For underwater wireless communication systems, acoustic waves are used as the primary carrier due to their relatively low absorption in underwater environments. However, acoustic waves have low propagation speed and a very limited frequency band.
- Electromagnetic (EM) waves. The use of EM waves in the radio frequency band has several advantages over acoustic waves, mainly faster velocity and high operating frequency (resulting in higher bandwidth). The key limitation of using EM waves for underwater communication is the high attenuation due to the conductive nature of seawater [255].
- Optical waves. Using optical waves for communication obviously has a big advantage in data rate. However, there are a couple of disadvantages for optical communication in water. Firstly, optical signals are rapidly absorbed in water. Secondly, optical scattering caused by suspended particles and plankton is significant. Thirdly, the high level of ambient light in the upper part of the water column is another adverse effect for using optical communication.

Apparently, each of the three physical waves as wireless information carrier has its own advantages and disadvantages. For a more intuitive comprehension, we summarize the major characteristics of acoustic, electromagnetic and optical carriers in Table 1.1. Acoustic waves propagate well in seawater and can reach a far distance. This justifies using acoustic waves for most underwater wireless communications.

Table 1.1 Comparison of acoustic, EM and optical waves in seawater environments

	Acoustic	Electromagnetic	Optical
Nominal speed (m/s)	~ 1500	$\sim 33\,333\,333$	$\sim 33\,333\,333$
Power loss	relatively small	large	$\propto$ turbidity
Bandwidth	$\sim$ kHz	$\sim$ MHz	$\sim$ 10–150 MHz
Frequency band	$\sim$ kHz	$\sim$ MHz	$\sim 10^{14}$–10^{15} Hz
Antenna size	$\sim$ 0.1 m	$\sim$ 0.5 m	$\sim$ 0.1 m
Effective range	$\sim$ km	$\sim$ 10 m	$\sim$ 10–100 m

Source: Liu 2008 [255], Table 2, p. 984. Reproduced with permission of Wiley.

1.1.3 Underwater Systems and Networks

Along with the tremendous scientific and technology advances in last several decades, a wide range of underwater exploration and applications have emerged. The scientific exploration spans across multiple disciplines, such as physical oceanography, marine biology, and deep sea archaeology (e.g., discovery of the wreck of the *Titanic*). Environmental applications involve studies in pollution monitoring, climate change, and global warming. Commercial applications of underwater technologies can be found in, e.g., offshore oil/gas field monitoring, fishery industries, and treasure discovery. Military applications of underwater technologies include tactical surveillance in coastal areas, harbors and ports etc.

In recent years, development of underwater vehicles of various sizes and capabilities, such as sea gliders and autonomous underwater vehicles (AUVs), has enabled underwater applications without human interaction. For example, sea gliders can be deployed in lakes or oceans to collect data samples of water over a large time period, and then send the data back to a control center for scientific studies. A fleet of underwater vehicles can form an underwater network, in which vehicles can collaborate to accomplish predetermined tasks. As more intelligent systems are deployed in underwater applications, the need of communications and networking keeps growing.

1.2 UWA Channel Characteristics

Given the complexity of underwater acoustic medium and the low propagation speed of sound in water, the underwater acoustic channel is commonly regarded as one of the most challenging channels for communication. Next we will look into several distinguishing characteristics of underwater acoustic channels. Comparisons between the underwater acoustic channel and the terrestrial radio channel are made along with the descriptions of underwater acoustic channel characteristics.

1.2.1 Sound Velocity

The extremely slow propagation speed of sound through seawater is an important factor that differentiates it from electromagnetic propagation. The speed of sound in water depends on the water properties of temperature, salinity and pressure; illustrative plots of the three parameters as functions of water depth are shown in Figure 1.1 [305, Chap. 9]. A typical speed of sound in water near the ocean surface is about 1520 m/s, which is more than 4 times faster than the speed of sound in air, but five orders of magnitude smaller than the speed of light. The speed of sound in water grows with increasing water temperature, increasing salinity and increasing depth. Approximately, the sound speed increases 4.0 m/s for water temperature rising 1°C. When salinity increases one practical salinity unit (PSU), the sound speed in water increases to 1.4 m/s. As the depth of water (therefore also the pressure) increases to 1 km, the sound speed increases roughly to 17 m/s. It is noteworthy to point out the above assessments are only for rough quantitative or qualitative discussions, and the variations in sound speed for a given property are not linear in general.

A typical sound speed profile as a function of depth in deep water, is shown in Figure 1.2 [305, Chap. 9]. Depending on the depth, the profile can be divided into four layers.

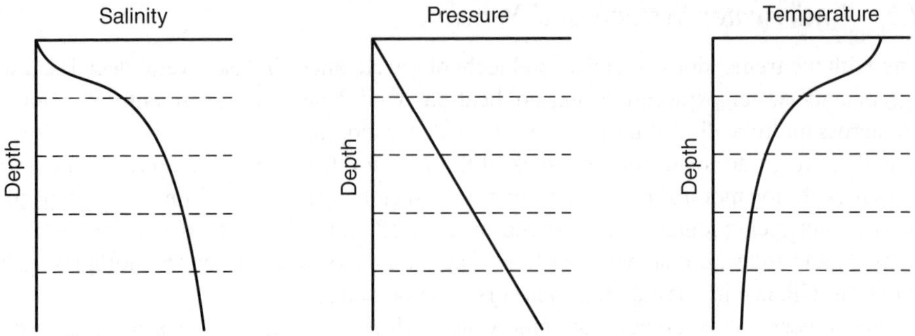

Figure 1.1 Variations of environmental parameters as functions of depth.

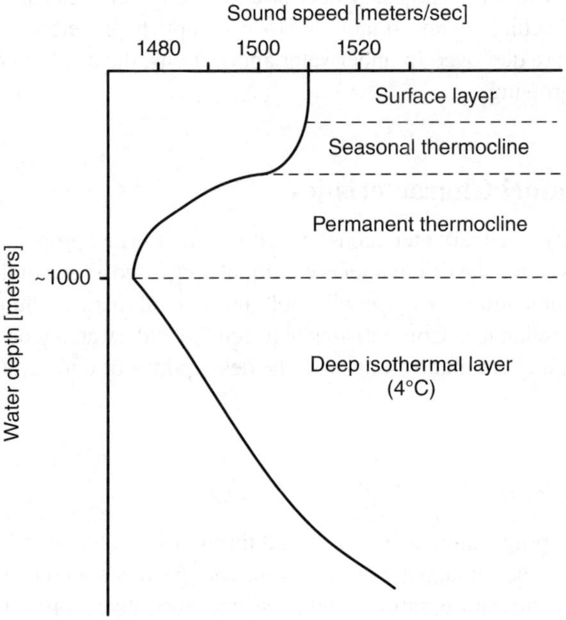

Figure 1.2 Typical sound speed profile in deep water.

- *Surface layer.* The surface layer usually has a water depth of a few tens of meters. Due to the mixing effect of wind, both temperature and salinity in this layer tend to be homogeneous, which leads to a constant sound velocity. This layer is also called a *mixed layer*.
- *Seasonal and permanent thermocline layers.* In the *thermocline* layers, the water temperature decreases as the water depth grows; as illustrated in Figure 1.1. In these two layers, the effect of increases in pressure and salinity cannot compensate the effect of temperature decrease. Therefore, there is a negative gradient of the sound speed profile in depth. In the seasonal thermocline layer, the negative gradient varies with seasons, while it is less seasonal in the permanent thermocline layer.

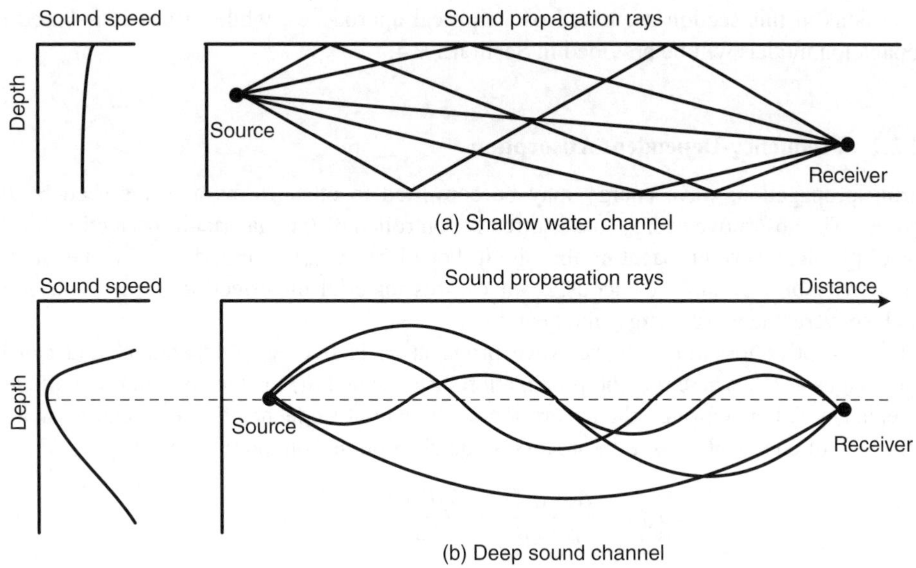

Figure 1.3 Multiple propagation paths.

- *Deep isothermal layer.* The water temperature is nearly constant around 4°C. The sound speed is therefore mainly determined by the water pressure, which leads to a positive gradient of sound speed in depth.

According to Snell's law, a ray of sound bends toward the direction of low propagation speed. In shallow water, the sound speed is usually constant throughout the water column. The acoustic signal usually propagates along straight lines, as illustrated in Figure 1.3(a). The sound speed profile of deep water channels diversifies the sound propagation paths. In particular notice that there is a minimal sound speed at a particular water depth (named the *channel axis*) between the permanent thermocline layer and the deep isothermal layer. For an acoustic signal transmitted at the channel axis, a ray of sound will be bent downward when propagating to the permanent thermocline layer and bent upward when propagating to the isothermal layer, thus being trapped within the two layers without interacting with the sea surface and bottom, as illustrated in Figure 1.3(b). This type of channel is called the *deep sound channel*, and the corresponding propagation is called SOFAR (for SOund Fixing And Ranging). An interesting phenomenon of SOFAR propagation is that a path traveling a longer distance could have a shorter travel time. Due to the refraction caused by inhomogeneous sound speed, there exist both shadow zones and convergence zones in the acoustic field, where a *shadow zone* denotes an area which cannot be penetrated by direct sound paths, and a *convergence zone* denotes an area which is insonified intensively by a bundle of sound paths.

1.2.2 Propagation Loss

There are three primary mechanisms of energy loss during the propagation of acoustic waves in water: (i) absorptive loss, (ii) geometric spreading, and (iii) scattering loss. Note that the discussions in this section are based on empirical approaches, while discussions based on propagation models will be provided in Section 1.2.4.

1.2.2.1 Frequency-Dependent Absorption

During propagation, wave energy may be converted to other forms and absorbed by the medium. The absorptive energy loss is directly controlled by the material imperfection for the type of physical wave propagating through it. For EM waves, the imperfection is the electric conductivity of seawater. For acoustic waves, this material imperfection is the inelasticity, which converts the wave energy into heat.

The absorptive loss for acoustic wave propagation is frequency-dependent, and can be expressed as $e^{\alpha(f)d}$, where d is the propagation distance and $\alpha(f)$ is the absorption coefficient at frequency f. For seawater, the absorption coefficient at frequency f in kHz can be written as the sum of chemical relaxation processes and absorption from pure water [4, 266]:

$$\alpha(f) = \frac{A_1 P_1 f_1 f^2}{f_1^2 + f^2} + \frac{A_2 P_2 f_2 f^2}{f_2^2 + f^2} + A_3 P_3 f^2, \tag{1.1}$$

where the first term on the right side is the contribution from boric acid, the second term is from the contribution of magnesium sulphate, and the third term is from the contribution of pure water; A_1, A_2, and A_3 are constants; the pressure dependencies are given by parameters P_1, P_2 and P_3; and the relaxation frequencies f_1 and f_2 are for the relaxation process in boric acid and magnesium sulphate, respectively. Please refer to [266, Chap. 2] for formulations of the coefficients A_1, A_2, A_3, P_1, P_2, P_3, f_1, and f_2 as functions of temperature, salinity and water depth.

In underwater acoustic communications, Thorp's formula can be used as a simplified absorption model for frequencies less than 50 kHz,

$$\alpha(f) = \frac{0.11 f^2}{1 + f^2} + \frac{44 f^2}{4100 + f^2} + 2.75 \times 10^{-4} f^2 + 0.003 \tag{1.2}$$

where f denotes frequency in kHz [51, 195].

1.2.2.2 Geometric Spreading Loss

Geometric spreading is the local power loss of a propagating acoustic wave due to energy conservation. When an acoustic impulse propagates away from its source with longer and longer distance, the wave front occupies larger and larger surface area. Hence, the wave energy in each unit surface (also called *energy flow*) becomes less and less. For the spherical wave generated by a point source, the power loss caused by geometric spreading is proportional to the square of the distance. On the other hand, the cylindrical waves generated by a very long line source, the power loss caused by geometric spreading is proportional to the distance. For a practical underwater setting, the geometric spreading is a hybrid of spherical and cylindrical spreading, with the power loss to be proportional to d^β, where β is between 1, for cylindrical spreading, and 2, for spherical spreading [397]. Provided that the sound propagation in real channels can hardly be classified into either of the two spreading models, a practical value of the spreading exponent can be taken as $\beta = 1.5$. Note that geometric spreading is frequency-independent.

1.2.2.3 Scattering Loss

Scattering is a general physical process in which the incident wave is reflected by irregular surfaces in many different directions. The sound scattering in underwater environments can be attributed to the nonuniformities in the water column and interactions of acoustic waves with nonideal sea surfaces and bottoms. Obstacles in the water column include point targets such as fish and plankton, and scattering volumes such as fish schools and bubble clouds. The corresponding scattering loss depends on the acoustic wavelength and target size. In particular, the scattering loss increases as the acoustic wavelength decreases. The scattering property of sea surface and bottom is mainly determined by the interface roughness. High interface roughness induces large spatial energy dispersion. The roughness of sea surface is due to the capillary waves caused by wind, the amplitude of which ranges from centimeters to meters (e.g., swells). The roughness of sea bottom depends on the geology, including e.g., the roughness of rocks, sand ripples, and organisms in sediments. The amplitude of the roughness also varies from centimeters to meters. Similar to the target scattering in the water column, scattering loss at sea surface and bottom is also frequency-dependent.

In real environments, the two types of scattering processes coexist. For example, in the presence of a high wind speed, the wind-generated waves increases the roughness of sea surface, and breaking waves can create bubble clouds of a large size. Both types of scattering losses happen when the acoustic wave interacts with both sea surface and bubble clouds. Moreover, the wind-generated waves become moving reflectors of acoustic waves, thus introducing energy dispersion not only in the spatial domain but also in the frequency domain.

1.2.2.4 Propagation Loss Parametrization

Denote ξ as the scattering loss. For an acoustic signal at frequency f, the attenuation after propagating a distance of d can be formulated as

$$P_{\text{att}}(f,d) = \xi d^{\beta} e^{\alpha(f)d}. \tag{1.3}$$

Different from the propagation loss in the terrestrial radio channel which only has spreading loss with an exponent $2 \sim 6$ [145], the propagation loss in underwater acoustic channels is spreading-loss dominant in the near-distance transmission and absorption-loss dominant in the long-distance transmission.

1.2.3 Time-Varying Multipath

An acoustic wave can reach a certain point through multiple paths. In a shallow water environment, where the transmission distance is larger than the water depth, wave reflections from the surface and the bottom generate multiple arrivals of the same signal. In deep water applications, surface and bottom reflections may be neglected. However, the wave refractions due to the spatially varying sound speed cause significant multipath phenomena.

Assume that there are N_{pa} paths, and let ξ_p denote the scattering loss, d_p the propagation distance and τ_p the propagation delay of the pth path. Then the pass loss along the pth path can be written as

$$P_{\text{att}}(f,d_p) = \xi_p d_p^{\beta} e^{\alpha(f)d_p} \tag{1.4}$$

which combines the effects of spreading loss, absorptive loss, and scattering loss. For a channel which is *time-invariant* within a certain time interval, the channel transfer function at frequency f can be described as

$$H(f) = \sum_{p=1}^{N_{\text{pa}}} \frac{1}{\sqrt{P_{\text{att}}(f, d_p)}} e^{-j2\pi f \tau_p}. \tag{1.5}$$

The channel transfer function in (1.5) reveals that the overall channel attenuation is dependent not only on the distance, but also on the frequency. Since $\alpha(f)$ increases as f increases, high frequency waves will be considerably attenuated within a short distance, while low frequency acoustic waves can travel far. As a result, the bandwidth is extremely limited for long-range applications, while for short-range applications, several tens of kHz bandwidth could be available (a thorough study on the relationship between bandwidth (capacity) and distance is reported in [362]).

1.2.3.1 Large Delay Spread

The channel delay spread is defined as the maximal difference in the times-of-arrival of channel paths,

$$D_\tau := \max\{|\tau_p - \tau_q|\}, \qquad \forall p, q. \tag{1.6}$$

The slow speed of acoustic waves and significant multipath phenomena cause very large channel delay spread. For example, two physical arrivals that differ 15 meters in path length lead to an arrival time difference of 10 ms (here we assume the propagation speed of sound is 1500 m/s). In shallow water, the typical delay spread is around several tens of milliseconds; an example is shown in Figure 1.4(a), but occasionally delay spread can be as large as 100 ms [208]. In deep water, the delay spread can be of the order of seconds. For underwater acoustic communications, the large delay spread leads to severe intersymbol interference due to the waveform time-dispersion (also called *time-spreading*).

1.2.3.2 Large Doppler Spread

Time variability is one of the most challenging features of underwater acoustic channels. Due to medium instability, such as the current-induced platform motion and wind-generated waves as time-varying reflectors, different propagation paths could have different time-variabilities. For example, the direct path without reflections could be very stable, while the sea surface reflected paths could have time variations incurred by the motion of surface waves. The different time variabilities lead to different Doppler scaling effects or Doppler shifts of the transmitted signal.

Denote v_p as the Doppler rate of the pth path, namely the change rate of the propagation length of the pth path. The channel Doppler rate spread is defined as the maximal difference of the Doppler rates of channel paths,

$$D_{\text{d}} = \max\left\{\frac{|v_p - v_q|}{c}\right\}, \qquad \forall p, q \tag{1.7}$$

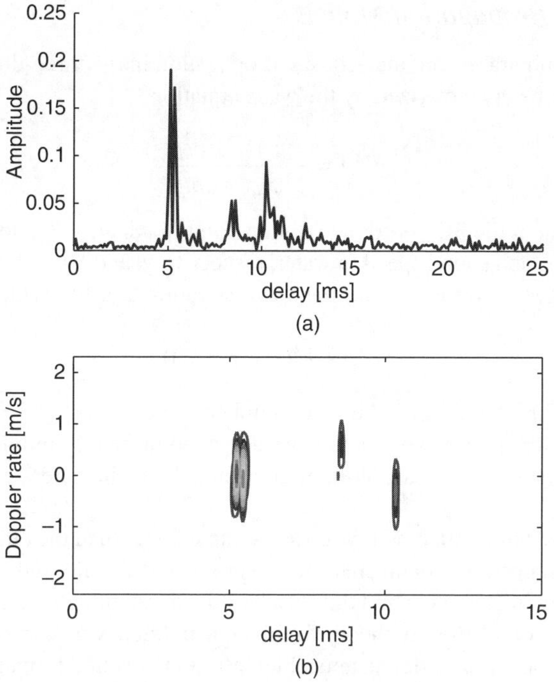

Figure 1.4 Channel profiles from a shallow water stationary experiment of SPACE'08. (a) an example channel impulse response as a function of delay; (b) an example channel scattering function on the delay-Doppler plane.

where c is the sound speed in water. The slow propagation speed of sound introduces large Doppler spread or shifts. For example, consider $\upsilon = 1.5$ m/s and $f_c = 30$ kHz, where υ is the rate of change of the propagation path length (e.g, the platform velocity), and f_c is the system center frequency. The Doppler frequency shift at f_c is given by $f_d = \upsilon/cf_c = 30$ Hz.

On the outset, large Doppler spread results in a reduction in the *channel coherence time* (the time period when the channel can be viewed as static) or an apparent increase in the rate of channel fluctuation [316]. The large Doppler spread causes severe interference among different frequency components of the signal (also referred to as *frequency-spreading*). An example of the channel scattering function is shown in Figure 1.4(b).

1.2.3.3 Sparsity of Channel Paths

Despite the large delay spread and Doppler rate spread, multiple paths in the underwater acoustic channel tend to be sparse, with channel energy concentrated in a few paths; see Figure 1.4. Although the joint presence of large delay and Doppler rate spread entails a complex communication channel, the multipath sparsity is one key feature to be exploited in communication system design.

1.2.4 Acoustic Propagation Models

Given environmental parameters, the acoustic propagation in the three-dimensional underwater environment can be characterized by the wave equation

$$\nabla^2 p = \frac{1}{c^2(x,y,z)} \frac{\partial^2 p}{\partial t^2} \tag{1.8}$$

where (x, y, z) represent the 3D coordinate of one point in water, p, t, and $c(x, y, z)$ denote the sound pressure, time, and sound speed in water, respectively, and ∇^2 is the Laplacian operator. For a sinusoidal wave of frequency f_0, the wave equation can be written as the Helmholtz equation,

$$\nabla^2 p + k^2(x,y,z)p = 0 \tag{1.9}$$

where $k(x, y, z) := 2\pi f_0/c(x, y, z)$ is the wave number.

Despite of simplicity of the wave equation, finding its solution is a complicated task. Depending on applications, several typical solutions are available to characterize the acoustic field.

- Ray theory: By assuming that the phase varies much faster than the amplitude, this method takes the three-dimensional sound pressure as a product of an amplitude function and a phase function which are independent and thus can be solved individually. This above assumption limits the application of the ray theory to the high-frequency systems. Compared to other methods, the ray theory provides an appealing intuition of sound propagation. A commonly used code is the Bellhop ray tracing program [313].
- Normal mode solutions: This method provides an exact solution of the wave equation, but is restricted to the horizontally stratified channel which only has variation of sound velocity in the depth direction and assumes a flat and horizontal bottom. Hence, the solution is range independent. This method is often used in the time-reversal processing and the matched field processing. One general code is KRAKEN based on the KRAKEN normal mode model [312].
- Wave number integration: Similar to the normal mode solution, this method assumes a stratified channel, and computes the acoustic field using wave number integration. In particular, using the fast Fourier transform (FFT), the fast field program (FFP) can directly evaluate the integral solution to obtain a numerical solution of the wave equation. Although this method can provide accurate solution, it fails to provide physical interpretation of acoustic fields relative to the other three methods. An example of the FFP program is OASES (for Ocean Acoustic and Seismic Exploration Synthesis) [335].
- Parabolic approximation: Considering only the forward propagation direction, this method approximates the Helmholtz equation in (1.9) by a parabolic equation (PE) which can be evaluated numerically. A large number of PE approximations have been developed since the 1970s [195]. The PE method is suitable for calculating acoustic field in a range-dependent environment. In various PE codes, the bottom topography and surface roughness can be accounted for. An example of the PE codes is MMPE based on the Monterey-Miami PE model [344]. A recent application to high frequency acoustic transmission can be found in [336].

Notice that the acoustic field is to be described with a resolution on the level of wavelength. The latter three methods are mainly suitable to the low-frequency domain in which the acoustic field is stable for observation. For low-frequency communications, the Parabolic approximation can be used to simulate the underwater acoustic channel, while the ray-tracing theory is commonly adopted for high-frequency systems. Please refer the textbook [195] for detailed presentations of the theory and to the website [161] for variants and program codes of the above four solutions.

1.2.5 Ambient Noise and External Interference

Noise is used to denote a signal that distorts the desired ones. Depending on applications, underwater acoustic noise consists of different components. Specific to the underwater acoustic communication system, the acoustic noise can be grouped into two categories: ambient noise and external interference.

Ambient noise is one kind of background noise which comes from a myriad of sources. The common sources of ambient noise in water include volcanic and seismic activities, turbulence, surface shipping and industrial activities, weather processes such as wind-generated waves and rain, and thermal noise [266]. Due to the multiple sources, ambient noise can be approximated as Gaussian, but it is not white. The level of underwater ambient noise may have large fluctuations upon a change with time, location or depth. For short-range acoustic communication, the level of ambient noise may be well below the desired signal. For long-range or covert acoustic communication, the noise level would be a limiting factor for communication performance.

External interference is an interfering signal which is recognizable in the received signal. Corresponding sources include marine animals, ice cracking, and acoustic systems working in the same environment. For example, snapping shrimp in warm water and ice cracking in polar regions generate impulsive interferences [78]. Sonar operations could occasionally happen at the same time with communications, creating an external interference which is highly structured [422]. Relative to ambient noise, external interferences are neither Gaussian nor white. The presence of this kind of noises may cause highly dynamic link error rate or even link outage.

It should be noted that the noise level is highly frequency-dependent. The noise power spectrum density almost monotonically decreases as frequency increases, until up to about 100 kHz when terminal noise becomes dominant. Thus, when selecting a suitable frequency band for communication, besides the frequency-dependent path loss as shown in (1.3), noise should also be taken into account [316, 362].

1.3 Passband Channel Input–Output Relationship

A diagram for the transmitter and receiver in the presence of underwater acoustic channels is shown in Figure 1.5. Since acoustic signals have low frequency, the passband samples $\tilde{x}[n]$ are often directly generated. After digital-to-analog (D/A) conversion, the passband signal $\tilde{x}(t)$ is amplified, and passed to matching circuits, matched to the transducer. At the receiver side, the weak signal is increased in level by a pre-amplifier, filtered by a simple bandpass filter, and sampled at the passband. From the signal processing point of view, the channel includes the imperfections of the transmitter and receiving circuits. All the modules that are lumped together are called channel between $\tilde{x}(t)$ and $\tilde{y}(t)$.

This book establishes the channel input–output relationship directly in the passband. This allows us to capture the wideband channel effect: (a) the propagation effect is frequency

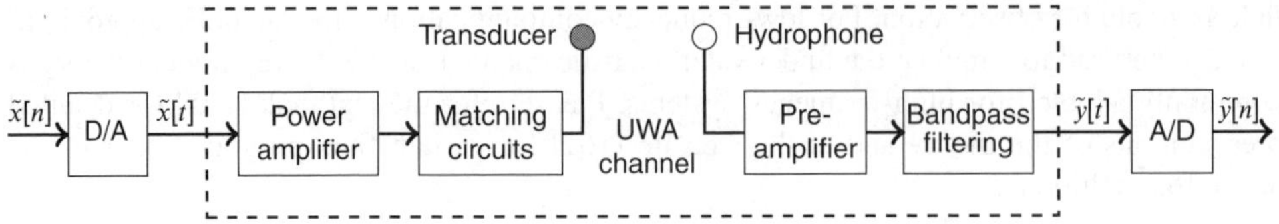

Figure 1.5 The UWA system in the passband.

dependent, (b) each transducer has its own transmit voltage response (TVR), and the matching is not uniform in the signal band, and (c) the Doppler distortion is frequency dependent. In all these considerations, the absolute values of the frequency band do matter.

Assume that the channel is linear time-varying channel. Then we can represent it by $h(t; \tau)$. From the signal processing perspective, the overall channel between the transmitter and the receiver is

$$\begin{aligned}\tilde{y}(t) &= \tilde{x}(t) \star h(t; \tau) + \tilde{n}(t) \\ &= \int h(t; \tau)x(t - \tau)d\tau + \tilde{n}(t) \end{aligned} \tag{1.10}$$

where $\star$ denotes the convolution operation.

1.3.1 Linear Time-Varying Channel with Path-Specific Doppler Scales

The channel $h(t; \tau)$ is general with no specified structure. From the system identification point of view, we need to parameterize the channel. We start with the assumption that the channel consists of $N_{\rm pa}$ discrete paths,

$$h(t; \tau) = \sum_{p=1}^{N_{\rm pa}} A_p(t)\delta(\tau - \tau_p(t)) \tag{1.11}$$

where $A_p(t)$ and $\tau_p(t)$ are the time-varying amplitude and delay for the pth path, respectively.

For a short block of length $T_{\rm bl}$, one can assume that $A_p(t)$ and $\tau_p(t)$ as slowly varying. For this, one can adopt the following assumptions

- AS1): The amplitude is constant within a short block

$$A_p(t) = A_p. \tag{1.12}$$

- AS2): The delay variation within one block can be approximated by a first-order polynomial

$$\tau_p(t) \approx \tau_p - a_p t, \qquad t \in [0, T_{\rm bl}], \tag{1.13}$$

where τ_p is the initial delay and a_p is first order derivative of $\tau_p(t)$.

The parameter a_p is often termed the Doppler scaling factor. Based on assumptions AS1) and AS2), we have a time varying channel with different Doppler scales on different paths as

$$h(t;\tau) = \sum_{p=1}^{N_{\rm pa}} A_p \delta\left(\tau - (\tau_p - a_p t)\right) \tag{1.14}$$

The received passband signal is related to the transmitted passband signal as

$$\tilde{y}(t) = \sum_{p=1}^{N_{\rm pa}} A_p \tilde{x}\left(\left(1 + a_p\right) t - \tau_p\right) + \tilde{w}(t) \tag{1.15}$$

where the equivalent noise $\tilde{w}(t)$ contains both ambient and model-mismatch noises as

$$\tilde{w}(t) = \tilde{n}(t) + \underbrace{\tilde{x}(t) \star h(t;\tau) - \sum_{p=1}^{N_{\rm pa}} A_p \tilde{x}\left(\left(1 + a_p\right) t - \tau_p\right)}_{\text{signal-dependent model mismatch noise}} \tag{1.16}$$

Note that a physical channel might not be able to be represented exactly, but it can be approximated by (1.14) from a signal processing point of view.

Equations (1.14) and (1.15) are the foundations used by the receiver designs in this book, where the input output relationship is parameterized by $N_{\rm pa}$ triplets $\{A_p, a_p, \tau_p\}$; see Figure 1.6 for illustrations on the general and special cases.

The following special cases are often used.

1.3.2 Linear Time-Varying Channels with One Common Doppler Scale

Assume that all the paths have the same Doppler scale factor. The channel is simplified to

$$h(t;\tau) = \sum_{p=1}^{N_{\rm pa}} A_p \delta\left(\tau - \left(\tau_p - at\right)\right). \tag{1.17}$$

This channel is parameterized by $N_{\rm pa}$ pairs $\{A_p, \tau_p\}$ plus the common Doppler scale. A common Doppler scale can be readily removed through a resampling operation.

1.3.3 Linear Time-Invariant Channel

Assume that all the paths are stable with no delay variations. The channel is simplified to

$$h(\tau) = \sum_{p=1}^{N_{\rm pa}} A_p \delta\left(\tau - \tau_p\right). \tag{1.18}$$

The channel is parameterized by $N_{\rm pa}$ pairs $\{A_p, \tau_p\}$.

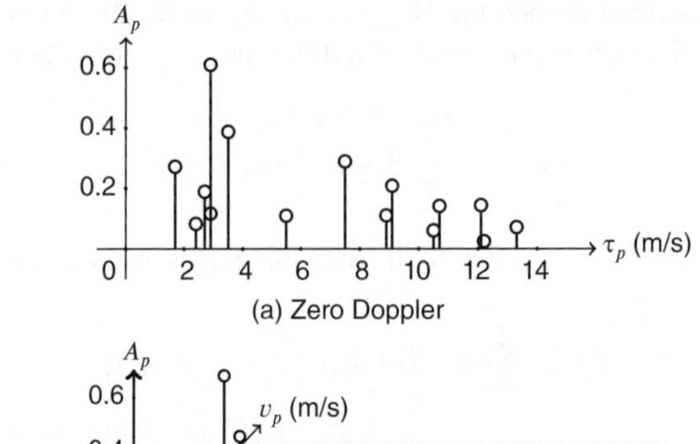

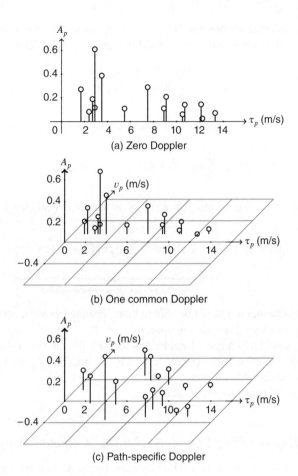

Figure 1.6 Illustration of three channel models. υ_p denotes the Doppler speed associated with the pth path.

1.3.4 Linear Time-Varying Channel with Both Amplitude and Delay Variations

One can certainly extend the work to be more general, by using a Taylor expansion on the amplitude and delay respectively. For example, the approximation on the amplitude can be the N_{amp}-th order. The approximation on the delay can be on the N_{delay}-th order.

- AS1) The amplitude variation within one block can be approximated by a polynomial up to order N_{amp}:

$$A_p(t) \approx A_p^{(0)} - A_p^{(1)} t + \frac{1}{2} A_p^{(2)} t^2 + \cdots + \frac{(-1)^{N_{\mathrm{amp}}}}{N_{\mathrm{amp}}!} A_p^{(N_{\mathrm{amp}})} t^{N_{\mathrm{amp}}}$$

$$= \sum_{n=0}^{N_{\mathrm{amp}}} \frac{(-1)^n}{n!} A_p^{(n)} t^n \tag{1.19}$$

- AS2) The delay variation within one block can be approximated by a polynomial up to order N_{delay}:

$$\tau_p(t) \approx a_p^{(0)} - a_p^{(1)} t + \frac{1}{2} a_p^{(2)} t^2 + \cdots + \frac{(-1)^{N_{\text{delay}}}}{N_{\text{delay}}!} a_p^{(N_{delay})} t^{N_{\text{delay}}}$$
$$= \sum_{n=0}^{N_{\text{delay}}} \frac{(-1)^n}{n!} a_p^{(n)} t^n \qquad (1.20)$$

Channel parameterization based on these two polynomials is quite general, with special cases available in the literature.

Setting $N_{\text{delay}} = 1$ in AS2), the multipath channel is approximated as:

$$h(t; \tau) \approx \sum_{p=1}^{N_{\text{pa}}} \left(\sum_{n=0}^{N_{\text{amp}}} \frac{(-1)^n}{n!} A_p^{(n)} t^n \right) \delta\left(\tau - \left(\tau_p - a_p t\right)\right). \qquad (1.21)$$

Receiver designs based on up to the second-order polynomial amplitude fitting and up to the first-order polynomial delay fitting have been reported in [442].

1.3.5 *Linear Time-Varying Channel with Frequency-Dependent Attenuation*

In a practical system as shown in Figure 1.5, the transmitter voltage response (TVR) is usually not constant due to imperfect circuit matching to the transducer across all the frequency band. Meanwhile, the signal attenuation in underwater acoustic channels is frequency-dependent. One could extend the channel in (1.11) as

$$h(t; \tau) = \sum_{p=1}^{N_{\text{pa}}} A_p(t) \gamma_p \left(\tau - \tau_p(t)\right) \qquad (1.22)$$

where $\gamma_p(t)$ represents a combined effect of TVR and the frequency-dependent acoustic channel attenuation. In practical systems, the TVR can be measured [34], and the propagation pattern can be determined analytically or experimentally. Recent experimental results in [404, 405] have further confirmed the wideband nature of the underwater acoustic channels. Incorporating the frequency-dependent templates into practical receiver designs is yet to be demonstrated.

1.4 Modulation Techniques for UWA Communications

The main techniques used in underwater acoustic communications are: frequency hopped FSK, direct sequence spread spectrum, single carrier transmission, sweep-spread carrier modulation, and multicarrier modulation.

1.4.1 *Frequency Hopped FSK*

In FSK modulation, information bits are used to select the carrier frequencies of the transmitted signal. The receiver compares the measured power at different frequencies to infer what has

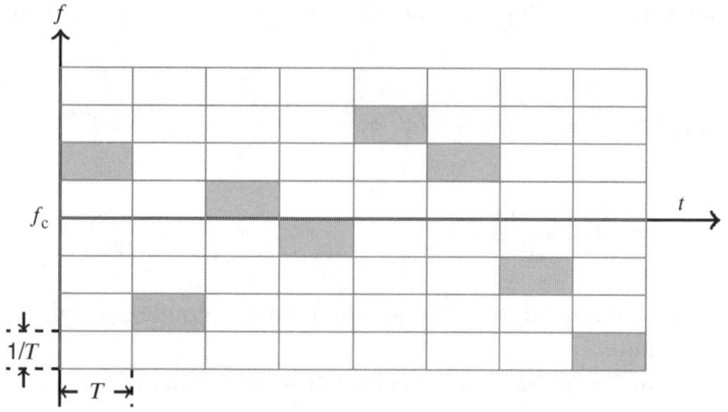

Figure 1.7 An illustration of frequency hopped FSK.

been sent. Using only an energy detector at the receiver, this scheme bypasses the need for channel estimation, and is thus robust to channel variations. However, guard bands are needed to avoid the interference caused by frequency-spreading, and guard intervals are needed to avoid the interference caused by time-spreading.

Frequency hopped (FH) FSK avoids the waiting of the channel clearing corresponding to the previous symbol, by hopping to a different frequency, as illustrated in Figure 1.7. The transmitted signal in passband is

$$\tilde{x}(t) = 2\Re\left\{\sum_{i=-\infty}^{\infty} e^{j2\pi f(i;s[i])(t-iT)} g(t - iT)\right\} \tag{1.23}$$

where T is the time duration of each tone with a frequency support of $1/T$, the tone of the ith symbol is determined by $f(i; s[i])$ which is a function of both the time slot index i and the data symbol $s[i]$, and $g(t)$ is the pulse shaper. Within a bandwidth B, the total number of tones that can be used is BT. Since BT is larger than one, there is a gain on noise suppression on the order of BT, a benefit from frequency-hopping spread spectrum. Due to the bandwidth expansion via frequency hopping, the overall bandwidth efficiency is low, typically less than 0.1 bits/sec/Hz.

1.4.2 Direct Sequence Spread Spectrum

In DSSS modulation, a narrowband waveform of bandwidth W is spread to a large bandwidth B before transmission. This is achieved by multiplying each symbol with a spreading code of length $N = \lfloor B/W \rfloor$, and transmitting the resulting sequence at a high rate as allowed by bandwidth B. For coherent DSSS, the baseband signal is

$$x(t) = \sum_{i=\infty}^{\infty} s[i] \sum_{n=0}^{N-1} c[i; n]p(t - (iN + n)T_c) \tag{1.24}$$

where $s[i]$ is the information-bearing symbol, $c[i; n]$ is the chip sequence for the ith symbol, T_c is the chip duration, and $p(t)$ is the pulse shaper on the chip level. The corresponding passband signal is

$$\tilde{x}(t) = 2\Re\{x(t)e^{j2\pi f_c t}\}. \tag{1.25}$$

Multiple arrivals at the receiver side can be separated via the de-spreading operation which suppresses the time-spreading induced interference, thanks to the nice auto-correlation properties of the spreading sequence. Channel estimation and tracking are needed as phase-coherent modulation is used to map information bits to symbols before spreading [128, 443].

For noncoherent DSSS, information bits can be used to select different spreading codes to be used, and the receiver compares the amplitudes of the outputs from different matched filters, with each one matched to one choice of spreading code. This avoids the need for channel estimation and tracking.

1.4.3 Single Carrier Modulation

One major step towards high rate communication is single carrier transmission of information symbols from constellations such as phase-shift-keying (PSK) and quadrature-amplitude-modulation (QAM) [367]. With symbols $s[i]$ and pulse shaping filter $p(t)$, the transmitted signal is

$$x(t) = \sum_{i=-\infty}^{\infty} s[i]p(t - iT), \tag{1.26}$$

where T is the symbol period. The corresponding passband signal can be similarly obtained as in (1.25).

The channel introduces intersymbol interference (ISI) due to multipath propagation. When data symbols are transmitted at a high rate, the same physical channel leads to more channel taps in the discrete-time equivalent model. Advanced signal processing at the receiver side is used to suppress the interference; this process is termed *channel equalization*. Although widely used for slowly-varying multipath channels in radio applications, channel equalization for fast-varying underwater channel is a significant challenge.

There are various receiver designs developed for single carrier transmissions.

- The canonical receiver in [367] successfully combined a second-order phase-locked-loop to track channel phase variations with an adaptive decision feedback equalizer to suppress intersymbol interference. Multichannel DFE is adopted when there are multiple receiving elements [366].
- In the time-reversal approach, the signals from multiple elements are combined, followed by a single channel equalizer, where the equalizer can be linear or based on decision feedback [108, 148, 350, 355, 451].
- The complexity of time-domain equalization grows quickly as the number of channel taps increases, which will eventually limit the rate increase for single-carrier phase-coherent transmission. A frequency-domain equalization approach can effectively deal with channels with a large number of taps [468].

Iterative channel equalization and data decoding can be carried out for decoding performance improvement, which leads to the so-called Turbo equalization.

1.4.4 Sweep-Spread Carrier (S2C) Modulation

In all the previous modulation schemes, the carrier frequency stays constant for the whole data burst, or at least within each symbol duration as in FH-FSK. A new signaling method based on the implementation of a sweep-spread carrier, which entails rapid fluctuation of carrier frequency, has been proposed in [207]. Let f_L and f_H denote the lower and higher ends of the signal band, and T_{sw} be the sweep time. The frequency variation rate is then $m = (f_H - f_L)/T_{sw}$. The sweep-spread carrier consists of a succession of sweeps as [207]

$$c(t) = \exp\left[j2\pi\left(f_L\bar{t} + \frac{1}{2}m\bar{t}^2\right)\right], \quad \bar{t} = t - \left\lfloor \frac{t}{T_{sw}} \right\rfloor T_{sw}. \tag{1.27}$$

The baseband signal $s(t)$, which could be coherent or differentially modulated, is converted to the passband through

$$\tilde{x}(t) = 2\Re\{s(t)c(t)\}. \tag{1.28}$$

The receiver carried out the carrier demodulation through the multiplication of the received signal with an appropriately varying gradient-heterodyne signal having the same sweep cycle and the same slope of the frequency variation [207]. In the presence of a multipath fading channel, the signals with different arrival times lead to different residual frequencies after carrier demodulation. Bandpass filtering is used to separate different arrival paths. Signals along stable paths are typically selected for data demodulation. This way, the interference from unstable paths is suppressed. Note that a signal bandwidth several times larger than the symbol rate is often used to ensure a large frequency variation rate m for a good multipath resolution; hence, this method can be regarded as one special form of spread spectrum communication, which entails some noise reduction capabilities.

1.4.5 Multicarrier Modulation

The idea of multicarrier modulation is to divide the available bandwidth into a large number of subbands, where each subband has its own (sub)carrier. Within each band, the symbol rate is reduced with an increased symbol duration, so that intersymbol interference can be less severe, which helps to simplify the receiver complexity of channel equalization. There are many variants of multicarrier modulation. One way to characterize them is to check whether the subbands are overlapping or nonoverlapping, as illustrated in Figure 1.8.

Due to the existence of guard bands between neighboring subbands in the multicarrier approach with nonoverlapping subbands, bandpass filtering can be used to separate the signals from different subbands. Hence, this approach is essentially a frequency-division-multiplexing (FDM) approach, by splitting a large bandwidth into smaller pieces. Within each band, one can adopt signaling schemes such as M-ary frequency-shift keying, single carrier transmission, or another multicarrier modulation with overlapping subcarriers in a small bandwidth.

Orthogonal frequency-division multiplexing (OFDM) is one prevailing example of multicarrier modulation with overlapping subcarriers. The waveform is carefully designed to maintain orthogonality even after propagating over a long multipath channel to eliminate the need for an equalizer [44, 418]. Precisely due to this advantage, OFDM has prevailed in recent

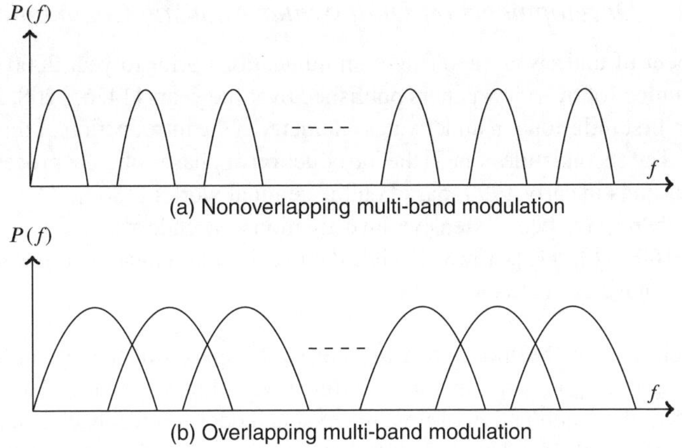

Figure 1.8 Illustration of multicarrier modulation techniques. $P(f)$ denotes the power spectrum density.

broadband wireless radio applications, including digital audio/video broadcasting, wireless local/metropolitan area networks, and fourth generation cellular networks. Filterbank based approaches also belong to the category of multicarrier modulation with overlapping subbands, making extensions to the Fourier bases used in OFDM.

The fast variations of underwater acoustic channels entail large Doppler spread which introduces significant interference among OFDM subcarriers. Signal processing specialized to underwater acoustic channels are needed to make OFDM work in underwater environments.

1.4.6 Multi-Input Multi-Output Techniques

A wireless system that employs multiple transmitters and multiple receivers is referred to as a multiple-input multiple-output (MIMO) system. It has been shown that the channel capacity in a scattering-rich environment increases linearly with $\min(N_t, N_r)$, where N_t and N_r are the numbers of transmitters and receivers, respectively [123, 379]. Such a drastic capacity increase does not incur a penalty on precious power and bandwidth resources; but instead it comes from the utilization of spatial dimension virtually creating parallel data pipes. Hence, MIMO modulation is a promising technology to offer yet another fundamental advance on high data rate underwater acoustic communication [209, 210]. MIMO has been applied in both single carrier transmission and multicarrier transmission.

MIMO introduces additional interference among parallel data streams from different transmitters. Also, each receiver has more channels to estimate, which requires more overhead spent on training symbols. For fast varying underwater channels, the number of transmitters might not be large for best rate-and-performance tradeoff. In addition to co-located antennas, distributed MIMO is also possible if clustered single-transmitter nodes could cooperate [398]. Certainly, implementation of distributed MIMO needs to address challenging practical issues such as node synchronization and cooperation.

1.4.7 Recent Developments on Underwater Acoustic Communications

The development of underwater acoustic communications prior to year 2000 has been summarized in a number of overview papers published over the years [14, 65, 208, 360]. It is often viewed that the first milestone in underwater telemetry is the introduction of digital techniques in early 1980s. The second milestone is the introduction of phase coherent processing for single carrier transmissions in early 1990, based on the seminal work in [367].

Since 2000, there have been extensive investigations on underwater acoustic communications [15, 79, 160, 343, 448]. Towards high data rate communications, the following three major directions have been pursued.

- Various receivers have been designed to improve the performance of single carrier transmissions. Example approaches include the time reversal processing in [107, 121, 330, 354, 355, 449, 450], the combination of time reversal and decision feedback equalization (DFE) in [108, 148, 346, 349, 350, 353, 355, 356, 451], the frequency domain equalization in [468], and the joint channel estimation and data detection approach in [249, 411].
- Multicarrier modulation has been successfully applied to underwater acoustic channels. For coherent demodulation, example approaches include the block-by-block based OFDM receivers in [149, 201, 232, 235, 257, 375, 457] and the adaptive OFDM receiver in [361, 363]. Explicit intercarrier interference (ICI) cancellation is one key element for enhanced receiver performance [38, 180, 392]. Noncoherent on-off keying and differential encoding have been explored in [140] and [12, 365] for OFDM systems, respectively. Adaptive modulation and coding have been recently studied for underwater OFDM [324, 413]. Variants of the multicarrier modulation other than OFDM have also been pursued [185, 287].
- MIMO techniques have attracted a lot of attention to increase the spectrum efficiency. For MIMO single carrier transmissions, example receivers include the multi-channel DFE based approach [209, 210, 331, 369], time-reversal combined with single channel DFE [347, 348, 352, 357], frequency domain equalization [460, 461], successive interference cancellation [250, 251], and iterative (Turbo) equalizations [325, 374, 377, 415, 416]. For MIMO OFDM, example approaches include the block-by-block receiver design in [172, 181, 233] and the adaptive receiver in [67, 364]. Recent receiver designs for MIMO systems with transmissions from multiple spatially distributed users can be found in [82, 83, 84, 183, 391, 423].

1.5 Organization of the Book

This book is solely focused on OFDM and MIMO OFDM for underwater acoustical channels. It is worthwhile to point out that different modulation schemes have their strengths in different situations. Generally speaking, FHSS and DSSS are good candidates for low data rates and robust operations. For low-medium rates and longer range, single carrier transmission is a suitable choice. S2C lies between DSSS and single carrier, depending on the spreading factor. For short-range and large data rates over long channels, OFDM has its competitive advantage. An intuitive illustration is shown in Figure 1.9, where the boundary and the overlapping regions should not be interpreted quantitatively.

There are many books available on OFDM and MIMO OFDM in wireless radio channels. However, as pointed out in this chapter, the challenges of radio and underwater acoustic channels are drastically different, and the receiver designs for underwater acoustic OFDM

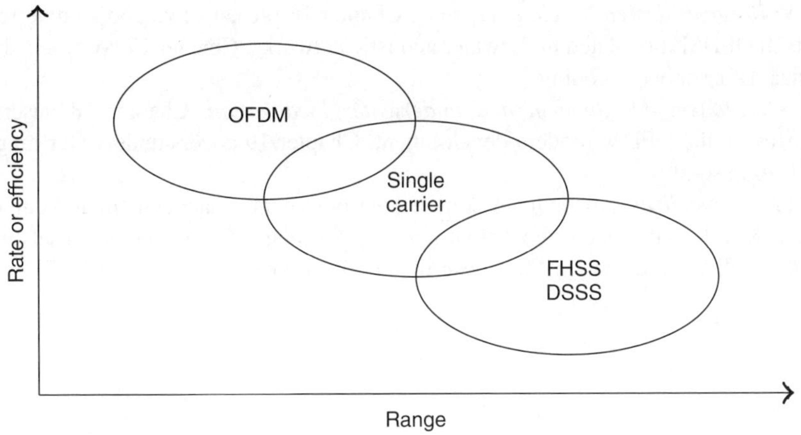

Figure 1.9 Illustration of relative merits of different techniques.

have some unique features. Specifically, the receiver design adopts a novel signal processing based model to parameterize UWA channels in (1.14), which is richer than that sufficient for radio channels. This book necessarily uses the passband formulation rather than the baseband formulation in existing books. This captures the wideband nature of underwater acoustic channels. Channel sparsity is a key element of the book, where the latest advances in compressive sensing have been incorporated into the channel estimation module. In addition, explicit intercarrier-interference consideration is adopted in this book, and iterative receiver processing is a central piece.

The chapters and appendices are arranged in the following order:

- *Part I: Basics.* Chapter 2 presents the basic principle of OFDM modulation and demodulation. Chapter 3 presents nonbinary low-density parity-check (LDPC) coded OFDM, where the LDPC codes are used within numerical and experimental results in this book. Chapter 4 discusses a property regarding to the peak-to-average-power-ratio of a transmitted OFDM signal.
- *Part II: Receiver components.* Chapter 5 presents an overview of OFDM transceiver design for point-to-point communications. Chapter 6 details the synchronization and Doppler estimation modules. Chapter 7 presents the channel and noise variance estimation algorithms. Chapter 8 contains the data detection algorithms under different channel input–output relationships.
- *Part III: Single-transmitter communications.* Chapter 9 develops a block-by-block progressive receiver along with its performance results. Chapter 10 describes a block-to-block adaptive receiver with clustered channel adaptation. Chapter 11 deals with channels having significantly separated multipath clusters, a concern in deep water horizontal communications. Chapter 12 presents an OFDM receiver in the presence of external interference.
- *Part IV: Multi-input multi-output communications.* Chapter 13 presents receiver design for co-located MIMO OFDM. Chapter 14 deals with distributed MIMO OFDM, with quasi-synchronous reception. Chapter 15 presents a receiver design for completely asynchronous multiple user communication.

- *Part V: Receiver design in relay networks.* Chapter 16 presents two cooperative relay protocols in OFDM modulated underwater acoustic networks. Chapter 17 considers the use of physical layer network coding.
- *Part VI: Modem development and underwater localization.* Chapter 18 describes the advances on the OFDM modem development. Chapter 19 covers underwater ranging and localization solutions.
- *Part VII: Appendixes.* Appendix A contains various sparse channel estimation algorithms. Appendix B describes the setup of two major experiments, from which data sets were collected to validate many algorithms described in this book.

2

OFDM Basics

We illustrate the basic ideas of OFDM transmission and reception in a time invariant channel $h(t)$. Assume that $h(t)$ has nonzero support within the interval $[0, T_{\text{ch}}]$. The channel frequency response is expressed as

$$H(f) = \int_0^{T_{\text{ch}}} h(t)e^{-j2\pi ft}dt. \tag{2.1}$$

Denote $\tilde{x}(t)$ as the transmitted passband signal. The received passband signal after channel transmission is

$$\begin{aligned}\tilde{y}(t) &= h(t) \star \tilde{x}(t) + \tilde{n}(t),\\ &= \int_0^{T_{\text{ch}}} h(\tau)\tilde{x}(t-\tau)d\tau + \tilde{n}(t),\end{aligned} \tag{2.2}$$

where $\tilde{n}(t)$ is the additive noise.

Given the dispersive nature of the channel, intersymbol interference appears for serially transmitted data streams. Compared to the canonical single-carrier serial transmission scheme, OFDM is a block transmission scheme, which partitions information symbols into blocks, and a guard interval is inserted between consecutive blocks before transmission. There are two different types of guard intervals: one is padding zeros at the end of an OFDM symbol, and the other is inserting a cyclic prefix in the front of an OFDM symbol. Accordingly, OFDM signaling has two popular formats: zero-padded (ZP) OFDM and cyclic-prefixed (CP) OFDM.

Prior to the discussion of OFDM signaling properties, we first specify the OFDM signal generation and reception both ZP- and CP-OFDM in Sections 2.1 and 2.2, respectively.

2.1 Zero-Padded OFDM

2.1.1 Transmitted Signal

Consider one ZP-OFDM block from the transmission illustrated in Figure 2.1. Let T denote the basic OFDM symbol duration, which dictates a subcarrier spacing of $1/T$. Assume K

OFDM for Underwater Acoustic Communications, First Edition. Shengli Zhou and Zhaohui Wang.
© 2014 John Wiley & Sons, Ltd. Published 2014 by John Wiley & Sons, Ltd.

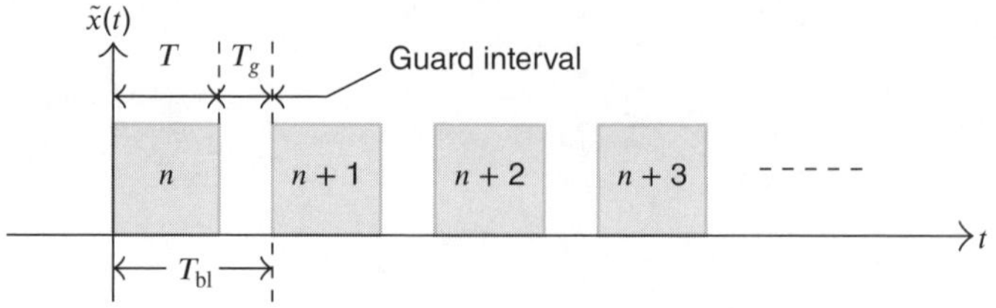

Figure 2.1 An illustration of ZP-OFDM blocks.

subcarriers in total. The kth subcarrier is located at the frequency

$$f_k = f_c + \frac{k}{T}, \qquad k = -\frac{K}{2}, \ldots, \frac{K}{2} - 1, \tag{2.3}$$

where f_c is the center frequency. Let T_g denote the zero-padding interval, and T_{bl} the total block duration corresponding to one OFDM block which includes both windowing operation and guard interval.

Let $s[k]$ denote the symbol to be transmitted on the kth subcarrier, where some of the symbols might be zeros. Let S_A denote the set of active subcarriers which deliver nonzero symbols. One OFDM symbol in the baseband is

$$x(t) = \sum_{k \in S_A} s[k] e^{j2\pi \frac{k}{T} t} g(t), \quad t \in [0, T_{bl}] \tag{2.4}$$

where $g(t)$ is the pulse shaping filter. After baseband-to-passband upshifting, the passband signal is given by

$$\begin{aligned} \tilde{x}(t) &= \sum_{k \in S_A} s[k] e^{j2\pi f_k t} g(t) + \sum_{k \in S_A} s^*[k] e^{-j2\pi f_k t} g(t) \\ &= 2\Re \left\{ \sum_{k \in S_A} s[k] e^{j2\pi f_k t} g(t) \right\}, \quad t \in [0, T_{bl}]. \end{aligned} \tag{2.5}$$

Since for any real signal $\tilde{X}(-f) = \tilde{X}^*(f)$, we focus on the Fourier transform at the positive frequency range with $f > 0$, and ignore the negative frequency part in our presentation. The Fourier transform of $\tilde{x}(t)$ at the positive frequency range is

$$\tilde{X}(f) = \sum_{k \in S_A} s[k] G(f - f_k), \quad f > 0 \tag{2.6}$$

where $G(f)$ is the Fourier transform of $g(t)$.

As illustrated in Figure 2.2, the orthogonality of subcarriers in OFDM imposes a key property of $g(t)$, which is expressed as

$$G(0) = 1, \text{ and } G\left(\frac{k}{T}\right) = 0, \quad \forall k \neq 0. \tag{2.7}$$

With such a property, the sample at f_m is

$$\tilde{X}(f_m) = \sum_{k \in S_A} s[k] G(f_m - f_k) = s[m], \tag{2.8}$$

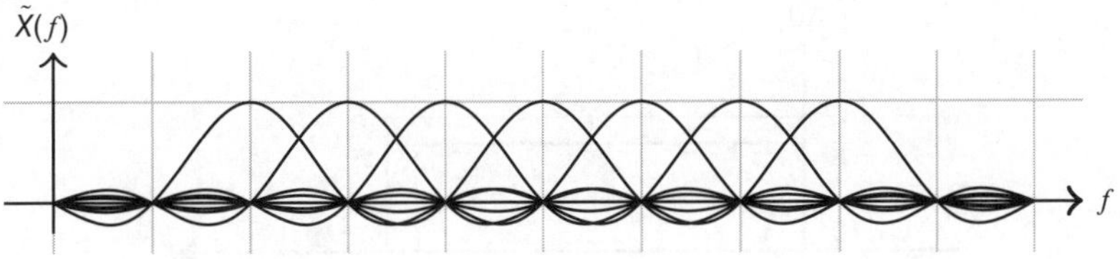

Figure 2.2 An illustration of ZP-OFDM waveform in the frequency domain.

which reveals that the symbols are separated in the frequency domain, and orthogonality is embedded in the transmitted waveform.

There are several possible choices of the pulse shaping filter which satisfy (2.7).

2.1.1.1 Rectangular Window

The popular choice of the pulse shaping filter is a rectangular window,

$$g_{\text{rec}}(t) = \begin{cases} \frac{1}{T}, & t \in [0, T] \\ 0, & \text{otherwise} \end{cases} \tag{2.9}$$

with the Fourier transform expressed as

$$G_{\text{rec}}(f) = \frac{\sin(\pi f T)}{\pi f T} e^{-j\pi f T}. \tag{2.10}$$

The duration for each ZP-OFDM block is $T_{\text{bl}} = T + T_{\text{g}}$.

2.1.1.2 Raised Cosine Window

Raised cosine filters are good choices for the purpose of reducing sidelobes of $G(f)$ [318]. With T denoting the basic OFDM symbol duration, and β denoting the roll-off factor, the raised-cosine window is give by

$$g_{\text{rc}}(t) = \begin{cases} \frac{1}{T}, & t \in [\beta T, T] \\ \frac{1}{2T}\left[1 + \cos\left(\frac{\pi}{\beta T}\left(\left|t - \frac{1+\beta}{2}T\right| - \frac{1-\beta}{2}T\right)\right)\right], & t \in [0, \beta T) \cup (T, (1+\beta)T] \\ 0, & \text{otherwise.} \end{cases} \tag{2.11}$$

Its Fourier transform is

$$G_{\text{rc}}(f) = \frac{\sin(\pi f T)}{\pi f T} \cdot \frac{\cos(\pi \beta f T)}{1 - 4\beta^2 f^2 T^2} e^{-j\pi f(1+\beta)T}. \tag{2.12}$$

When $\beta = 0$, $g_{\text{rc}}(t)$ in (2.11) reduces to the rectangular window $g_{\text{rec}}(t)$ in (2.9). The duration for each ZP-OFDM block is $T_{\text{bl}} = (1 + \beta)T + T_{\text{g}}$. Figures 2.3(a) and 2.3(b) demonstrate the rectangular and raised cosine pulse shapes and the amplitude of the corresponding frequency transform, respectively.

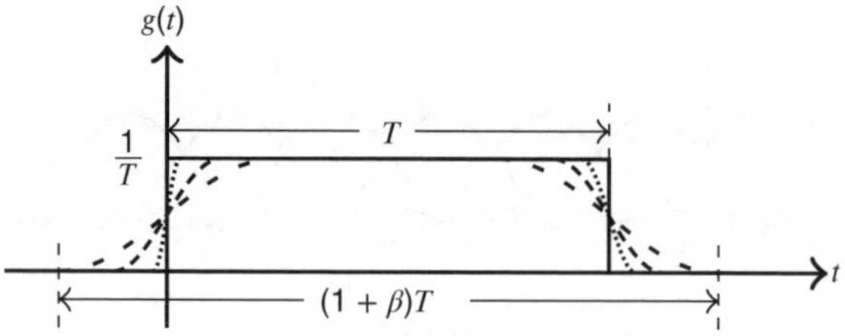

(a) Time-domain waveform

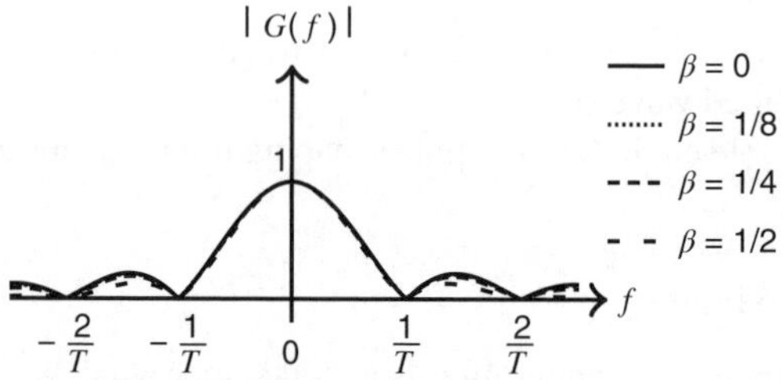

(b) Frequency response

Figure 2.3 An illustration of rectangular and raised cosine windows.

2.1.2 *Receiver Processing*

Let the ZP-OFDM signal pass through the channel as described in (2.2). The Fourier transform of the received signal $\tilde{y}(t)$ is

$$\tilde{Y}(f) = \int_0^{T_{\mathrm{bl}}} \tilde{y}(t)e^{-j2\pi ft}\,dt \tag{2.13}$$

Note that a time-domain convolution amounts to a frequency-domain multiplication. We have

$$\tilde{Y}(f) = H(f)\tilde{X}(f) + \tilde{N}(f), \tag{2.14}$$

where $\tilde{N}(f)$ is the Fourier transform of $\tilde{n}(t)$.

The receiver draws samples at specific frequency points. The frequency-domain sample at the subcarrier frequency f_m is

$$y[m] = \tilde{Y}(f)\big|_{f=f_m}. \tag{2.15}$$

Similarly define the noise sample as $\eta[m] = \tilde{N}(f)|_{f=f_m}$. Substituting (2.6) into (2.14), the frequency sample at the mth subcarrier can be formulated as

$$\begin{aligned} y[m] &= H(f_m)\sum_{k\in S_{\mathrm{A}}} s[k]G(f_m - f_k) + \eta[m] \\ &= H(f_m)s[m] + \eta[m]. \end{aligned} \tag{2.16}$$

Clearly, a time dispersive channel in the continuous time is converted to parallel flat-fading (nondispersive) channel in the frequency domain. The embedded orthogonality in the transmission is kept even after a dispersive channel.

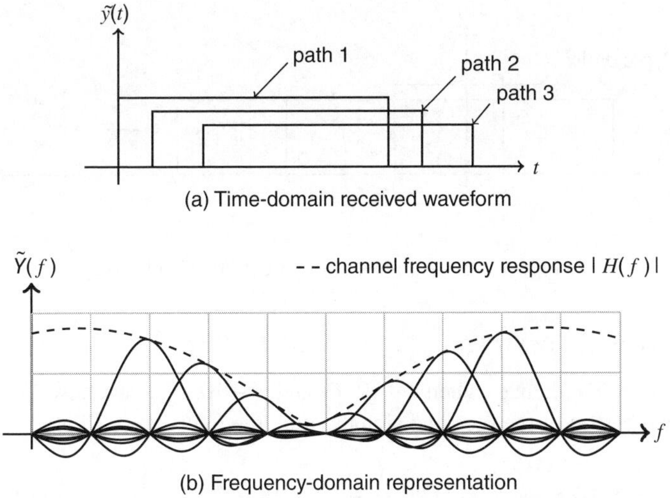

Figure 2.4 Illustration of the received signals in the time and frequency domain; noise not included.

An illustration is shown in Figure 2.4, in a channel with several paths. There is no intercarrier interference (ICI), with each symbol experiencing an amplitude attenuation and a phase rotation. Avoiding ISI and ICI in a time-invariant dispersive channel is the unique advantage of OFDM.

2.2 Cyclic-Prefixed OFDM

2.2.1 Transmitted Signal

Consider one CP-OFDM block from the transmission illustrated in Figure 2.5. Let T_{cp} denote the length of the CP, and define a rectangular window of length $T_{\mathrm{cp}} + T$ as

$$q(t) = \begin{cases} \frac{1}{T} & t \in [-T_{\mathrm{cp}}, T], \\ 0 & \text{otherwise.} \end{cases} \tag{2.17}$$

The baseband signal is

$$x_{\mathrm{cp}}(t) = \sum_{k \in \mathcal{S}_{\mathrm{A}}} s[k] e^{j2\pi \frac{k}{T} t} q(t) \tag{2.18}$$

One can verify $x_{\mathrm{cp}}(t) = x_{\mathrm{cp}}(t + T)$, when $t \in [-T_{\mathrm{cp}}, 0]$. The passband signal is

$$\tilde{x}_{\mathrm{cp}}(t) = 2\Re \left\{ \sum_{k \in \mathcal{S}_{\mathrm{A}}} s[k] e^{j2\pi f_k t} q(t) \right\}, \tag{2.19}$$

where $s[k]$ and $\mathcal{S}_{\mathrm{A}}$ are as defined before.

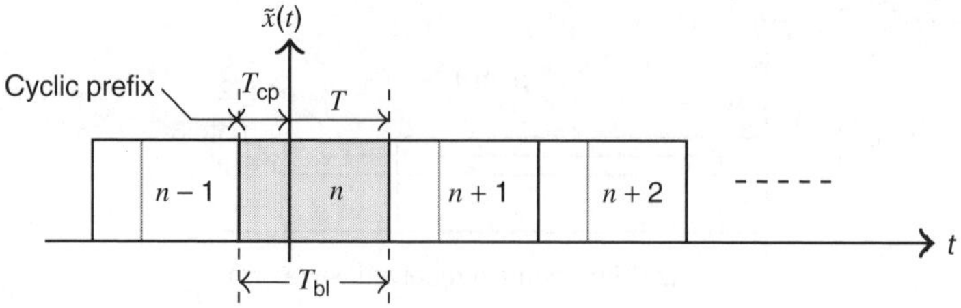

Figure 2.5 An illustration of CP-OFDM blocks.

2.2.2 Receiver Processing

Assume that $T_{\rm cp} > T_{\rm ch}$, hence, when $t \in [0, T]$ and $\tau \in [0, T_{\rm ch}]$, we have $q(t-\tau) = 1/T$. A closed-form expression of the received signal within the interval [0, T] is

$$\begin{aligned}\tilde{y}_{\rm cp}(t) &= \int_0^{T_{\rm ch}} h(\tau) 2\Re\left\{\frac{1}{T}\sum_{k\in S_{\rm A}} s[k]e^{j2\pi f_k(t-\tau)}\right\} d\tau + \tilde{n}(t) \\ &= 2\Re\left\{\frac{1}{T}\sum_{k\in S_{\rm A}} H(f_k)s[k]e^{j2\pi f_k t}\right\} + \tilde{n}(t).\end{aligned} \tag{2.20}$$

Performing the Fourier transform on the truncated waveform of duration T, we have

$$\begin{aligned}\tilde{Y}_{\rm cp}(f) &= \int_0^{T} \tilde{y}_{\rm cp}(t)e^{-j2\pi ft}dt \\ &= \sum_{k\in S_A} H(f_k)s[k]G_{\rm rec}(f-f_k) + \tilde{N}(f), \quad f > 0\end{aligned} \tag{2.21}$$

Draw the samples on the subcarrier frequencies, we have

$$\begin{aligned}y_{\rm cp}[m] &= \tilde{Y}_{\rm cp}(f)\Big|_{f=f_m} \\ &= H(f_m)s[m] + \eta[m].\end{aligned} \tag{2.22}$$

Again, the symbols are separated at different subcarrier frequencies. The orthogonality is preserved after the dispersive channel.

Notice that due to different signaling formats within the guard interval, the receiver processing of ZP-OFDM and CP-OFDM differs in that the Fourier transform of ZP-OFDM is performed within the interval [0, $T_{\rm bl}$] as in (2.13), while it is performed within the interval [0, T] for CP-OFDM as in (2.21).

2.3 OFDM Related Issues

2.3.1 ZP-OFDM versus CP-OFDM

Due to the dispersive nature of channels, the guard interval is necessary to avoid intersymbol interference in OFDM block transmissions. Comparing (2.16) and (2.22), both ZP- and CP-OFDM yield an identical input–output relationship in time-invariant channels.

A majority of textbooks use CP-OFDM as the basis. In this book, we mainly use ZP-OFDM for illustration. The derivations for CP-OFDM can be carried out similarly. For underwater applications, the guard interval could be very long. Zero padding will save the transmission power relative to CP, and reduces the duty cycle for a practical transducer. It also turns out that the derivation of ZP-OFDM is simpler than that for CP-OFDM. For example, the window effect can be easily incorporated in ZP-OFDM, while less convenient for CP-OFDM. Also, all the received samples can be kept in ZP-OFDM, but the CP-OFDM has to throw away the samples in the CP portion, as a result, ZP-OFDM can have more information gleaned from the received signal.

2.3.2 *Peak-to-Average-Power Ratio*

Due to the summation operation in (2.5), OFDM signals have large peak-to-average-power ratio (PAPR) relative to single-carrier transmissions. A mathematical formulation of PAPR is given as

$$\text{PAPR} := \frac{\max(|\tilde{x}(t)|^2)}{\mathbb{E}[|\tilde{x}(t)|^2]}. \tag{2.23}$$

PAPR reduction is an important issue in practical OFDM systems. Chapter 4 will be dedicated to the PAPR control of OFDM, with several practical strategies discussed.

2.3.3 *Power Spectrum and Bandwidth*

Assume that there are many ZP-OFDM blocks transmitted in sequence and that the information symbols are uncorrelated and have zero mean. The power spectral density (PSD) of the ZP-OFDM signal at the positive frequency range can be easily obtained as

$$\text{PSD}_{\text{zp-ofdm}}(f) = \frac{E_s}{T_{\text{bl}}} \sum_{k \in S_A} |G(f - f_k)|^2, \quad f > 0 \tag{2.24}$$

where $E_s := \mathbb{E}[|s[m]|^2]$ denotes the average symbol energy at each subcarrier. Figure 2.6 provides an illustration of the PSD of ZP-OFDM transmissions.

Assuming that there are a few null subcarriers placed on the end of the frequency band, the bandwidth is roughly defined as

$$B = \frac{K}{T}. \tag{2.25}$$

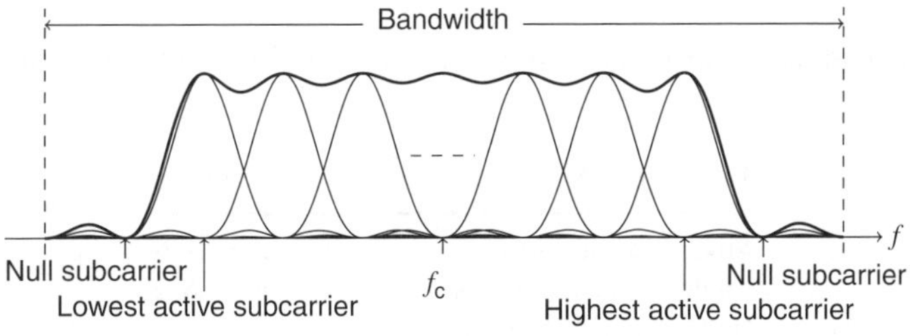

Figure 2.6 Illustration of ZP-OFDM power spectrum density.

A more careful definition of bandwidth, such as 3-dB bandwidth, can be computed by knowing the exact locations of the null subcarriers on the edges of the frequency band.

2.3.4 Subcarrier Assignment

There are three types of subcarriers in OFDM transmissions, which are used for different purposes.

- Null subcarriers are used at the edge of the frequency band to prevent leakage. Also, null subcarriers are mixed with active subcarriers to facilitate Doppler estimation and noise variance estimation.
- Pilot subcarriers are used for channel estimation.
- Data subcarriers are used to carry information symbols.

In this book, a lot of experimental results are carried out using the subcarrier assignment described in Table 2.1. Out of a total of $K = 1024$ subcarriers, $K_\mathrm{p} = 256$ pilot subcarriers are uniformly spaced; among $K_\mathrm{n} = 96$ null subcarriers, 24 null subcarriers are on each side, and 48 null subcarriers are randomly inserted in the middle, and the remaining $K_\mathrm{d} = 672$ subcarriers are used as data subcarriers.

2.3.5 Overall Data Rate

Let us compute the overall data rate considering various overheads; the details of channel coding and constellation mapping will be provided in Chapter 3. Let the coding rate be r. The spectral efficiency in terms of bits per second per Hz is

$$\alpha = r\log_2 M \frac{T}{T_\mathrm{bl}} \frac{K_\mathrm{d}}{K}, \tag{2.26}$$

where M is the size of the constellation for mapping information bits into symbols. The achieved data rate in bits per second is

$$R = \alpha B. \tag{2.27}$$

It can be simply computed as

$$R = \frac{rK_\mathrm{d} \log_2 M}{T_\mathrm{bl}}, \tag{2.28}$$

by counting the number of information bits over one ZP-OFDM block duration.

Table 2.1 One example subcarrier assignment

Number of subcarriers	$K = 1024$
Number of data carriers	$K_\mathrm{d} = 672$
Number of pilot carriers	$K_\mathrm{p} = 256$
Number of null subcarriers	$K_\mathrm{n} = 96$

2.3.6 Design Guidelines

Finding suitable T and T_g is important for an OFDM system design.

- On the one hand, using a large symbol duration T increases the the spectral efficiency. On the other hand, there could be more channel variations within a large symbol duration.
- On the one hand, a small T_g is desirable for lowering the overhead to achieve a large spectral efficiency. On the other hand, T_g shall be larger than T_{ch} to avoid interblock interference.

The selection of T and T_g depends on the channel characteristics such as delay spread and the channel coherence time. Once the subcarrier spacing $1/T$ is specified, the number of subcarriers can be found based on the desired signal bandwidth and the transducer characteristics.

2.4 Implementation via Discrete Fourier Transform

We have used the continuous time waveform for illustration of the orthogonality property of OFDM modulation. Now let us consider the digital implementation.

Let f_s denote the sampling rate, which has to be higher than twice of the highest frequency of the signal component $f_{K/2-1}$. For ease of implementation, one can choose

$$f_s = \lambda B \tag{2.29}$$

where λ is the ratio of the passband sampling rate and the baseband sampling rate. There are $N_{bl} = T_{bl}/T_s$ samples to generate, where $T_s = 1/f_s$ is the sampling step size.

The first step is to construct a complex baseband signal as

$$u(t) = \sum_{k=-K/2}^{K/2-1} s[k]e^{j2\pi \frac{k}{T} t}. \tag{2.30}$$

This is a periodic signal with period T. So it suffices to generate the samples within only one period $[0, T]$. There are $N = T/T_s$ samples, as denoted by

$$u[n] = \sum_{k=-K/2}^{K/2-1} s[k]e^{j2\pi \frac{kn}{\lambda K}} = \sum_{k=0}^{\lambda K-1} \bar{s}[k]e^{j2\pi \frac{kn}{\lambda K}}, \tag{2.31}$$

where the equivalent sequence $\bar{s}[k]$ is defined as

$$\bar{s}[k] = \begin{cases} s[k], & k = 0, \dots, \frac{K}{2} - 1 \\ 0, & k = K, \dots, \lambda K - \frac{K}{2} - 1 \\ s[k - \lambda K], & k = \lambda K - \frac{K}{2}, \dots, \lambda K - 1. \end{cases} \tag{2.32}$$

From (2.31), the sequence $u[n]$ can be generated by a DFT of size λK applied on the input sequence $\bar{s}[k]$.

$$u[n] = \text{IDFT}\{\bar{s}[k]\}, \quad n, k = 0, \dots, \lambda K - 1. \tag{2.33}$$

Once $u[n]$ is obtained, the sequence $\tilde{x}[n]$ can be simply obtained as

$$\tilde{x}[n] = 2\Re\{u[n \bmod N]e^{j2\pi n f_c T_s} g(nT_s)\}, \quad n = 0, \dots, N_{bl} - 1 \tag{2.34}$$

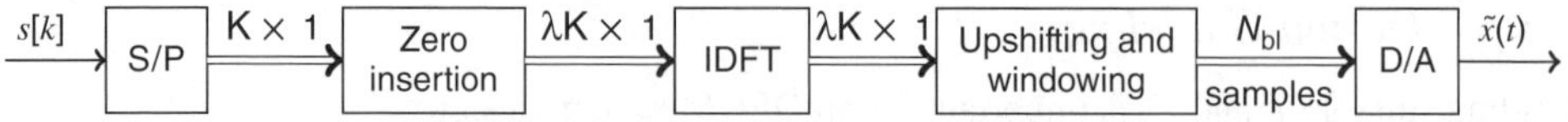

Figure 2.7 The transmitter implementation.

This real sequence is passed to the D/A converter, amplified by a power amplifier, and transmitted through the transducer. The illustration is shown in Figure 2.7.

It is desirable to choose λK to be a power of 2 so that an FFT can be applied. If λK is not a power of 2, one can simply choose an FFT of a larger size by filling zeros at the edges of the frequency band.

The digital implementation of the receiver in the presence of a time-invariant channel is as follows: (1) Taking the time-domain samples at a sampling rate f_s. (2) Downshifting the passband sequence to the baseband. (3) Low pass filtering and downsampling the baseband sequence to the sampling rate B. (4) Overlap-and-add followed by DFT for ZP-OFDM, or just simply DFT for CP-OFDM where the FFT size is K. More details on the receiver implementation for underwater acoustic channels will be available in later chapters.

2.5 Challenges and Remedies for OFDM

To make OFDM modulation successful in a practical underwater system, the following three issues must be adequately addressed.

- Plain (or uncoded) OFDM has poor performance in fading channels, since it does not exploit the multipath diversity inherent to the channel.
- OFDM is sensitive to intercarrier interference (ICI) caused by channel variations. Underwater channels vary fast due to the large ratio of the platform motion relative to the sound propagation speed. Even with stationary transmitters and receivers, significant ICI could still exist due to wave action and water motion.
- OFDM transmission has large peak-to-average power ratio (PAPR). The required large power backoff reduces the transmission range.

Countermeasures need to be developed to address these challenges.

- Among other possible choices, diversity combining and channel coding are two effective approaches that can drastically improve the system performance. Multichannel reception introduces both power gain and diversity gain, through the use of multiple receiving elements. Channel coding introduces both diversity gain and coding gain through the use of redundant signaling.
- Since ICI is inevitable in fast-varying channels, it has to be explicitly dealt with. This book provides an in-depth treatment on advanced receiver design for OFDM systems in the presence of fast fading channels.
- The transmitter will apply PAPR reduction as much as possible. Although OFDM has high PAPR, it is an ideal choice for short range, high data rate applications. Note that some systems may not be peak power constrained, but instead average power constrained.

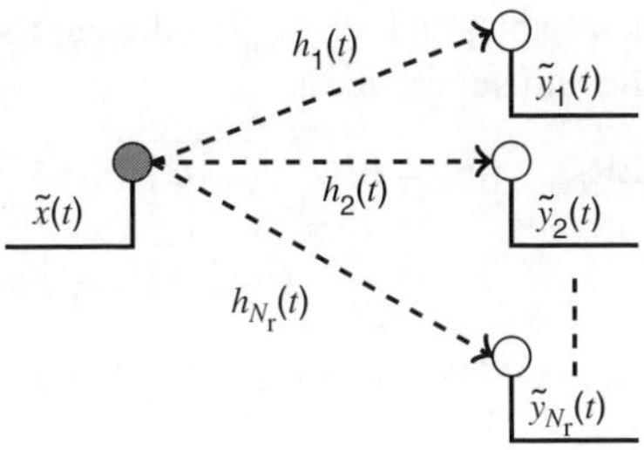

Figure 2.8 A system with one transmitter and N_r receivers.

2.5.1 Benefits of Diversity Combining and Channel Coding

Here let us illustrate the performance gains of diversity combining and channel coding. Consider a system with N_r receivers as shown in Figure 2.8. The system model in (2.16) can be recast as

$$y_\nu[m] = H_\nu[m]s[m] + \eta_\nu[m], \quad \nu = 1, \dots, N_r \tag{2.35}$$

where $H_\nu[m] := H_\nu(f_m)$ is the channel gain on the mth subcarrier.

On each subcarrier, maximum ratio combining (MRC) is applied, leading to a symbol estimate:

$$\hat{s}[m] = \frac{\sum_{\nu=1}^{N_r} H_\nu^*[m]y_\nu[m]}{\sum_{\nu=1}^{N_r} |H_\nu[m]|^2}. \tag{2.36}$$

Define the symbol energy $\sigma_s^2 = \mathbb{E}[|s[m]|^2]$ and noise variance $\sigma_\eta^2 = \mathbb{E}[|\eta[m]|^2]$. The instantaneous SNR at the mth subcarrier is

$$\gamma[m] = \left(\sum_{\nu=1}^{N_r} |H_\nu[m]|^2\right)\frac{\sigma_s^2}{\sigma_\eta^2} \tag{2.37}$$

with the average SNR defined as $\bar{\gamma}[m] = \mathbb{E}[\gamma[m]]$.

First, let us consider *uncoded* transmissions. Assuming a BPSK symbol on each data subcarrier, and that there is no fading, i.e., $\gamma[m] = \bar{\gamma}[m]$, the average bit error rate (BER) on the mth subcarrier is:

$$\text{BER}_{\text{awgn}}[m] = Q\left(\sqrt{2\bar{\gamma}[m]}\right), \tag{2.38}$$

where

$$\begin{aligned} Q(x) &= \int_x^\infty \frac{1}{\sqrt{2\pi}} e^{-t^2/2} dt \\ &= \frac{1}{\pi}\int_0^{\pi/2} \exp\left(-\frac{x^2}{2\sin^2\theta}\right) d\theta. \end{aligned} \tag{2.39}$$

With fading, $\gamma[m]$ is random variable, and let $f_{\gamma[m]}(\gamma)$ denote its probability density function (pdf). The average BER on the mth subcarrier is

$$\begin{aligned}\overline{\mathrm{BER}}_{\mathrm{fading}}[m] &= \mathbb{E}[Q(\sqrt{2\gamma[m]})] \\ &= \int_0^{\infty} Q(\sqrt{2\gamma}) f_{\gamma[m]}(\gamma) d\gamma.\end{aligned} \tag{2.40}$$

The average BER of the uncoded OFDM system can be obtained by taking the average of the BER on all the data subcarriers.

Now consider channel coding is applied. In practical OFDM systems, coding is applied across data subcarriers. For theoretical computation, let us consider channel coding applied across time, and assume that long codewords and Gaussian signaling are used. Without fading, the capacity on the mth subcarrier is

$$C_{\mathrm{awgn}}[m] = \log_2(1 + \overline{\gamma}[m]). \tag{2.41}$$

With fading, the ergodic capacity on the mth subcarrier is

$$\begin{aligned}\overline{C}_{\mathrm{fading}}[m] &= \mathbb{E}[\log_2(1 + \gamma[m])] \\ &= \int_0^{\infty} \log_2(1 + \gamma) f_{\gamma[m]}(\gamma) d\gamma.\end{aligned} \tag{2.42}$$

The total ergodic capacity per OFDM symbol is the sum of the capacities on all the data subcarriers.

The average BER and ergodic capacity of fading channels can be readliy evaluated once the fading characteristic is known.

Example 2.1

We assume that $H_\nu[m]$ is complex Gaussian with zero mean and unit variance, for all subcarriers at all receive elements. Thus, $|H[m]|$ is Rayleigh distributed with unit variance, and the same average SNR $\overline{\gamma}[m]$ on all the subchannels is the same

$$\overline{\gamma} = N_{\mathrm{r}} \frac{\sigma_s^2}{\sigma_\eta^2} \tag{2.43}$$

Since $\gamma[m]$ is Chi-square distributed with $2N_{\mathrm{r}}$ degrees of freedom, its pdf is

$$f_{\gamma[m]}(\gamma) = \left(\frac{N_{\mathrm{r}}}{\overline{\gamma}}\right)^{N_{\mathrm{r}}} \frac{\gamma^{N_{\mathrm{r}}-1}}{\Gamma(N_{\mathrm{r}})} \exp\left(-\frac{N_{\mathrm{r}}\gamma}{\overline{\gamma}}\right). \tag{2.44}$$

The average BER in (2.40) can be evaluated in a single integral as [342]:

$$\overline{\mathrm{BER}}_{\mathrm{fading}}[m] = \frac{1}{\pi} \int_0^{\pi/2} \left(1 + \frac{\overline{\gamma}}{N_{\mathrm{r}} \sin^2\theta}\right)^{-N_{\mathrm{r}}} d\theta. \tag{2.45}$$

Figure 2.9 plots the uncoded BER curves as a function of the average SNR $\overline{\gamma}$ (in decibels) for both AWGN and Rayleigh fading channels. At uncoded BER 10^{-3}, the SNR gap is about 17 dB if no diversity combining is used. This comparison demonstrates that *fading channel drastically affects the uncoded performance.*

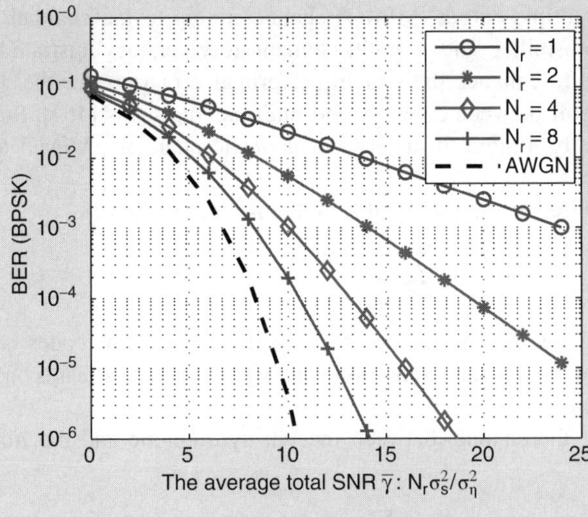

Figure 2.9 Comparison of the bit error rates in AWGN and Rayleigh fading channels, BPSK constellation.

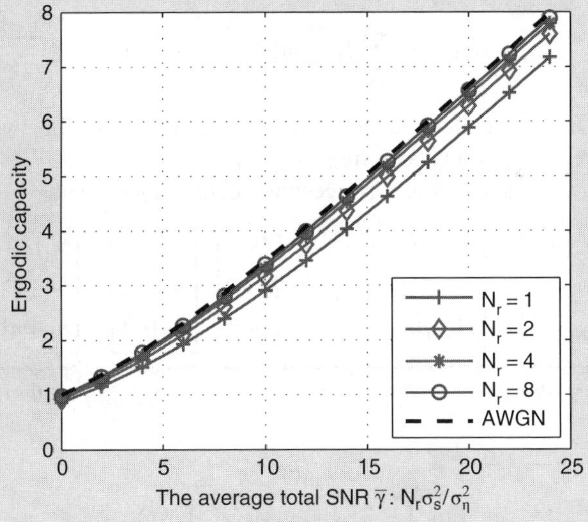

Figure 2.10 Comparison of the capacities of the AWGN and Rayleigh fading channels.

Figure 2.10 plots the capacity curves of the AWGN and Rayleigh fading channels as a function of the average SNR $\bar{\gamma}$ (in decibels), on each OFDM subcarrier. At data rate of 2 bits per symbol, the SNR gap is only about 2 dB even with $N_r = 1$, much smaller than the gap as shown in Figure 2.9. This example illustrates that channel coding is a crucial component for improving the performance of OFDM in fading channels.

2.6 MIMO OFDM

A multiple-input multiple-output (MIMO) channel refers to the channel in a system with a multi-element transmitter and a multi-element receiver, as illustrated in Figure 2.11. MIMO can be used to enhance the system performance or improve the data rate. There is a fundamental tradeoff between diversity and multiplexing gains [467]. Since the frequency band is fundamentally limited in underwater acoustic channels, MIMO techniques appear especially attractive.

Define the transmitted passband signal from the μth transmitter as

$$\tilde{x}_\mu(t) = \sum_{k\in S_A} s_\mu[k]e^{j2\pi f_k t}g(t), \quad t \in [0, T_{\rm bl}], \tag{2.46}$$

The symbols $s_\mu[k]$ could be coded using the so-called space time codes or space frequency codes. One simple method is spatial multiplexing, where the data streams on different transmit elements are independent.

Let $h_{\nu,\mu}(t)$ denote the channel between the νth hydrophone and the μth transducer. The received signal is

$$\tilde{y}_\nu(t) = \sum_{\mu=1}^{N_t} h_{\nu,\mu}(t) \star \tilde{x}_\mu(t) + \tilde{n}_\nu(t), \tag{2.47}$$

where $\tilde{n}_\nu(t)$ is the additive noise on the νth receiver. The received samples in the frequency domain are

$$y_\nu[m] = \sum_{\mu=1}^{N_t} H_{\nu,\mu}[m]s_\mu[m] + \eta_\nu[m], \tag{2.48}$$

where $H_{\nu,\mu}[m] := H_{\nu,\mu}(f_m)$ is the frequencey response of the channel between the μth transmitter and the νth receiver, evaluated at frequency f_m.

Stacking frequency measurements at N_r receiving hydrophones yields

$$\underbrace{\begin{bmatrix} y_1[m] \\ \vdots \\ y_{N_r}[m] \end{bmatrix}}_{:=\mathbf{y}[m]} = \underbrace{\begin{bmatrix} H_{1,1}[m] & \cdots & H_{1,N_t}[m] \\ \vdots & \ddots & \vdots \\ H_{N_r,1}[m] & \cdots & H_{N_r,N_t}[m] \end{bmatrix}}_{:=\mathbf{H}[m]} \underbrace{\begin{bmatrix} s_1[m] \\ \vdots \\ s_{N_t}[m] \end{bmatrix}}_{:=\mathbf{s}[m]} + \underbrace{\begin{bmatrix} \eta_1[m] \\ \vdots \\ \eta_{N_r}[m] \end{bmatrix}}_{:=\boldsymbol{\eta}[m]} \tag{2.49}$$

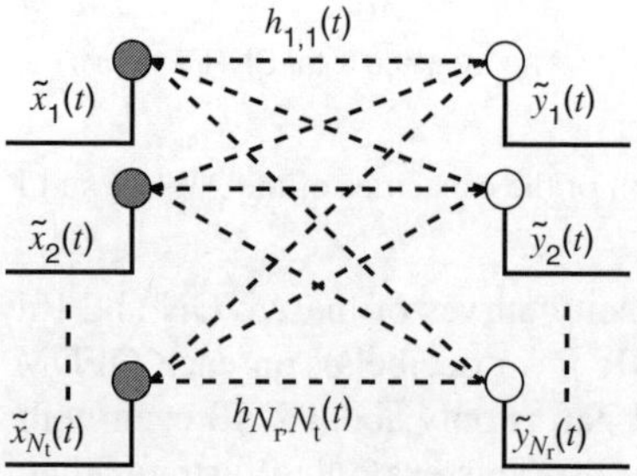

Figure 2.11 A system with N_t transmitters and N_r receivers.

which can be compactly expressed as

$$\mathbf{y}[m] = \mathbf{H}[m]\mathbf{s}[m] + \boldsymbol{\eta}[m], \qquad m = -\frac{K}{2}, \cdots, \frac{K}{2} - 1. \tag{2.50}$$

Assuming that $\boldsymbol{\eta}[m]$ is complex Gaussian distributed with zero mean and covariance matrix $\sigma_\eta^2 \mathbf{I}_{N_r}$, the channel capacity at the mth subcarrier conditional on $\mathbf{H}[m]$ is expressed as

$$C(N_t, N_r|\mathbf{H}[m]) = \log_2 \det \left(\mathbf{I}_{N_r} + \frac{\sigma_{\text{total}}^2}{N_0 N_t} \mathbf{H}[m]\mathbf{H}^H[m] \right) \tag{2.51}$$

where $\sigma_{\text{total}}^2 = \sum_{\mu=1}^{N_t} \mathbb{E}[|s_\mu[m]|^2]$ is the transmission energy across all N_t elements. Averaged over all possible channel realizations, the ergodic capacity on the mth subcarrier is [379]:

$$\overline{C}(m; N_t, N_r) = \mathbb{E}_{\{\mathbf{H}[m]\}}[C(N_t, N_r|\mathbf{H}[m])]. \tag{2.52}$$

Example 2.2

Now let us assume that all the entries of $\mathbf{H}[m]$ are i.i.d. Gaussian distributed with zero mean and unit variance. Define $\overline{\gamma}$ as the average SNR at each receive antenna, which equals $\sigma_{\text{total}}^2/\sigma_\eta^2$ in this example. In the high SNR regime, the ergodic capacity can be approximated as

$$\overline{C}(m; N_t, N_r) \approx \min\{N_t, N_r\} \log_2(\overline{\gamma}), \tag{2.53}$$

which shows a linear growth with respect to the minimum number of transmit and receive elements. Assuming an identical number of transmit and receive elements, $N_t = N_r = N$, Figure 2.12 depicts the ergodic capacity at each subcarrier with different values of receiver SNR. A considerable capacity growth can be observed by increasing the number of transmit and receive elements.

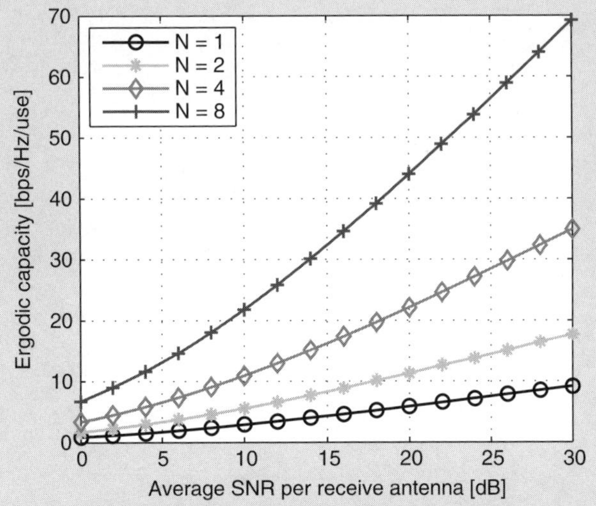

Figure 2.12 The ergodic capacity in MIMO transmissions, $N_t = N_r = N$.

There are more channels to estimate at each receiver in a MIMO system. The main challenge of realizing MIMO in underwater communications is that the channels change very fast and channel estimation may require substantial overhead, which can compromise the benefits of MIMO. Also, the transducers at the transmitter side are often more costly than the hydrophones at the receiver side.

2.7 Bibliographical Notes

The idea of OFDM dates back to 1960s. The first use of overlapping but noninterferencing subchannels was due to Chang in 1966 [68]. Weinstein and Ebert proposed to use the discrete Fourier transform for baseband modulation and demodulation in 1971 [432]. Peled and Ruiz proposed the use of cyclic prefix to achieve orthogonality after the channel dispersion in 1980 [306]. The first commercial success of OFDM is in the application of asymmetrical digital subscribe lines (ADSL) in early 1990s. Now, OFDM has a wideapread use in modern communication systems, with prominent examples such as WiFi (IEEE 802.11 a/g/n), WiMAX (IEEE 802.16), digital audio and video broadcasting (DAB/DVB), and fourth generation (4G) cellular systems. A historic review of OFDM is recently provided by Weinstein, 2009 [431].

Influential tutorial-style papers on OFDM include [44, 120, 333, 418]. A number of books on OFDM technology are available in the market, including Bingham, 2000 [45], Hanzo, Munster, Choi and Keller, 2003 [157], Prasad, 2004 [315], Bahai, Saltzberg, and Ergen, 2004 16, Li and Stuber (eds), 2006 [243], Chiueh and Tsai, 2007 [81], Fazel and Kaiser, 2008 [118], Narasimhamurthy, Banavarand, and Tepedelenlioglu, 2010 [291]. The application of OFDM in optical communication is described by Shieh and Djordjevic, 2009 [340].

3

Nonbinary LDPC Coded OFDM

Channel coding is an integral part of a digital communication system. Various topics regarding to channel coding have been covered by existing books. In this chapter, we first go over two basic formats of coupling channel coding with OFDM modulation. Then, we will focus on nonbinary low-density-parity-check (LDPC) codes, which are used in the simulation and experimental results contained in this book.

3.1 Channel Coding for OFDM

Figure 3.1 shows a classic diagram of a digital communications system in wireless channels, where information bits pass through source coding and channel coding modules before being modulated for transmission. Channel coding is introduced to protect information bits from errors after transmission through communication channels, and the bit-to-symbol mapping are used to increase the transmission spectral efficiency. Figures 3.2 and 3.3 show typical one-dimensional and two-dimensional constellations widely used in wireless systems, including binary phase-shift keying (BPSK), M-ary pulse amplitude modulation (PAM), quadrature phase shift keying (QPSK) and M-ary quadrature amplitude modulation (QAM).

Stemming from Shannon's information theory [337], significant progress on channel coding has been made. Meanwhile, OFDM has emerged as an effective modulation technique for high-rate communications. Next, we will give a brief overview on channel coding, coded modulation, and discuss several channel coding schemes widely used in OFDM systems.

3.1.1 Channel Coding

To combat the imparity of communication channels, channel coding allows *error detection and correction* at receiver side by introducing *redundancy* into transmitted symbols. As illustrated in Figure 3.1, the information sequence from source can be divided into multiple blocks, and each block is a k-tuple vector, denoted by $\mathbf{u} = [u_0, \cdots, u_{k-1}]^T$ and termed as message block. Channel coding projects each message block into a larger space which is spanned by n-tuple vectors, and the corresponding vector $\mathbf{v} = [v_0, \cdots, v_{n-1}]^T$ is called a *codeword*. The corresponding code is referred to as (n, k) code, with the *code rate* defined as $r := k/n$. The primary

OFDM for Underwater Acoustic Communications, First Edition. Shengli Zhou and Zhaohui Wang.
© 2014 John Wiley & Sons, Ltd. Published 2014 by John Wiley & Sons, Ltd.

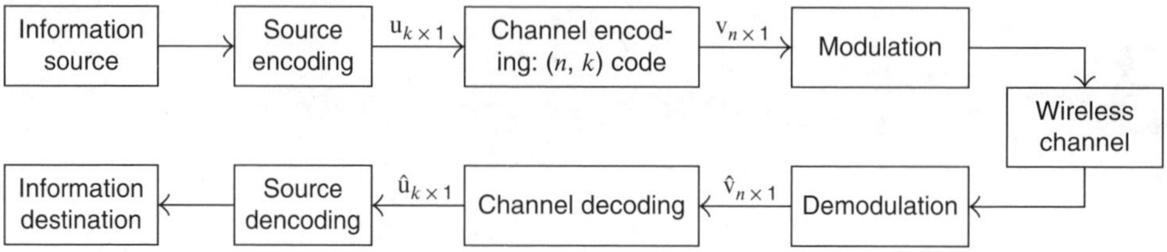

Figure 3.1 A schematic diagram of wireless communication system.

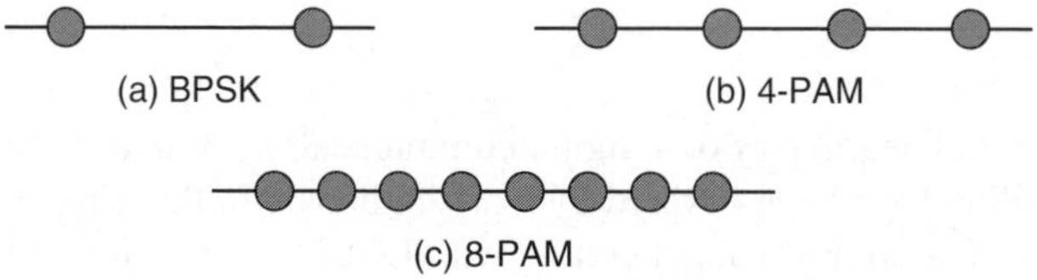

Figure 3.2 One-dimensional constellations: BPSK, 4-PAM, 8-PAM.

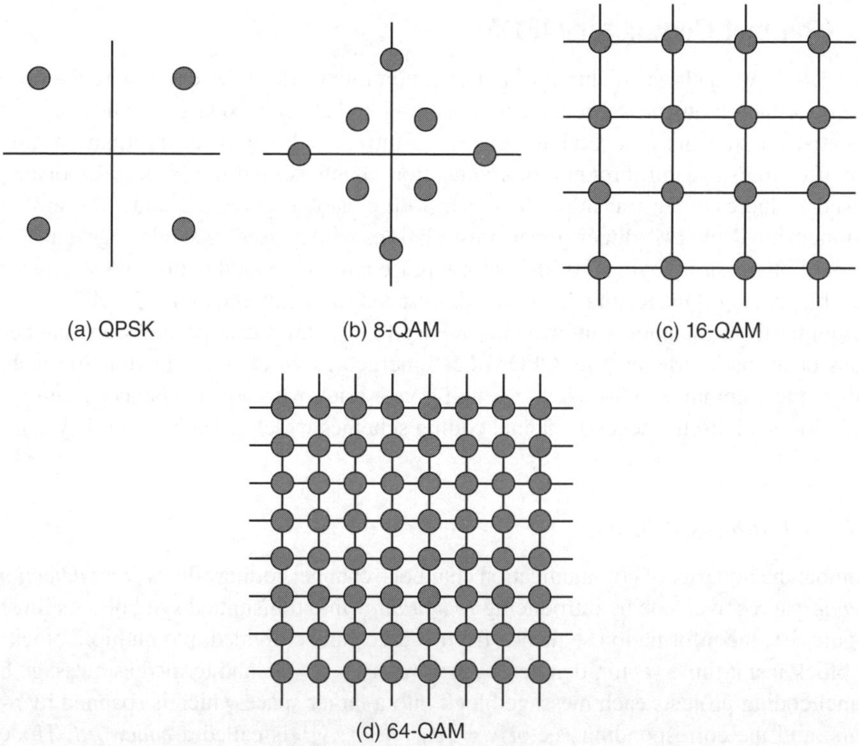

Figure 3.3 Two-dimensional constellations: QPSK, 8-QAM, 16-QAM, 64-QAM.

task of channel coding is to design mapping and demapping strategies between the space spanned by the k-tuple message vectors and that spanned by n-tuple vectors, which yields desirable error detection and correction capabilities at moderate computation complexity.

There are two structurally different types of codes: block codes and convolutional codes (CC). In block channel coding, each message block is encoded independently by multiplying the message block with a *generator matrix*, while in convolutional coding, each message block is encoded through convolutionary operations between message blocks and predefined generator sequences of length m, leading to the obtained codeword as a function of the current message block and previous m message blocks. Classical block coding schemes include Hamming codes, Bose-Chaudhuri-Hocquenghem (BCH) codes, Reed-Solomon codes (RS), cyclic codes, etc. The modern coding theory is defined by two types of codes: turbo codes and low-density parity-check (LDPC) codes, both Shannon capacity-approaching. In particular, turbo codes are generated with two or multiple convolutional encoders which are concatenated according to certain structures. LDPC codes are defined by *sparse* parity-check matrices, which facilitates code design and implementation on graphs [137, 269].

Channel encoding and decoding are performed over finite fields, which are referred to as Galois fields, in memory of the French mathematician Evariste Galois (1811–1832). A Galois field of size q is denoted as GF(q), where q is often chosen as $q = 2^p$. The simplest field is GF(2), which only has two elements $\{0, 1\}$. Accordingly, elements in both $\mathbf{u}$ and $\mathbf{v}$ are therefore binary bits, and the corresponding codes are called binary codes. For GF(q) with $q = 2^p$ and $p > 1$, each element in both $\mathbf{u}$ and $\mathbf{v}$ is formed by a sequence of symbols, each represented by p binary bits. From pure coding point of view, nonbinary coding can have better performance than binary coding, especially when the code block length is small and the code rate is high.

3.1.2 Coded Modulation

The concept that channel coding and modulation in Figure 3.1 shall be taken as one entity creates a research area called *coded modulation* [275]. Four important approaches are: trellis coded modulation (TCM), block coded modulation (BCM), bit interleaved coded modulation (BICM), and nonbinary coded modulation (NCM).

Different coded modulation strategies have applications in different scenarios. The key idea of TCM is to encode message blocks onto an expanded modulation constellation set [396]. It is effective for the AWGN channel, and has been widely used in wireline systems. Through a multilevel coding method that uses multiple error-correcting codes of different error correction capabilities, BCM is mainly used for unequal error protection [189]. In BICM, the information bit sequence is firstly encoded into a coded bit sequence, and then interleaved before mapped to information symbols. Compared with other schemes, BICM can effectively counteract the fading property of wireless channels [52]. The binary coded modulation involves constellation mapping that converts bits into symbols. As shown in two examples of coded modulation in Figures 3.2 and 3.3, each symbol in a constellation of size 2^b carries b coded bits. Different ways of labeling the symbols with bit indices may lead to different receiver performance. In practice, natural mapping, Gray mapping, and set-partitioning based mapping have been studied.

At the expense of design complexity, nonbinary coded modulation has an advantage over binary coded modulation that the nonbinary codeword can match very well with the underlying constellations, avoiding bit-to-symbol conversion at the transmitter and symbol-to-bit conversion at the receiver [31]. Suppose that a constellation size of 2^b will be used. To couple nonbinary coding with constellation mapping, it is desirable to match the field order with the constellation size, i.e., $p = b$. This way, one element in GF(q) is mapped to one point in the signal constellation. In occasions when b is small, it may be preferable to choose $p > b$. Then, it is convenient to choose $\frac{p}{b}$ as an integer, and map each element in GF(q) to $\frac{p}{b}$ symbols drawn from the constellation. In both scenarios, constellation labeling does not affect the receiver performance.

3.1.3 Coded OFDM

As indicated in Section 2.5, channel coding is important for OFDM systems to address the fading property of wireless channels. Figures 3.4 and 3.5 depict a binary coded BICM-OFDM system and a nonbinary coded OFDM system, respectively.

Over the years, there have been many binary codes developed for binary coded OFDM systems. Several examples are as follows. Convolutional codes have been widely used in wireless OFDM systems. For example, a 64-state rate-$\frac{1}{2}$ convolutional code, has widespread use in WiFi (IEEE 802.11a/g) systems, where the Viterbi algorithm is used for decoding. RS codes are often used in combination with CC, where a high rate RS code is used as an outer code and a CC is used as an inner code. The bursty errors from the Viterbi decoder can be effectively mitigated by the RS decoder. Turbo codes and LDPC codes are more powerful channel coding schemes, which have been adopted into recent wireless standards, such as IEEE 802.16. Iterative decoding algorithms are used for decoding.

In the literature, both nonbinary Turbo codes and nonbinary LDPC codes have been constructed. In this book, we focus on nonbinary LDPC codes and their application to underwater OFDM systems.

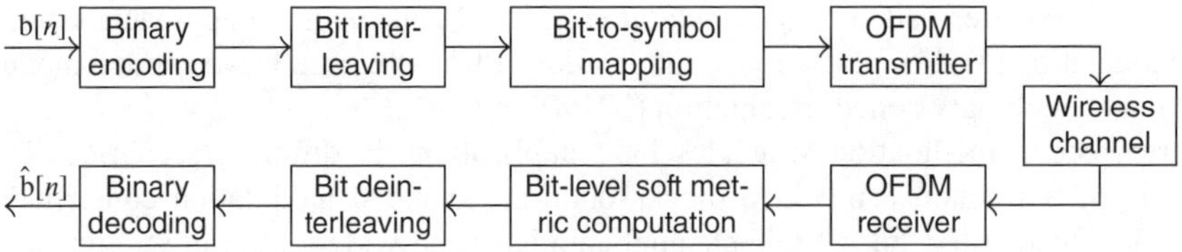

Figure 3.4 A schematic block diagram of a bit-interleaved coded OFDM system.

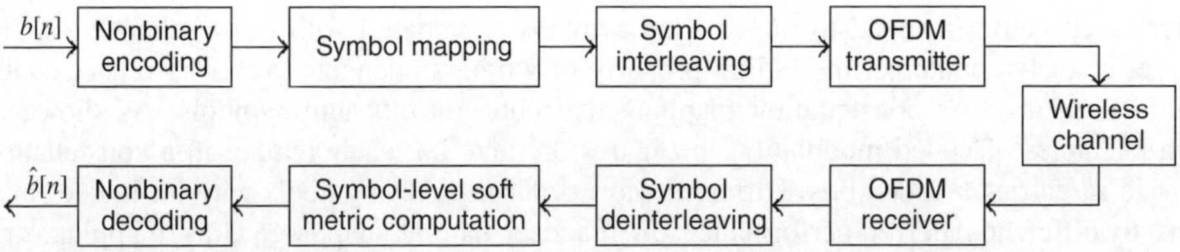

Figure 3.5 A schematic block diagram of a nonbinary coded OFDM system.

3.2 Nonbinary LDPC Codes

Consider a high order Galois field of GF(q), and let $\{\alpha_0 = 0, \alpha_1, \ldots, \alpha_{q-1}\}$ denote elements in GF(q). Nonbinary LDPC code is a linear block code defined over GF(q). Define a codeword consisting of N coded symbols as

$$\mathbf{c} = \left[c[0] \;\; \ldots \;\; c[N-1]\right]^{\mathrm{T}}, \tag{3.1}$$

where $c[n] \in \mathrm{GF}(q)$. The codeword $\mathbf{c}$ is a valid codeword if and only if it satisfies the parity check constraint

$$\mathbf{Hc} = \mathbf{0}, \tag{3.2}$$

where $\mathbf{H}$ denotes the parity check matrix[1]. For an LDPC code, the matrix $\mathbf{H}$ has low density (or percentage) of nonzero entries. For binary codes over GF(2), the nonzero entries of $\mathbf{H}$ can only have value 1. For nonbinary codes over GF(q), the nonzero entries of $\mathbf{H}$ take values from $\{\alpha_1, \ldots, \alpha_{q-1}\}$. With a parity check matrix $\mathbf{H}$ of size $M \times N$, the designed code rate of an LDPC code is

$$r = \frac{N-M}{N}. \tag{3.3}$$

The column and row weights of $\mathbf{H}$ are defined as the number of nonzero entries in the column and row, respectively. An LDPC code whose $\mathbf{H}$ has fixed column weight and fixed row weight is called a regular code. Otherwise, it is called an irregular code.

Example 3.1

Consider a code of $N = 12$ symbols $[c[0], \ldots, c[11]]^{\mathrm{T}}$ from GF(8), where any valid codeword satisfies six parity check equations as

$$\alpha_5 c[0] + \alpha_1 c[3] + \alpha_1 c[6] + \alpha_2 c[9] = 0, \tag{3.4}$$

$$\alpha_3 c[1] + \alpha_3 c[4] + \alpha_5 c[7] + \alpha_3 c[10] = 0, \tag{3.5}$$

$$\alpha_1 c[0] + \alpha_5 c[5] + \alpha_2 c[7] + \alpha_7 c[11] = 0, \tag{3.6}$$

$$\alpha_5 c[2] + \alpha_6 c[4] + \alpha_3 c[6] + \alpha_7 c[11] = 0, \tag{3.7}$$

$$\alpha_2 c[2] + \alpha_3 c[5] + \alpha_1 c[8] + \alpha_5 c[9] = 0, \tag{3.8}$$

$$\alpha_2 c[1] + \alpha_4 c[3] + \alpha_2 c[8] + \alpha_4 c[11] = 0. \tag{3.9}$$

The parity check matrix $\mathbf{H}$ then has $M = 6$ rows and $N = 12$ columns, as depicted in Figure 3.6(a). The code rate is $\frac{1}{2}$. Each column has weight 2 and each row has weight 4, and hence it is a regular code. An LDPC code can be represented by a Tanner graph, as illustrated in Figure 3.6(b), where variable nodes (VND) correspond to the columns of $\mathbf{H}$ and the check nodes (CND) correspond to rows of $\mathbf{H}$.

[1] The $\mathbf{H}$ matrix in this chapter stands for the parity check matrix for an LDPC code, while it stands for a channel matrix in other chapters.

$$
\mathbf{H} = \begin{array}{c} \\ 1 \\ 2 \\ 3 \\ 4 \\ 5 \\ 6 \end{array}
\begin{array}{c}
\begin{array}{cccccccccccc} 1 & 2 & 3 & 4 & 5 & 6 & 7 & 8 & 9 & 10 & 11 & 12 \end{array} \\
\begin{bmatrix}
5 & 0 & 0 & 1 & 0 & 0 & 1 & 0 & 0 & 2 & 0 & 0 \\
0 & 3 & 0 & 0 & 3 & 0 & 0 & 5 & 0 & 0 & 3 & 0 \\
1 & 0 & 0 & 0 & 0 & 5 & 0 & 2 & 0 & 0 & 7 & 0 \\
0 & 0 & 5 & 0 & 6 & 0 & 3 & 0 & 0 & 0 & 0 & 7 \\
0 & 0 & 2 & 0 & 0 & 3 & 0 & 0 & 1 & 5 & 0 & 0 \\
0 & 2 & 0 & 4 & 0 & 0 & 0 & 0 & 2 & 0 & 0 & 4
\end{bmatrix}
\end{array}
$$

(a) Parity check matrix

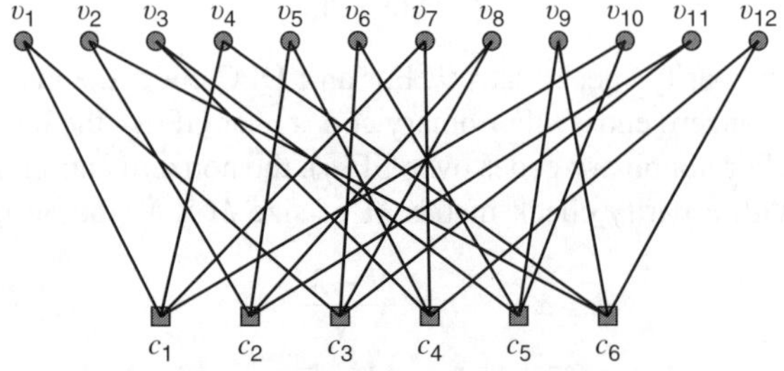

(b) Tanner graph representation

Figure 3.6 An example code with $M = 6$, $N = 12$ over GF(8).

The performance of nonbinary LDPC codes depend on various factors. In general, the larger the field size and the larger the block length, the better the performance for nonbinary LDPC codes. Column and row weight distributions can be optimized to improve the performance. It turns out that the relationship of the performance with the mean column weight is not monotonic. In general, as the field size increases, the density shall get smaller. In this book, we focus on two particular code designs, one being the regular cycle code suitable for a large q and the other being irregular nonbinary LDPC codes suitable for a small or moderate q.

3.2.1 Nonbinary Regular Cycle Codes

An LDPC code whose $\mathbf{H}$ has fixed column weight $j = 2$ is called a cycle code [200]. The parity check matrices of cycle codes have the the lowest density, as each coded symbol needs to at least participate two parity check equations. Cycle codes can be regular or irregular depending on whether the row weight is fixed or not.

Here we focus on regular cycle codes, assuming that the row weight is fixed to be d. Counting the number of nonzeros entries in $\mathbf{H}$, we have $Md = 2N$. The designed code rate of regular cycle codes is restricted to

$$r = \frac{N - M}{N} = \frac{d - 2}{d}, \tag{3.10}$$

where d is an integer. For example, r can be $\frac{1}{3}, \frac{1}{2}, \frac{3}{5}, \frac{2}{3}, \frac{5}{7}, \frac{3}{4}, \ldots, \frac{7}{8}, \ldots, \frac{15}{16}$, etc.

The check matrix $\mathbf{H}$ of a regular cycle code is well structured. Specifically, the parity check matrix $\mathbf{H}$ of any regular cycle code can be put into *a concatenation form of row-permuted block-diagonal matrices* after row and column permutations if d is even, or, if d is odd and the code's associated graph contains at least one spanning subgraph that consists of disjoint edges [176]. For convenience, let us only state the result here when d is even.

Let us use the notation of $\mathbf{H}_1 \cong \mathbf{H}_2$ to denote the equivalence of two matrices $\mathbf{H}_1$ and $\mathbf{H}_2$ such that one can be transformed to the other simply through row and column permutations. For any regular cycle GF(q) code with row weight $d = 2v$, its parity check matrix $\mathbf{H}$ of size $M \times N$ has the equivalent form

$$\mathbf{H} \cong \left[\overline{\mathbf{H}}_1 \;\; \mathbf{P}_2\overline{\mathbf{H}}_2 \;\; \ldots \;\; \mathbf{P}_v\overline{\mathbf{H}}_v \right], \tag{3.11}$$

where $\mathbf{P}_i$ is an $M \times M$ permutation matrix, and $\overline{\mathbf{H}}_i$ is of size $M \times M$, $1 \le i \le v$. The matrix $\overline{\mathbf{H}}_i$ has an equivalent block-diagonal form

$$\overline{\mathbf{H}}_i \cong \text{diag}\left(\left[\tilde{\mathbf{H}}^{\text{cir}}_{i,1} \;\; \tilde{\mathbf{H}}^{\text{cir}}_{i,2} \;\; \ldots \;\; \tilde{\mathbf{H}}^{\text{cir}}_{i,Q_i} \right] \right), \tag{3.12}$$

where the matrix $\tilde{\mathbf{H}}^{\text{cir}}_{i,l}$ is of size $\kappa_{i,l} \times \kappa_{i,l}$ that satisfies $M = \sum_{l=1}^{Q_i} \kappa_{i,l}$ and has an equivalent form

$$\tilde{\mathbf{H}}^{\text{cir}} = \begin{bmatrix} \zeta_1 & 0 & 0 & \ldots & \beta_\kappa \\ \beta_1 & \zeta_2 & 0 & \ldots & 0 \\ 0 & \beta_2 & \zeta_3 & \ldots & 0 \\ \vdots & \vdots & \ddots & \ddots & \vdots \\ 0 & \ldots & 0 & \beta_{\kappa-1} & \zeta_\kappa \end{bmatrix} \tag{3.13}$$

with all ζ_i and β_i being nonzero entries from GF(q).

Example 3.2

The $\mathbf{H}$ matrix of the regular cycle code in Figure 3.6 has $d = 4$. Applying Theorem 1, one can obtain the two submatrices $\overline{\mathbf{H}}_1$ and $\overline{\mathbf{H}}_2$ in the block diagonal form as shown in Figure 3.7. The permutation matrix

$$\mathbf{P}_2 = \begin{bmatrix} 1&0&0&0&0&0 \\ 0&0&0&0&0&1 \\ 0&0&1&0&0&0 \\ 0&0&0&0&1&0 \\ 0&1&0&0&0&0 \\ 0&0&0&1&0&0 \end{bmatrix} \tag{3.14}$$

will permute the rows of $\overline{\mathbf{H}}_2$, before appending it after $\overline{\mathbf{H}}_1$. It can be verified $[\overline{\mathbf{H}}_1, \mathbf{P}_2\overline{\mathbf{H}}_2]$ is equivalent to the $\mathbf{H}$ matrix in Figure 3.6 after row and column permutations.

3.2.2 *Nonbinary Irregular LDPC Codes*

Cycle codes over large Galois fields (e.g., $q \ge 64$) can achieve near-Shannon-limit performance [169]. However, cycle codes over small to moderate Galois fields (e.g., $4 \le q \le 32$) suffer from performance loss due to a "tail" in the low weight regime of the distance

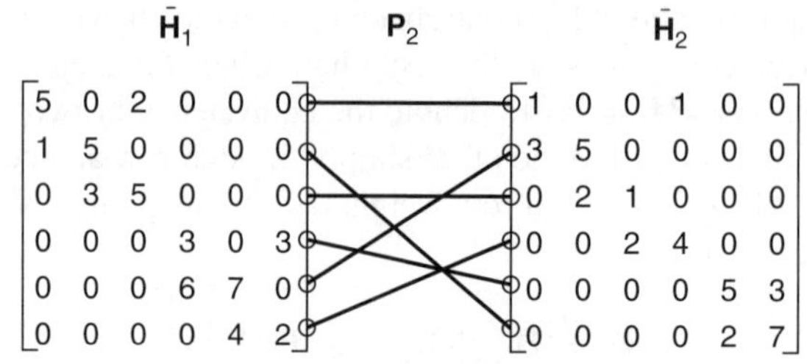

Figure 3.7 The equivalent form for a regular cycle code.

spectrum [169]. Irregular weight distribution can be used to improve the performance of nonbinary LDPC codes. A simple strategy has been proposed in [175] that improves the code performance while retaining the benefits of a regular cycle code as much as possible.

The approach is to replace a portion of weight-2 columns of $\mathbf{H}$ of a cycle code by columns of weight $t > 2$, (e.g., $t = 3$ or $t = 4$). Let N_1 columns of $\mathbf{H}$ have weight 2 and N_2 columns have weight t. The mean column weight is

$$\eta = \frac{2N_1 + tN_2}{N}. \tag{3.15}$$

The matrix $\mathbf{H}$ can be arranged as

$$\mathbf{H} = [\mathbf{H}_1 \mid \mathbf{H}_2] \tag{3.16}$$

where $\mathbf{H}_1$ contains all weight-2 columns and $\mathbf{H}_2$ contains all weight-t columns. Clearly, $\mathbf{H}_1$ is of size $M \times N_1$ and $\mathbf{H}_2$ is of size $M \times N_2$.

Note that $\mathbf{H}_1$ corresponds to the check matrix of a general cycle code. To maximally benefit from the structure of regular cycle codes presented in Section 3.2.1, it is desirable to make $\mathbf{H}_1$ be as close to a regular cycle code as possible. Specifically, one can split the matrix as

$$\mathbf{H} = [\mathbf{H}_{1a} \mid \mathbf{H}_{1b} \mid \mathbf{H}_2] \tag{3.17}$$

where the matrix $\mathbf{H}_{1a}$ is of size $M \times N_{1a}$ and the matrix $\mathbf{H}_{1b}$ is of size $M \times N_{1b}$. The number N_{1a} is the largest integer not greater than N_1 that can render $d_{1a} = \frac{2N_{1a}}{M}$ an integer, that is, $\mathbf{H}_{1a}$ is the largest sub-matrix of $\mathbf{H}_1$ that could be made regular, with row weight d_{1a}. If $N_{1a} = N_1$, then $N_{1b} = 0$. As such, $\mathbf{H}_1$ itself is regular, which is a special case.

In a nutshell, the proposed nonbinary irregular LDPC codes try to make a large portion of the check matrix come from a regular cycle code. This way, many benefits of a regular cycle code can be retained.

3.3 Encoding

Encoding has been an important issue for LDPC codes. Encoding of a general LDPC code can be done in almost linear time instead of quadratic time on the block length [270].

The proposed codes in Sections 3.2.1 and 3.2.2 can be encoded in linear time in parallel as follows. Assume $N_1 \geq M$, then $N_{1a} \geq M$. Since $\mathbf{H}_{1a}$ is regular, it can be decomposed as in (3.11). Hence, the first $M \times M$ submatrix of $\mathbf{H}$ can be made to have the form in (3.12). Let us denote it as $\overline{\mathbf{H}}_1$, and split $\mathbf{H}$ as $\mathbf{H} = [\overline{\mathbf{H}}_1 \mid \mathbf{H}^{\perp}]$ where $\mathbf{H}^{\perp}$ is of size $M \times (N - M)$. Make sure that $\overline{\mathbf{H}}_1$ has full rank by suitable choices of its nonzero entries. Partition the codeword $\mathbf{c}$ into two parts as $\mathbf{c} = [\mathbf{p}^{\mathrm{T}}, \mathbf{d}^{\mathrm{T}}]^{\mathrm{T}}$ where $\mathbf{p}$ is of length M. Let $\mathbf{p}$ contain the parity symbols and $\mathbf{d}$ contain the information symbols. A valid codeword satisfies $\mathbf{Hc} = \mathbf{0}$, which implies that

$$\overline{\mathbf{H}}_1\mathbf{p} = -\mathbf{H}^{\perp}\mathbf{d}. \tag{3.18}$$

From (3.12), $\overline{\mathbf{H}}_1$ can be made block diagonal as $\mathrm{diag}(\tilde{\mathbf{H}}^{\mathrm{cir}}_{1,1}, \ldots, \tilde{\mathbf{H}}^{\mathrm{cir}}_{1,Q_1})$. According to the sizes of $\{\tilde{\mathbf{H}}^{\mathrm{cir}}_{1,l}\}_{l=1}^{Q_1}$, let us partition $\mathbf{p}$ and the right-hand side of (3.18) into Q_1 pieces as

$$\mathbf{p} = \begin{bmatrix}\mathbf{p}_1^{\mathrm{T}} & \cdots & \mathbf{p}_{Q_1}^{\mathrm{T}}\end{bmatrix}^{\mathrm{T}}, \tag{3.19}$$

$$-\mathbf{H}^{\perp}\mathbf{s} = \begin{bmatrix}\mathbf{b}_1^{\mathrm{T}} & \cdots & \mathbf{b}_{Q_1}^{\mathrm{T}}\end{bmatrix}^{\mathrm{T}}, \tag{3.20}$$

respectively. Computation of $\mathbf{p}$ requires solving the following Q_1 equations

$$\tilde{\mathbf{H}}^{\mathrm{cir}}_{1,i}\mathbf{p}_i = \mathbf{b}_i, \quad 1 \le i \le Q_1. \tag{3.21}$$

Now, let us describe how to solve an equation in the general form of $\tilde{\mathbf{H}}^{\mathrm{cir}}\mathbf{x} = \mathbf{b}$, where $\mathbf{p} = [p_1, p_2, \ldots, p_k]^{\mathrm{T}}$, $\mathbf{b} = [b_1, b_2, \ldots, b_k]^{\mathrm{T}}$, and $\tilde{\mathbf{H}}^{\mathrm{cir}}$ has the structure in (3.13). Let us rewrite the relationship using the matrix-vector notation as

$$\begin{bmatrix}\zeta_1 & & & & \beta_k\\ \beta_1 & \zeta_2 & & & \\ & \beta_2 & \ddots & & \\ & & \ddots & \zeta_{k-1} & \\ & & & \beta_{k-1} & \zeta_k\end{bmatrix}\begin{bmatrix}p_1\\ p_2\\ \vdots\\ p_k\end{bmatrix} = \begin{bmatrix}b_1\\ b_2\\ \vdots\\ b_k\end{bmatrix}. \tag{3.22}$$

An LU decomposition leads to

$$\begin{bmatrix}1 & & & & \\ \gamma_1 & 1 & & & \\ & \gamma_2 & \ddots & & \\ & & \ddots & 1 & \\ & & & \gamma_{k-1} & 1\end{bmatrix}\underbrace{\begin{bmatrix}1 & & & u_1\\ & 1 & & u_2\\ & & \ddots & \vdots\\ & & 1 & u_{k-1}\\ & & & u_k\end{bmatrix}\underbrace{\begin{bmatrix}\zeta_1 & & & \\ & \zeta_2 & & \\ & & \ddots & \\ & & & \zeta_k\end{bmatrix}\begin{bmatrix}p_1\\ p_2\\ \vdots\\ p_k\end{bmatrix}}_{:=[x_1, x_2, \cdots, x_k]^{\mathrm{T}}}}_{:=[z_1, z_2, \cdots, z_k]^{\mathrm{T}}} = \begin{bmatrix}b_1\\ b_2\\ \vdots\\ b_k\end{bmatrix} \tag{3.23}$$

where the coefficients are defined as

$$\gamma_i = \zeta_i^{-1}\beta_i, \quad i = 1, \ldots, k \tag{3.24}$$

$$u_i = -\gamma_1 \cdots \gamma_i \gamma_k, \quad i = 1, \ldots, k-1, \tag{3.25}$$

$$u_k = (1 + \gamma_1\gamma_2 \cdots \gamma_k). \tag{3.26}$$

The solution to the above question is the following algorithm [179].

1. $z_1 = b_1; z_i = \gamma_{i-1} z_{i-1} + b_i, i = 2, 3, \ldots, k$;
2. $x_k = (1 + \gamma_1 \gamma_2 \ldots \gamma_k)^{-1} z_k$;
   $x_i = z_i - \gamma_1 \gamma_2 \ldots \gamma_{i-1} \gamma_k x_k, i = 1, 2, \ldots, k-1$;
3. $p_i = \zeta_i^{-1} x_i, i = 1, 2, \ldots, k$.

Assume that the coefficients have been stored before computing. The computation complexity is $2(k-1)$ additions, $2(k-1)$ multiplications, and $k+1$ divisions over GF(q).

Note that solving those Q_1 equations in (3.21) can be done in parallel. The overall complexity of solving the equation (3.18) is about $2M$ additions, $2M$ multiplications, and M divisions over GF(q), when some coefficients are precomputed. Fast and parallel encoding is quite desirable especially when the block length is large, or, when multiple rounds of encoding is needed for the proposed OFDM peak-to-average-power-ratio reduction as will be detailed in Section 4.2.2.

3.4 Decoding

Decoding of LDPC codes can use standard sum-product algorithms (SPA) and its low-complexity variants. When applying a standard SPA as described in [93], the complexity is $O(q^2)$. An important variation is to use FFT based Q-ary SPA (FFT-QSPA) [351], whose complexity is $O(q \log q)$. Using the log-domain representation, multiplication operations are replaced by additions [440]. In this chapter, we present the log-domain SPA, and an approximation called Max-Log-SPA, or Min-Sum, following the lines in [440].

The decoder has the following four key steps, as illustrated in Figure 3.8.

S1. Initialize the decoder with soft metrics, in the form of the log-likelihood-ratio vector (LLRV), from the demodulator block of the receiver.

S2. Variable to check node update. Define the set $\mathcal{M}(n)$ as all the check nodes associated with the variable node n. Each variable node n sends updated information to all the check nodes in $\mathcal{M}(n)$.

S3. Check to variable node update. Define the set $\mathcal{N}(m)$ as all the variable nodes associated with the check node m. Each check node m sends updated information to all the variable nodes in $\mathcal{N}(m)$.

S4. Make a tentative decision of the codeword as $\hat{\mathbf{c}}$. If $\mathbf{H}\hat{\mathbf{c}} = 0$, the decoder declares success and exits with the correct decision. If not, it returns to S2 for the next round of iteration. The decoder stops and outputs soft information, once the maximum number of iterations have been reached.

We next specify the details of these steps.

3.4.1 *Initialization*

How to compute the soft information in different contexts will be the subject of Chapter 8 on data detection. Here we present a simple scenario to illustrate the concept. Assume that the equivalent channel input–output model after symbol deinterleaving is:

$$y[n] = g[n]\phi(c[n]) + \eta[n], \quad n = 0, \ldots, N-1, \tag{3.27}$$

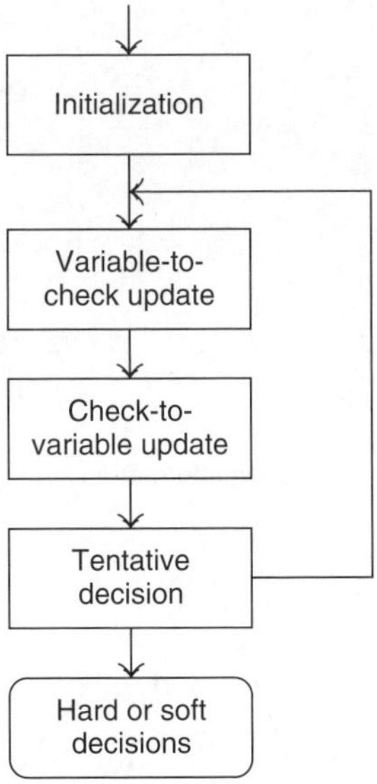

Figure 3.8 The key steps in an iterative decoder for LDPC codes.

where $y[n]$ is the receiver output, $g[n]$ is the equivalent channel gain for the nth symbol, $\phi(\cdot)$ denotes the mapping from a coded symbol to a constellation symbol, and $\eta[n]$ is the equivalent noise.

Assume that $\eta[n]$ has variance σ^2 per real and imaginary dimension. The likelihood can be computed as

$$\Pr(c[n] = \alpha_i) = \frac{1}{2\pi\sigma^2} \exp\left(\frac{-|y[n] - g[n]\phi(\alpha_i)|^2}{2\sigma^2} \right), \tag{3.28}$$

for $n = 0, \ldots, N-1$, and $i = 0, \ldots, q-1$. The log-likelihood-ratio vector (LLRV) of $c[n]$ is defined as

$$\mathbf{L}^{\mathrm{ch}}[n] = \left[L_0^{\mathrm{ch}}[n] \; L_1^{\mathrm{ch}}[n] \; \ldots \; L_{q-1}^{\mathrm{ch}}[n] \right]^{\mathrm{T}}, \tag{3.29}$$

where

$$L_i^{\mathrm{ch}}[n] := \ln \frac{\Pr(c[n] = \alpha_i)}{\Pr(c[n] = 0)}. \tag{3.30}$$

From (3.28), we have

$$\begin{aligned} L_i^{\mathrm{ch}}[n] &= -\frac{1}{2\sigma^2} \left(\left| y[n] - g[n]\phi(\alpha_i) \right|^2 - |y[n] - g[n]\phi(0)|^2 \right) \\ &= \frac{1}{\sigma^2} \Re\{y^*[n]g[n](\phi(\alpha_i) - \phi(0))\} - \frac{1}{2\sigma^2} |g[n]|^2 (|\phi(\alpha_i)|^2 - |\phi(0)|^2) \end{aligned} \tag{3.31}$$

3.4.2 Variable-to-Check-Node Update

As illustrated in Figurer 3.9(a), The variable node n sends a message to the check node m, which summarizes the *a priori* probabilities from the channel and the beliefs from all the

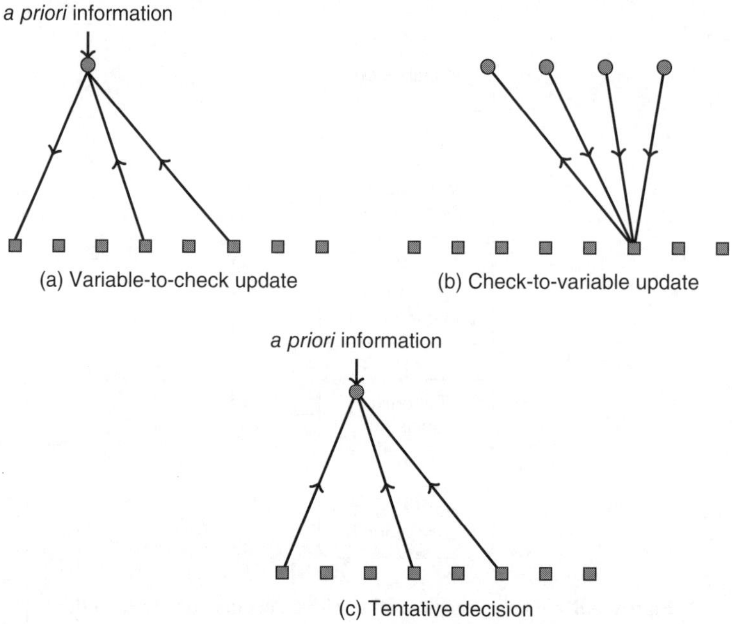

Figure 3.9 An illustration of LDPC decoding process.

check nodes in $\mathcal{M}(n)$ excluding the check node m itself,

$$\mathbf{L}^{\text{VND}}[m,n] = \mathbf{L}^{\text{ch}}[n] + \sum_{i \in \mathcal{M}(n)\backslash m} \mathbf{L}^{\text{CND}}[i,n], \tag{3.32}$$

where $\mathbf{L}^{\text{VND}}[m,n]$ denotes the LLRV sent from the nth variable node to the mth check node, and $\mathbf{L}^{\text{VND}}[i,n]$ denotes the LLRV sent from the ith check node to the nth variable node. Both $\mathbf{L}^{\text{VND}}[m,n]$ and $\mathbf{L}^{\text{VND}}[i,n]$ are similarly defined as $\mathbf{L}^{\text{ch}}[n]$ in (3.29).

3.4.3 Check-to-Variable-Node Update

At the mth check node, the parity check equation is

$$\sum_{j \in \mathcal{N}(m)} H_{m,j} c[j] = 0. \tag{3.33}$$

Hence, the variable node n can be expressed as

$$c[n] = \sum_{j \in \mathcal{N}(m)\backslash n} H_{m,n}^{-1} H_{m,j} c[j]. \tag{3.34}$$

With this relationship, the LLRV for $c[n]$ can be obtained based on the knowledge of LLRVs for other symbols $c[j]$, where $j \in \mathcal{N}(m)\backslash n$, as illustrated in Figure 3.9(b).

Note that the addition in (3.34) is done in the finite field. By defining intermediate variables, the LLRV updates are carried out repeatedly using the updating rules of adding two variables.

In particular, assume A_1, A_2 as constants in GF(q), and v_1 and v_2 as variables in GF(q). The LLRV for the sum $A_1v_1 + A_2v_2$ can be computed as follows:

$$\begin{aligned} L(A_1v_1 + A_2v_2 = \alpha_i) &= \ln \frac{P(A_1v_1 + A_2v_2 = \alpha_i)}{P(A_1v_1 + A_2v_2 = 0)} \\ &= \ln \frac{\sum_{x\in \mathrm{GF}(q)} P(v_1 = x)P(v_2 = A_2^{-1}(\alpha_i + A_1x))}{\sum_{x\in \mathrm{GF}(q)} P(v_1 = x)P(v_2 = A_2^{-1}A_1x)} \\ &= \ln \frac{\sum_{x\in \mathrm{GF}(q)} \frac{P(v_1=x)P(v_2=A_2^{-1}(\alpha_i+A_1x))}{P(v_1=0)P(v_2=0)}}{\sum_{x\in \mathrm{GF}(q)} \frac{P(v_1=x)P(v_2=A_2^{-1}A_1x)}{P(v_1=0)P(v_2=0)}} \\ &= \ln \left(\sum_{x\in \mathrm{GF}(q)} e^{L(v_1=x)+L(v_2=A_2^{-1}(v_2\alpha_i+A_1x))} \right) \\ &\quad - \ln \left(\sum_{x\in \mathrm{GF}(q)} e^{L(v_1=x)+L(v_2=A_2^{-1}(A_1x))} \right) \end{aligned} \tag{3.35}$$

One basic operation in (3.35) is $\ln(e^x + e^y)$. Define

$$\max{}^*(x, y) = \ln(e^x + e^y) = \max(x, y) + \ln(1 + e^{-|x-y|}), \tag{3.36}$$

which can be conveniently implemented by using a look up table for the second term. The computation in (3.35) can be accomplished by repeatedly using the $\max^*(\cdot)$ operation as

$$\max{}^*(x, y, z) = \max{}^*(\max{}^*(x, y), z). \tag{3.37}$$

The $\max^*(\cdot)$ operation can be replaced by the $\max(\cdot)$ operation, and corresponding algorithm is called Max-log-SPA or Min-Sum [440]. The update rule is simplified as

$$\begin{aligned} L(A_1v_1 + A_2v_2 = \alpha_i) \approx &\max_{x\in \mathrm{GF}(q)} (L(v_1 = x) + L(v_2 = A_2^{-1}(v_2\alpha_i + A_1x))) \\ &- \max_{x\in \mathrm{GF}(q)} (L(v_1 = x) + L(v_2 = A_2^{-1}(A_1x))). \end{aligned} \tag{3.38}$$

The advantage is that the LLRV can be now arbitrarily scaled in the Max-log-SPA decoder. Hence, the noise variance in (3.31) can be set to an arbitrary value.

Note that the presentation here focuses on one individual symbol from the check equation. Efficient implementation exists to update all the variable nodes connected to one check node with shared intermediate computations [439].

3.4.4 *Tentative Decision and Decoder Outputs*

As shown in Figure 3.9(c), the decoder collects all the information from the channel and the check nodes to compute the LLRV corresponding to the variable node n as

$$\mathbf{L}^{\mathrm{app}}[n] = \mathbf{L}^{\mathrm{ch}}[n] + \sum_{m\in \mathcal{M}(n)} \mathbf{L}^{\mathrm{CND}}[m, n]. \tag{3.39}$$

Hard decision is made on the symbol by symbol basis as

$$\hat{c}[n] = \arg\max_{\alpha_i} L_i^{\text{app}}[n]. \tag{3.40}$$

If $\mathbf{H}\hat{\mathbf{c}} = \mathbf{0}$, then the decoder declares success. If it has reached the maximum number of iterations, the decoder stops and reports the soft symbol estimates, which may be used as additional inputs for channel estimation and data detection. The soft symbol estimates can be computed based on either the *a posteriori* probability (APP) information $\mathbf{L}^{\text{app}}[n]$, or the extrinsic information. For ease of reference by later chapters, we spell out different soft symbol estimates explicitly as follows.

- Soft information based on the APP information. First convert the LLRs into probabilities as:

$$P_0^{\text{app}}[n] = \frac{1}{1 + \sum_{i=1}^{q-1} e^{L_i^{\text{app}}[n]}}, \tag{3.41}$$

$$P_i^{\text{app}}[n] = P_0[n] e^{L_i^{\text{app}}[n]}, \quad i = 1, \ldots, q-1. \tag{3.42}$$

The soft symbol estimate and its corresponding variance are computed as

$$\bar{s}^{\text{app}}[n] = \sum_{\alpha_i \in S} \phi(\alpha_i) P_i^{\text{app}}[n] \tag{3.43}$$

$$\text{Var}\left(\bar{s}^{\text{app}}[n]\right) = \left(\sum_{\alpha_i \in S} |\phi(\alpha_i)|^2 \cdot P_i^{\text{app}}[n] \right) - |\bar{s}^{\text{app}}[n]|^2. \tag{3.44}$$

- Soft information based on the extrinsic information. The extrinsic information is

$$\mathbf{L}^{\text{ext}}[n] = \mathbf{L}^{\text{app}}[n] - \mathbf{L}^{\text{ch}}[n], \tag{3.45}$$

which leads to the corresponding probabilities

$$P_0^{\text{ext}}[n] = \frac{1}{1 + \sum_{i=1}^{q-1} e^{L_i^{\text{ext}}[n]}}, \tag{3.46}$$

$$P_i^{\text{ext}}[n] = P_0^{\text{ext}}[n] e^{L_i^{\text{ext}}[n]}, \quad i = 1, \ldots, q-1. \tag{3.47}$$

The soft symbol estimate and its corresponding variance are then computed as

$$\bar{s}^{\text{ext}}[n] = \sum_{\alpha_i \in S} \phi(\alpha_i) P_i^{\text{ext}}[n] \tag{3.48}$$

$$\text{Var}(\bar{s}^{\text{ext}}[n]) = \left(\sum_{\alpha_i \in S} |\phi(\alpha_i)|^2 P_i^{\text{ext}}[n] \right) - |\bar{s}^{\text{ext}}[n]|^2. \tag{3.49}$$

3.5 Code Design

Two steps are needed to design nonbinary LDPC codes. The first step is to design the code structure that specifies the locations of nonzero entries in the parity check matrix. The second step is to determine the nonzero entries of the parity check matrix. We next discuss some design methodologies corresponding to different cases.

3.5.1 Design of Regular Cycle codes

There are three methods outlined in [176] on the design of the code structure for regular cycle codes.

1. One can utilize known regular graphs with good properties, such as the Ramanujan graphs [134]. Ref. [169] first utilized this kind of promising graphs to construct cycle codes. Later, Ref. [177] showed through simulations that these cycle codes can achieve performance within 1 dB away from the corresponding Shannon limits, including codes of rate $1/2$, $2/3$ and $3/4$. However, good known graphs may be very limited in the number of code choices.
2. One can resort to computer search algorithms. Computer search based algorithms have been widely adopted to construct LDPC codes [170, 272]. Among them the progressive edge-growth (PEG) [170] algorithm has been shown efficient and feasible for constructing LDPC codes with short code lengths and high rates as well as LDPC codes with long code lengths. Although the original PEG algorithm aims to construct a bipartite Tanner graph, the same principle of PEG can be adopted to construct the associated graphs for regular cycle codes [176].
3. Based on the structures presented in Section 3.2.1, one can construct good regular associated graphs through carefully designing interleavers.

Once the nonzero locations of the check matrix are decided, one starts with the selection of nonzero entries. References [177, 213, 277, 314] have addressed the issue for the selection of nonzero entries for cycle codes. Essentially, resolvable cycles [179] with short length correspond to low-weight codewords, which may induce undetected errors during the decoding process [177, 213, 314]. Therefore, to lower the error floor, it becomes desirable to make all cycles irresolvable, especially those with short lengths.

3.5.2 Design of Irregular LDPC Codes

Section 3.2.2 presented an irregular LDPC code whose check matrix has only column weights 2 and t. The proposed design steps in [175] are as follows.

- **Step** 1: Specify the structure of $\mathbf{H}_{1a}$.
  Construct a cycle code of fixed row weight d_{1a} using the design methodologies presented in Section 3.5.1.
- **Step** 2: Specify the structure of $\mathbf{H}_{1b}$ and $\mathbf{H}_2$.
  Apply the progressive-edge-growth (PEG) algorithm [170] to attach N_{1b} columns of weight 2 and N_2 columns of weight t to the matrix $\mathbf{H}_{1a}$. This way, the structure of $\mathbf{H}$ in (3.17) is established.
- **Step** 3: Specify the nonzero entries of $\mathbf{H}_1$.
  Note that the sub-matrix $\mathbf{H}_1 = [\mathbf{H}_{1a} \mid \mathbf{H}_{1b}]$ can be regarded as a check matrix of a cycle code. Hence, we can apply the design criterion of [176] to choose appropriate nonzero entries for $\mathbf{H}_1$ to make as many as possible short length cycles of the associated graph [179] of $\mathbf{H}_1$ irresolvable.
- **Step** 4: Specify the nonzero entries of $\mathbf{H}_2$.
  The nonzero entries of $\mathbf{H}_2$ are generated randomly with a uniform distribution over the set GF$(q)\backslash 0$.

One key design issue of the proposed irregular LDPC codes is to decide the average column weight, which has a considerable impact on the performance, as illustrated in the following example.

Example 3.3

Figure 3.10 compares the performance of irregular LDPC codes over GF(16) with different mean column weights. All the codes have rate 1/2 and codeword length of 1008 bits that correspond to 252 GF(16) symbols. BPSK modulation is used on the binary input AWGN

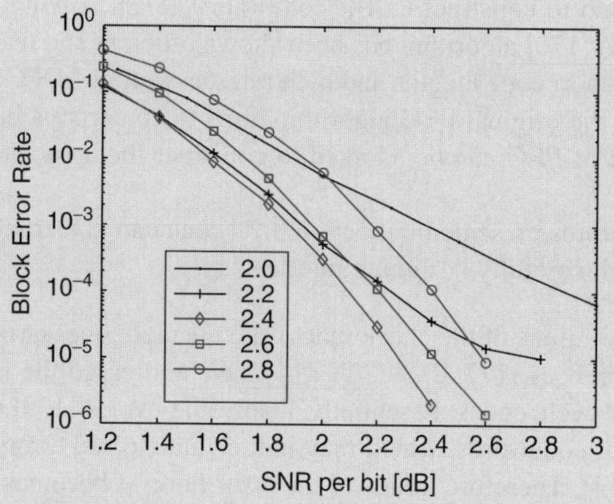

Figure 3.10 Performance comparison of irregular codes over GF(16) with different mean column weight; $t = 3$, $r = 1/2$ and the codeword length is 1008 bits, i.e., 252 GF(16) symbols.

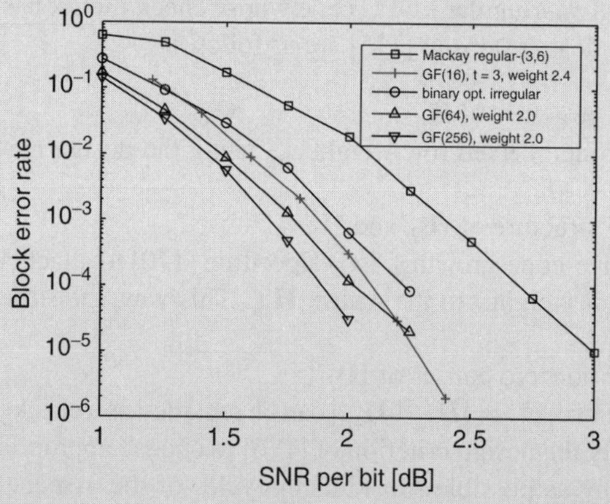

Figure 3.11 Performance comparison of irregular codes over GF(16) with an optimized binary irregular LDPC code; $r = 1/2$ and the codeword length is 1008 bits, which correspond to 252 GF(16) symbols, 168 GF(64) symbols, and 126 GF(256) symbols.

channel and the decoder has a maximum of 80 iterations. One can observe from Figure 3.10 that the performance curves of the codes with $\eta = 2.0$ and $\eta = 2.2$ level off above 10^{-5} due to the contribution from the probability of undetected errors. This is not the case if $\eta \geq 2.4$. Actually no undetected errors have been observed for $\eta \geq 2.4$ in the simulations. Another interesting observation is that as η increases from 2.4 to 2.6 and 2.8, the code performance degrades. Therefore, the code with $\eta = 2.4$ is the best one for this particular example.

Figure 3.11 shows the performance comparison between the irregular LDPC codes over GF(16) and an optimized binary irregular LDPC code. The performance of Mackay's (3,6)-regular code and cycle codes over GF(64) and GF(256) of the same block length are also included (see more details on these codes' parameters in [176]). It can be seen from Figure 3.11 that by adopting an irregular column weight distribution, the code's performance can be greatly improved without having to use a very large q; this is desirable from both the complexity and the constellation matching perspectives.

3.5.3 Quasi-Cyclic Nonbinary LDPC codes

Quasi-Cyclic (QC) LDPC codes are appealing to practical systems as the quasi-cyclic structure of the parity check matrix allows for linear time encoding using only shift registers, renders efficient routing for decoding implementation, and enables the storage of the coding matrix with only a few memory units.

A binary QC-LDPC code is obtained as follows. First, construct a mother matrix, or base matrix, $\mathbf{M}(\mathbf{H})$, which has a small size $m \times n$. Then apply a cyclic expansion operation, that is, replacing the entries "0" and "1" in $\mathbf{M}(\mathbf{H})$ with zero sub-matrices of size $L \times L$ and circulant permutation sub-matrices of size $L \times L$, respectively. Specifically, let $\mathbf{P}$ be an $L \times L$ permutation matrix as

$$\mathbf{P} = \begin{bmatrix} 0 & 1 & 0 & \dots & 0 \\ 0 & 0 & 1 & \dots & 0 \\ \vdots & \vdots & \vdots & \ddots & \vdots \\ 0 & 0 & 0 & \dots & 1 \\ 1 & 0 & 0 & \dots & 0 \end{bmatrix}. \tag{3.50}$$

For a finite a, $\mathbf{P}^a$ denotes a circulant permutation sub-matrix of size $L \times L$ which is obtained by cyclically shifting the identity matrix $\mathbf{I}_L$ to the right by a times if $a > 0$, or to the left by $-a$ times if $a < 0$. For simple notation, $\mathbf{P}^\infty$ denotes the zero matrix of size $L \times L$. A parity check matrix of size $mL \times nL$ for a binary QC-LDPC code is then obtained as [124, 289]

$$\mathbf{H} = \begin{bmatrix} \mathbf{P}^{a_{11}} & \mathbf{P}^{a_{12}} & \dots & \mathbf{P}^{a_{1n}} \\ \mathbf{P}^{a_{21}} & \mathbf{P}^{a_{22}} & \dots & \mathbf{P}^{a_{2n}} \\ \vdots & \vdots & \ddots & \vdots \\ \mathbf{P}^{a_{m1}} & \mathbf{P}^{a_{m2}} & \dots & \mathbf{P}^{a_{mn}} \end{bmatrix} \tag{3.51}$$

where the shift offset value $a_{ij} \in \{-(L-1), \dots, L-1, \infty\}$, $i = 1, 2, \dots, m$, $j = 1, 2, \dots, n$. Designing a binary QC-LDPC code amounts to the selection of the shift offset values $\{a_{i,j}\}$.

Now, one needs to replace each "1" entry in $\mathbf{H}$ by an element from GF(q)\0 to obtain a nonbinary QC-LDPC code. In each nonbinary circulant permutation matrix, let ρ_{ij} denote the nonzero element in the first row of $\mathbf{P}^{a_{ij}}$, which could be randomly drawn from GF(q)\0, the nonzero elements for the remaining rows of $P^{a_{ij}}$ are obtained by multiplying the one in the row above it by α^{λ}, where α is a primitive element of GF(2^p) and λ is an integer. To qualify the codes as quasi-cyclic codes, the nonzero element in the first row needs to be made equal to the nonzero element in the last row multiplied by α^{λ} [248, 308]. One example is to choose the smallest λ for any given L such that $\lambda L = \gamma(2^p - 1)$ where γ is an integer [173].

Example 3.4

Here we show one example of rate compatible nonbinary QC-LDPC design [173], where the mother matrices for eight different rates are constructed as shown in Figure 3.12.

This set of QC-LDPC codes falls into the family of rate-compatible codes. Here, the rate compatibility means that lower rate codes are obtained by shortening higher-rate codes, i.e., some information symbols of higher-rate codes are set to be zeros for lower-rate codes (refereed to as information nulling [384]). The mother matrix is of size 6×54 when the code rate is $r = 8/9$. When the rate changes to $r = 1/2, 2/3, 3/4, 5/6$, the mother matrix corresponds to the first 12, 18, 24, and 36 columns of $\mathbf{M}(\mathbf{H})$ in (3.12), respectively. Note that all columns of $\mathbf{M}(\mathbf{H})$ have either weight 2 or weight $t = 3$. The mean column weights for rates $1/2, 2/3, 3/4, 5/6$, and $8/9$ are 2.5, 2.667, 2.75, 2.8333 and 2.8333, respectively. The 6 columns inside the box marked on $\mathbf{M}(\mathbf{H})$, having columns weights of 2, correspond to parity check symbols and can be used to facilitate linear time encoding for all codes along the lines in Section 3.3.

For cyclic expansion, the shift offset values are designed to be

$$a_{ij} = r_{ij} l_j, \tag{3.52}$$

where r_{ij} is drawn from a difference set $\{0, 1, 3, 7, 12, \dots\}$ sequentially for each nonzero entry in the j-th column of $\mathbf{M}(\mathbf{H})$ and $\{l_j\}$'s are obtained via computer search [173]:

$l_1 - l_{10}$	$l_{11} - l_{20}$	$l_{21} - l_{30}$	$l_{31} - l_{40}$	$l_{41} - l_{50}$	$l_{51} - l_{54}$
4,0,3,-1,3,-7, -7,29,10,-21	9,25,5,-4,6, -1,7,1,-2,5	13,0,10,-8, 2,-2,1,1,2,3	2,2,1,-2,4, -3,1,-4,8,-5	6,-2,9,3,0,5, -1,-3,10,-8	4,4,1,-1

Two sets of codes with $L = 30$ and $L = 60$ are constructed over GF(8), where the nonzero entry on the first row of each circulant matrix was randomly chosen. With $L = 30$ the constructed codes over GF(8) have block lengths of $1080, 1620, 2160, 3240$ and 4860 bits, respectively. With $L = 60$ the constructed codes over GF(8) have block lengths of $2160, 3240, 4320, 6480$ and 9720 bits, respectively. Figure 3.13(a) shows the performance of these codes over the AWGN channel and Figure 3.13(b) shows the performance of these codes over the i.i.d. Rayleigh fading channel, where BPSK is used.

As shown in Figure 3.13, there is a coding gain of 0.2 to 0.5 dB at BLER of 10^{-4} when the block lengths get doubled.

- The Shannon limits for rates $1/2, 2/3, 3/4, 5/6$ and $8/9$ over the BPSK AWGN channel are 0.188, 1.0, 1.6, 2.4 and 3.1 dB, respectively. It can be seen from Figure 3.13(a) that at BLER of 10^{-4} the five codes with $L = 60$ are about 1.6, 1.3, 1.1, 0.9 and 0.7 dB from the Shannon limit of the AWGN channel.
- The Shannon limits for rates $1/2, 2/3, 3/4, 5/6$ and $8/9$ over the BPSK Rayleigh fading channel are $1.83, 3.65, 4.9, 6.7$ and 8.6 dB, respectively. It can be seen from Figure 3.13(b) that at BLER of 10^{-4} the five codes with $L = 60$ are about 2.2, 1.95, 1.9, 1.9 and 1.7 dB from the Shannon limit of the Rayleigh fading channel.

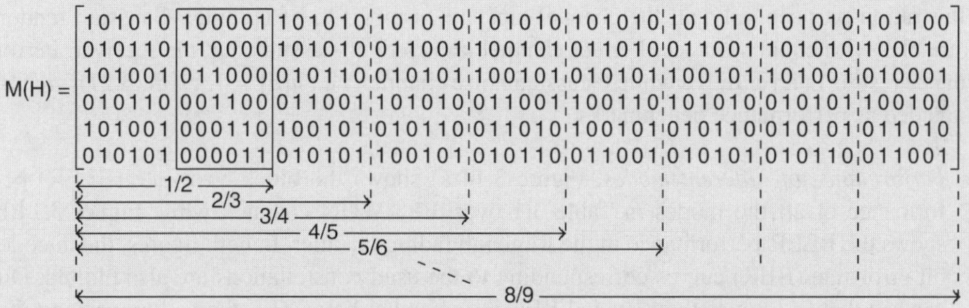

Figure 3.12 The first $6d$ columns form the PCM of size $6 \times 6d$ for the code of rate $r = (d-1)/d$, $d = 2, 3, 4, 6, 9$.

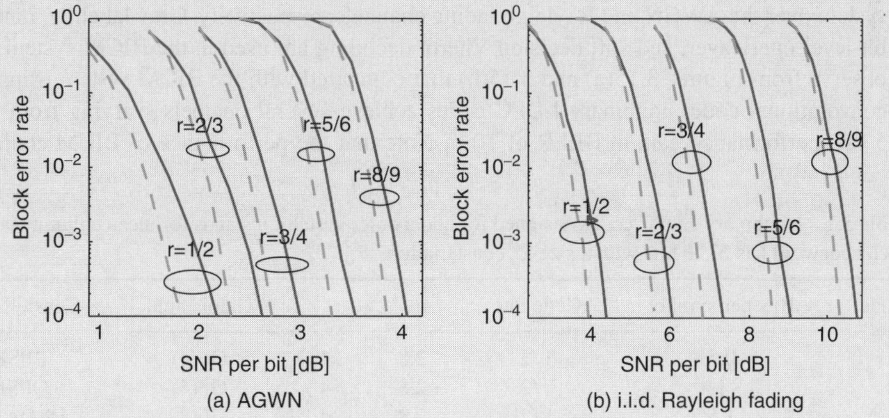

Figure 3.13 Performance of rate-compatible nonbinary QC-LDPC codes of various lengths. Solid curves correspond to $L = 30$ whereas dashed curves correspond to $L = 60$.

3.6 Simulation Results of Coded OFDM

This section presents some Monte Carlo simulation results of coded OFDM. The OFDM parameters are selected as in Table 2.1, where $K_\mathrm{d} = 672$ subcarriers are used for data transmission out of a total of $K = 1024$ subcarriers. When constructing the nonbinary LDPC codes, the coding alphabet is matched to the modulation alphabet, i.e., $p = b$. Using each OFDM block to carry exactly one codeword, the codeword length is 672 symbols. Seven modulation-coding pairs have been constructed in [175], as shown in Table 3.1, with the bandwidth efficiency ranging from 0.5 to 5 bits/symbol. For LDPC codes over GF(64), regular cycle codes as described in Section 3.5.1 are used [176]. For LDPC codes over GF(q) where $q < 64$, near regular construction as described in Section 3.5.2 are used.

Assume a signal bandwidth $B = 12\,\mathrm{kHz}$, the subcarrier spacing is $B/K = 11.72\,\mathrm{Hz}$, and the OFDM symbol duration is $T = 85.33\,\mathrm{ms}$. Consider a Rayleigh fading multipath channel delay spread of 10 ms, which will lead to a discrete-time with 120 channel taps in the baseband. For illustration purpose only, all the channel taps are assumed to be complex Gaussian random variables with equal variance. Perfect channel knowledge is assumed at the receiver. Performance results in an additive white Gaussian noise channel, i.e., $g[n] = 1, \forall n$ in (3.27), are also included as performance benchmarks.

- *Performance of different modes*. Figure 3.14(a) shows the block-error-rate (BLER) performance of all the modes in Table 3.1 over the AWGN channel, while Figure 3.14(b) shows the BLER performance in the Rayleigh fading channel. In both figures, the uncoded bit-error-rate (BER) curves corresponding to the used constellations are also plotted. One can see that as long as the uncoded BER is somewhat below 0.1, the coding performance improves drastically. Indeed, the performance curves are very steep in the *waterfall* region.
- *Comparison with convolutional codes based BICM*. Figures 3.15(a) and 3.15(b) show the performance comparisons between a BICM system [52, 459] based on a 64-state rate-1/2 convolutional code with the generator (133, 171) and the proposed nonbinary LDPC coding system over the AWGN and Rayleigh fading channels, respectively. Gray labeling, random bit-level interleaver, and soft decision Viterbi decoding are used in the BICM system. We observe from Figures 3.15(a) and 3.15(b) that compared with the BICM system using the convolutional code, nonbinary LDPC codes achieve several decibels (varying from 2 to 5 dB) performance gain at BLER of 10^{-2}. Note that the performance of BICM could be

Table 3.1 Nonbinary LDPC codes designed for underwater system. η stands for mean column weight. Each codeword has $672b$ bits with a size-2^b constellation

Mode	Bits per symbol	Code rate	η	t	Galois field	Constellation
1	0.5	1/2	2.8	4	GF(4)	BPSK
2	1	1/2	2.8	4	GF(4)	QPSK
3	1.5	1/2	2.8	4	GF(8)	8-QAM
4	2	1/2	2.3	3	GF(16)	16-QAM
5	3	1/2	2.0	-	GF(64)	64-QAM
6	4	2/3	2.0	-	GF(64)	64-QAM
7	5	5/6	2.0	-	GF(64)	64-QAM

Source: Huang 2008 [175], Table I, p. 1691. Reproduced with permission of IEEE.

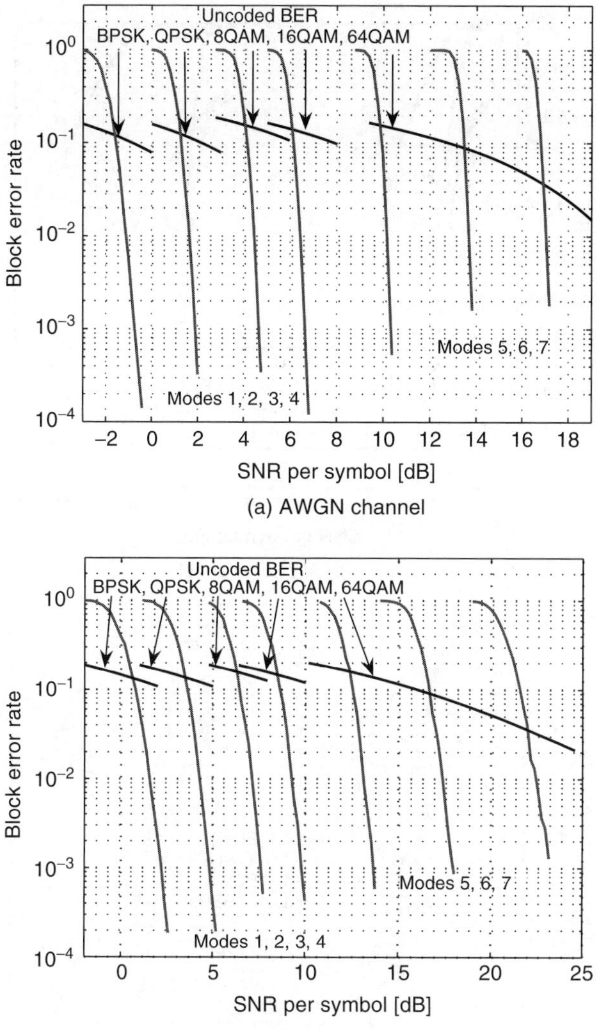

(a) AWGN channel

(b) Rayleigh fading channel

Figure 3.14 BLER performance of different modes over the AWGN and Rayleigh fading channels.

considerably improved by using more powerful binary codes such as turbo codes and binary LDPC codes, and through iterative constellation demapping [242]. The comparisons here provide a basic reference.

3.7 Bibliographical Notes

Channel coding is an area extensively studied ever since Shannon's classic paper in 1948 [337]. Channel coding is often covered in popular textbooks on digital communication systems, e.g.,

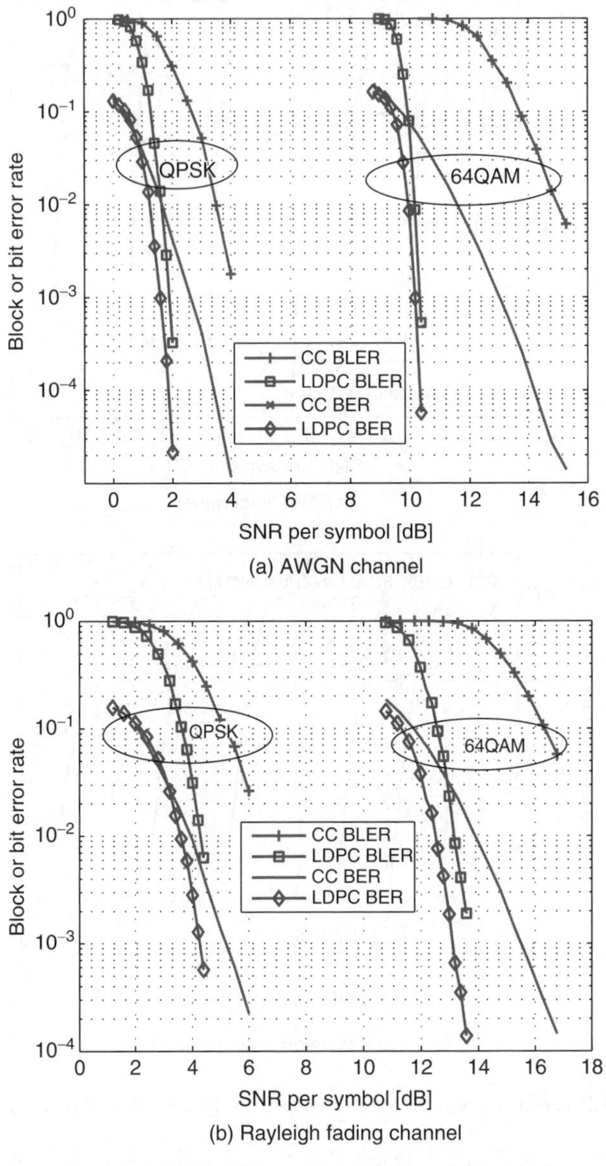

Figure 3.15 Comparison between LDPC and convolutional codes of rate $1/2$ under different modulation over the AWGN and the Rayleigh fading channels.

Proakis and M. Salehi, 2008 [318], Benedetto and Biglieri, 1999 [30], or on information theory and learning algorithms, e.g., MacKay, 2003 [270]. There are also many books dedicated to channel coding, e.g., Blahut, 1985 [46], Wicker, 1995 [433], Lin and Costello, 2004 [247], Richardson and Urbanke, 2008 [328].

Classic coding theory often covers topics such as Hamming Codes, BCH codes, Reed-Solomon codes, Reed-Muller codes, and convolutional codes. The new era of modern coding theory was marked by the invention of Turbo codes by Berrou, Glavieux, and Thitimajshima in 1993 [42]. Following the invention of LDPC codes by Gallager in 1960s [137] and its rediscovery by MacKay in 1996 [269], major developments on LDPC codes are the extension to irregular LDPC codes [261, 262] and the extension to higher order Galois field [93].

4

PAPR Control

Due to the superposition of data symbols on a large number of subcarriers, an OFDM waveform has a large peak-to-average power ratio (PAPR). Since nonlinear amplification causes inter modulation among subcarriers and undesired out-of-band radiation, the power amplifier at the transmitter must operate with a large backoff to limit nonlinear distortion. The severity of the PAPR issue in practical systems depends on the system operating conditions.

- In short range underwater acoustic applications, the average power matters more than the peak power as the power amplifier is often not working in the full power mode. As long as the transmitter circuits can handle occasionally large peaks, the PAPR issue is not a major concern.
- In long range underwater acoustic applications, the power amplifier might work in the full power mode. With the peak power limited by the hardware, a signal with large PAPR leads to small average power output, reducing the transmission range. In such scenarios, the PAPR issue is a major concern.

From the implementation point of view, more bits are needed for quantization to synthesize a signal with large PAPR, which adds to the transmitter complexity.

Next, we explore the PAPR issue of OFDM modulation, and then present three simple yet effective PAPR reduction methods.

4.1 PAPR Comparison

Let $\tilde{x}(t)$ denote one OFDM block in the passband, as in (2.5). The PAPR is defined as

$$\text{PAPR} := \frac{\max\left(|\tilde{x}(t)|^2\right)}{\mathbb{E}[|\tilde{x}(t)|^2]}. \tag{4.1}$$

The PAPR is a random variable, with its value decided by the realization of data and pilot symbols on OFDM subcarriers. In addition to the mean and variance, the complementary cumulative distribution function (CCDF) is often used as one performance metric:

$$\text{CCDF}(x) := \Pr(\text{PAPR} > x). \tag{4.2}$$

OFDM for Underwater Acoustic Communications, First Edition. Shengli Zhou and Zhaohui Wang.
© 2014 John Wiley & Sons, Ltd. Published 2014 by John Wiley & Sons, Ltd.

Now let us evaluate the PAPR of OFDM blocks. First, take samples from the continuous-time signal

$$\tilde{x}[n] = \tilde{x}(t)|_{t=\frac{n}{f_s}}, \tag{4.3}$$

where f_s is the *passband* sampling rate. Using a reasonably large sampling rate, there is a good match between the continuous-time and the discrete-time formulation,

$$\text{PAPR} \approx \frac{\max\left(|\tilde{x}[n]|^2\right)}{\mathbb{E}[|\tilde{x}[n]|^2]}. \tag{4.4}$$

As a performance benchmark, we consider a discrete-time sequence of length N, where all elements are independent and identically distributed (i.i.d.) following a standard normal distribution $\mathcal{N}(0,1)$. The CCDF of the PAPR of such a white noise sequence is

$$\Pr(\text{PAPR} > x) = 1 - (1 - 2Q(x))^N \tag{4.5}$$

$$\approx 2NQ(x), \tag{4.6}$$

where the Q-function is defined as

$$Q(x) := \frac{1}{\sqrt{2\pi}} \int_x^{\infty} e^{-\frac{t^2}{2}}\, dt. \tag{4.7}$$

Example 4.1

With a system bandwidth $B = 6\,\text{kHz}$, center frequency $f_c = 17\,\text{kHz}$, and sampling rate $f_s = 96\,\text{kHz}$, Figure 4.1 shows the PAPR distribution of OFDM modulation as a function of the number of subcarriers K. Corresponding to different lengths of the OFDM symbol $N = Kf_s/B$, the PAPR curves of Gaussian noise in (4.5) are also plotted. As the number of subcarriers grows, an increase in PAPR is observed, and the PAPR characteristic of the passband OFDM samples approaches that of Gaussian noise.

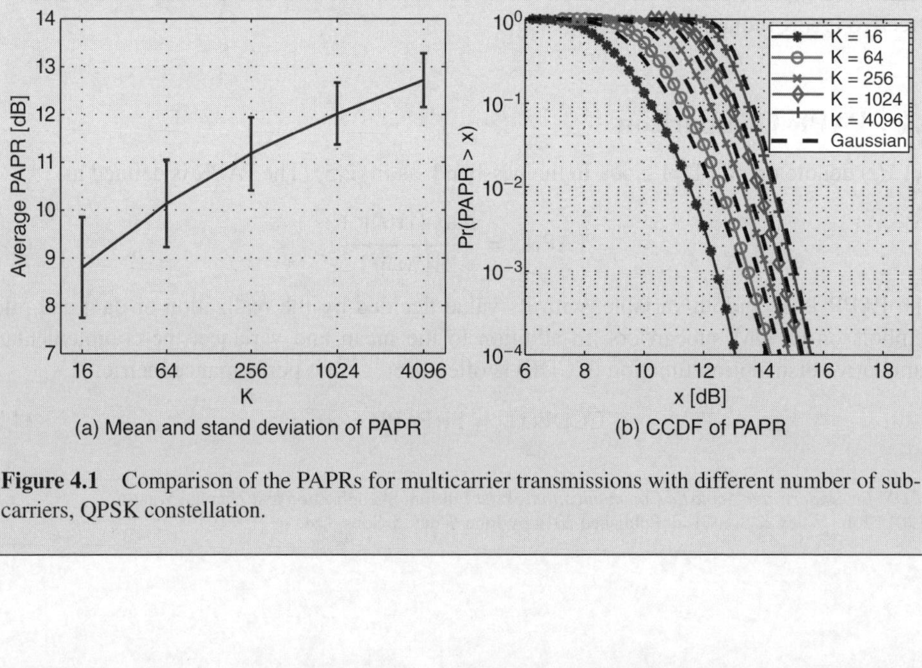

Figure 4.1 Comparison of the PAPRs for multicarrier transmissions with different number of subcarriers, QPSK constellation.

Example 4.2

Now compare the PAPRs of OFDM and single-carrier transmissions. The number of subcarriers is $K = 1024$ for OFDM modulation. The single-carrier waveform is generated according to (1.26), using a raised cosine pulse shaping filter with 25% excess bandwidth. Figure 4.2 shows the PAPR properties of OFDM and single-carrier transmissions. Clearly, the PAPR of OFDM signal is not sensitive to the symbol constellation, in contrast to single-carrier transmission. There is a significant gap in PAPR between OFDM and a single-carrier signal. However, this gap decreases as the symbol constellation size increases. At the CCDF value of 10^{-2}, the gap decreases from 6 dB to 4 dB when the constellation changes from QPSK to 16-QAM.

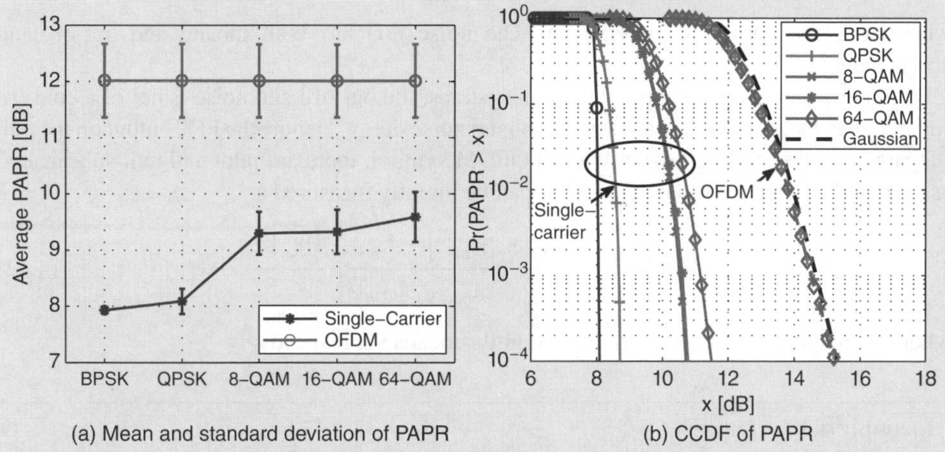

Figure 4.2 Comparison of the PAPRs between OFDM and single-carrier transmissions.

4.2 PAPR Reduction

Next, we present three PAPR reduction approaches: clipping, selective mapping, and peak reduction subcarriers. The first approach introduces signal distortion, while the other two are distortionless at the expense of some signaling overhead.

4.2.1 Clipping

The transmitter can simply clip the signal before amplification, which provides effective PAPR reduction and has no side information needed. Since it is a nonlinear process, clipping may cause significant inband distortion and out-of-band noise.

We here illustrate the clipping effect using a simple model called soft limiter (SL). Operating in the passband, the SL output is related to its input as

$$\tilde{y}(t) = \begin{cases} -A, & \tilde{x}(t) \leq -A \\ \tilde{x}(t), & |\tilde{x}(t)| < A \\ A, & \tilde{x}(t) \geq A. \end{cases} \tag{4.8}$$

Define the clipping ratio Γ as

$$\Gamma := \frac{A}{\sqrt{P_{\text{in}}}}. \tag{4.9}$$

where P_{in} is the power of the input signal before clipping.

Clearly, the clipping noise

$$\tilde{e}(t) = \tilde{y}(t) - \tilde{x}(t) \tag{4.10}$$

is signal dependent. If the clipping ratio is high, clipping may only occur infrequently. In such a case, the clipping noise is more of an impulse-like noise. On the other hand, one can decompose the SL output to two uncorrelated signal components as

$$\tilde{y}(t) = \alpha\tilde{x}(t) + \tilde{\eta}(t). \tag{4.11}$$

where α is the correlation coefficient. The noise $\tilde{\eta}(t)$ has both inband and out-of-band components.

For underwater acoustic communication systems, the out-of-band noise is not of a concern so far. Let us measure the variance of the inband noise. Let y_m denote the FFT output on the mth subcarrier corresponding to $\tilde{y}(t)$. In each OFDM symbol, there are pilot and null subcarriers, as specified in Chapter 2. The pilot SNR can be directly measured as

$$\text{Pilot SNR} = \frac{\mathbb{E}_{m\in\mathcal{S}_{\text{P}}}\left[|y_m|^2\right] - \mathbb{E}_{m\in\mathcal{S}_{\text{N}}}\left[|y_m|^2\right]}{\mathbb{E}_{m\in\mathcal{S}_{\text{N}}}\left[|y_m|^2\right]}, \tag{4.12}$$

where $\mathcal{S}_{\text{P}}$ and $\mathcal{S}_{\text{N}}$ are the sets of pilot and null subcarriers, respectively.

Example 4.3

Using the same system parameters as in Example 1, Figure 4.3 shows the measured pilot SNR as a function of the clipping ratio with both $K = 512$ and $K = 1024$. The pilot SNR measured at the transmitter serves as an upper limit of its counterpart measured at the receiver. Clearly there is a tradeoff: A smaller clip ratio reduces PAPR, and hence improves

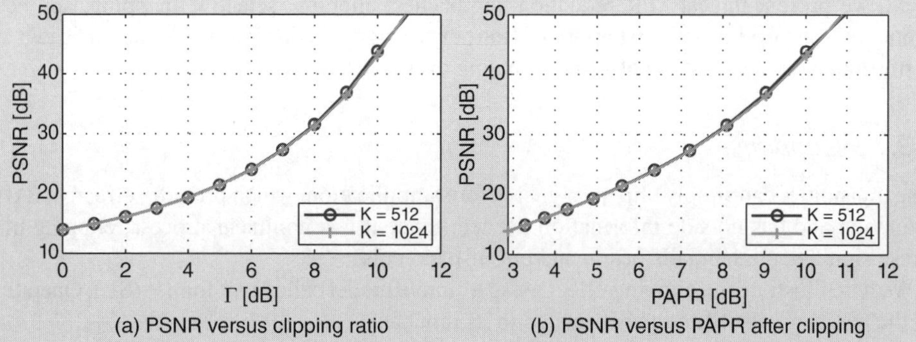

Figure 4.3 The measured pilot SNR and PAPR with different values of the clipping ratio; QPSK is used; however, other constellations from BPSK to 64-QAM lead to the same curves.

the transmission power, but leads to a reduction on the pilot SNR. Depending on the ambient noise level, an optimal clipping ratio exists to maximize the signal to noise ratio of the received signal where the noise includes both the clipping noise and the ambient noise.

4.2.2 Selective Mapping

The idea of selective mapping (SLM) was presented in different forms in [279, 401], and [26]. The key idea is that by introducing extra freedom, the transmitter generates a set of sufficiently different candidate signals which all include the same information. The one with the lowest PAPR is selected for transmission.

Denote L as the number of candidate signals available, and denote the PAPR value of the ith candidate signal as PAPR_i. The PAPR of the SLM output has a CCDF expressed as

$$\Pr\left(\min_{i=1,\ldots,L} \mathrm{PAPR}_i > x\right) = \left(\Pr(\mathrm{PAPR}_i > x)\right)^L, \tag{4.13}$$

which shows that the SLM method can reduce the PAPR effectively as L increases.

There are different ways of implementing the SLM idea. Phase rotations are applied on the modulation symbols in [26, 279], where different scramblers are used on the information bits in [401]. In these results, side information on which signal candidate has been chosen needs to be transmitted. This causes signaling overhead. In addition, side information has high importance and has to be strongly protected.

In the modified approach [50], prior to scrambling and channel encoding, additional bits which are used to select different scrambling code patterns, are inserted to the information bits; as illustrated in Figure 4.4. This way, the side information bits are contained in the data and do not need separate encoding.

Following the SLM principle in [50], the fact that the generator matrix an LDPC code has high density has been used in [175] to reduce PAPR through multiple times of LDPC encoding. The transmitter operates as follows.

- For each set of information bits to be transmitted within one OFDM block, reserve z bits for the PAPR reduction purpose.
- For each choice of the values of these z bits, carry out LDPC encoding and OFDM modulation, and calculate the PAPR.
- Out of 2^z candidates, select the OFDM symbol with the lowest PAPR for transmission.

Due to the nonsparseness of generator matrix, a single bit change will lead to a drastically different codeword after LDPC encoding [270]. (For high rate codes such as 3/4, nonsystematic LDPC codes are preferred relative to systematic codes [175]). Since z is very small, e.g., $z = 2$,

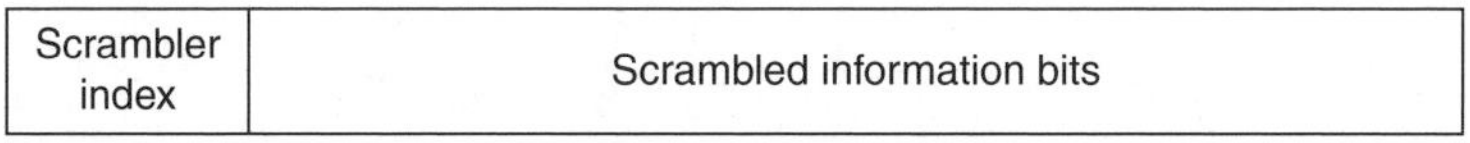

Figure 4.4 The scrambled information bits to be encoded by a channel encoder. The decoded bits will be scrambled using a descrambler as specified by the scrambler index.

or $z = 4$, the reduction on transmission rate is negligible. At the receiver side, those z bits are simply dropped after channel decoding. This is aligned with the principle in [50], but bypasses the scrambling operation at the transmitter and the descrambling operation at the receiver.

Example 4.4

Using the same parameters as in Example 1, and pilot allocation specified in Table 2.1, Figure 4.5 shows the PAPR CCDF curves for different values of z, in which mode 2 specified in Table 3.1 is used for LDPC encoding. Using a nonbinary LDPC code with 4 bits overhead can decrease the PAPR by 2.2 dB relative to the case with no overhead at the CCDF value of 10^{-2}. As z increases, there is a diminishing return, suggesting that initial reduction of PAPR through SLM is very effective, while further reduction will get more difficult. With $z = 4$ bits, the PAPR of OFDM blocks is essentially varying within the range of 10.5 dB and 11.5 dB.

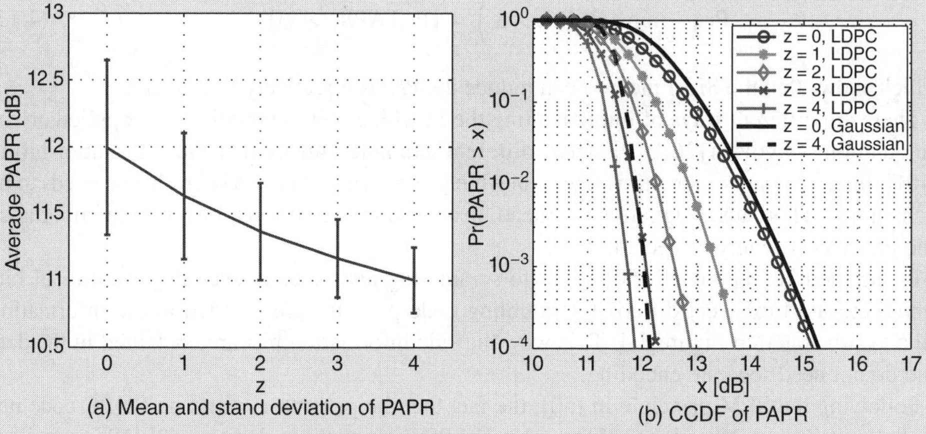

Figure 4.5 PAPR of OFDM with different overhead levels, QPSK constellation.

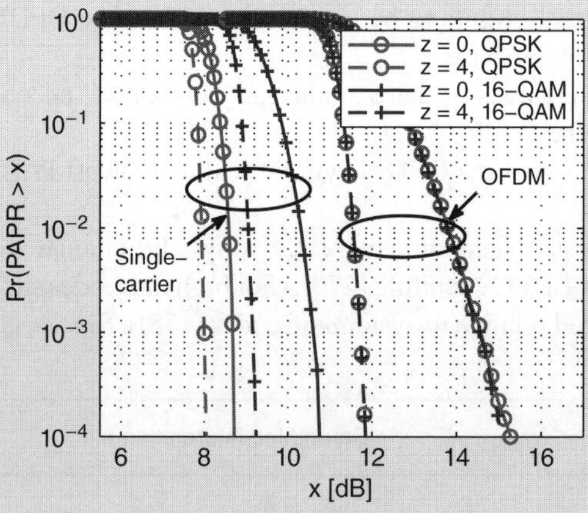

Figure 4.6 Comparison of SC and OFDM with or without PAPR control; $z = 4$ in the SLM method.

The SLM principle can be applied to single-carrier transmissions as well. Figure 4.6 compares the PAPR of OFDM with single-carrier transmissions, both having $K = 1024$ information symbols in each data block. After PAPR control, the gaps between OFDM and single-carrier transmissions at the CCDF of 10^{-2} are only 3.5 dB and 2.5 dB, for QPSK and 16-QAM, respectively.

4.2.3 Peak Reduction Subcarriers

The idea here is that a set of subcarriers are not used to transmit useful information, but reserved to reduce the peak power. Let $\mathcal{S}_\mathrm{A}$ denote the set of active subcarriers in an OFDM symbol, and $\mathcal{S}_\mathrm{PRC}$ as the set of peak reduction carriers (PRC) [380]. Corresponding to the OFDM expression in (2.5), the resulting signal is

$$\tilde{x}(t) = 2\Re\left\{\sum_{k\in\mathcal{S}_\mathrm{A},k\notin\mathcal{S}_\mathrm{PRC}} s[k]e^{j2\pi f_k t}g(t) + \sum_{\ell\in\mathcal{S}_\mathrm{PRC}} p[\ell]e^{j2\pi f_\ell t}g(t)\right\}. \tag{4.14}$$

The symbols $\{p[\ell], \ell \in \mathcal{S}_\mathrm{PRC}\}$ do not need to be drawn from the same constellations as information symbols $s[k]$, and can be arbitrary.

First the locations of the peak reduction subcarriers have to determined beforehand. Contiguous subcarriers often provides worse results in comparison with randomly distributed subcarriers. Once the locations are determined, the symbols $p[\ell]$ will be optimized for each individual block. Optimal and some suboptimal algorithms can be found in e.g., [380].

For underwater acoustic communication systems, so far there is no regulation on out-of-band radiation. Also, a typical transducer has a rather narrow frequency band, and can serve as a filter naturally. Hence, the PRC do not need to be within the signal band, and the choice is termed as out-of-band tone insertion (OTI) in [329]. The transmitted signal is

$$\tilde{x}(t) = 2\Re\left\{\sum_{k\in\mathcal{S}_\mathrm{A}} s[k]e^{j2\pi f_k t}g(t) + \sum_{f_\ell\notin\mathcal{B}} p[\ell]e^{j2\pi f_\ell t}g(t)\right\}, \tag{4.15}$$

where $\mathcal{B} := [f_\mathrm{c} - B/2, f_\mathrm{c} + B/2]$ denotes the useful signal frequency band. Extending the tone reservation technique to the out-of-band carriers, the PAPR reduction is achieved without a loss in the data rate, but with slight increase on the average power consumption.

4.3 Bibliographical Notes

The PAPR of OFDM symbols is a random variable. The number of signals with PAPR much greater than a typical value is small, but the number of signals with PAPR much smaller than the typical value is also small. Excluding only those signals with high peaks, it requires relatively little effort to achieve the typical PAPR, and it is hard to decrease the PAPR further [253].

PAPR reduction is an area that has been extensively investigated, where various methods have been proposed for DSL and wireless OFDM systems. In addition to the clipping [8, 98, 99, 241, 300], selective mapping [26, 163, 279, 401], and peak reduction subcarriers [230, 329, 380] approaches described in this chapter, other methods include coding methods [167], trellis shaping [162, 298], tone injection [220, 380, 381], active constellation extension [221], and partial transmit sequences [288, 382]. These methods can be evaluated and compared based on different metrics such as the PAPR reduction capability, the distortion on the signals, the overhead on the data rate, the complexity of implementation, and whether the method requires some side information to be delivered to the receiver. A comprehensive treatment can be found in the research monograph by Litsyn, 2007 [253].

5

Receiver Overview and Preprocessing

Chapters 5 to 17 contain the major body of this book–OFDM receiver design for underwater acoustic communications. One prominent feature of receiver algorithm design in underwater communications is that the receiving algorithms are heavily tailored to the characteristics of underwater acoustic channels, computational capability of system hardware, and desired system performance. Given large dynamics of an underwater acoustic environment and high cost of the system hardware, the primary goal of the receiving algorithm design is to achieve *efficient* and *reliable* communications with *low computational cost*, whereas a tradeoff always exists between system performance and algorithm complexity. To suit the necessity of various communication applications, both low-complexity and high-complexity receiving algorithms have their own merits. Meanwhile, reducing computational complexity while maintaining reliable communication performance becomes a major theme of advanced receiver design.

Compared to terrestrial radio communications, one unique feature of underwater acoustic communications is that communication channels in different applications could exhibit drastically different characteristics, and hence a universal receiver design might not exist. As a result, different communication applications have to be examined individually. Several typical communication scenarios will be considered in the following chapters of this book: (i) shallow water acoustic communications in Chapters 9, 10 and 13, (ii) deep water horizontal acoustic communications in Chapter 11, (iii) acoustic communications with external interference in Chapter 12, (iv) distributed multiuser acoustic communications in Chapters 14 and 15, and (v) underwater relay communications in Chapters 16 and 17. Before moving to the overall OFDM receiver design in above communication scenarios, the common structural components in the OFDM receivers will be discussed in Chapters 5–8, and the major task of OFDM receiver design for specific communication channels becomes addressing challenges posed by the channel to particular receiver components.

In this chapter, we will have an overview on the OFDM receiver structure for three types of channel settings: (i) the single-input single-output (SISO) channel, (ii) the single-input multi-output (SIMO) channel, and (iii) the multi-input multi-output (MIMO) channel. We start with a brief description on the overall OFDM receiver structure, then provide detailed discussion on the receiver preprocessing which converts the analogous time-domain waveform to the

OFDM for Underwater Acoustic Communications, First Edition. Shengli Zhou and Zhaohui Wang.
© 2014 John Wiley & Sons, Ltd. Published 2014 by John Wiley & Sons, Ltd.

frequency-domain discrete samples. After the receiver preprocessing, the input–output relationship between frequency samples and transmitted data symbols in the above three types of channel settings will be presented, which lay down the foundation for OFDM receiver design in following chapters. Then, based on two data decoding philosophies, a general overview on OFDM receivers will be presented. We also look into the system model for CP-OFDM transmissions, and pinpoint the subtle difference between ZP-OFDM and CP-OFDM receiver designs. The discussion applies to a variety of communication scenarios we will look into in later chapters. In the last part of this chapter, numerical results on the OFDM receiver theoretical performance bound will be presented, which will invoke resonance with the performance of practical receivers to be covered in later chapters.

5.1 OFDM Receiver Overview

A typical OFDM frame structure in each transmission usually includes a detection preamble and a synchronization preamble followed by a number of OFDM symbols; as shown in Figure 5.1. Correspondingly, the OFDM receiver in UWA communications consists of several components as shown in Figure 5.2.

- *Preamble detection.* This module keeps monitoring the environment to detect the arrival of useful signal, or being more specific, the arrival of the detection preamble. Once the signal is detected, the receiver starts recording the useful signal for processing.
- *Synchronization and Doppler scale estimation.* Based on the recorded waveform, this module estimates the time-of-arrival of useful signal τ_0 and the Doppler scaling factor a caused by platform mobility. These two parameters are used in the subsequent processing,

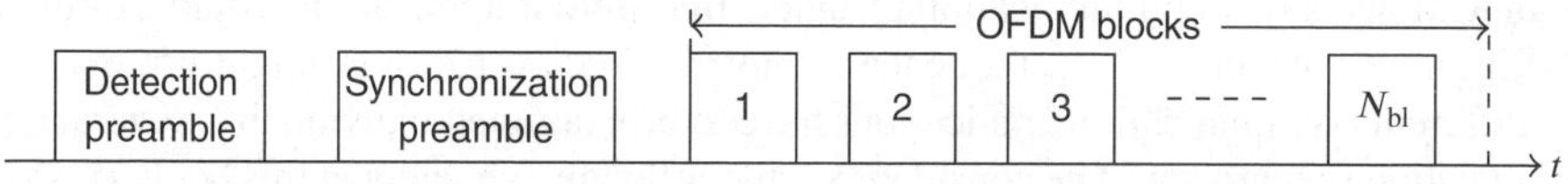

Figure 5.1 An example frame structure of OFDM transmission.

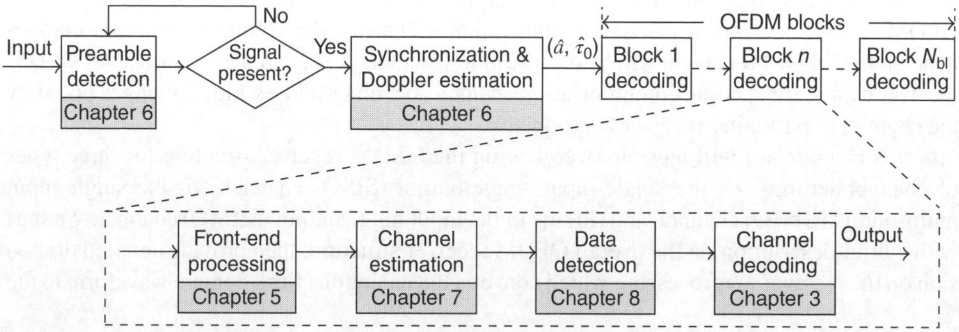

Figure 5.2 Diagram of typical receiver processing modules.

including the Doppler scaling effect compensation and the OFDM signal partition for block processing.

- *Receiver processing*. Based on the estimated time-of-arrival of useful signal, the received OFDM signal is usually truncated into individual processing units according to the transmitted OFDM block structure. Information symbols within each received OFDM block will be recovered with an identical processing flow as indicated in Figure 5.2.

Among the above three modules, the receiver processing module is the major component of an OFDM receiver. Taking the recorded time-domain waveform of one particular block as input, this module deciphers the transmitted information bits carried by this block. Typical processing units in this module are illustrated in Figure 5.2, including:

- Receiver preprocessing. At each receiving element, the receiver preprocessing is conducted to convert the recorded time-domain signal to frequency-domain discrete samples. During this process, the Doppler scaling effect caused by platform mobility will be compensated.
- Channel estimation. In most scenarios, the communication channel is unknown to receiver. Prior to parameterizing the channel distortion effect on the transmitted signal, estimation of channel parameters at each receiving element is necessary.
- Data detection. With the discrete frequency samples and channel estimates at all the receiving elements, information symbols in the transmitted signal are estimated. The reliability information on information bits or symbols obtained.
- Channel decoding. Channel coding is an indispensable component in modern communication systems to approach the Shannon capacity. Taking the data detection results as input, channel decoder deciphers the transmitted information bits within each OFDM block.

Inspired by the turbo decoding algorithm [42], iterative processing among the above components is often adopted to improve system performance through back and forth message passing: parameters estimated in the preceding iteration can be used as known parameters in the current iteration [103, 439]. For example, the estimated data symbols from the channel decoder can be fed back to facilitate channel estimation, which in turn improves data detection and channel decoding performance. We will defer detailed description on the iterative OFDM receiver design to Chapter 9 and beyond.

5.2 Receiver Preprocessing

5.2.1 Receiver Preprocessing

The receiver preprocessing consists of Doppler compensation and time-to-frequency domain conversion. This leads to measurements in the frequency domain, based on which channel estimation, data detection, and channel decoding are carried out.

5.2.1.1 Two-Step Doppler Compensation

Let a and $\hat{a}$ denote the accurate and the estimated mean Doppler scaling factor, respectively. A two-step operation can be performed within each OFDM block to convert the continuous passband signal $\tilde{y}(t)$ to frequency domain samples at all subcarriers. Due to possible inaccuracy in the Doppler scale factor estimation, the channel Doppler scaling effect on the received passband signal $\tilde{y}(t)$ is compensated in two steps.

(i) Main Doppler scale compensation. The main Doppler effect can be removed through a resampling operation of $\tilde{y}(t)$ with a resampling factor $(1+\hat{a})$, leading to the resampled signal $\tilde{y}\left(\frac{t}{1+\hat{a}}\right)$.
(ii) Residual Doppler shift compensation. After resampling, the effect of residual Doppler shift can be approximately viewed as due to carrier frequency offset. Assume that $\hat{\epsilon}$ is the estimated residual Doppler shift, the Doppler shift compensation is performed to obtain

$$\tilde{z}(t) = \tilde{y}\left(\frac{t}{1+\hat{a}}\right) e^{-j2\pi\hat{\epsilon}t} \tag{5.1}$$

5.2.1.2 Time-to-Frequency Conversion

The frequency-domain measurement at the mth subcarrier is obtained as

$$\begin{aligned} z[m] &= \int_0^{T+T_g} \tilde{z}(t) e^{-j2\pi f_m t} dt \\ &= \int_0^{T+T_g} \tilde{y}\left(\frac{t}{1+\hat{a}}\right) e^{-j2\pi(f_m+\hat{\epsilon})t} dt \\ &= (1+\hat{a})\tilde{Y}((1+\hat{a})(f_m+\hat{\epsilon})), \end{aligned} \tag{5.2}$$

which indicates that the mth frequency-domain measurement corresponds to the frequency sample of $\tilde{Y}(f)$ at the one particular frequency point $(1+\hat{a})(f_m+\hat{\epsilon})$.

5.2.2 *Digital Implementation*

5.2.2.1 Passband-to-Baseband Downshifting

Define

$$K_{\text{bl}} := K + (T_g/T)K. \tag{5.3}$$

The passband continuous-time signal $\tilde{y}(t)$ is sampled as

$$\tilde{y}[n] = \tilde{y}(t)|_{t=nT_{s,p}}, \qquad n = 0, \cdots, \lambda_p K_{\text{bl}} - 1 \tag{5.4}$$

where $T_{s,p} := 1/f_{s,p}$ denotes the sampling interval, and the passband sampling rate is $f_{s,p} = \lambda_p B$ with λ_p being the ratio between the passband sampling rate and the system bandwidth (c.f. Section 2.4).

The baseband samples can be obtained by passing the passband samples through a low bandpass filter,

$$\text{LPF}\{\tilde{y}[n]e^{-j2\pi f_c n T_{s,p}}\}, \tag{5.5}$$

for $n = 0, \cdots, \lambda_p K_{\text{bl}} - 1$, where LPF stands for the low bandpass filtering. For computational efficiency, a downsampling operation is usually adopted after the passband-to-baseband downshifting. We denote $f_{s,b} := \lambda_b B$ as the baseband sampling rate, with a sampling interval $T_{s,b} := 1/f_{s,b}$. The baseband sample is denoted as $y[n]$, with $n = 0, \cdots, \lambda_b K_{\text{bl}} - 1$.

5.2.2.2 Method 1: Interpolation and FFT

Based on the discrete time-domain sample $y[n]$, the frequency-domain sample $z[m]$ can be obtained via a two-step operation: Doppler-scale compensation and time-to-frequency transformation. The mean Doppler-scale compensation can be achieved via a sampling-rate conversion, and the time-to-frequency transform can be accomplished via the fast Fourier transform (FFT). We next illustrate the two-step approach in detail.

Denote $y_{\mathrm{r}}(t)$ as the baseband signal of $\tilde{y}_r(t) = \tilde{y}(t/(1+\hat{a}))$. We have

$$y_{\mathrm{r}}(t) = y\left(\frac{t}{1+\hat{a}}\right) e^{-j2\pi \hat{a} f_{\mathrm{c}} \frac{t}{1+\hat{a}}}, \qquad t \in [0, T_{\mathrm{bl}}]. \tag{5.6}$$

According to the sampling theory [317, Chapter 6], the continuous-time baseband signal $y(t)e^{-j2\pi\hat{a}f_{\mathrm{c}}t}$ can be expressed as

$$y(t)e^{-j2\pi\hat{a}f_{\mathrm{c}}t} = \sum_{n=-\infty}^{\infty} y[n]e^{-j2\pi\hat{a}f_{\mathrm{c}}nT_{\mathrm{s,b}}} p(t - nT_{\mathrm{s,b}}) \tag{5.7}$$

where $p(t)$ is the interpolation function for signal reconstruction,

$$p(t) = \frac{\sin(\pi t/T_{\mathrm{s,b}})}{\pi t/T_{\mathrm{s,b}}}. \tag{5.8}$$

Substituting (5.7) into (5.6) yields

$$y_{\mathrm{r}}(t) = \sum_{n=-\infty}^{\infty} y[n]e^{-j2\pi\hat{a}f_{\mathrm{c}}nT_{\mathrm{s,b}}} p\left(\frac{t}{1+\hat{a}} - nT_{\mathrm{s,b}}\right). \tag{5.9}$$

Denote $y_{\mathrm{r}}[l]$ as the discrete sample of $y_{\mathrm{r}}(t)$ at a rate $f_{\mathrm{s,b}}$

$$y_{\mathrm{r}}[l] = y_{\mathrm{r}}(t)|_{t=lT_{\mathrm{s,b}}} \tag{5.10}$$

which can be formulated as

$$y_{\mathrm{r}}[l] = \sum_{n=-\infty}^{\infty} y[n]e^{-j2\pi\hat{a}f_{\mathrm{c}}nT_{\mathrm{s,b}}} p\left(\frac{lT_{\mathrm{s,b}}}{1+\hat{a}} - nT_{\mathrm{s,b}}\right). \tag{5.11}$$

Define

$$\overline{T}_{\mathrm{s,b}} = \frac{T_{\mathrm{s,b}}}{1+\hat{a}}. \tag{5.12}$$

We rewrite (5.11) as

$$y_{\mathrm{r}}[l] = \sum_{n=-\infty}^{\infty} y[n]e^{-j2\pi\hat{a}f_{\mathrm{c}}nT_{\mathrm{s,b}}} p(l\overline{T}_{\mathrm{s,b}} - nT_{\mathrm{s,b}}) \tag{5.13}$$

which shows that the mean Doppler compensation can be achieved by changing the sampling rate of the discrete time-domain samples $y[n]e^{-j2\pi\hat{a}f_{\mathrm{c}}nT_{\mathrm{s,b}}}$ from $1/T_{\mathrm{s,b}}$ to a new sampling rate $1/\overline{T}_{\mathrm{s,b}}$.

There are many approaches for signal sampling-rate conversion. Here we present two approaches applicable in two different scenarios.

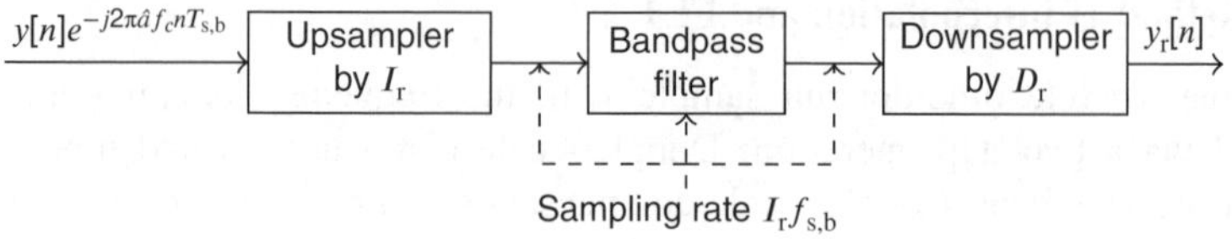

Figure 5.3 Sampling-rate conversion via interpolation and decimation.

- Large Doppler scale factor. For large values of $\hat{a}$, one canonical procedure for sampling-rate conversion is based on signal interpolation and decimation [317, Chapter 11] [338]. We express the ratio $\overline{T}_{s,b}/T_{s,b}$ as a rational number

$$\frac{\overline{T}_{s,b}}{T_{s,b}} = \frac{D_r}{I_r} = \frac{1}{1+\hat{a}}, \tag{5.14}$$

where D_r and I_r are relative prime integers. The sampling-rate conversion can be achieved by an interpolator of a factor I_r and a decimator of a factor D_r, as shown in Figure 5.3. A bandpass filter is introduced to avoid aliasing in the interpolation and decimation process. A polyphase filter structure can be used to implement the interpolator and decimator with high computational efficiency [317, Chapter 11].

In some scenarios, I_r and D_r could be too large for implementation. One can approximate $1/(1+\hat{a})$ in (5.14) using smaller values of I_r and D_r at the expense of signal distortion.
- Small Doppler scale factor. Note that in practical underwater communication systems, the Doppler scaling factor $\hat{a}$ can be very small. For example, with a platform speed $v_0 = 1.5$ m/s, the Doppler scale is $a = 0.001$. The interpolation and decimation factors are $I_r = 1001$ and $D_r = 1000$, which require a very high computational load. Hence, in the scenario with small values of $\hat{a}$, direct calculation of (5.11) tends to be more efficient.

Notice that the operation in (5.9) can be interpreted as a convolution between the discrete sequence $y[n]e^{-j2\pi\hat{a}f_c nT_{s,b}}$ and a continuous signal $p(t/(1+\hat{a}))$. Corresponding to the *windowing* method in the finite-impulse-response (FIR) filter design, the Doppler compensated signal $y_r(t)$ can be closely reconstructed by replacing the infinite summation in (5.9) by a finite summation in the practical implementation. To accommodate various values of $\hat{a}$, the discrete samples of $p(t)$ can be pre-computed at a very high sampling rate, stored in a lookup table, and interpolation can be used to calculate values of $p(t)$ at arbitrary points. For extremely small values of $\hat{a}$, a very high sampling rate of $p(t)$ is necessary to achieve a reasonable reconstruction accuracy, which could incur a very large storage requirement. More detailed studies on the interpolation-based sampling-rate conversion can be found from the CCRMA group in Stanford University [66].

Once $y_r[l]$ is obtained, the residual Doppler shift can be compensated by multiplying $y_r[l]$ with $e^{-j2\pi\hat{\epsilon}lT_{s,b}}$. The frequency sample $z[m]$ can then be obtained via a $\lambda_b K$-point FFT.

5.2.2.3 Method 2: Chirp z-Transform

Compared to the two-step approach, an alternative approach to the Doppler-scale compensation operates in the frequency domain using the chirp z-transform (CZT) [317, Chapter 8] [323].

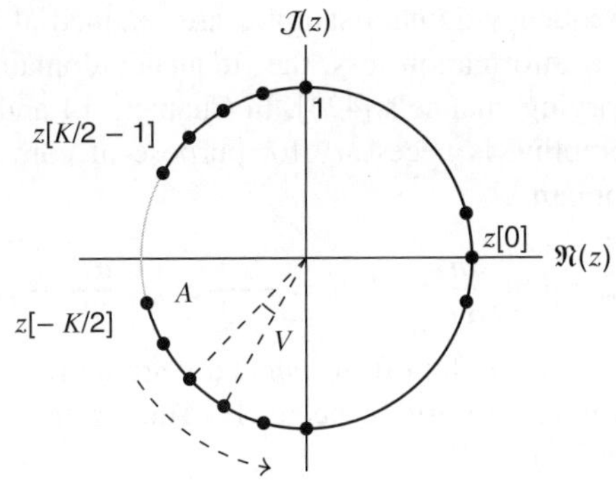

Figure 5.4 Illustration of CZT-based frequency-domain sampling.

With the baseband sample $y[n]$, a direct calculation of (5.2) is

$$z[m] = (1+\hat{a}) \sum_{n=0}^{\lambda_b K_{bl}-1} y[n] e^{-j2\pi\left[(1+\hat{a})\left(\frac{m}{T}+\hat{\epsilon}\right)+\hat{a}f_c\right]\frac{n}{\lambda_b B}}. \tag{5.15}$$

For illustration convenience, we redefine the frequency sample index $\ell = m + K/2$, with $\ell = 0, 1, \cdots, K-1$. The frequency sample in (5.15) can be reformulated as

$$z[\ell - K/2] = (1+\hat{a}) \sum_{n=0}^{\lambda_b K_{bl}-1} y[n] e^{-j2\pi\left[(1+\hat{a})\left(\frac{\ell}{T}-\frac{K}{2T}\right)+(1+\hat{a})\hat{\epsilon}+\hat{a}f_c\right]\frac{n}{\lambda_b B}}. \tag{5.16}$$

Define

$$A := e^{-j2\pi\left[\frac{(1+\hat{a})\hat{\epsilon}+\hat{a}f_c}{\lambda_b B}-\frac{1+\hat{a}}{2\lambda_b}\right]}, \qquad V := e^{j2\pi(1+\hat{a})\frac{1}{\lambda_b K}}. \tag{5.17}$$

We have (5.16) rewritten as

$$z[\ell - K/2] = \sum_{n=0}^{\lambda_b K_{bl}-1} y[n] A^{-n} V^{-nl}, \tag{5.18}$$

for $\ell = 0, \cdots, K-1$. From (5.18), we can see that the discrete sample $z[\ell - K/2]$ corresponds to sampling the z-transform of sequence $y[n]$ over an arc of unit circle: the starting point of the arc is defined by A, while the sampling interval over the arc is determined by V, as illustrated in Figure 5.4. The transform in (5.18), corresponding to the chirp z-transform, can be evaluated via FFT [317, Chapter 8] [323].

Relative to the first implementation method which performs sampling-rate conversion and time-to-frequency transform individually, the CZT-based method yields frequency samples directly based on discrete time-domain samples, hence achieves a higher accuracy without making any approximation for complexity reduction. Meanwhile, leveraging FFT, the CZT-based approach may require a lower complexity compared to the step-by-step implementation.

5.2.3 Frequency-Domain Oversampling

Notice that for each ZP-OFDM block, a total of K_{bl} time-domain samples are obtained with a baseband sampling rate, which contain all useful information about the current block, while there are only $K < K_{\rm bl}$ frequency-domain samples are retained after the time-to-frequency transform. Hence, to avoid information loss, the frequency-domain oversampling is necessary, especially in time-varying channels [424]. In Chapters 14 and 15, we will see that the frequency-domain oversampling is necessary for purpose of converting samples from frequency domain to time domain.

Define

$$\check{f}_{\check{m}} = f_{\rm c} + \frac{\check{m}}{\alpha T}, \qquad \check{m} = -\frac{\alpha K}{2}, \dots, \frac{\alpha K}{2} - 1, \tag{5.19}$$

where $\alpha > 1$ is an integer oversampling factor, $\check{m}$ and k are indices of the oversampled measurements and the physical subcarriers, respectively. The frequency-domain samples at the frequency oversampling point $\check{f}_{\check{m}}$ is

$$z[\check{m}] = (1+\hat{a})\tilde{Y}\left((1+\hat{a})(\check{f}_{\check{m}} + \hat{\epsilon})\right). \tag{5.20}$$

Similar to the frequency-domain sampling with a sampling rate of $1/T$, the discrete samples in (5.20) can be obtained via the two implementation methods discussed above. In particular for the two-step implementation approach, an αK-point FFT operation will be performed after padding $(\alpha K - K_{\rm bl})$ zeros to the baseband samples after Doppler shift compensation.

5.3 Frequency-Domain Input–Output Relationship

5.3.1 Single-Input Single-Output Channel

We consider a path-based channel with path-specific Doppler scales defined in (1.14). Assume that the channel consists of $N_{\rm pa}$ discrete paths, with a triplet (A_p, τ_p, a_p) representing the amplitude, initial delay and Doppler rate of the pth path, respectively. Denote $\tilde{x}(t)$ as the transmitted passband signal. The received signal is expressed as

$$\tilde{y}(t) = \sum_{p=1}^{N_{\rm pa}} A_p \tilde{x}((1+a_p)t - \tau_p) + \tilde{n}(t). \tag{5.21}$$

The two-step Doppler compensation in (5.1) leads to

$$\tilde{z}(t) = \sum_{p=1}^{N_{\rm pa}} A_p \tilde{x}\left(\frac{1+a_p}{1+\hat{a}}t - \tau_p\right) e^{-j2\pi\hat{\epsilon}t} + \tilde{w}(t), \tag{5.22}$$

where the Doppler compensated noise is denoted by

$$\tilde{w}(t) = \tilde{n}\left(\frac{t}{1+\hat{a}}\right) e^{-j2\pi\hat{\epsilon}t} \tag{5.23}$$

Based on the transmitted signal formula in (2.6) and the time-to-frequency conversion in (5.2), the input–output relationship between the discrete frequency samples and the transmitted symbols is expressed as

$$z[m] = \sum_{k=-K/2}^{K/2-1} H[m,k]s[k] + w[m], \quad m = -\frac{K}{2}, \dots, \frac{K}{2} - 1 \tag{5.24}$$

where $w[m]$ is the noise sample in the frequency domain, and $H[m,k]$ is the ICI coefficient which specifies the contribution of the symbol transmitted on the kth subcarrier to the output on the mth subcarrier. The ICI coefficient can be represented by the N_{pa} path parameters as

$$H[m,k] = \sum_{p=1}^{N_{\text{pa}}} \xi_p e^{-j2\pi \frac{m}{T}\overline{\tau}_p} G\left(\frac{f_m + \hat{\epsilon}}{1+b_p} - f_k\right), \tag{5.25}$$

with

$$b_p := \frac{a_p - \hat{a}}{1+\hat{a}}, \qquad \xi_p := \frac{A_p}{1+b_p} e^{-j2\pi(f_c+\hat{\epsilon})\overline{\tau}_p}, \qquad \overline{\tau}_p := \frac{\tau_p}{1+b_p}. \tag{5.26}$$

Using a matrix-vector notation, we can rewrite (5.24) as

$$\underbrace{\begin{bmatrix} z[-\frac{K}{2}] \\ \vdots \\ z[\frac{K}{2}-1] \end{bmatrix}}_{:=\mathbf{z}} = \underbrace{\begin{bmatrix} H[-\frac{K}{2},-\frac{K}{2}] & \cdots & H[-\frac{K}{2},\frac{K}{2}-1] \\ \vdots & \ddots & \vdots \\ H[\frac{K}{2}-1,-\frac{K}{2}] & \cdots & H[\frac{K}{2}-1,\frac{K}{2}-1] \end{bmatrix}}_{:=\mathbf{H}} \underbrace{\begin{bmatrix} s[-\frac{K}{2}] \\ \vdots \\ s[\frac{K}{2}-1] \end{bmatrix}}_{:=\mathbf{s}} + \underbrace{\begin{bmatrix} w[-\frac{K}{2}] \\ \vdots \\ w[\frac{K}{2}-1] \end{bmatrix}}_{:=\mathbf{w}} \tag{5.27}$$

which can be compactly expressed as

$$\mathbf{z} = \mathbf{H}\mathbf{s} + \mathbf{w}. \tag{5.28}$$

The channel mixing matrix $\mathbf{H}$ is specified by N_{pa} triplets $\{\xi_p, \overline{\tau}_p, b_p\}$,

$$\mathbf{H} = \sum_{p=1}^{N_{\text{pa}}} \xi_p \boldsymbol{\Lambda}(\overline{\tau}_p)\boldsymbol{\Gamma}(b_p, \hat{\epsilon}) \tag{5.29}$$

where $\boldsymbol{\Lambda}(\tau)$ is a $K \times K$ generic diagonal matrix with the mth diagonal entry,

$$[\boldsymbol{\Lambda}(\tau)]_{m,m} = e^{-j2\pi \frac{m}{T}\tau}, \tag{5.30}$$

and $\boldsymbol{\Gamma}(b,\epsilon)$ is a $K \times K$ generic matrix with the (m,k)th entry,

$$[\boldsymbol{\Gamma}(b,\epsilon)]_{m,k} = G\left(\frac{f_m + \epsilon}{1+b} - f_k\right). \tag{5.31}$$

5.3.2 *Single-Input Multi-Output Channel*

For a SIMO channel with N_r receiving elements as shown in Figure 2.8, denote $N_{\text{pa},\nu}$ as the number of paths between the transmitter and the νth receiving element, and the triplet $(A_{\nu,p}, \tau_{\nu,p}, a_{\nu,p})$ is similarly defined as (A_p, τ_p, a_p). With a transmitted signal $\tilde{x}(t)$, the received signal at the νth element is

$$\tilde{y}_\nu(t) = \sum_{p=1}^{N_{\text{pa},\nu}} A_{\nu,p}\tilde{x}((1+a_{\nu,p})t - \tau_{\nu,p}) + \tilde{n}_\nu(t). \tag{5.32}$$

Based on the transmitted signal formula in (2.6) and the time-to-frequency conversion in (5.2), the frequency measurement in the mth subcarrier can be expressed as

$$z_\nu[m] = \sum_{k=-K/2}^{K/2-1} H_{\nu,\mu}[m,k]s[k] + w_\nu[m], \quad m = -\frac{K}{2}, \ldots, \frac{K}{2} - 1 \tag{5.33}$$

where the channel coefficient $H_\nu[m,k]$ is similarly defined as $H[m,k]$ in (5.25),

$$H_\nu[m,k] = \sum_{p=1}^{N_{\mathrm{pa},\nu}} \xi_{\nu,p} e^{-j2\pi \frac{m}{T}\bar{\tau}_{\nu,p}} G\left(\frac{f_m + \hat{\epsilon}_\nu}{1 + b_{\nu,p}} - f_k\right), \tag{5.34}$$

with $\{\xi_{\nu,p}, \bar{\tau}_{\nu,p}, b_{\nu,p}\}$ similarly defined as $\{\xi_p, \bar{\tau}_p, b_p\}$ in (5.26) based on $\{A_{\nu,p}, \tau_{\nu,p}, a_{\nu,p}\}$.

Hence, a compact expression of the discrete input–output relationship is

$$\mathbf{z}_\nu = \sum_{\mu=1}^{N_t} \mathbf{H}_{\nu,\mu}\mathbf{s} + \mathbf{w}_\nu, \tag{5.35}$$

for $\nu = 1, \cdots, N_r$, with

$$\mathbf{H}_\nu = \sum_{p=1}^{N_{\mathrm{pa},\nu}} \xi_{\nu,p} \mathbf{\Lambda}(\bar{\tau}_{\nu,p}) \mathbf{\Gamma}(b_{\nu,p}, \hat{\epsilon}_\nu). \tag{5.36}$$

The input–output relationship in (5.35) can also be put in a matrix form as

$$\begin{bmatrix} \mathbf{z}_1 \\ \vdots \\ \mathbf{z}_{N_r} \end{bmatrix} = \begin{bmatrix} \mathbf{H}_1 \\ \vdots \\ \mathbf{H}_{N_r} \end{bmatrix} \mathbf{s} + \begin{bmatrix} \mathbf{w}_1 \\ \vdots \\ \mathbf{w}_{N_r} \end{bmatrix}. \tag{5.37}$$

5.3.3 Multi-Input Multi-Output Channel

For a MIMO channel with N_t transmitters and N_r receiving elements as shown in Figure 2.11, denote $N_{\mathrm{pa},\nu,\mu}$ as the number of paths between the μth transmitter and the νth receiving element, and the triplet $(A_{\nu,\mu,p}, \tau_{\nu,\mu,p}, a_{\nu,\mu,p})$ is similarly defined as (A_p, τ_p, a_p). Define $\tilde{x}_\mu(t)$ as the transmitted signal from the μth transmitter. The received signal at the νth receiving element is

$$\tilde{y}_\nu(t) = \sum_{\mu=1}^{N_t} \sum_{p=1}^{N_{\mathrm{pa},\nu,\mu}} A_{\nu,\mu,p} \tilde{x}\left((1 + a_{\nu,\mu,p})t - \tau_{\nu,\mu,p}\right) + \tilde{n}_\nu(t). \tag{5.38}$$

Based on the transmitted signal formula in (2.6) and the time-to-frequency conversion in (5.2), the frequency measurement in the mth subcarrier can be expressed as

$$z_\nu[m] = \sum_{\mu=1}^{N_t} \sum_{k=-K/2}^{K/2-1} H_{\nu,\mu}[m,k]s_\mu[k] + w_\nu[m], \quad m = -\frac{K}{2}, \ldots, \frac{K}{2} - 1 \tag{5.39}$$

where the channel coefficient $H_{\nu,\mu}[m,k]$ is similarly defined as $H[m,k]$ in (5.25),

$$H_{\nu,\mu}[m,k] = \sum_{p=1}^{N_{\mathrm{pa},\nu,\mu}} \xi_{\nu,\mu,p} e^{-j2\pi \frac{m}{T}\overline{\tau}_{\nu,\mu,p}} G\left(\frac{f_m + \hat{\epsilon}_\nu}{1+b_{\nu,\mu,p}} - f_k\right), \tag{5.40}$$

with $\{\xi_{\nu,\mu,p}, \overline{\tau}_{\nu,\mu,p}, b_{\nu,\mu,p}\}$ similarly defined as $\{\xi_p, \overline{\tau}_p, b_p\}$ in (5.26) based on $\{A_{\nu,\mu,p}, \tau_{\nu,\mu,p}, a_{\nu,\mu,p}\}$.

Hence, a compact expression of the discrete input–output relationship is

$$\mathbf{z}_\nu = \sum_{\mu=1}^{N_\mathrm{t}} \mathbf{H}_{\nu,\mu}\mathbf{s}_\mu + \mathbf{w}_\nu, \tag{5.41}$$

with channel mixing matrices

$$\mathbf{H}_{\nu,\mu} = \sum_{p=1}^{N_{\mathrm{pa},\nu,\mu}} \xi_{\nu,\mu,p}\mathbf{\Lambda}(\overline{\tau}_{\nu,\mu,p})\mathbf{\Gamma}(b_{\nu,\mu,p}, \hat{\epsilon}_\nu), \tag{5.42}$$

for $\nu = 1, \cdots, N_\mathrm{r}$, and $\mu = 1, \cdots, N_\mathrm{t}$. The input–output relationship in (5.41) can also be put into a matrix form as

$$\begin{bmatrix} \mathbf{z}_1 \\ \vdots \\ \mathbf{z}_{N_\mathrm{r}} \end{bmatrix} = \begin{bmatrix} \mathbf{H}_{1,1} & \cdots & \mathbf{H}_{1,N_\mathrm{t}} \\ \vdots & \ddots & \vdots \\ \mathbf{H}_{N_\mathrm{r},1} & \cdots & \mathbf{H}_{N_\mathrm{r},N_\mathrm{t}} \end{bmatrix} \begin{bmatrix} \mathbf{s}_1 \\ \vdots \\ \mathbf{s}_{N_\mathrm{t}} \end{bmatrix} + \begin{bmatrix} \mathbf{w}_1 \\ \vdots \\ \mathbf{w}_{N_\mathrm{r}} \end{bmatrix}. \tag{5.43}$$

5.3.4 Channel Matrix Structure

We take the SISO channel as an example to illustrate different structures of the channel matrix in both time-invariant and time-varying environments. From (5.29) we see that the channel matrix structure is determined by the matrix $\mathbf{\Gamma}(b_p, \hat{\epsilon})$ defined in (5.31). For a rectangular pulse shaping window in (2.9), we have the following observations on the channel matrix structure.

- In the time-invariant channel defined in (1.18) with $b_p = 0$ and $\hat{\epsilon} = 0$, we have

$$[\mathbf{\Gamma}(b_p, \hat{\epsilon})]_{m,k} = \begin{cases} 1, & \forall m = k \\ 0, & \forall m \neq k \end{cases} \tag{5.44}$$

  which leads to a diagonal channel matrix $\mathbf{H}$ with the mth diagonal element expressed as

$$H[m] = \sum_{p=1}^{N_\mathrm{pa}} \xi_p e^{-j2\pi \frac{m}{T}\overline{\tau}_p}, \tag{5.45}$$

  indicating that the orthogonality of subcarriers is well preserved in the received signal.
- In the time-varying channel defined in (1.14) with $b_p \neq 0$ and $\hat{\epsilon} \neq 0$, we have

$$[\mathbf{\Gamma}(b_p, \hat{\epsilon})]_{m,k} \neq 0, \qquad \forall m, k \tag{5.46}$$

  which leads to a full channel matrix, with its off-diagonal elements characterizing ICI in the received signal.

5.4 OFDM Receiver Categorization

Based on the input-output relationship in (5.28), (5.37) and (5.43), two types of OFDM receivers exist for block processing:

- The ICI-ignorant receiver which takes the potential ICI in the received signal as ambient noise. This type of receivers applies to the time-invariant channel, or the time-varying channel with computational capability constraint at receiver.
- The ICI-aware receiver which addresses ICI explicitly in receiver processing. This type of receivers applies to the time-varying channel with a reasonable computational capability at receiver.

For the OFDM transmission with multiple blocks, two methodologies exist for block-level processing:

- The block-by-block processing which decodes received blocks individually. This type of receivers assumes a large channel variation from one block to the next, having robust performance in underwater communication systems.
- The block-to-block processing where the processing results of preceding blocks (e.g., the channel estimate) is used for decoding the current block. This type of receivers enjoys a high system throughput by reducing the number of pilots for channel estimation, but relying on the channel coherence across blocks.

We next give a brief description about the above receiver design and processing methods.

5.4.1 ICI-Ignorant Receiver

Assuming the absence of ICI, the channel matrix **H** is taken as a diagonal matrix. The input–output relationship of the three transmission systems is simplified as

- SISO channel

$$z[m] = H[m]s[m] + \eta[m] \tag{5.47}$$

  where the equivalent noise is

$$\eta[m] = \sum_{k\neq m} H[m,k]s[k] + w[m]. \tag{5.48}$$

- SIMO channel

$$z_\nu[m] = H_\nu[m]s[m] + \eta_\nu[m], \qquad \nu = 1, \cdots, N_r \tag{5.49}$$

  where the equivalent noise is

$$\eta_\nu[m] = \sum_{k\neq m} H_\nu[m,k]s[k] + w_\nu[m], \qquad \nu = 1, \cdots, N_r \tag{5.50}$$

- MIMO channel

$$z_\nu[m] = \sum_{\mu=1}^{N_t} H_{\nu,\mu}[m]s_\mu[m] + \eta_\nu[m], \qquad \nu = 1, \cdots, N_r \tag{5.51}$$

where the equivalent noise is

$$\eta_\nu[m] = \sum_{\mu=1}^{N_t} \sum_{k \neq m} H_{\nu,\mu}[m,k]s_\mu[k] + w_\nu[m], \qquad \nu = 1, \cdots, N_r \tag{5.52}$$

In (5.47), (5.49) and (5.51), the noise term $\eta[m]$ or $\eta_\nu[m]$ is formed by the ambient noise and the ignored ICI. Based on the above input–output relationship, ICI-ignorant receiver processing can be performed. With the frequency measurements at the pilot subcarriers, either the least-squares based method or the compressive-sensing based sparse channel estimation method can be adopted to estimate the channel path parameters, which are then used to reconstruct the channel coefficients at the data subcarriers. The symbol detection is individually performed for each data subcarrier, which is followed by channel decoding to recover information bits.

5.4.2 ICI-Aware Receiver

As discussed in Section 5.3.4, the channel matrix $\mathbf{H}$ is full in time-varying channels. However, the ICI dispersion described by (5.31) reveals that the ICI at one subcarrier is mainly contributed by subcarriers within its near neighborhood. For a reasonable receiver processing complexity, a banded-ICI assumption is usually adopted [185, 196, 232, 332], by assuming that each symbol only affects its D direct neighbors on each side, i.e.,

$$[\boldsymbol{\Gamma}(b_p, \hat{\epsilon})]_{m,k} \simeq 0, \qquad \forall |m-k| > D \tag{5.53}$$

which leads to

$$H[m,k] \simeq 0, \qquad \forall |m-k| > D \tag{5.54}$$

As illustrated in Figure 5.5, a banded channel matrix $\mathbf{H}_D$ can be defined to approximate the full channel matrix $\mathbf{H}$, with its (m,k)th element formulated as

$$H_D[m,k] = \begin{cases} H[m,k], & |m-k| \le D, \\ 0, & |m-k| > D. \end{cases} \tag{5.55}$$

With the band-limited ICI assumption of the channel matrix, the effective system model becomes

- SISO channel

$$z[m] = \sum_{k=m-D}^{m+D} H[m,k]s[k] + \eta[m] \tag{5.56}$$

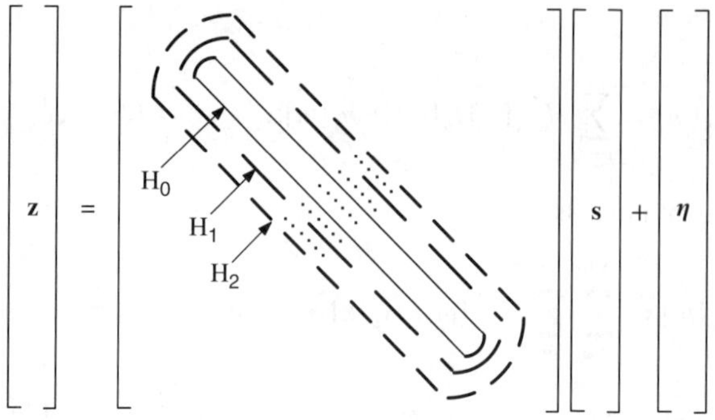

Figure 5.5 The equivalent system model with different ICI depths.

where the equivalent noise is

$$\eta[m] = \sum_{|k-m|>D} H[m,k]s[k] + w[m]. \tag{5.57}$$

- SIMO channel

$$z_\nu[m] = \sum_{k=m-D}^{m+D} H_\nu[m,k]s[k] + \eta_\nu[m] \qquad \nu = 1,\cdots,N_\mathrm{r} \tag{5.58}$$

where the equivalent noise is

$$\eta_\nu[m] = \sum_{|k-m|>D} H_\nu[m,k]s[k] + w_\nu[m], \qquad \nu = 1,\cdots,N_\mathrm{r} \tag{5.59}$$

- MIMO channel

$$z_\nu[m] = \sum_{\mu=1}^{N_\mathrm{t}} \sum_{k=m-D}^{m+D} H_{\nu,\mu}[m,k]s_\mu[k] + \eta_\nu[m], \qquad \nu = 1,\cdots,N_\mathrm{r} \tag{5.60}$$

where the equivalent noise is

$$\eta_\nu[m] = \sum_{\mu=1}^{N_\mathrm{t}} \sum_{|k-m|>D} H_{\nu,\mu}[m,k]s_\mu[k] + w_\nu[m], \qquad \nu = 1,\cdots,N_\mathrm{r} \tag{5.61}$$

Building upon the above input–output relationship, the receiver processing with an explicit consideration of ICI can be performed for information bits recovery. To reconstruct the channel matrix based on the frequency measurements at both pilot and null subcarriers, a sparse channel estimator can be adopted. Based on the estimated channel matrix **H**, multiple symbol detection algorithms can be employed to estimate the transmitted symbol, such as the maximum *a posteriori* probability (MAP) algorithm, the trellis-based message passing algorithm, and the linear minimum mean squared error (LMMSE) algorithm. The transmitted information bits are then recovered by a channel decoder.

5.4.3 Block-by-Block Processing

The block-by-block processing takes each received OFDM block as one processing unit, and performs channel estimation, symbol detection and channel decoding within each block; as illustrated in Figure 5.2. The processing procedure ignores the channel coherence property across blocks, hence has robust performance in channels with large variations.

5.4.4 Block-to-Block Processing

Relative to the block-by-block processing, the block-to-block processing exploits the processing results of preceding blocks to facilitate the decoding process of the current block. Depending on communication scenarios, different processing results from preceding blocks can be used. For example, in Chapter 10, we will see that in shallow water communications with a large channel coherence time, channel estimates from preceding blocks can be used for channel estimation in the current block for systems with a small number of pilot subcarriers. In Chapter 11, we will see that in the deep water horizontal communications, data decoding results of preceding and succeeding blocks will be used for interblock interference cancellation at the current block.

5.4.5 Discussion

With the above receiver processing procedures, there are four different combinations of receiver structures. The desirable combination depends on the channel characteristics, receiver computational capability, and targeted performance of one particular communication system. In the following chapters, different combinations will be used for in different application scenarios.

5.5 Receiver Performance Bound with Simulated Channels

Define σ_s^2 as the average symbol energy, and σ_w^2 as the ambient noise variance. The average SNR of the received OFDM signal is

$$\bar{\gamma} = \frac{\sigma_s^2}{\sigma_w^2} \tag{5.62}$$

Let R denote the data rate in the unit of bits/symbol. Assuming that the symbols are Gaussian distributed and perfect channel state information (CSI) is available at receiver, the outage probability with full ICI equalization is formulated as

$$p_{\text{out}}(\bar{\gamma}, R) = \Pr\left\{ \log_2 \left| \mathbf{I}_K + \bar{\gamma}\mathbf{H}\mathbf{H}^{\text{H}} \right| < R \right\}. \tag{5.63}$$

For the time-varying channel with a banded assumption of ICI, the signal-to-interference-and-noise ratio (SINR) is defined as

$$\bar{\gamma}_D := \frac{\sigma_s^2}{\sigma_w^2 + \frac{1}{K}\text{tr}(\mathbf{H}_D^-(\mathbf{H}_D^-)^{\text{H}})\sigma_s^2} \tag{5.64}$$

with $\mathbf{H}_D^- := \mathbf{H} - \mathbf{H}_D$. Note that the residual ICI is colored. The outage probability is thus upper bounded by

$$p_{\text{out}}(\bar{\gamma}, R, D) \leq \Pr\left\{\log_2\left|\mathbf{I}_K + \bar{\gamma}_D \mathbf{H}_D \mathbf{H}_D^{\mathrm{H}}\right| < R\right\}. \tag{5.65}$$

5.5.1 *Simulating Underwater Acoustic Channels*

Based on the path-based channel model in Section 1.3, we simulate the underwater acoustic channel by randomly generating N_{pa} paths individually. Specifically, the amplitude, delay and Doppler scale of each path are generated as follows.

- *Delay*: The interarrival time of paths is generated following an exponential distribution with mean $\Delta\tau$, which leads to an average channel delay spread $N_{\text{pa}}\Delta\tau$.
- *Amplitude*: The path amplitude is generated according a Rayleigh distribution with the average power exponentially decreasing with delay. We denote ΔP_{pa} dB as the average path-power difference between the beginning and the end of the guard interval of length T_{g}.
- *Doppler scale*: We define v_0 m/s as the relative platform moving speed between the transmitter and receiver, and v_p as the speed of the pth path. To simulate a time-varying channel, the speed of each path is drawn from a uniform distribution within the interval $[v_0 - \sqrt{3}\sigma_{\text{v}}, v_0 + \sqrt{3}\sigma_{\text{v}}]$ m/s which leads to a standard deviation of σ_{v} m/s. Hence, the maximum Doppler rate of each path is $\sqrt{3}\sigma_{\text{v}} f_{\text{c}}/c$ where $c = 1500$ m/s denotes the nominal sound speed in water.

In the following chapters, the underwater acoustic channel will be simulated according to the above specification. Two examples of the simulated channel profiles are shown in Figure 1.6, with channel parameters: $N_{\text{pa}} = 15$, $\Delta\tau = 1$ ms, $\Delta P_{\text{pa}} = 20$ dB, $T_{\text{g}} = 24.6$ ms, and $v_0 = 0$ m/s and $\sigma_{\text{v}} = 0$ m/s for the time-invariant channel, $v_0 = 0.2$ m/s and $\sigma_{\text{v}} = 0$ m/s for the channel with a common Doppler, and $v_0 = 0$ m/s and $\sigma_{\text{v}} = 0.25$ m/s for the channel with path-specific Doppler scales.

5.5.2 *ICI Effect in Time-Varying Channels*

We use a numerical example to gain insights on the ICI level in the time-varying channel with different path-speed variances. Consider a stationary system with center frequency $f_{\text{c}} = 13$ kHz, symbol duration $T = 104.86$ ms, guard interval $T_{\text{g}} = 24.6$ ms, and subcarrier spacing $\Delta f = 9.54$ Hz. The underwater acoustic channel is assumed consisting of $N_{\text{pa}} = 15$ paths, and simulated with parameters $\Delta\tau = 1$ ms, $\Delta P_{\text{pa}} = 20$ dB, $v_0 = 0$ m/s and a varying σ_{v}.

Figure 5.6(a) shows the average ICI power $\mathbb{E}\{|H[m, m-D]|^2\}$ normalized by the desired signal power $\mathbb{E}\{|H[m, m]|^2\}$, as a function of the relative subcarrier index for various standard deviation of path speed σ_{v}. The ICI coefficients are calculated based on full CSI. As expected, the average ICI power decreases as the ICI index increases. Most of the symbol energy concentrates around the neighborhood of the desired subcarrier, while the ICI energy increases with σ_{v}. Hence, it is necessary to increase the considered ICI depth when the path-speed variation σ_{v} increases.

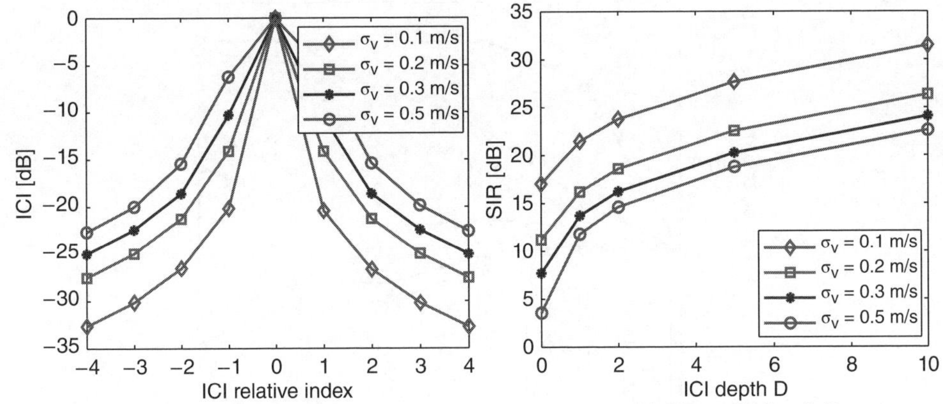

(a) Normalized ICI power *vs* relative subcarrier index

(b) SIR *vs* ICI Depth *D*

Figure 5.6 Illustration on ICI in time-varying channels.

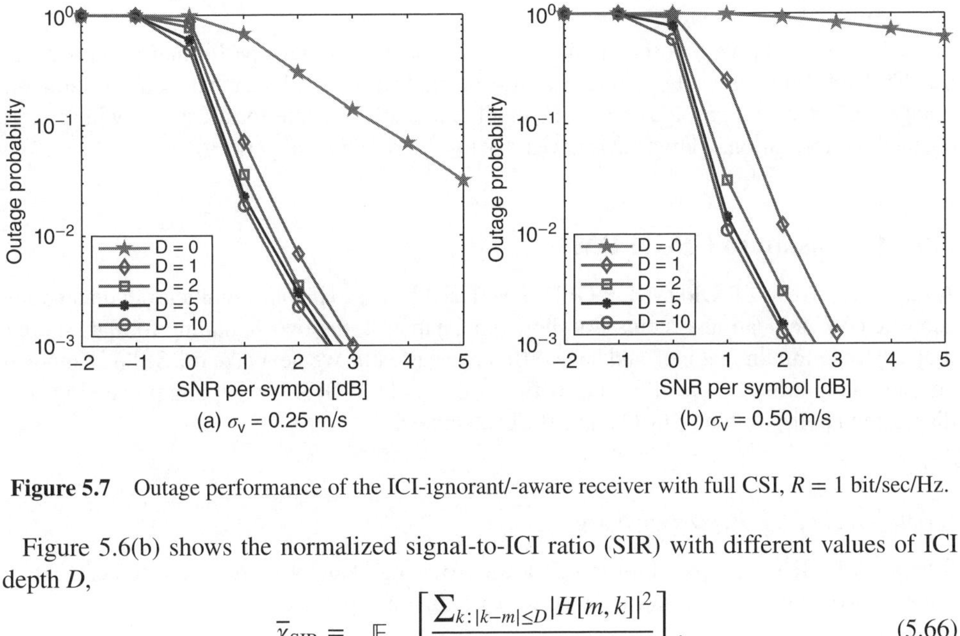

(a) $\sigma_v = 0.25$ m/s

(b) $\sigma_v = 0.50$ m/s

Figure 5.7 Outage performance of the ICI-ignorant/-aware receiver with full CSI, $R = 1$ bit/sec/Hz.

Figure 5.6(b) shows the normalized signal-to-ICI ratio (SIR) with different values of ICI depth D,

$$\overline{\gamma}_{\text{SIR}} = \mathop{\mathbb{E}}_{m \in \mathcal{S}_{\text{all}}} \left[\frac{\sum_{k:|k-m| \le D} |H[m,k]|^2}{\sum_{k:|k-m| > D} |H[m,k]|^2} \right]. \tag{5.66}$$

One can see that SIR increases as the considered ICI depth increase, which again reveals that a larger ICI depth is desirable as the channel path-speed variation increases.

5.5.3 *Outage Performance of SISO Channel*

With identical system and channel settings in Section 5.5.2, Figures 5.7 and 5.8 show the outage performance of a SISO channel in time-varying channels under different path-speed

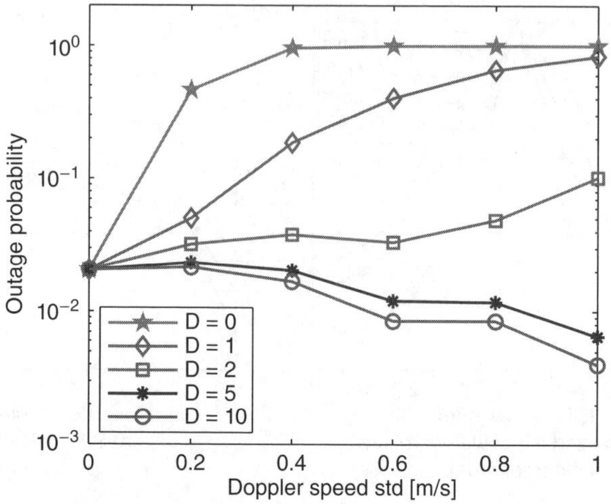

Figure 5.8 Outage performance of the ICI-aware receiver with full CSI, SNR = 1 dB.

variations. Agreeing with observations in Figure 5.6(b), the outage performance improves as the ICI depth D increases. For a time-varying channel with a small channel path variation, the outage performance converges fast with small values of D, while for a channel with a large channel path variation, a larger D is necessary to achieve an ideal performance.

5.6 Extension to CP-OFDM

As an alternative to ZP-OFDM, CP-OFDM has also been used in many multicarrier underwater acoustic communication systems. As shown in Chapter 2, the two signalling formats share a majority of common features, and have differences as well. We next take the SISO channel as an example to illustrate the difference between ZP- and CP-OFDM receiving processing. The discussion applies to both SIMO and MIMO channels.

5.6.1 Receiver Preprocessing

Similar to (5.21), after passing through a time-varying channel with path-specific Doppler scales, the CP-OFDM waveform at receiver side can be expressed as

$$\tilde{y}_{\rm cp}(t) = \sum_{p=1}^{N_{\rm pa}} A_p \tilde{x}_{\rm cp}\left((1+a_p)t - \tau_p\right) + \tilde{n}_{\rm cp}(t), \tag{5.67}$$

where $\tilde{n}_{\rm cp}(t)$ is the ambient noise, and the transmitted signal $\tilde{x}_{\rm cp}(t)$ is described in (2.19).

During the receiver preprocessing, the two-step Doppler compensation approach for ZP-OFDM can be applied directly to CP-OFDM. Different from the time-to-frequency conversion of ZP-OFDM in (5.2), the frequency-domain samples of CP-OFDM after removing the cyclic prefix at the front of each block, can be obtained as

$$z_{\text{cp}}[m] = \int_0^T \tilde{y}_{\text{cp}}\left(\frac{t}{1+\hat{a}}\right) e^{-j2\pi(f_m+\hat{\epsilon})t} dt. \tag{5.68}$$

5.6.2 *Frequency-Domain Input–Output Relationship*

Based on (2.19) and (5.67), one can reformulate (5.68) as

$$z_{\text{cp}}[m] = \sum_{k=-K/2}^{K/2-1} H_{\text{cp}}[m,k]s_{\text{cp}}[k] + w_{\text{cp}}[m], \qquad m = -\frac{K}{2}, \cdots, \frac{K}{2} - 1 \tag{5.69}$$

where $w_{\text{cp}}[m]$ corresponds to the noise sample in frequency domain, and the channel coefficient $H_{\text{cp}}[m,k]$ is expressed as

$$H_{\text{cp}}[m,k] = \sum_{p=1}^{N_{\text{pa}}} \xi_{\text{cp},p} e^{-j2\pi\frac{k}{T}\tau_p} G_{\text{rec}}(f_m + \hat{\epsilon} - (1+b_p)f_k) \tag{5.70}$$

with $G_{\text{rec}}(f)$ denoting the Fourier transform of the rectangular pulse shaping window in (2.10), b_p being defined in (5.26), and

$$\xi_{\text{cp},p} := A_p e^{-j2\pi f_c \tau_p}. \tag{5.71}$$

Stacking frequency samples into a vector yields the matrix-vector expression,

$$\mathbf{z}_{\text{cp}} = \mathbf{H}_{\text{cp}}\mathbf{s}_{\text{cp}} + \mathbf{w}_{\text{cp}} \tag{5.72}$$

where the channel matrix can be expressed as

$$\mathbf{H}_{\text{cp}} = \sum_{p=1}^{N_{\text{pa}}} \xi_{\text{cp},p}\boldsymbol{\Gamma}(b_p, \hat{\epsilon})\boldsymbol{\Lambda}(\tau_p). \tag{5.73}$$

Comparing (5.24) with (5.69), one can see that CP-OFDM has an identical input–output relationship structure as ZP-OFDM, except a slight difference between the channel coefficient expressions in (5.25) and (5.70), which leads to different orders of $\boldsymbol{\Gamma}(\cdot)$ and $\boldsymbol{\Lambda}(\cdot)$ in (5.29) and (5.73).

Despite the difference, discussions on the channel matrix structure for ZP-OFDM in Section 5.3.4 carries over to the channel matrix $\mathbf{H}_{\text{cp}}$ in (5.73). Therefore, the receiver processing techniques for one signalling format can be easily extended to its alternative with slight modifications [37, 35].

5.7 Bibliographical Notes

Resampling is one module that is common to underwater sonar and communication applications. The two-step preprocessing, resampling followed by residual Doppler compensation, was specifically proposed for OFDM receivers in [235]. The formulation of the channel input–output relationship for a multipath channel with path-specific Doppler scales was presented in [38] for ZP-OFDM and in [35, 36] for CP-OFDM.

6

Detection, Synchronization and Doppler Scale Estimation

In practical communication systems, a vital component of receiver design is to detect the arrival of the useful signal from the transmitter. As illustrated in Figure 5.1, once the useful signal is detected, the receiver starts recording the waveform, and employs an appropriate synchronization algorithm to identify the starting point of the useful signal from the recorded waveform. Note that in the presence of platform motions, the transmitted signal is usually dilated or compressed at the receiver side. Synchronization and Doppler scale estimation therefore often operate together.

Three key components in the OFDM receiver design will be discussed in this chapter.

- *Preamble detection.* As revealed in Figure 5.1, a preamble is usually transmitted prior to OFDM blocks for incoming signal detection. Depending on the preamble structure, two popular methods exist. The first method is to perform cross correlation between the received signal and the transmitted preamble which is known to the receiver. In this method, Doppler-insensitive waveforms are usually adopted to account for channel variations. The second method is to embed certain structure in the transmitted preamble, and signal detection can be accomplished by monitoring the structure of the incoming signal at the receiver side. In this method, the receiver can be ignorant of the transmitted preamble and only needs to know the preamble structure. In this chapter, both methods will be discussed.
- *Synchronization.* It is well-known that detection and parameter estimation go hand in hand. The signaling formats for detection apply well to the time-of-arrival and Doppler scale estimation. Moreover, for ZP-OFDM data symbols which multiplex data, pilot and null subcarriers, synchronization and Doppler scale estimation can be achieved using the measurements at pilot and null subcarriers.
- *Residual Doppler shift estimation.* With the estimated Doppler scale factor, resampling the received signal converts the frequency-dependent Doppler scaling effect to an approximately frequency-independent Doppler shift which can be taken as the carrier frequency offset (CFO), a term which is commonly used in radio communications. Note that slight inaccuracy of Doppler scale estimation could incur severe phase rotation of the resampled

OFDM for Underwater Acoustic Communications, First Edition. Shengli Zhou and Zhaohui Wang.
© 2014 John Wiley & Sons, Ltd. Published 2014 by John Wiley & Sons, Ltd.

samples in the time domain. The residual Doppler shift estimation and compensation are vital for successful data decoding.

For clarity of presentation, this chapter focuses on detection, synchronization and Doppler scale estimation in the underwater acoustic OFDM system in a linear time-varying channel with a common Doppler scale factor

$$h(t;\tau) = \sum_{p=1}^{N_{\rm pa}} A_p \delta(\tau - (\tau_p - at)). \tag{6.1}$$

The developed methods can be applied to general channels with path-specific Doppler scales, e.g., (1.14), where the Doppler scale estimate can be approximately viewed as the dominant or the mean Doppler scale of the channel.

The layout of this chapter is as follows.

- In Section 6.1, based on Doppler-insensitive or Doppler-sensitive waveforms, several cross-correlation based methods for detection, synchronization and Doppler scale estimation.
- In Section 6.2, the detection, synchronization and Doppler scale estimation schemes are developed for a specifically designed waveform, which consists of two identical OFDM symbols preceded by a cyclic prefix and is referred to as CP-OFDM.
- In Section 6.3, the ZP-OFDM block structure is exploited for synchronization and Doppler scale estimation. The Doppler scales estimation performance of CP-OFDM and ZP-OFDM are compared in simulations.
- In Section 6.5, several design examples for practical systems are presented.
- In Section 6.6, the residual Doppler shift estimation methods for ZP-OFDM are discussed and tested in simulation.

6.1 Cross-Correlation Based Methods

6.1.1 Cross-Correlation Based Detection

6.1.1.1 Matched Filter Processing

Denote H_0 as the hypothesis that a useful signal is absent from the incoming waveform, and H_1 as the hypothesis that a useful signal is present. For a time-varying channel depicted in (1.14), the received waveforms under the two hypotheses are expressed as

$$\tilde{y}(t) = \begin{cases} \tilde{n}(t) & \mathrm{H}_0 : \text{no useful signal} \\ \sum_{p=1}^{N_{\rm pa}} A_p x((1+a)t - \tau_p) + \tilde{n}(t) & \mathrm{H}_1 : \text{with useful signal} \end{cases} \tag{6.2}$$

where $\tilde{n}(t)$ is the ambient noise. After downshifting the passband waveform to baseband and lowpass filtering (LPF), the baseband waveform $y(t) := \mathrm{LPF}\{\tilde{y}(t)e^{-j2\pi f_c t}\}$ can be formulated as

$$y(t) = \begin{cases} n(t) & \mathrm{H}_0 : \text{no useful signal} \\ \sum_{p=1}^{N_{\mathrm{pa}}} A_p e^{j2\pi f_c(at-\tau_p)} x((1+a)t-\tau_p) + n(t) & \mathrm{H}_1 : \text{with useful signal} \end{cases} \quad (6.3)$$

Taking the transmitted waveform in baseband $x(t)$ as a local template, the matched filter output of the received waveform can be formulated as

$$r_{\mathrm{MF}}(\tau) = \int x^*(t) y(t+\tau) dt. \quad (6.4)$$

Substituting (6.3) into (6.4) yields

$$r_{\mathrm{MF}}(\tau) = \begin{cases} r_{xn}(t) & \mathrm{H}_0 : \text{no useful signal} \\ \sum_{p=1}^{N_{\mathrm{pa}}} A_p e^{j2\pi f_c(at-\tau_p)} r_{xx}((1+a)t-\tau_p) + r_{xn}(t) & \mathrm{H}_1 : \text{with useful signal} \end{cases} \quad (6.5)$$

where $r_{xx}(t)$ is the auto-correlation function of $x(t)$, and $r_{xn}(t)$ is the correlation between $x(t)$ and the ambient noise $n(t)$. Let $r_{\mathrm{MF}}[n]$ denote the discrete samples of $r_{\mathrm{MF}}(t)$ with a baseband sampling rate $f_{\mathrm{s,\,b}}$,

$$r_{\mathrm{MF}}[n] = r_{\mathrm{MF}}(t)|_{t=\frac{n}{f_{\mathrm{s,b}}}}. \quad (6.6)$$

Assume that the correlation sample $r_{xn}[n]$ follows a complex Gaussian distribution, i.e., $r_{xn}[n] \sim \mathcal{CN}(0, \sigma_{\mathrm{w}}^2)$. Define the normalized matched filter output squared as

$$c_{\mathrm{MF}}[n] := \frac{|r_{\mathrm{MF}}[n]|^2}{\sigma_{\mathrm{w}}^2} \quad (6.7)$$

which follows a noncentrally chi-squared distribution under H_0.

In the time-varying channel conditions, Doppler-insensitive waveforms are desirable for signal detection. Two popular waveforms widely used in radar and sonar systems are as follows.

Example 6.1

The linear frequency-modulated (LFM) waveform is a *Doppler-shift* insensitive waveform, which is often used in narrowband systems. The LFM waveform is formulated as

$$\tilde{x}_{\mathrm{LFM}}(t) = 2\cos\left[2\pi\left(f_0 t + \frac{1}{2} m t^2\right)\right] \quad (6.8)$$

where f_0 is the starting frequency, and m is a frequency sweeping parameter. The instantaneous frequency is

$$f_{\mathrm{LFM}}(t) = \frac{d}{dt}\left(f_0 t + \frac{1}{2} m t^2\right) = f_0 + mt \quad (6.9)$$

which is linearly proportional to time index t. Given a center frequency f_c, the baseband waveform is formulated as

$$x_{\mathrm{LFM}}(t) = \exp\left\{j2\pi\left[(f_0 - f_c)t + \frac{1}{2} m t^2\right]\right\}. \quad (6.10)$$

The narrowband ambiguity function is defined as

$$\chi_{\text{NB}}(\tau,f) = \int_{-\infty}^{\infty} x(t)x^*(t-\tau)e^{j2\pi ft}dt. \tag{6.11}$$

The absolute value and the contour of the narrowband ambiguity function of an LFM waveform are shown in Figure 6.1.

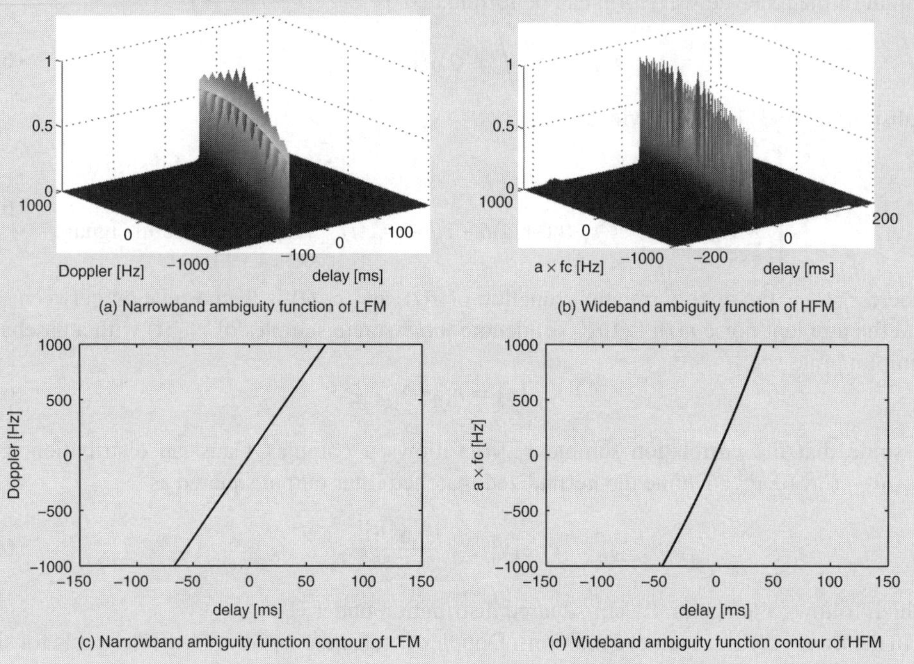

Figure 6.1 Ambiguity functions of LFM and HFM waveforms, $f_c = 10\,\text{kHz}$, $B = 6\,\text{kHz}$ and $T = 200\,\text{ms}$.

Example 6.2

The hyperbolic frequency-modulated (HFM) waveform is a *Doppler-scale* insensitive waveform, which is often used in wideband systems. The HFM waveform is formulated as

$$\tilde{x}_{\text{HFM}}(t) = 2\cos\left[\frac{2\pi}{k}\ln(1+kf_0t)\right] \tag{6.12}$$

where k is a design parameter depending on the available frequency band. The instantaneous frequency is

$$f_{\text{HFM}}(t) = \frac{d}{dt}\left[\frac{1}{k}\ln(1+kf_0t)\right] = \frac{1}{\frac{1}{f_0}+kt}. \tag{6.13}$$

The instantaneous period of the signal, which is the inverse of the instantaneous frequency, is proportional to time index t. The HFM waveform is therefore also referred to as the linear period modulation (LPM) waveform. Given a center frequency f_c, the baseband waveform is formulated as

$$x_{\mathrm{HFM}}(t) = \exp\left\{j2\pi\left[\frac{1}{k}\ln(1+kf_0t) - f_c t\right]\right\}. \tag{6.14}$$

The wideband ambiguity function which is defined as

$$\chi_{\mathrm{WB}}(\tau, a) = \sqrt{|a|}\int_{-\infty}^{\infty} x(t)x^*(a(t-\tau))dt. \tag{6.15}$$

The absolute value and the contour of the wideband ambiguity function of a HFM waveform are shown in Figure 6.1.

6.1.1.2 Test Statistics

Based on the matched filter output, two test statistics are widely used for radar and sonar detection.

- Thresholding method

$$\max\left\{|r_{\mathrm{MF}}[n]|^2\right\} \underset{\mathrm{H}_1}{\overset{\mathrm{H}_0}{\lessgtr}} \Gamma_{\mathrm{th}}. \tag{6.16}$$

- Constant false alarm rate (CFAR) test

$$\max\{c_{\mathrm{MF}}[n]\} \underset{\mathrm{H}_1}{\overset{\mathrm{H}_0}{\lessgtr}} \Gamma_{\mathrm{th}}. \tag{6.17}$$

  As indicated by its name, through constantly measuring the noise variance, the test statistic in (6.17) can achieve a constant probability of false alarm.

In the wireless channel with multiple paths, the two methods declare detection based on the signal propagating along the strongest path, thus incurring performance degradation in the presence of dense paths and impulsive interferences, especially in underwater environments. To accommodate multiple propagation paths, two sequential detectors based on the cumulative sum of the log-likelihood ratio (LLR) of $c_{\mathrm{MF}}[n]$ can be used. The LLR of $c_{\mathrm{MF}}[n]$ is formulated as

$$\mathrm{L}(c_{\mathrm{MF}}[n]) = \ln\frac{f(c_{\mathrm{MF}}[n] \mid \mathrm{H}_1)}{f(c_{\mathrm{MF}}[n] \mid \mathrm{H}_0)}. \tag{6.18}$$

- Sequential probability ratio test (SPRT): By approximating that $\{c_{\mathrm{MF}}[n]\}$ are i.i.d., the SPRT can be used for signal detection, where the cumulative sum of LLR is formulated as

$$\mathrm{T}_{\mathrm{SPRT}}[n] = \mathrm{T}_{\mathrm{SPRT}}[n-1] + \mathrm{L}(c_{\mathrm{MF}}[n]). \tag{6.19}$$

A stopping rule of SPRT is defined as

$$\phi[n] = \begin{cases} 0 & \Gamma_{\text{th}}^{(\text{L})} < \text{T}_{\text{SPRT}}[n] \le \Gamma_{\text{th}}^{(\text{H})} \\ 1 & \text{T}_{\text{SPRT}}[n] > \Gamma_{\text{th}}^{(\text{H})} \text{ or } \text{T}_{\text{SPRT}}[n] \le \Gamma_{\text{th}}^{(\text{L})} \end{cases} \tag{6.20}$$

and a decision rule is defined as

$$\delta[n] = \begin{cases} 0 & \text{T}_{\text{SPRT}}[n] < \Gamma_{\text{th}}^{(\text{L})} \\ 1 & \text{T}_{\text{SPRT}}[n] > \Gamma_{\text{th}}^{(\text{H})} \\ \text{undefined} & \text{else.} \end{cases} \tag{6.21}$$

Decision can be made based on $\delta[\tilde{n}]$, where $\tilde{n}$ is the index of the first sample that $\phi[n] = 1$. The two thresholds $\Gamma_{\text{th}}^{(\text{L})}$ and $\Gamma_{\text{th}}^{(\text{H})}$ are determined based on the probability of detection P_{D} and probability of false alarm P_{FA} [341, 410]:

$$\Gamma_{\text{th}}^{(\text{L})} = \frac{1 - P_{\text{D}}}{1 - P_{\text{FA}}}, \qquad \Gamma_{\text{th}}^{(\text{H})} = \frac{P_{\text{D}}}{P_{\text{FA}}}. \tag{6.22}$$

Note that except the calculation of LLR, SPRT is distribution free, which makes SPRT very appealing in the scenario with dependent and nonidentically distributed measurement samples.

- Page test: The cumulative sum of LLR in the Page test is defined as

$$\text{T}_{\text{Page}}[n] = \max\left\{0, \text{T}_{\text{Page}}[n-1] + \text{L}\left(c_{\text{MF}}[n]\right)\right\}, \tag{6.23}$$

and a decision of H_1 is made if

$$\text{T}_{\text{Page}}[n] > \Gamma_{\text{th}} \tag{6.24}$$

A careful inspection on (6.19) and (6.23) reveals that Page test based sequential detector can be taken as a serial execution of multiple SPRTs, hence achieves faster detection of any change occurring in the data sequence [2].

Note that the LLR computation in (6.18) requires the probability density function of the normalized matched filter output in the presence and absence of the useful signal. To avoid such a requirement in the Page test, a locally optimal detector nonlinearity is often used [205]. For the noncentrally chi-squared distributed signal $c_{\text{MF}}[n]$, the locally optimal nonlinearity is

$$\text{L}(c_{\text{MF}}[n]) = c_{\text{MF}}[n] - b \tag{6.25}$$

where b is a false alarm inhibiting bias. Appropriate choice of b in underwater acoustic applications can be found in [2]. The bias and the thresholds need to be fine tuned by practical engineers to achieve a tradeoff between low false alarm rate and high probability of detection.

6.1.2 *Cross-Correlation Based Synchronization and Doppler Scale Estimation*

The synchronization preamble is used to estimate the time-of-arrival and the Doppler scale factor of the useful signal. Two types of waveforms are available for the above purpose.

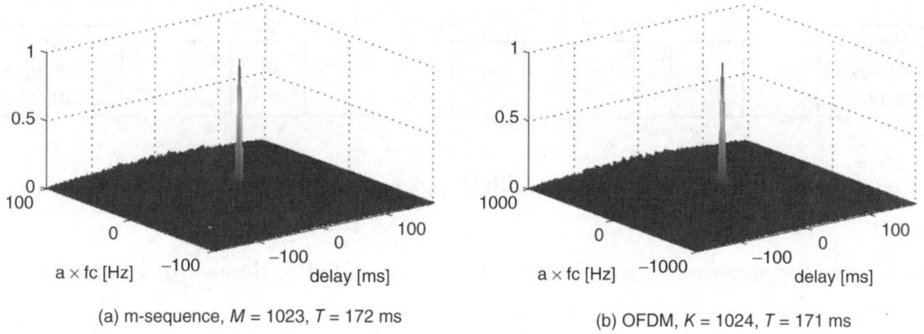

Figure 6.2 The wideband ambiguity functions of m-sequence and OFDM waveforms, $f_c = 10\,\text{kHz}$ and $B = 6\,\text{kHz}$.

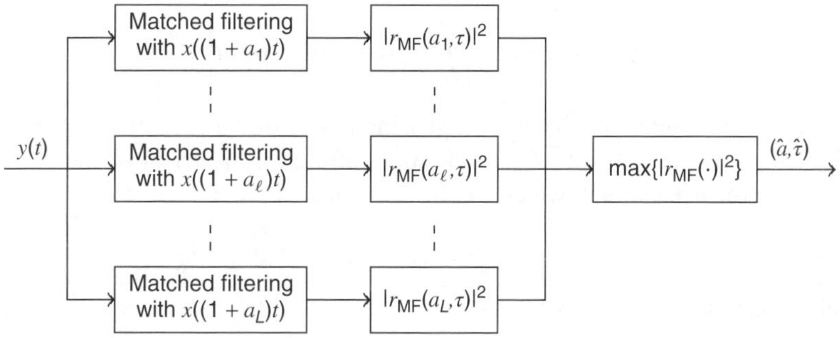

Figure 6.3 A bank of cross-correlation branches for time synchronization and Doppler scale estimation.

6.1.2.1 Doppler-Sensitive Waveform

In radar and sonar applications, Doppler sensitive waveforms are usually adopted for joint time-of-arrival and Doppler scale estimation. Typical Doppler sensitive waveforms include m-sequence coded waveforms, Costas waveforms, and OFDM waveforms; see the thumbtack wideband ambiguity functions of the Doppler-sensitive waveforms in Figure 6.2.

Denote T_0 as the time duration of the synchronization preamble. Joint estimates of the time-of-arrival and the Doppler scale factor can be obtained as

$$(\hat{a}, \hat{\tau}) = \arg\max_{a,\tau} \left| \int_0^{T_0} y(t+\tau)x^*\left((1+a)t\right) e^{-j2\pi a f_c t} dt \right|. \tag{6.26}$$

The above operation can be implemented via a bank of cross-correlators as shown in Figure 6.3, where the branch yielding the largest peak provides the needed Doppler scale estimate and location of the strongest path.

6.1.2.2 Doppler-Insensitive Preamble and Postamble

Besides the Doppler-sensitive waveforms, the Doppler-insensitive waveforms can also be used for joint time-of-arrival and Doppler scale factor estimation. An example of the signal design is

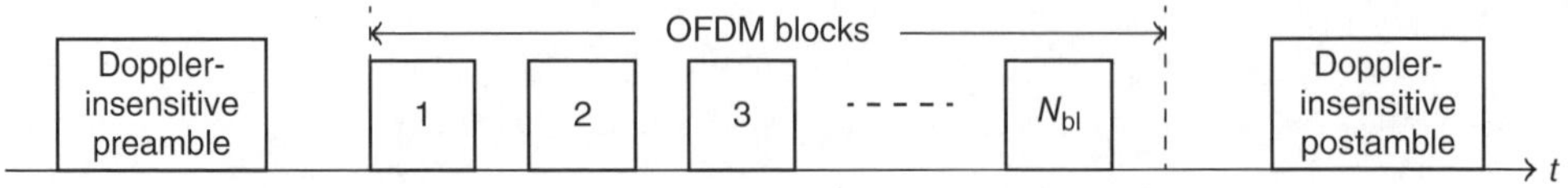

Figure 6.4 OFDM frame structure with Doppler-insensitive preamble and postamble.

to transmit two Doppler-insensitive waveforms before and after the useful signal transmission, respectively; as shown in Figure 6.4 [338].

The time-of-arrivals of the preamble and postamble can be estimated with two matched filters,

$$\hat{\tau}_{\text{pre}} = \arg\max_{\tau} \left| \int_0^{T_0} y(t+\tau)x^*_{\text{pre}}(t)\, dt \right|, \tag{6.27}$$

$$\hat{\tau}_{\text{post}} = \arg\max_{\tau} \left| \int_0^{T_0} y(t+\tau)x^*_{\text{post}}(t)\, dt \right|. \tag{6.28}$$

Denote T_{tp} as the time interval between the starting points of the preamble and postamble in the transmitted signal, and denote T_{rp} as the time-difference-of-arrival of the preamble and postamble in the received signal, which can be estimated as

$$\hat{T}_{\text{rp}} = \hat{\tau}_{\text{post}} - \hat{\tau}_{\text{pre}}. \tag{6.29}$$

The Doppler scale factor can be estimated as

$$\hat{a} = \frac{T_{\text{tp}}}{\hat{T}_{\text{rp}}} - 1. \tag{6.30}$$

Compared to the Doppler sensitive waveform, the two-LFM signaling format in Figure 6.4 enjoys a much lower complexity, but suffers a processing delay as it requires the whole data burst to be buffered before parameter estimation, hence is not appropriate for online receiver processing.

6.1.2.3 Precise Timing

Note that in underwater acoustic channels with multiple paths, the early arrival paths are not necessarily the strongest paths. Due to channel fading, the strongest path can appear at random positions within the delay spread, which is undesirable for the task of data block partitioning after the synchronization. In contrast, the position of the first path is more stable. Rather than taking the strongest path at the matched filter output as the time-of-arrival of useful signal, a precise timing procedure is necessary to locate the first arrival in the presence of multiple paths. Joint timing and channel estimation methods for fine timing have been investigated in [40, 97, 226, 246, 281, 282]. Note that for communication purposes, OFDM does not require precise timing. Small or moderate offsets in the time domain translate to phase shifts in the frequency domain, which can be estimated by pilot tones, or bypassed by differential modulation. For this reason heuristic precise timing can be achieved based on the sequential test statistics, such as the SPRT and Page tests which are described in Section 6.1.1.

6.2 Detection, Synchronization and Doppler Scale Estimation with CP-OFDM

The cross-correlation based detection and parameter estimation have been widely used in the radar and sonar literature. Compared to those applications, communication systems usually operate in the high signal-to-noise regime. This allows diversity in the waveform and algorithm design, especially tailored to the particular characteristics of communication channels, such as multipath and fading effects. In this section we present detection, synchronization and Doppler scale estimation for a specifically designed waveform that consists of two identical OFDM symbols preceded by a cyclic prefix (CP).

Note that OFDM symbols are Doppler sensitive. The cross-correlation based method discussed in Section 6.1.2 applies to the designed CP-OFDM waveform by taking one OFDM symbol as the signal template. This section presents the algorithm design exploiting the embedded signal structure.

6.2.1 CP-OFDM Preamble with Self-Repetition

Consider the CP-OFDM preamble structure in Figure 6.5, which consists of two identical OFDM symbols of length T_0 and a cyclic prefix of length $T_{\rm cp}$ in front, with an embedded structure

$$x_{\rm cp}(t) = x_{\rm cp}(t + T_0), \qquad -T_{\rm cp} \leq t \leq T_0. \tag{6.31}$$

Let B denote the system bandwidth, and define $K_0 := BT_0$ as the number of subcarriers. The baseband CP-OFDM signal is

$$x_{\rm cp}(t) = \sum_{k=-K_0/2}^{K_0/2-1} d[k] e^{j2\pi \frac{k}{2T_0}} q(t), \qquad t \in [-T_{\rm cp}, 2T_0] \tag{6.32}$$

where $d[k]$ is the transmitted symbol on the kth subcarrier, with $k \in S := \{-K_0/2, \cdots, K_0/2 - 1\}$, and $q(t)$ is a pulse shaping window,

$$q(t) = \begin{cases} \frac{1}{T_0}, & t \in [-T_{\rm cp}, 2T_0] \\ 0, & \text{elsewhere.} \end{cases} \tag{6.33}$$

The passband signal can be obtained as $\tilde{x}_{\rm cp}(t) = 2\Re\{x_{\rm cp}(t)e^{j2\pi f_c t}\}$, where f_c is the center frequency.

After transmitting the passband signal $\tilde{x}_{\rm cp}(t)$ through the multipath channel defined in (6.1), the received passband signal is

$$\tilde{y}(t) = \sum_{p=1}^{N_{\rm pa}} A_p \tilde{x}_{\rm cp}(t - (\tau_p - at)) + \tilde{n}(t). \tag{6.34}$$

Define the channel transfer function

$$H(f) := \sum_{p=1}^{N_{\rm pa}} A_p e^{-j2\pi f \tau_p} \tag{6.35}$$

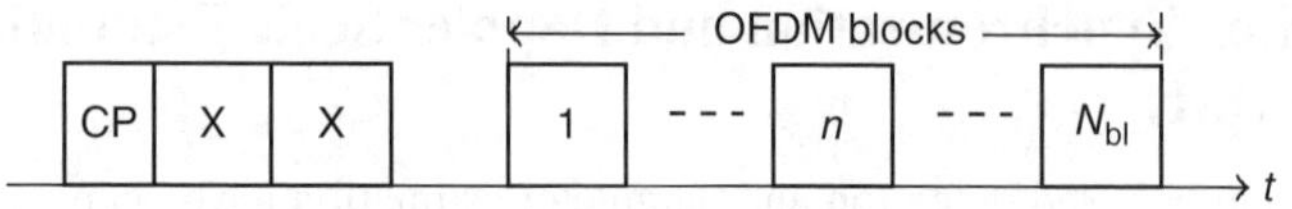

Figure 6.5 A CP-OFDM preamble which consists of two identical OFDM symbols and a cyclic prefix (CP), precedes the data transmission which uses zero padding.

and the frequency response on the kth subcarrier is

$$H_k = H(f_k). \tag{6.36}$$

Denote $\tau_{\max}$ as the maximum delay of multipath arrivals, and define

$$\mathcal{T}_{\text{cyclic}} := \left[-\frac{T_{\text{cp}} - \tau_{\max}}{1+a}, \frac{2T_0}{1+a}\right].$$

Conversion of the passband signal $\tilde{y}(t) = 2\Re\{y(t)e^{j2\pi f_c t}\}$ to baseband leads to

$$\begin{aligned} y(t) &= \sum_{k\in\mathcal{S}} H_k d[k] e^{j2\pi\left(\frac{k}{T_0}+af_k\right)t} + n(t) \\ &= e^{j2\pi a f_c t} \sum_{k\in\mathcal{S}} H_k d[k] e^{j2\pi\frac{k}{T_0}(1+a)t} + n(t), \qquad t \in \mathcal{T}_{\text{cyclic}} \end{aligned} \tag{6.37}$$

where $n(t)$ is the noise at baseband.

For the channel depicted in (6.1) where all paths have a common Doppler scale factor a, the the received waveform in (6.37) has an embedded structure as

$$y(t) = e^{-j2\pi\frac{a}{1+a}f_c T_0} y\left(t + \frac{T_0}{1+a}\right), \qquad -\frac{T_{\text{cp}} - \tau_{\max}}{1+a} \le t \le \frac{T_0}{1+a} \tag{6.38}$$

which has a repetition period $T_0/(1+a)$ regardless of the channel amplitudes.

6.2.2 *Self-Correlation Based Detection, Synchronization and Doppler Scale Estimation*

By exploiting the structure in (6.38), the self correlation of the two repetitions is

$$r_{\text{cp}}(a, \tau) = \int_0^{\frac{T_0}{1+a}} y(t+\tau) y^*\left(t + \tau + \frac{T_0}{1+a}\right) dt, \tag{6.39}$$

which does not require the knowledge of the channel and data symbols. Define a test statistic as

$$M(a, \tau) = \frac{r_{\text{cp}}(a, \tau)}{\sqrt{\int_0^{\frac{T_0}{1+a}} |y(t+\tau)|^2 dt \cdot \int_0^{\frac{T_0}{1+a}} \left| y^*\left(t+\tau+\frac{T_0}{1+a}\right)\right|^2 dt}}. \tag{6.40}$$

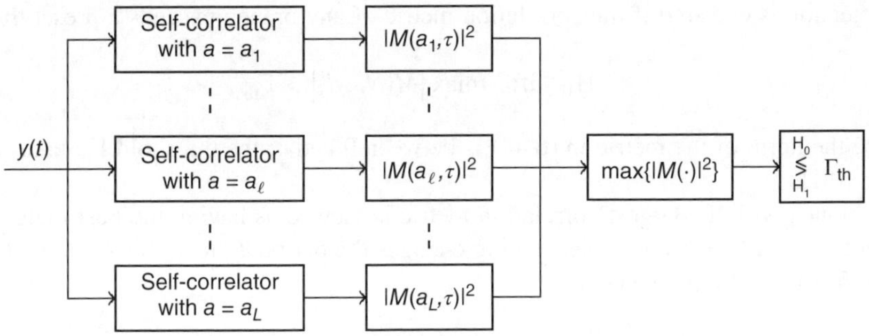

Figure 6.6 Self-correlation bank for CP-OFDM preamble based detection.

The useful signal detection can be achieved via

$$\max_{a,\tau} \{|M(a,\tau)|\} \underset{H_1}{\overset{H_0}{\lessgtr}} \Gamma_{\text{th}}. \tag{6.41}$$

The time-of-arrival and the Doppler scale factor of the CP-OFDM symbol in the received signal can be jointly estimated via

$$(\hat{a}, \hat{\tau}) = \arg\max_{a,\tau} |M(a,\tau)|. \tag{6.42}$$

This method can be implemented with a bank of self-correlators as shown in Figure 6.6, where each branch calculates the correlation result with one tentative value of the Doppler scale factor [274]. The quantization step size of a depends on the desired Doppler scale estimation resolution and the receiver computational capability. We defer our discussion on the needed Doppler scale resolution to Section 6.6.

6.2.3 *Implementation*

We now present implementation of the receiver processing based on the sampled baseband signal. The baseband signal is usually oversampled at a multiple of the system bandwidth $\lambda_b B$:

$$y[n] = y(t)|_{t=\frac{n}{\lambda_b B}}, \tag{6.43}$$

where λ_b is an integer oversampling factor. The receiver processing includes the following steps.

1. Each of the L branches calculates a correlation metric with one candidate value of the window size N_ℓ. The window size N_ℓ shall be close to $\lambda_b K_0$, which is the number of samples of one OFDM symbol when no Doppler scaling occurs. For each delay d, the correlation metric corresponding to the window size N_ℓ is

$$M(N_\ell, d) = \frac{\sum_{i=d}^{d+N_\ell-1} y^*[i]\, y[i+N_\ell]}{\sqrt{\sum_{i=d}^{d+N_\ell-1} |y[i]|^2 \cdot \sum_{i=d}^{d+N_\ell-1} |y[i+N_\ell]|^2}}, \tag{6.44}$$

for $\ell = 1, \ldots L$.

2. A detection is declared if the correlation metric of any branch exceeds a preset threshold Γ_{th}:
$$\text{H}_1 \quad \text{if:} \quad \max_{\ell} |M(N_\ell, d)| > \Gamma_{\text{th}} \tag{6.45}$$
Since the norm of the metric in (6.44) is between 0 and 1, the threshold Γ_{th} takes a value from [0,1].
3. The branch with the largest correlation metric is viewed as having the best match on the repetition length. Since Doppler scaling changes the period T_0 to $T_0/(1+a)$, the Doppler scale factor can be estimated as
$$\hat{a} = \frac{\lambda_{\text{b}} K_0}{\hat{N}} - 1, \quad \text{where } \hat{N} = \arg\max_{\{N_\ell\}} |M(N_\ell, d)| \tag{6.46}$$
The speed estimate follows as
$$\hat{v} = c\hat{a}, \tag{6.47}$$
where c is the speed of sound in water.
The Doppler scale estimate can be refined using a multi-grid approach for the implementation:
(a) *Coarse-grid search for detection.* Only a few parallel self-correlators are used to monitor the incoming data. This helps to reduce the receiver complexity.
(b) *Fine-grid search for data demodulation.* After a detection is declared, a set of parallel self-correlators with better Doppler scale resolution is used only on the captured preamble. Fine-grid search is centered around the Doppler scale estimate from the coarse-grid search.
Instead of a multi-grid search, one may also use an interpolation based approach to improve the estimation accuracy beyond the step size. For example, consider the technique from [192], which is usually used in spectral peak location estimation based on a limited number of DFT samples. After coarse- or fine-grid search, let $|X_k|$ denote the amplitude from the branch with the largest correlation output, and $|X_{k-1}|$ and $|X_{k+1}|$ be the amplitudes from the left and right neighbors. Let Δa denote the grid spacing. The quadratic interpolation
$$\delta = \frac{|X_{k+1}| - |X_{k-1}|}{4|X_k| - 2|X_{k-1}| - 2|X_{k+1}|}\Delta a \tag{6.48}$$
can be used to estimate an offset δ of the Doppler scale deviating from the strongest branch towards the second strongest branch.
4. Synchronization is performed on the branch that yields the maximum correlation metric. After the maximum is determined, the start of transmission can be selected as suggested in [334]; starting from the peak the 80% "shoulders" are found (first sample of this correlator branch before and after the peak that is less than 80% of the peak) and the middle is chosen as synchronization point. This is beneficial, since due to the CP structure the correlation metric has a plateau around the peak [334].

6.2.3.1 Doppler Scale Resolution

Since the window size N_ℓ is an integer, the minimum step size on the Doppler scale is $1/(\lambda_{\text{b}} K_0)$. To improve the Doppler scale resolution, the receiver operates on the oversampled baseband signal. The oversampling factor depends on the needed Doppler scale resolution and the parameter K_0. The maximum value of λ_b is the ratio of the passband sampling rate to the baseband signal bandwidth, which is typically less than 50 in underwater applications.

6.2.3.2 Implementation Complexity

The three sliding summations in (6.44) can be computed recursively. For example, define $\psi(N_\ell, d) = \sum_{i=d}^{d+N_\ell-1} y^*[\ell]y[i+N_\ell]$. Instead of summing over N_ℓ multiplications, one can use

$$\psi(N_\ell, d+1) = \psi(N_\ell, d) + y^*[d+N_\ell]y[d+2N_\ell] - y^*[d]y[d+N_\ell], \tag{6.49}$$

which amounts to two complex multiplications and two complex additions for each update. Hence, for each delay d, the metric $M(N_\ell, d)$ in (6.44) can be computed by no more than seven complex multiplications, six complex additions, one square root operation, and one division. Note that the complexity does not depend on the window size.

6.2.3.3 Fine Timing

With the estimated Doppler scale $\hat{a}$, the receiver can resample the preamble. This way, the *wideband* channel effect of frequency-dependent Doppler shifts can be reduced to the *narrowband* channel effect of frequency-independent Doppler shifts [235]. The fine-timing algorithms developed for cross-correlation based synchronization methods in Section 6.1.2 can be applied on the resampled preamble. This way, the *first* path instead of the *strongest* path can be synchronized [40].

6.3 Synchronization and Doppler Scale Estimation for One ZP-OFDM Block

To effectively deal with fast channel variations, the ZP-OFDM signaling format multiplexing pilot and null subcarriers with data subcarriers is often adopted. This format allows synchronization and Doppler scale estimation within each ZP-OFDM block [412].

Corresponding to (2.4), the baseband transmitted ZP-OFDM signal is

$$x_{zp}(t) = \sum_{k\in S_A} s[k]e^{j2\pi \frac{k}{T}t} g(t), \qquad t \in [0, T_{bl}] \tag{6.50}$$

where $g(t)$ is the pulse-shaping window. After transmitting the ZP-OFDM symbol through a multipath channel defined in (6.1), the passband signal $\tilde{y}(t)$ reaches the receiver, with the baseband signal obtained as $y(t) = \text{LPF}\{\tilde{y}(t)e^{-j2\pi f_c t}\}$. The availability of null subcarriers, pilot subcarriers, and data subcarriers will be exploited next for Doppler scale estimation.

6.3.1 *Null-Subcarrier based Blind Estimation*

Assume that coarse synchronization is available from the preamble. After truncating each ZP-OFDM block from the received signal, the total of energy measurements at null subcarrier frequencies is used as a metric for the Doppler scale estimation

$$\hat{a} = \arg\min_{a} \sum_{k \in \mathcal{S}_\mathrm{N}} \left| \int_0^{T+T_\mathrm{g}} y\left(\frac{t}{1+a}\right) e^{-j2\pi a f_\mathrm{c} t} e^{-j2\pi \frac{k}{T} t} dt \right|^2, \tag{6.51}$$

where a one-dimensional grid search can be used. For each tentative a, a resampling operation is carried out, which is followed by a fast Fourier transform.

6.3.2 Pilot-Aided Estimation

As introduced above, a set of subcarriers in $\mathcal{S}_\mathrm{P}$ is dedicated to transmit pilot symbols. Hence, the transmitted waveform $x_{\mathrm{zp}}(t)$ is partially known, containing

$$x_{\mathrm{pilot}}(t) = \sum_{k \in \mathcal{S}_\mathrm{P}} s[k] e^{j2\pi \frac{k}{T} t} g(t), \qquad t \in [0, T_{\mathrm{bl}}]. \tag{6.52}$$

The joint time-of-arrival and Doppler scale estimation is achieved via

$$(\hat{a}, \hat{\tau}) = \arg\max_{a,\tau} \left| \int_0^{\frac{T}{1+a}} y(t+\tau) x^*_{\mathrm{pilot}}((1+a)t - \tau)\, e^{-j2\pi a f_\mathrm{c} t} dt \right| \tag{6.53}$$

which can be implemented via a bank of cross-correlators.

6.3.3 Decision-Aided Estimation

For the OFDM transmission with multiple blocks, the Doppler estimated in one block can be used for the resampling operation of the next block if there is small Doppler variation across blocks. After the decoding operation the receiver can reconstruct the transmitted time-domain waveform, by replacing $s[k]$ by its estimate $\hat{s}[k]$, $\forall k \in \mathcal{S}_D$ in (6.50). Denote the reconstructed waveform as $\hat{x}_{\mathrm{zp}}(t)$.

Similar to the pilot-aided method, the decision-aided method performs the joint time-of-arrival and Doppler scale estimation via

$$(\hat{a}, \hat{\tau}) = \arg\max_{a,\tau} \left| \int_0^{\frac{T}{1+a}} y(t+\tau) \hat{x}^*_{\mathrm{zp}}((1+a)t - \tau)\, e^{-j2\pi a f_\mathrm{c} t} dt \right| \tag{6.54}$$

which can, again, be implemented via a bank of cross-correlators. The estimated $\hat{a}$ can be used for the resampling operation of the next block.

Relative to the pilot-aided method, the decision-aided method leverages the estimated information symbols, thus is expected to achieve a better estimation performance. Assuming that all the information symbols have been successfully decoded, the decision-aided method enjoys a $10\log_{10}((|\mathcal{S}_P| + |\mathcal{S}_\mathrm{D}|)/|\mathcal{S}_\mathrm{P}|)$ dB power gain relative to the pilot-aided method.

6.4 Simulation Results for Doppler Scale Estimation

In this section, the Doppler scale estimation performance of the CP-OFDM waveform and that of the ZP-OFDM waveforms are compared numerically. Specific to the CP-OFDM,

Table 6.1 CP/ZP OFDM parameters in simulations

System parameters	CP-OFDM	ZP-OFDM
Center frequency	13 kHz	13 kHz
Bandwidth	4.88 kHz	4.88 kHz
# of subcarriers	512	1024
Frequency spacing	9.54 Hz	4.77 Hz
Time duration	104.86 ms	209.72 ms
Guard interval	100 ms	40.3 ms

the cross-correlation based method for Doppler sensitive waveforms in Section 6.1 and the self-correlation based method in Section 6.2 will be examined. For the ZP-OFDM waveform, the three methods discussed in Section 6.3 will be considered.

The OFDM parameters are summarized in Table 6.1. For CP-OFDM, the data symbols at all the 512 subcarriers are randomly drawn from a QPSK constellation. For ZP-OFDM, out of 1024 subcarriers, there are $|\mathcal{S}_{\rm N}| = 96$ null subcarriers with 24 on each edge of the signal band for band protection and 48 evenly distributed in the middle for the carrier frequency offset estimation; $|\mathcal{S}_{\rm P}| = 256$ are pilot subcarriers uniformly distributed among the 1024 subcarriers, and the remaining are $|\mathcal{S}_{\rm D}| = 672$ data subcarriers for delivering information symbols. The pilot symbols are drawn randomly from a QPSK constellation. The data symbols are encoded with a rate-$1/2$ nonbinary LDPC code and modulated by a QPSK constellation.

Two types of channels are considered. One is a single-path channel defined in (1.18) with $N_{\rm pa} = 1$, and the other is a multipath channel defined in (1.17) where all paths have one common Doppler scale factor, and $N_{\rm pa} = 15$. For the multipath channel, the interarrival times of paths are assumed to follow an exponential distribution with a mean of 1 ms. The amplitudes of paths are Rayleigh distributed with the average power decreasing exponentially with the delay, where the difference between the beginning and the end of the guard time is 20 dB.

A common Doppler scale a in these two types of channels is associated to a Doppler speed v (with unit m/s) through

$$a = v/c, \tag{6.55}$$

where the Doppler speed v is assumed to follow a uniform distribution within [−4.5, 4.5] m/s, and $c = 1500$ m/s is the sound speed in water.

With the ground truths of v and a, the root-mean-squared-error (RMSE) of the estimated Doppler speed is adopted as the performance metric,

$$\text{RMSE} = \sqrt{\mathbb{E}[|\hat{v} - v|^2]} = \sqrt{\mathbb{E}[|(\hat{a} - a)c|^2]}, \tag{6.56}$$

which has the unit m/s.

6.4.1 RMSE Performance with CP-OFDM

For the single-path channel, Figure 6.7 shows the RMSE performance of two estimation methods at different SNR levels. One can see a considerable gap between the self-correlation method and the cross-correlation method, while in the medium to high SNR region, both methods can provide a reasonable performance to facilitate receiver decoding.

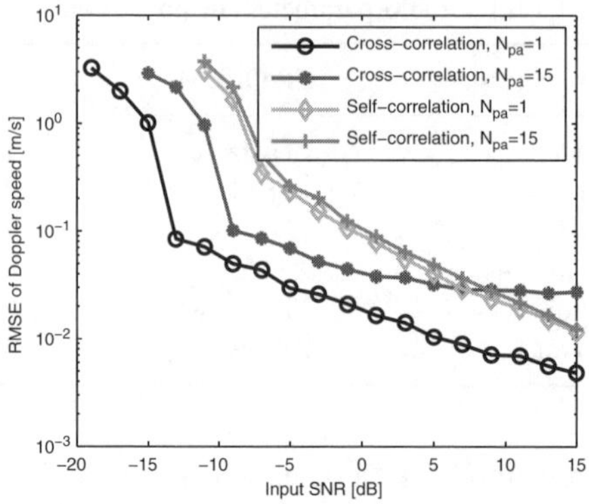

Figure 6.7 CP-OFDM Doppler speed estimation in one path and multipath one common Doppler scale channel.

For the multipath channel with a single Doppler speed, Figure 6.7 shows the RMSE performance of two estimation methods. One can see that the cross-correlation method outperforms the self-correlation method considerably in the low SNR region. However, the former suffers an error floor in the high SNR region, while the latter does not.

Relative to the RMSE performance in the single-path channel, a considerable performance degradation can be observed for the cross-correlation method in the multipath channel, whereas the performance of the self-correlation method is quite robust. The reason for the difference lies in the capability of the self-correlation method to collect the energy from all paths for Doppler scale estimation, while the cross-correlation method aims to get the Doppler scale estimate from only one path, the strongest path.

6.4.2 RMSE Performance with ZP-OFDM

Figure 6.8 shows the RMSE performance of three estimation methods for ZP-OFDM in the single-path channel. In the low SNR region, one can see that the decision-aided method is the best, while the null-subcarrier based blind method is the worst. With the ratio $(|\mathcal{S}_{\rm D}| + |\mathcal{S}_{\rm P}|)/|\mathcal{S}_{\rm P}| \approx 4$, the pilot-aided method is expected to suffer around 6 dB performance loss relative to the decision-aided method, due to a lower matched-filtering gain. In the medium and high SNR region, the pilot-aided method suffers an error floor due to the interference from the data subcarriers, and the null-subcarrier based blind method gets good estimation performance. The Cramer-Rao lower bound (CRLB) with a known waveform is also included as the performance benchmark, whose derivation can be carried out similar to [129, 297].

Figure 6.9 shows the RMSE performance of the three methods in the multipath channel with a single Doppler speed. Again, one can see that in the low SNR region, the decision-aided method has the best performance, while the null-subcarrier based blind method is the worst. As opposed to the performance in the single-path channel, the decision-aided method has an error floor in the high SNR region, since it only picks up the maximum correlation peak of one path. On the other hand, the null-subcarrier method has robust performance in the presence of multiple paths.

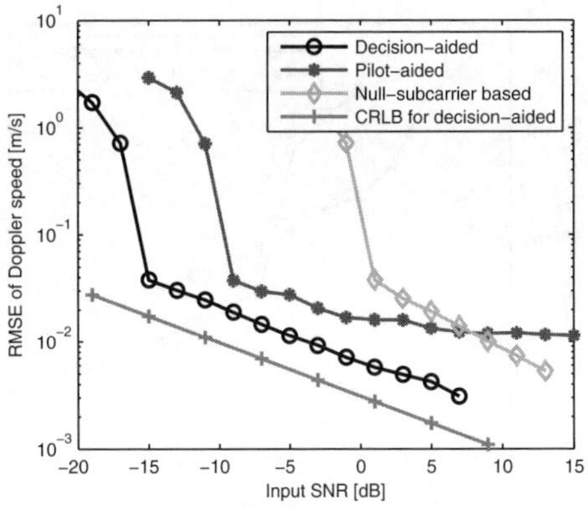

Figure 6.8 ZP-OFDM Doppler speed estimation in one path with delay channel. The CRLB with all data known is included as a benchmark.

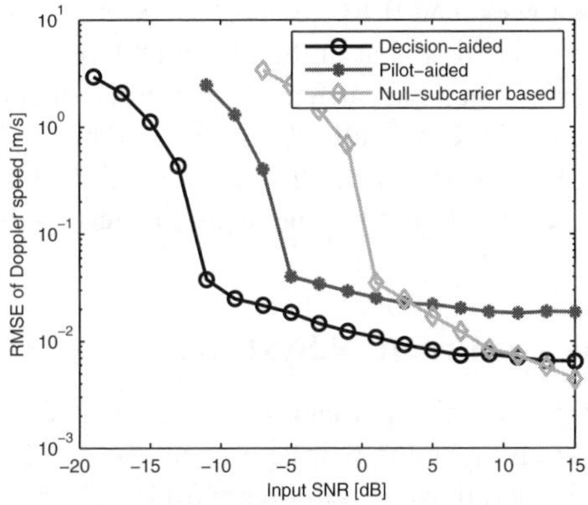

Figure 6.9 ZP-OFDM Doppler speed estimation in multipath one Doppler offset channel.

6.4.3 Comparison of Blind Methods of CP- and ZP-OFDM

The self-correlation method for the CP-OFDM preamble is closely related to the null-subcarrier based blind method for ZP-OFDM. This can be easily verified by rewriting (6.32) as

$$x_{\mathrm{cp}}(t) = \sum_{k=-K_0}^{K_0-1} d[k]e^{j2\pi\frac{k}{2T_0}t}q(t), \qquad t \in [-T_{\mathrm{cp}}, 2T_0] \tag{6.57}$$

where $d[k] = 0$ when k is odd and $d[k] = s[k/2]$ when k is even. The cyclic repetition pattern in (6.31) is generated by placing zeros on all odd subcarriers in a long OFDM symbol of duration $2T_0$. Hence, the self-correlation implementation could be replaced by the null-subcarrier based implementation for the CP-OFDM preamble.

Figure 6.10 shows the performance comparison between the blind method for ZP-OFDM and that for CP-OFDM in the multipath channel with one Doppler scale factor, respectively. At low

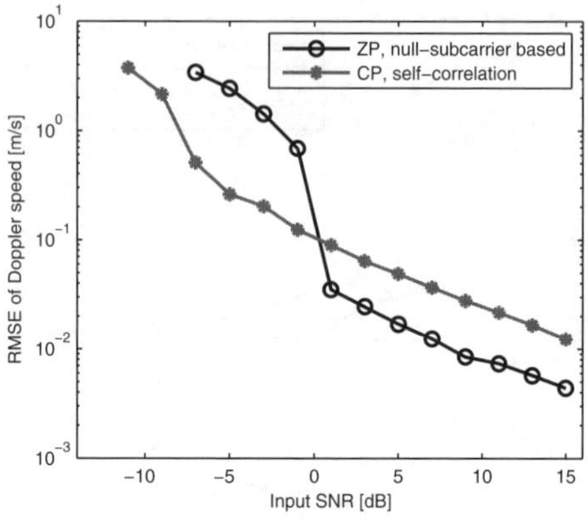

Figure 6.10 Null subcarrier based method in ZP-OFDM and CP-OFDM.

SNR, typically when it is lower than 0 dB, the null-subcarrier based method in CP-OFDM system has a better performance than that in the ZP-OFDM system, which is due to the fact that CP-OFDM system has 512 null subcarriers, far more than that the 96 null subcarriers in the ZP-OFDM block. At high SNR, the null subcarrier based method in ZP-OFDM has better performance. A possible reason is that the null subcarriers in ZP-OFDM are distributed with an irregular pattern, which could outperform the regular pattern in the CP-OFDM preamble.

6.5 Design Examples in Practical Systems

Note that waveforms for synchronization can also be applied for detection. The detection preamble and synchronization preamble will be a combination of the waveforms and techniques, as listed in Table 6.2 and illustrated in Figure 6.11.

To further elaborate, several design examples for practical systems are shown in Figure 6.12. In example 1, the LFM/HFM preamble can be used for detection, and the synchronization and Doppler scale estimation can be performed for each individual ZP-OFDM data block; in examples 2 and 3, both the m-sequence and CP-OFDM preamble can be used for all purposes, i.e., detection, synchronization and Doppler scale estimation; in example 4, the LFM/HFM preamble can be used for both signal detection and synchronization, and coupled with the LFM/HFM postamble, the Doppler scale estimate can be obtained; in examples 5 and 6, the

Table 6.2 Typical waveforms and techniques for detection and synchronization

Waveform property	Waveform	Parameters	Processing techniques	Complexity
Doppler-insensitive	LFM, HFM	$\hat{\tau}$	Matched filtering	Low
Doppler-sensitive	m-seq., OFDM	$(\hat{\tau}, \hat{a})$	Bank of matched filters	High
Pulse combination	LFM, HFM	$(\hat{\tau}, \hat{a})$	Matched filtering	Low
Repetition	CP-OFDM	$(\hat{\tau}, \hat{a})$	Bank of self-correlators	Medium

τ: time of arrival, and a: Doppler scale factor

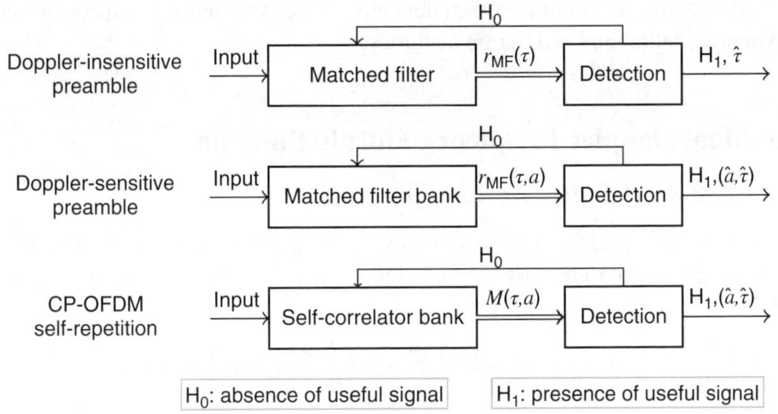

Figure 6.11 Detection and synchronization diagram for different preambles.

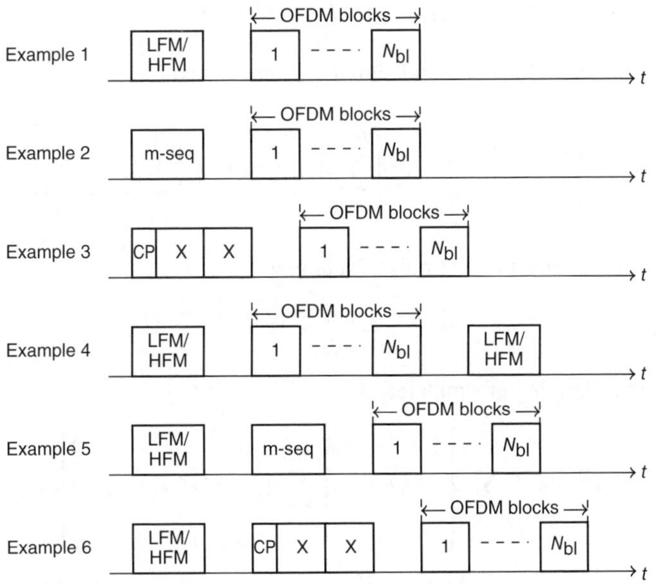

Figure 6.12 Design options of detection and synchronization preambles in a practical system.

LFM/HFM signal can be used for detection, and the m-sequence and the CP-OFDM preamble can be used for synchronization and Doppler scale estimation.

Compared with other examples, example 1 requires synchronization and Doppler scale estimation at each data block, incurring a very high computational complexity, and example 4 relying on a batch operation can only be used for the offline processing. Compared to examples 2 and 3, although examples 5 and 6 require two individual preambles for detection and synchronization, the computational complexity of detection is much lower, hence they are more suitable for online implementation.

In practical systems, the optimal design depends on the system operation constraints and the desired synchronization and estimation accuracy.

6.6 Residual Doppler Frequency Shift Estimation

6.6.1 System Model after Resampling

As illustrated in Section 5.2, in the preprocessing step the receiver resamples the received OFDM block at passband to remove the main Doppler scaling effect using an estimated resampling factor $\hat{a}$

$$\tilde{y}_{\rm r}(t) = \tilde{y}\left(\frac{t}{1+\hat{a}}\right) = \sum_{p=1}^{N_{\rm pa}} A_p \tilde{x}\left(\frac{1+a}{1+\hat{a}}t - \tau_p\right) + \tilde{w}(t), \tag{6.58}$$

where $\tilde{w}(t)$ is the resampled noise.

The baseband signal can be obtained after a passband-to-baseband downshifting and a low bandpass filtering (LPF),

$$y_{\rm r}(t) = {\rm LPF}\left\{e^{-j2\pi f_{\rm c} t}\tilde{y}_{\rm r}(t)\right\}. \tag{6.59}$$

After some manipulation, the baseband signal $y_{\rm r}(t)$ can be expressed as

$$y_{\rm r}(t) = e^{j2\pi \frac{a-\hat{a}}{1+\hat{a}} f_{\rm c} t} \sum_{k\in S_{\rm A}} \left\{ s[k] e^{j2\pi k\Delta f \frac{1+a}{1+\hat{a}} t} \left[\sum_{p=1}^{N_{\rm pa}} A_p e^{-j2\pi f_k \tau_p} g\left(\frac{1+a}{1+\hat{a}}t - \tau_p\right)\right]\right\} + w(t), \tag{6.60}$$

where $w(t)$ is the baseband ambient noise. Note that

$$\frac{1+a}{1+\hat{a}} \approx 1. \tag{6.61}$$

The baseband signal can be reformulated as

$$y_{\rm r}(t) \approx e^{j2\pi \frac{a-\hat{a}}{1+\hat{a}} f_{\rm c} t} \sum_{k\in S_{\rm A}} \left\{ s[k] e^{j2\pi k\Delta f t} \left[\sum_{p=1}^{N_{\rm pa}} A_p e^{-j2\pi f_k \tau_p} g(t - \tau_p)\right]\right\} + w(t), \tag{6.62}$$

which reveals that the residual Doppler effect can be viewed as a frequency shift, which is identical for all subcarriers

$$\epsilon := \frac{a-\hat{a}}{1+\hat{a}} f_{\rm c}. \tag{6.63}$$

The residual Doppler shift therefore can be compensated via multiplying $y_{\rm r}(t)$ by $e^{-j2\pi\epsilon t}$, leading to

$$z(t) := e^{-j2\pi\epsilon t} y_{\rm r}(t). \tag{6.64}$$

The frequency measurement at the mth subcarrier can be obtained as

$$z[m] = \int_0^{T_{\rm g}+T} y_{\rm r}(t) e^{-j2\pi\epsilon t} e^{-j2\pi \frac{m}{T} t}\, dt. \tag{6.65}$$

6.6.2 Impact of Residual Doppler Shift Compensation

In this section, we analyze the impact of the residual Doppler shift compensation on the receiver decoding performance, which will help to specify the needed Doppler scale resolution in the Doppler scale estimators.

With the estimated residual Doppler shift estimate $\hat{\epsilon}$, plugging in $z(t)$ and carrying out the integration, the frequency measurement $z[m]$ can be rigorously formulated as

$$z[m] = H\left(\frac{1+\hat{a}}{1+a}(f_m + \hat{\epsilon})\right) \sum_{k \in S_A} s[k]\varrho_{m,k} + \eta[m], \tag{6.66}$$

where $H(f)$ is the channel frequency response defined in (6.35), $\eta[m]$ is additive noise, and

$$\varrho_{m,k} = \frac{1+\hat{a}}{1+a} G\left(\beta_{m,k}\right), \tag{6.67}$$

$$\beta_{m,k} = (k-m)\frac{1}{T} + \frac{(a-\hat{a})f_m - (1+\hat{a})\hat{\epsilon}}{1+a}, \tag{6.68}$$

where $G(f)$ is the Fourier transform of the pulse shaping filter $g(t)$. Defining the symbol energy as $\sigma_s^2 = \mathbb{E}\left[|s[k]|^2\right]$ and the noise variance as σ_η^2, the effective SNR on the mth subcarrier is

$$\gamma_m = \frac{|\varrho_{m,m}|^2 \sigma_s^2}{\dfrac{\sigma_\eta^2}{\left|H\left(\frac{1+\hat{a}}{1+a}(f_m + \hat{\epsilon})\right)\right|^2} + \sum_{k \neq m} |\varrho_{m,k}|^2 \sigma_s^2}. \tag{6.69}$$

The first term in the denominator is due to additive noise, while the second term is due to the self-interference aroused by the Doppler scale mismatch. Even when the additive noise diminishes, the effective SNR is bounded by

$$\gamma_m \leq \bar{\gamma}_m := \frac{|\varrho_{m,m}|^2}{\sum_{k \neq m} |\varrho_{m,k}|^2} \tag{6.70}$$

due to self-interference induced by Doppler scale mismatch.

The SNR upper bound is evaluated for two cases

- Case 1: No Doppler shift compensation by setting $\epsilon = 0$.
- Case 2: Ideal Doppler shift compensation where

$$\epsilon_{\text{opt}} = \frac{a-\hat{a}}{1+\hat{a}} f_c, \tag{6.71}$$

such that

$$\beta_{m,k} = (k-m)\frac{1}{T} + \frac{a-\hat{a}}{1+a}\frac{m}{T}. \tag{6.72}$$

For the first case, the leading term $(k-m)/T$ in $\beta_{m,k}$ is the frequency difference between the kth and the mth subcarriers, while the second term $\frac{a-\hat{a}}{1+a} f_m$ is the extra frequency shift. For the second case, the leading term $(k-m)/T$ in $\beta_{m,k}$ is the frequency difference between the

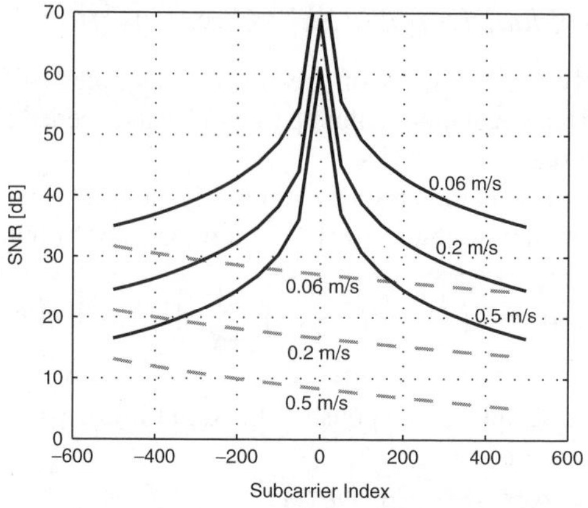

Figure 6.13 The SNR upper bound $\overline{\gamma}_m$ for $\hat{\epsilon} = \epsilon_{\text{opt}}$ (thick, full lines) and $\hat{\epsilon} = 0$ (thin, dashed lines) as a function of Δv, where $a - \hat{a} = \Delta v/c$.

kth and the mth subcarriers, while the second term $\frac{a-\hat{a}}{1+a} \cdot \frac{m}{T}$ is the extra frequency shift. Since f_m is much larger than $\frac{m}{T}$, the performance of Doppler shift compensation will improve. Consider an example of $f_c = 27\,\text{kHz}$ and $B/2 = 6\,\text{kHz}$, we have $f_m \in [21, 33]\,\text{kHz}$ and $\max_m \frac{m}{T} = 6\,\text{kHz}$. Hence, the accuracy of $(a - \hat{a})$ can be relaxed at least by four times to reach similar performance. This illustrates that Doppler shift compensation is one crucial step in receiver design [235].

The upper bound $\overline{\gamma}_m$ for $\hat{\epsilon} = 0$ and $\hat{\epsilon} = \epsilon_{\text{opt}}$ will be evaluated numerically. With system parameters $f_c = 27\,\text{kHz}$, $B = 12\,\text{kHz}$, and $K = 1024$, Figure 6.13 shows the bounds for these two cases respectively. Suppose that it is targeted to limit the self-noise to be at least 20 dB below the signal power. In the case $\hat{\epsilon} = 0$, Δv should be less than 0.06 m/s. While in the case $\hat{\epsilon} = \epsilon_{\text{opt}}$, the Δv can be as large as 0.3 m/s.

Figure 6.13 provides guidelines on the selection of the Doppler scale spacing of the parallel correlators. For example, assuming that the correlator branch closest to the true speed will yield the maximum metric, then with fine Doppler shift compensation $\hat{\epsilon} = \epsilon_{\text{opt}}$ the Doppler scale spacing can be set as 0.4 m/s (where Δv needs to be less than 0.2 m/s) to achieve an SNR upper bound of at least 25 dB. On the other hand, if an SNR upper bound of 15 dB is sufficient, the Doppler scale spacing could be as large as 1.0 m/s.

6.6.3 Two Residual Doppler Shift Estimation Methods

In this section, two residual Doppler shift estimation methods will be discussed [267, 454, 469]. The first method takes advantage of the fact that most OFDM systems have null subcarriers, and estimates the residual Doppler shift by minimizing the energy of frequency measurements at null subcarriers [267]. The second method exploits the non-Gaussianity of frequency measurements at all subcarriers [454].

6.6.3.1 The Null Subcarrier Based Method

Similar to the null subcarrier based method for Doppler scale estimation, the residual Doppler shift can be estimated as

$$\hat{\epsilon} = \arg\min_{\epsilon} \sum_{m \in S_N} \left| \int_0^{T+T_g} y_r(t) e^{-j2\pi\epsilon t} e^{-j2\pi \frac{m}{T} t} dt \right|^2 , \tag{6.73}$$

which can be solved using one dimensional search including coarse initial search followed by fine bi-sectional search [235, 445].

6.6.3.2 The Y-G Algorithm

The key idea of this algorithm is that the distribution of frequency measurements is more non-Gaussian when $\hat{\epsilon}$ approaches the optimal value ϵ_{opt}. The non-Gaussianity of a distribution can be measured by kurtosis. The objective function of the Y-G algorithm can be cast as

$$J(\epsilon) = \frac{\sum_{m \in S_A} \left| \int_0^{T+T_g} y_r(t) e^{-j2\pi\epsilon t} e^{-j2\pi \frac{m}{T} t} dt \right|^4}{\left[\sum_{m \in S_A} \left| \int_0^{T+T_g} y_r(t) e^{-j2\pi\epsilon t} e^{-j2\pi \frac{m}{T} t} dt \right|^2 \right]^2} . \tag{6.74}$$

With a sufficiently large number of data points, the objective function $J(\epsilon)$ resembles a period of the cosine function and $\hat{\epsilon} = \epsilon_{opt}$ will be a unique global minimum or maximum solution of the objective function. Instead of an exhaustive search, a closed-form estimate of $\hat{\epsilon}$ is available. Define two parameters

$$A := \frac{J(\frac{1}{4}\Delta f) + J(-\frac{1}{4}\Delta f)}{2} - J(0), \tag{6.75}$$

$$B := \frac{J(-\frac{1}{4}\Delta f) - J(\frac{1}{4}\Delta f)}{2}, \tag{6.76}$$

where $\Delta f = 1/T$ is the subcarrier spacing. The extreme point of the objective function $J(\hat{\epsilon})$ normalized by the subcarrier spacing can be obtained analytically as

$$\frac{\hat{\epsilon}}{\Delta f} = \begin{cases} \frac{1}{2\pi} \tan^{-1}\left(\frac{B}{A}\right) & \text{if } A > 0 \\ \frac{1}{2} + \frac{1}{2\pi} \tan^{-1}\left(\frac{B}{A}\right) & \text{if } A < 0, B \geq 0 \\ -\frac{1}{2} + \frac{1}{2\pi} \tan^{-1}\left(\frac{B}{A}\right) & \text{if } A < 0, B < 0. \end{cases} \tag{6.77}$$

6.6.4 *Simulation Results*

In this section, the two residual Doppler shift estimators will be compared in simulations.

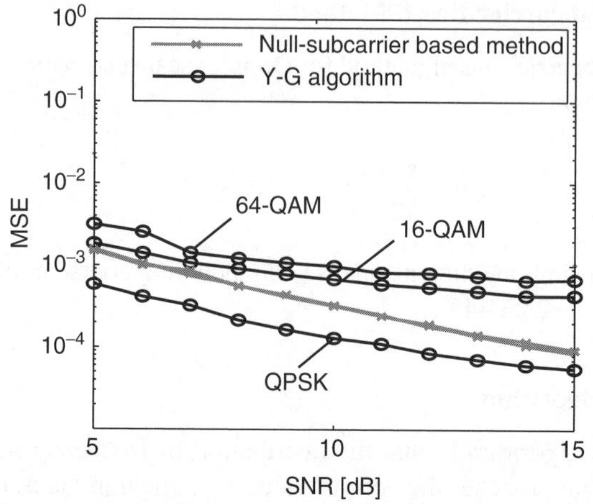

Figure 6.14 MSE comparison of two residual Doppler shift estimators.

6.6.4.1 Simulation Setup

The underwater acoustic channel is simulated according to the specifications in Section 5.5.1. The channel parameters are $N_{pa} = 10$, $\Delta\tau = 1$ ms, $\Delta P_{pa} = 20$ dB, $T_g = 24.6$ ms, and $v_0 = 0$ m/s. The OFDM parameters are identical to the parameters used in an experiment, which are specified in Table B.1.

Note that the residual Doppler shift in (6.62) is identical to the carrier frequency offset (CFO) in radio communications [267, 394]. To test the performance of the above two residual Doppler shift estimators, a Doppler shift is artificially added in the received signal before receiver processing; see (6.62). The artificial Doppler shift is generated uniformly and independently within $[-0.4, 0.4] \times \Delta f$ for each OFDM block.

6.6.4.2 Computational Complexity

For the null subcarrier based algorithm, a coarse initial search followed by a bisection search is performed [445]. The initial search range normalized by the subcarrier spacing Δf is from -0.4 to 0.4 with a step size 0.1. During the bi-sectional search, the step size relative to Δf decreases to 0.05, 0.025, 0.0125, and 0.00625, with two FFTs at each iteration searching the left and right points around the previously obtained CFO value. Thus the total computational complexity is $9 + 8 = 17$ K-point FFTs.

For the Y-G algorithm, the desired estimate is a global minimum of the objective function for all tested cases, thus the computational complexity is just that of three K-point FFTs.

6.6.4.3 MSE Comparison

Figure 6.14 shows the MSE performance of the two residual Doppler shift estimators with an artificially added Doppler shift using QPSK, 16-QAM, and 64-QAM constellations. It can be seen that the null subcarrier based method is very stable regardless of the adopted constellation.

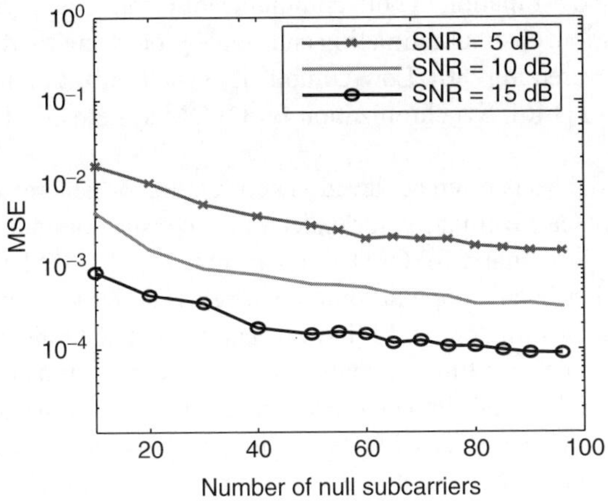

Figure 6.15 MSE performance with different numbers of null subcarriers.

In fact, the performance of the null subcarrier based method depends only on the desired resolution which is related to the computational complexity. The Y-G algorithm degrades as the constellation size increases from QPSK to 64-QAM. For QPSK modulation, the Y-G algorithm can outperform the null subcarrier based method whereas for 16-QAM and 64-QAM modulations the null-subcarrier based method outperforms the Y-G algorithm, especially in the high SNR range. However, the Y-G algorithm has the least computational complexity.

6.6.4.4 Effect of Number of Null Subcarriers

The effect of the number of null subcarriers in the null subcarrier based method is shown in Figure 6.15 for a 16-QAM constellation. The energy on a portion of the total 96 null subcarriers is used as the objective function for this purpose. Figure 6.15 shows that the performance of the null subcarrier based method degrades as the number of null subcarriers decreases. For this setting, the MSE improvement after more than 60 null subcarriers is minimal.

6.7 Bibliographical Notes

Detection, ranging, and range rate estimation are the topics which have been well-investigated in radar and sonar applications. The earliest collection of studies in series can be dated back to 1968 by H. Van Trees in [402], and the masterpiece by A. Papoulis in 1965 on probability, random variables and stochastic processes in [302]. Examples of the following publications include signal detection in non-Gaussian noise in [205], the recursive estimation algorithms in [358], and estimation for tracking and navigation in [22]. Classic texts for underwater telemetry include Urick, 1983 [397], Medwin and Clay, 1997 [277] and Lurton, 2010 [266].

The basic principles for detection and estimation in radar and sonar systems carry over to communication systems, whereas the latter operating in the high SNR regime allow diverse signal and algorithm design to accommodate properties of communication channels. Parallel studies on detection and estimation in both communication and radar systems are summarized in [453]. Synchronization, channel estimation and signal processing for digital communication receiver design are covered in [280]. Low-complexity synchronization for pilot-aided OFDM systems is discussed in [146]. Synchronization for OFDM systems in IEEE802. 11a, DVB-T is analyzed in [154].

Estimation of time delays is often achieved via correlation or generalized correlation methods [63, 214]. Doppler scale estimation for underwater acoustic communications can be found in e.g., [338, 452] and particularly for OFDM signals in [274, 412, 455]. Carrier frequency offset (CFO) estimation has been a popular topic for wireless OFDM systems. Existing methods include both data-aided estimators and blind estimators. The data-aided estimators, such as [285, 286, 334], rely on either training symbols or periodically transmitted pilots for CFO estimation. There is a rich set of blind estimators that exploit the signal structure or statistical characteristics of measurements for CFO estimation, such as the CP segment in CP-OFDM [228, 399], frequency measurements at null subcarriers [11, 85, 254, 267, 395], second-order statistics of received signal [133], the nonlinearity or high-order statistics of frequency measurements [264, 454], and relationships among time-domain oversampled measurements in CP-OFDM [71]. Comparison of the null subcarrier based method [267], the O-M algorithm [264] and the Y-G algorithm [454] in underwater acoustic OFDM systems can be found in [469].

7

Channel and Noise Variance Estimation

In this chapter, we focus on channel estimation within each received OFDM block. The underwater acoustic channel is well-known to consist of *sparsely* distributed propagation paths. Compared to the canonical least squares based method which estimates all the discretized channel samples within the channel delay spread, channel sparsity can be exploited to reduce the number of unknown parameters to estimate. This chapter treats sparse channel estimation for OFDM in both time-invariant and time-varying environments.

Sparse channel estimation has been underpinned by the striking development of *compressive sensing theory* in the last decade. To facilitate the development of sparse channel estimation in this chapter, the basics of compressive sensing are presented in Appendix A. The reader is encouraged to read this chapter and Appendix A in parallel for an in-depth understanding.

This chapter is organized as follows.

- In Section 7.1, to exploit the sparsity of underwater acoustic channels, a *dictionary* based formulation of the system input–output relationship is developed. It subsumes the classical input–output relationship formulation for the least squares (LS) based channel estimation, yet allows channel parameterization with a limited number of measurements.
- In Sections 7.2 and 7.3, building upon the dictionary-based formulation, an ICI-ignorant sparse channel estimator and an ICI-aware sparse channel estimator are derived for the single-input channel. A variety of practical issues, such as dictionary construction and pilot design, are also examined.
- In Section 7.4, two typical sparse recovery algorithms are briefly discussed.
- In Section 7.5, the sparse channel estimator for the single-input channel is extended to the multi-input channel, such as the multi-input multi-output (MIMO) channels where channel estimation is conducted at each receiving element individually. The developed sparse channel estimators will be widely used in the MIMO channels presented in Chapters 13–17.
- In Sections 7.6 and 7.7, noise variance estimation and noise pre-whitening in practical systems are examined. The estimated noise variance could be used in both symbol detection and channel decoding.

OFDM for Underwater Acoustic Communications, First Edition. Shengli Zhou and Zhaohui Wang.
© 2014 John Wiley & Sons, Ltd. Published 2014 by John Wiley & Sons, Ltd.

The sparse channel estimators presented in this chapter lay the foundation of channel estimation in the receivers developed in later chapters, such as the iterative receiver design in Chapter 9 where the data symbols estimated from preceding iterations are used for channel estimation refinement, and the block-to-block receiver processing in Chapter 10 in which channel estimates of preceding blocks are used for the channel estimation of the current block, thus maintaining a decent channel estimation performance with a reduced pilot overhead.

7.1 Problem Formulation for ICI-Ignorant Channel Estimation

7.1.1 *The Input–Output Relationship*

We index pilot subcarriers as the set $\mathcal{S}_P = \{q_1, \cdots, q_{K_P}\}$ with $K_P := |\mathcal{S}_P|$ subcarriers in total. Based on the input–output relationship in (5.45) and (5.47), stacking frequency measurements at pilot subcarriers into a vector yields

$$\underbrace{\begin{bmatrix} z[q_1] \\ \vdots \\ z[q_{K_P}] \end{bmatrix}}_{:=\mathbf{z}_P} = \underbrace{\begin{bmatrix} s[q_1] & & \\ & \ddots & \\ & & s[q_{K_P}] \end{bmatrix} \begin{bmatrix} e^{-j2\pi\frac{q_1}{T}\bar{\tau}_1} & \cdots & e^{-j2\pi\frac{q_1}{T}\bar{\tau}_{N_{\rm pa}}} \\ \vdots & \ddots & \vdots \\ e^{-j2\pi\frac{q_{K_P}}{T}\bar{\tau}_1} & \cdots & e^{-j2\pi\frac{q_{K_P}}{T}\bar{\tau}_{N_{\rm pa}}} \end{bmatrix}}_{:=\mathbf{A}} \underbrace{\begin{bmatrix} \xi_1 \\ \vdots \\ \xi_{N_{\rm pa}} \end{bmatrix}}_{:=\xi} + \underbrace{\begin{bmatrix} \eta[q_1] \\ \vdots \\ \eta[q_{K_P}] \end{bmatrix}}_{:=\eta_P}, \quad (7.1)$$

which can be written in the vector-matrix form

$$\mathbf{z}_P = \mathbf{A}\xi + \eta_P. \quad (7.2)$$

Should the receiver be aware of path delays, the channel estimation problem can be solved via the least squares (LS) approach,

$$\hat{\xi}_{\rm LS} = \arg\min_{\xi} ||\mathbf{z}_P - \mathbf{A}\xi||^2, \quad (7.3)$$

$$= (\mathbf{A}^{\rm H}\mathbf{A})^{-1}\mathbf{A}^{\rm H}\mathbf{z}_P = \mathbf{A}^{\dagger}\mathbf{z}_P \quad (7.4)$$

where $\mathbf{A}^{\dagger}$ denotes the pseudo-inverse of matrix $\mathbf{A}$. However, the number of paths and the path delays in (7.1) are usually not available at the receiver side.

7.1.2 *Dictionary Based Formulation*

Notice that the observed signal in (7.1) is a linear combination of an *unknown* number of structured signals, each defined by an *unknown* delay $\bar{\tau}_p$. The estimation problem can be reformulated by constructing a so-called dictionary, made of the signals parameterized by a representative selection of possible values of parameter $\bar{\tau}_p$. In this model, parameter sets not part of the solution will be assigned a zero weight coefficient, and weight coefficients of parameter sets within the combination basis will be selected to minimize the fitting error in (7.3).

Following the above line of thought, a representative set of $\bar{\tau}_p$ is chosen as

$$\bar{\tau} \in \left\{ 0, \frac{1}{\lambda_{\rm b}B}, \frac{2}{\lambda_{\rm b}B}, \cdots, \frac{N_{\rm de}-1}{\lambda_{\rm b}B} \right\}, \quad (7.5)$$

where λ_b is an oversampling factor, and the number of representative values is associated to the maximal path delay τ_{max} via

$$N_{de} := \left\lceil \frac{\tau_{max}}{\lambda_b B} \right\rceil, \tag{7.6}$$

with $\lceil \cdot \rceil$ denoting the ceiling operation. Without any prior information of channel path delays, discretization of $\overline{\tau}$ in (7.5) is based on the assumption that after synchronization all arriving paths fall into the guard interval, i.e., $\tau_{max} = T_g$.

One should note that the original $\overline{\tau}_p$ can in general take any *continuous* value, e.g., within $[0,\ T_g]$. So only as $\lambda_b \to \infty$ are we guaranteed to include the original $\overline{\tau}_p$ in the set (7.5). On the other hand for $\lambda_b > 1$ the difference between a column in matrix $\mathbf{A}$ corresponding to a particular $\overline{\tau}_p$ and the closest $\overline{\tau}$ diminishes rather quickly for any further increase in λ_b, more on that in Section 7.2.1.

For channels that vary slowly, Chapter 10 will show that prior knowledge on the rough locations of channel paths within the guard interval can be obtained, through preamble processing or channel estimates of preceding blocks. The representative values of $\overline{\tau}$ can be drawn from the possible *delay zones* of channel paths, leading to a reduced number of candidate paths and thus a decrease in the number of necessary measurements. We will defer detailed discussions on this topic to Chapter 10.

With the representative values of path delays in (7.5), the dictionary based formulation is cast as

$$\underbrace{\begin{bmatrix} z[q_1] \\ \vdots \\ z[q_{K_P}] \end{bmatrix}}_{:=\mathbf{z}_P} = \underbrace{\begin{bmatrix} s[q_1] & & \\ & \ddots & \\ & & s[q_{K_P}] \end{bmatrix} \begin{bmatrix} 1 & \cdots & e^{-j2\pi \frac{q_1}{K} \frac{(N_{de}-1)}{\lambda_b}} \\ \vdots & \ddots & \vdots \\ 1 & \cdots & e^{-j2\pi \frac{q_{K_P}}{K} \frac{(N_{de}-1)}{\lambda_b}} \end{bmatrix}}_{:=\mathbf{\Psi}} \underbrace{\begin{bmatrix} \xi_0 \\ \vdots \\ \xi_{N_{de}-1} \end{bmatrix}}_{:=\boldsymbol{\xi}} + \underbrace{\begin{bmatrix} \eta[q_1] \\ \vdots \\ \eta[q_{K_P}] \end{bmatrix}}_{:=\boldsymbol{\eta}_P} \tag{7.7}$$

which can be put in the vector-matrix form

$$\begin{aligned} \mathbf{z}_P &= \left[\boldsymbol{\psi}_0 \ \cdots \ \boldsymbol{\psi}_{p-1} \ \cdots \ \boldsymbol{\psi}_{N_{de}-1} \right] \boldsymbol{\xi} + \boldsymbol{\eta}_P \\ &= \mathbf{\Psi}\boldsymbol{\xi} + \boldsymbol{\eta}_P \end{aligned} \tag{7.8}$$

where $\boldsymbol{\psi}_{p-1}$ denotes the pth dictionary entry, and the size of the dictionary matrix $\mathbf{\Psi}$ is $K_P \times N_{de}$.

Once the estimate $\hat{\boldsymbol{\xi}}$ is obtained, the channel coefficients at all subcarriers can be reconstructed as

$$\hat{H}[m] = \sum_{p=0}^{N_{de}-1} \hat{\xi}_p e^{-j2\pi \frac{mp}{\lambda_b K}} \tag{7.9}$$

for $m = -K/2, \cdots, K/2-1$. Eq. (7.9) can be computed via the $\lambda_b K$-point FFT of $\hat{\boldsymbol{\xi}}$.

The meaning of converting the measurement representation from (7.1) to (7.7) is two-fold:

- It changes a small-size nonlinear estimation problem to a large-size linear estimation problem. Rather than estimating the N_{pa} path tuples $\{\xi_p, \overline{\tau}_p\}$ based on the nonlinear input–output relationship in (7.1), the channel estimation problem is reoriented as a combinatorial problem which estimates the linear combination coefficients in (7.8).

When the number of unknowns is no more than the number of pilots, i.e., $N_{\rm de} \le K_{\rm P}$, the LS estimate is well-posed,

$$\hat{\xi}_{\rm LS} = (\mathbf{\Psi}^{\rm H}\mathbf{\Psi})^{-1}\mathbf{\Psi}^{\rm H}\mathbf{z}_{\rm P}. \tag{7.10}$$

This corresponds to the canonical LS channel estimator widely used in the early days of OFDM. Given a fixed pilot overhead, it has been shown [294, 299] that for LS channel estimation, equally-spaced pilot subcarriers and equally-powered pilot symbols achieve the minimal mean-square channel estimation error. Assume a baseband sampling rate ($\lambda_{\rm b} = 1$), and define L as the number of channel samples, the input–output relationship can be put as

$$\begin{bmatrix} z[q_1] \\ \vdots \\ z[q_{K_{\rm P}}] \end{bmatrix} = \begin{bmatrix} s[q_1] & & \\ & \ddots & \\ & & s[q_{K_{\rm P}}] \end{bmatrix} \begin{bmatrix} 1 & \cdots & e^{-j2\pi\frac{q_1(L-1)}{K}} \\ \vdots & \ddots & \vdots \\ 1 & \cdots & e^{-j2\pi\frac{q_{K_{\rm P}}(L-1)}{K}} \end{bmatrix} \begin{bmatrix} \xi_0 \\ \vdots \\ \xi_{L-1} \end{bmatrix} + \begin{bmatrix} \eta[q_1] \\ \vdots \\ \eta[q_{K_{\rm P}}] \end{bmatrix} \tag{7.11}$$

where the LS estimate in (7.10) can be computed efficiently using a $K_{\rm P}$-point FFT.

- The dictionary based representation enables exploring sparsity of underwater acoustic channels for channel estimation, especially when the problem in (7.8) is ill-defined with $N_{\rm de} > K_{\rm P}$. Despite a large number of representative entries in $\mathbf{\Psi}$, the number of effective channel paths could be far less than the number of unknowns, i.e., $N_{\rm pa} \ll N_{\rm de}$. The estimation problem therefore can be formulated by constraining most of elements in the vector ξ to be zero.

  In this chapter, we will focus on sparse channel estimation based on compressive sensing techniques. A brief introduction of compressive sensing theory is included in Appendix A.

7.2 ICI-Ignorant Sparse Channel Sensing

When $K_{\rm P} \ll N_{\rm de}$, the problem in (7.8) is ill-defined in the least squares sense. Exploiting the sparsity of UWA channels, the $N_{\rm pa}$ paths can be identified by minimizing the number of channel paths subject to a constraint on the mean-squared fitting error. The estimation problem can be cast as

$$\ell_0\text{-norm:} \qquad \hat{\xi} = \arg\min_{\xi} ||\xi||_0 \qquad \text{s.t.} \qquad ||\mathbf{\Psi}\xi - \mathbf{z}_{\rm P}||^2 \le \delta \tag{7.12}$$

where the ℓ_0-norm of vector ξ is defined as the number of its nonzero elements, and δ is a design parameter characterizing the fidelity of the estimate to the observations in the least squares sense. Note that (7.12) is essentially a combinatorial problem with an unknown number of entries. Directly solving this problem is NP-hard.

Since the formulation in (7.12) is not amenable to efficient computation, one variant of (7.12) is to turn to convex relaxation by replacing the ℓ_0-norm by the ℓ_1-norm, so that linear programming techniques can be applied. The Lagrangian formulation is cast as

$$\ell_1\text{-norm:} \qquad \hat{\xi} = \arg\min_{\xi} \frac{1}{2}||\mathbf{\Psi}\xi - \mathbf{z}_{\rm P}||^2 + \zeta||\xi||_1 \tag{7.13}$$

where the ℓ_1-norm of ξ is defined as $||\xi||_1 := \sum_{p=0}^{N_{\rm de}-1} |\xi_p|$, and the factor ζ is a tuning parameter which controls the solution sparsity.

7.2.1   Dictionary Resolution versus Channel Sparsity

7.2.1.1   How Sparse is the Baseband Channel?

The channel representation with the baseband sampling rate ($\lambda_b = 1$) is commonly used in receiver design [389]. Although this representation can capture the full channel effect, it can only be treated as approximately sparse due to the available observations within a finite frequency band; as shown in Figure 7.1.

$$H(f) = \sum_{p=1}^{Npa} \xi_p e^{-j2\pi f\tau_p}$$

(a) Channel frequency response vs subcarrier index

$$h(\tau) = \sum_{p=1}^{Npa} \xi_p \delta(\tau - \tau_p)$$

(b) Channel impulse response vs delay τ in (ms $\cdot$ B)

$$h[\ell] = \sum_{k=-K/2}^{K/2-1} H(f_k) e^{j2\pi\frac{k\ell}{K}}$$

(c) Channel baseband impulse response vs sample ℓ

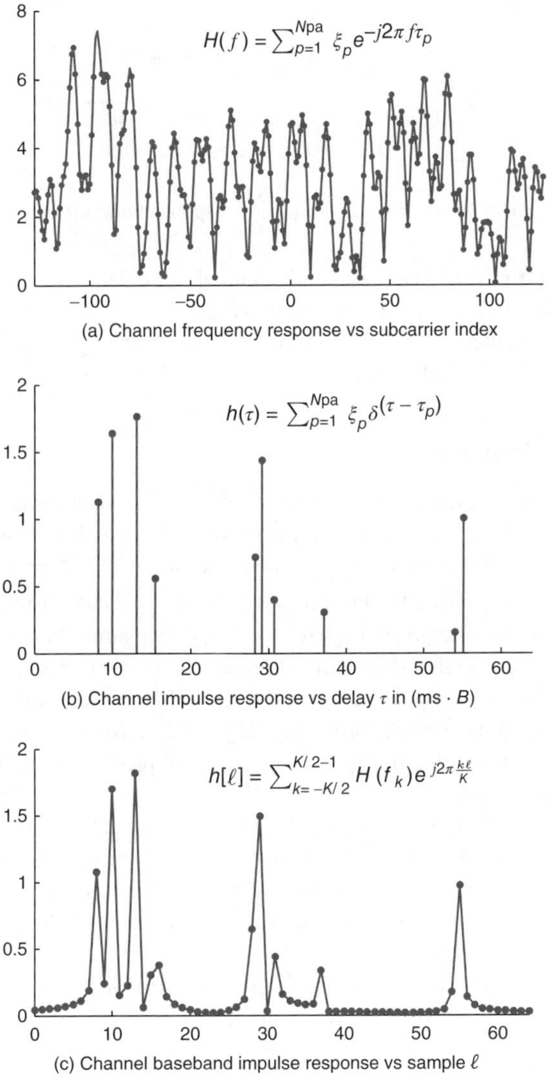

Figure 7.1   The channel frequency response $H(f)$ maps to the impulse response $h(\tau)$, but from a limited number of samples $H(f_k)$ only the baseband model $h[\ell]$ can be determined unambiguously; in this example there are $N_{pa} = 10$ discrete paths, and $K = 256$ frequency samples; all plots are magnitude only.

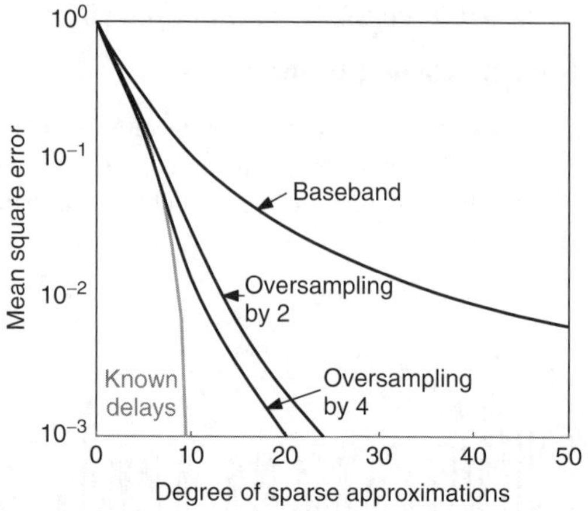

Figure 7.2 Approximation error of the frequency response with different degrees of sparsity.

To increase the sparsity of the baseband channel representation, one can increase the time-domain oversampling factor λ_b. In other words, the representative set of path delays in (7.5) can be constructed with a finer grained resolution $T_{s,b}$, which can more closely approximate the continuous path delays $\overline{\tau}_p$.

7.2.1.2 Numerical Example

Let us consider a simple example as shown in Figure 7.1, where the noiseless frequency measurements consist of the stacked frequency response $H(f_k)$ at K subcarriers, and each measurement is a linear combination of N_{pa} complex phases $e^{-j2\pi f_k \tau_p}$ with complex weighting coefficients ξ_p. The signal $H(f)$ is approximated using s basis vectors as $\hat{H}_s(f)$; the corresponding mean square error (MSE) is $\mathbb{E}_k[|H(f_k) - \hat{H}_s(f_k)|^2]$. Naturally when using a less sparse approximation (increasing s) the MSE will decrease. For example, for known delays the error will reach zero for $s = N_{pa}$. In general, how fast the MSE reduces with s will indicate how sparse the corresponding basis can approximate $H(f)$. While for the baseband model ($\lambda_b = 1$) there is a trivial way to determine the optimum s-sparse approximation, this is not the case for the redundant dictionaries.

In Figure 7.2 the MSE decreases similarly for all cases up to about 10^{-1}, this means that for this multipath channel about 90 % of the channel energy is concentrated in the ten strongest channel taps. On the other hand, the baseband model will need more than 30 nonzero channel taps to approximate the frequency response with a MSE of 10^{-2}, while using a redundant basis one needs about half that. This points towards an interesting fact, that the baseband channel taps are not approximately sparse in terms of a power law; if they would follow a power law, the slope in the plot would be constant, while in fact it levels off.

7.2.2 *Sparsity Factor*

Since the sparsity of UWA channel depends on the signal-to-noise ratio (SNR) at the receiver, it is suggested in [38] to choose the sparsity factor ζ in (7.13) as a function of SNR for each

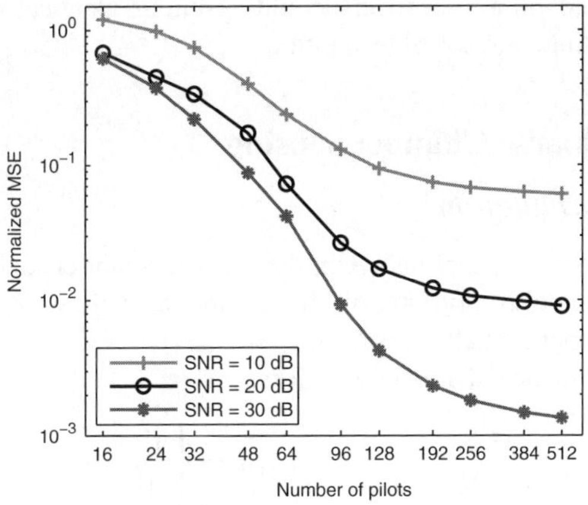

Figure 7.3 Normalized mean square error of channel frequency response estimate in the ICI-ignorant receiver with different number of randomly distributed pilot subcarriers. There are $N_{\text{pa}} = 15$ paths in simulation. An oversampling factor $\lambda_{\text{b}} = 2$ is used for channel estimation.

OFDM block as $\zeta = \kappa \|\mathbf{\Psi}^{\text{H}}\mathbf{z}_{\text{P}}\|_{\infty}/\sqrt{\text{SNR}}$, where κ is a constant, and $\|\mathbf{x}\|_{\infty}$ is the maximal absolute value of the elements in vector $\mathbf{x}$. Based on our simulation and experimental data decoding experience, a good choice of κ is 0.6. Please refer to [109, 136, 147] for theoretical discussions on the selection of ζ.

7.2.3 Number of Pilots versus Number of Paths

Provided the fact that exploiting channel sparsity enables a reduced number of measurements for channel estimation relative to that in the least squares sense, one question that arises naturally is how many measurements are needed to estimate a channel with a certain degree of sparsity. For the sensing problem with a well-conditioned sensing matrix, theoretical bounds on the number of measurements have been presented in [110].

For the sensing matrix with highly correlated entries, as $\mathbf{\Psi}$ in (7.7), the normalized channel estimation mean square error

$$\text{NMSE}(\hat{\xi}) = \frac{\sum_{k\in\mathcal{S}_{\text{all}}} \left|H(f_k) - \hat{H}(f_k)\right|^2}{\sum_{k\in\mathcal{S}_{\text{all}}} |H(f_k)|^2} \tag{7.14}$$

is presented numerically in Figure 7.3, where an OFDM signaling format of $K = 1024$ subcarriers is used, with different numbers of pilot subcarriers *randomly* distributed. At an SNR level of 20 dB, 64 pilots can achieve a fairly good estimation performance, with the normalized MSE around 0.07, and 128 pilots can further reduce the estimation error to 0.02.

Note that different from LS channel estimation where equally-spaced pilot subcarriers promise the minimal MSE, randomly distributed pilot subcarriers generate a sensing matrix with a better sensing property through reducing the coherence of sensing entries. One example is that when the number of pilots is less than the number of unknown channel samples, equally-spaced pilot subcarriers could result in identical entries in the sensing matrix, leading to multiple unresolvable solutions.

7.3 ICI-Aware Sparse Channel Sensing

7.3.1 Problem Formulation

In contrast to the diagonal channel matrix in the time-invariant channels, the channel matrix in the time-varying channel is a full matrix. ICI-aware channel estimation therefore relies on the frequency measurements at all subcarriers.

For illustration convenience, define two vectors,

$$[\mathbf{s}_{\mathrm{P}}]_k = \begin{cases} s[k] & k \in \mathcal{S}_{\mathrm{P}} \\ 0 & k \in \mathcal{S}_{\mathrm{D}} \cup \mathcal{S}_{\mathrm{N}}, \end{cases} \qquad [\mathbf{s}_{\mathrm{D}}]_k = \begin{cases} s[k] & k \in \mathcal{S}_{\mathrm{D}} \\ 0 & k \in \mathcal{S}_{\mathrm{P}} \cup \mathcal{S}_{\mathrm{N}}. \end{cases} \tag{7.15}$$

The frequency measurement vector in (5.28) can be reformulated as

$$\mathbf{z} = \mathbf{H}\mathbf{s}_{\mathrm{P}} + \mathbf{H}\mathbf{s}_{\mathrm{D}} + \mathbf{w}. \tag{7.16}$$

Since data symbols in $\mathbf{s}_{\mathrm{D}}$ are unknown at the receiver, the frequency components corresponding to the unknown data symbols can be combined with the ambient noise as one equivalent noise, which leads to

$$\mathbf{z} = \mathbf{H}\mathbf{s}_{\mathrm{P}} + \mathbf{v}_{\mathrm{D}}. \tag{7.17}$$

Substituting (5.29) into (7.17) yields

$$\mathbf{z} = \sum_{p=1}^{N_{\mathrm{pa}}} \xi_p \mathbf{\Lambda}(\overline{\tau}_p)\mathbf{\Gamma}(b_p, \hat{\epsilon})\mathbf{s}_{\mathrm{P}} + \mathbf{v}_{\mathrm{D}}. \tag{7.18}$$

7.3.2 ICI-Aware Channel Sensing

Given the nonlinearity between the frequency measurements and the path triplets $\{\xi_p, \overline{\tau}_p, b_p\}$ in (7.18), a linear estimator, such as the LS channel estimator in ICI-ignorant receiver processing, is not applicable. Similar to the sparse channel estimation in the ICI-ignorant receiver, to circumvent the difficulty incurred by the nonlinearity, a dictionary based representation of the frequency measurements can be used to convert the problem to a linear combinatorial problem.

We first define the representative sets of path parameters $(\overline{\tau}_p, b_p)$ as

$$\overline{\tau} \in \left\{0, \frac{1}{\lambda_{\mathrm{b}} B}, \frac{2}{\lambda_{\mathrm{b}} B}, \cdots, \frac{N_{\mathrm{de}} - 1}{\lambda_{\mathrm{b}} B}\right\}, \tag{7.19}$$

$$b \in \{-b_{\max}, -b_{\max} + \Delta b, \cdots, b_{\max}\}. \tag{7.20}$$

The discretization in $\overline{\tau}$ is identical to that used in the sparse ICI-ignorant channel measurement representation, with N_{de} tentative delays. For the Doppler scaling factors, it suffices to assume that they are spread around zero after a mean-Doppler scale compensation, and $b_{\max}$ can be chosen based on the assumed Doppler spread. With a grid resolution Δb, there are $N_{\mathrm{Dop}} := 2b_{\max}/(\Delta b) + 1$ tentative Doppler scale values. Hence, a total of $N_{\mathrm{de}} N_{\mathrm{Dop}}$ candidate paths will be included in the dictionary construction.

Define a vector formed by the amplitudes corresponding to all representative delays associated with the Doppler scale b_i,

$$\xi^{(i)} = \left[\xi_1^{(i)} \ \cdots \ \xi_{N_{\rm de}}^{(i)}\right]^{\rm T} \tag{7.21}$$

for $i = 1, \cdots, N_{\rm Dop}$. The $N_{\rm Dop}$ vectors are stacked into a long vector

$$\xi = \left[(\xi^{(1)})^{\rm T} \ \cdots \ (\xi^{(N_{\rm Dop})})^{\rm T}\right]^{\rm T}. \tag{7.22}$$

A linear formulation of the input–output expression in (7.18) is

$$\begin{aligned}\mathbf{z} &= \sum_{i=1}^{N_{\rm Dop}} \sum_{p=1}^{N_{\rm de}} \xi_p^{(i)} \boldsymbol{\Lambda}(\overline{\tau}_p) \boldsymbol{\Gamma}(b^{(i)}, \overline{\epsilon}) \mathbf{s}_{\rm P} + \boldsymbol{\eta}_D \\ &:= \boldsymbol{\Psi}\xi + \boldsymbol{\eta}_{\rm D}\end{aligned} \tag{7.23}$$

where the frequency measurement vector is presented as a linear combination of entries from the dictionary matrix $\boldsymbol{\Psi}$ with $N_{\rm de}N_{\rm Dop}$ columns. When $N_{\rm Dop} = 1$, the dictionary representation in (7.23) degrades to the dictionary representation for the ICI-ignorant channel estimation in (7.8); although there is a slight difference caused by the residual Doppler shift $\hat{\epsilon}$, the two representations are essentially identical.

Given an identical structure of (7.8) and (7.23), the sparse estimation problem can be formulated as

$$\ell_0\text{-norm:} \qquad \hat{\xi} = \arg\min_{\xi} ||\xi||_0 \qquad \text{s.t.} \qquad ||\boldsymbol{\Psi}\xi - \mathbf{z}||^2 \le \delta \tag{7.24}$$

and its convex relaxation by ℓ_1-norm is

$$\ell_1\text{-norm :} \qquad \hat{\xi} = \arg\min_{\xi} \frac{1}{2}||\boldsymbol{\Psi}\xi - \mathbf{z}||^2 + \zeta||\xi||_1 \tag{7.25}$$

Based on the estimated $\hat{\xi}$, the (m, k)th element of the channel matrix $\mathbf{H}$ can be reconstructed as

$$\hat{H}[m,k] = \sum_{i=1}^{N_{\rm Dop}} \sum_{p=1}^{N_{\rm de}} \hat{\xi}_p^{(i)} e^{-j2\pi \frac{m}{T}\overline{\tau}_p} G\left(\frac{f_m + \hat{\epsilon}}{1 + b^{(i)}} - f_k\right). \tag{7.26}$$

Since the channel estimation error is determined by the relationship between the sparse estimate and the channel coefficients as shown in (7.26), the estimation error on far off-diagonal values of $\mathbf{H}$ will surely be much larger than the actual values. Therefore, one can reduce the channel estimation error by approximating $\mathbf{H}$ as a banded matrix with D off-diagonals on each side. How many off-diagonals to keep will depend on the estimation accuracy of $\hat{\xi}$ and on the rate with which the magnitude of the off-diagonal values of $\mathbf{H}$ decrease. The banded structure of the reconstructed channel matrix resonates the banded assumption of the channel matrix in Section 5.4.2.

7.3.3 Pilot Subcarrier Distribution

As paths with nonzero Doppler scale b_p need to be identified based on their ICI pattern, pilots on adjacent subcarriers are necessary. Conversely if one selects pilots adjacent to data symbols, the ICI from these unknown symbols will be stronger. Therefore a random pilot assignment, as would be expected from compressed sensing theory, will very likely be suboptimal due to the specific structure of the dictionary $\boldsymbol{\Psi}$. A primary investigation on pilot placement for CP-OFDM [36] shows that a systematic pilot design which places pilots in equally-spaced clusters of alternating size, outperforms the random pilot assignment and the equally-spaced pilot assignment. The reason is explained by the necessity of sufficient amount of adjacent observations for ICI estimation.

Meanwhile, in iterative receiver processing, the data symbols estimated in the previous round can serve as pilot symbols for channel estimation in the current round, so that the ICI coefficient estimation is refined as iterations go on. A detailed description of iterative sparse channel estimation will be presented in Chapter 9.

7.3.4 Influence of Data Symbols

7.3.4.1 Small ICI Depth *D*

Note that the ICI energy decreases as the subcarrier distance increases; see Figure 5.5. In a scenario with small values of ICI depth D, channel estimation can be achieved based on frequency measurements within clusters of pilot subcarriers, by treating the ICI leakage from adjacent data subcarriers as ambient noise. The frequency measurement vector $\mathbf{z}$ in (7.24) and (7.25) will be replaced by the vector $\mathbf{z}_{\mathrm{P}}$ stacked by measurements at pilot subcarriers.

7.3.4.2 Large ICI Depth *D*

In scenarios with large ICI depth D, a pre-whitening procedure can be performed to reduce the effect of unknown data symbols on channel estimation.

Assuming $\mathbf{w} \sim \mathcal{CN}(\mathbf{0}, \sigma^2\mathbf{I})$, the covariance matrix of the equivalent noise vector $\boldsymbol{\eta}_{\mathrm{D}}$ is

$$\mathrm{Cov}(\boldsymbol{\eta}_{\mathrm{D}}) = \mathbb{E}(\mathbf{H}\boldsymbol{\Sigma}_{\mathrm{D}}\mathbf{H}^{\mathrm{H}}) + \sigma^2\mathbf{I} \tag{7.27}$$

with $\boldsymbol{\Sigma}_{\mathrm{D}}$ denoting the covariance matrix of $\mathbf{s}_{\mathrm{D}}$,

$$[\boldsymbol{\Sigma}_{\mathrm{D}}]_{k,k} = \begin{cases} 1 & k \in \mathcal{S}_{\mathrm{D}} \\ 0 & k \in \mathcal{S}_{\mathrm{P}} \cup \mathcal{S}_{\mathrm{N}}. \end{cases} \tag{7.28}$$

Assume $\mathbb{E}\{\mathbf{H}\mathbf{H}^{\mathrm{H}}\} \approx \overline{\gamma}\sigma^2\mathbf{I}$ in (7.27), where $\overline{\gamma}$ is the average SNR at the receiver. The covariance matrix of $\boldsymbol{\eta}_{\mathrm{D}}$ can be approximated by a diagonal matrix

$$\mathrm{Cov}(\boldsymbol{\eta}_{\mathrm{D}}) \approx \sigma^2 \left(\overline{\gamma}\boldsymbol{\Sigma}_{\mathrm{D}} + \mathbf{I}\right). \tag{7.29}$$

A pre-whitening operation can be performed prior to the channel estimation with a diagonal pre-whitening matrix

$$[\mathbf{W}]_{k,k} = \begin{cases} \sqrt{\frac{1}{\overline{\gamma}+1}} & k \in \mathcal{S}_{\mathrm{D}} \\ 1 & k \in \mathcal{S}_{\mathrm{P}} \cup \mathcal{S}_{\mathrm{N}}. \end{cases} \tag{7.30}$$

Multiplying both sides of (7.17) with $\mathbf{W}$ leads to

$$\mathbf{z}_{\mathrm{w}} = \mathbf{W}\mathbf{H}\mathbf{s}_{\mathrm{P}} + \boldsymbol{\eta}_{\mathrm{w}} \tag{7.31}$$

where the pre-whitened noise vector $\boldsymbol{\eta}_\mathrm{w} \sim \mathcal{CN}(\mathbf{0}, \sigma^2\mathbf{I})$. Substituting (5.29) into (7.31) yields

$$\mathbf{z}_\mathrm{w} = \sum_{p=1}^{N_\mathrm{pa}} \xi_p \mathbf{W}\boldsymbol{\Lambda}(\overline{\tau}_p)\boldsymbol{\Gamma}(b_p, \hat{\epsilon})\mathbf{s}_\mathrm{P} + \boldsymbol{\eta}_\mathrm{w}. \tag{7.32}$$

The ICI-aware channel estimation will be performed based on (7.32).

7.4 Sparse Recovery Algorithms

The problem defined by the ICI-ignorant/-aware channel estimation directly falls into a research area called *compressive sensing*. A variety of recovery algorithms have been developed to obtain the sparse solution. In Appendix A, we provide a brief introduction about compressive sensing theory and several typical sparse recovery algorithms which are computationally tractable for practical applications. A current summary of compressive sensing techniques can be found in, e.g., [110].

Algorithms that reconstruct a signal taking advantage of its sparse structure have been used well before the term compressive sensing was coined. The surprising discovery is that it can be shown that several of these algorithms will – under certain conditions – render the same solution as the combinatorial approach. These conditions largely amount to tighter constraints on the sparsity of $\boldsymbol{\xi}$ beyond identifiability. We briefly introduce the two main types of algorithms.

7.4.1 Matching Pursuit

In this type of approach the combinatorial problem is circumvented by heuristically choosing which values of $\boldsymbol{\xi}$ are nonzero and solving the resulting constrained least-squares problem. The most popular algorithms of this type are greedy algorithms, like Matching Pursuit (MP) or Orthogonal Matching Pursuit (OMP), that identify the nonzero elements of $\boldsymbol{\xi}$ in an iterative fashion.

A short algorithmic description of OMP would be:

1. Initialize the set of nonzero elements $\boldsymbol{\Omega}_0$ as empty, the observation residual as, $\mathbf{r} = \mathbf{z}$, and the iterative index $i = 0$.
2. Correlate all columns of $\boldsymbol{\Psi}$ with the residual $\mathbf{r}$, viz., $\boldsymbol{\Psi}^\mathrm{H}\mathbf{r}$, choose the largest element by magnitude and add its index p_i to the set of nonzero elements $\boldsymbol{\Omega}_i = \boldsymbol{\Omega}_{i-1} \cup \{p_i\}$.
3. With the constraint that only elements of $\boldsymbol{\xi}$ with indices in $\boldsymbol{\Omega}_i$ are nonzero, find an estimate $\hat{\boldsymbol{\xi}}_i$ that minimizes $||\mathbf{z} - \boldsymbol{\Psi}\hat{\boldsymbol{\xi}}||^2$.
4. Update the residual as $\mathbf{r} = \mathbf{z} - \boldsymbol{\Psi}\hat{\boldsymbol{\xi}}_i$.
5. Repeat steps (2)-(4) until either a known degree of sparsity s is reached or the norm of the residual $||\mathbf{r}||^2$ falls below a predetermined threshold.

This type of algorithm has been popular mainly because it can be easily implemented and has low computational complexity. However, it could fail to find the sparsest solution in certain scenarios.

7.4.2 ℓ_1-Norm Minimization

Although the ℓ_1-norm optimization problem in (7.13) and (7.25) has been used in various applications to promote sparse solutions in the past (see references in [56]), it is now largely popular under the name Basis Pursuit (BP), as introduced in [73]. While originally the term BP was used to designate the case of noiseless measurements and the qualifier Basis Pursuit De-Noising (BPDN) to refer to the case of noisy measurements [73], we will generally refer to both cases simply by BP. Under appropriate parameterization, BP can be shown to be equivalent to the LASSO algorithm which is well-known in statistics [385]; see Appendix A.2.2.

All these algorithms have in common that they lead to convex optimization problems, which can be solved efficiently using existing computational techniques. Here we consider the projected gradient methods. A framework of algorithms in this category is as follows.

(1) *Initialization.* Set iteration index $i = 0$, and an initial estimate ξ_0.
(2) *Solution for a separable approximation*
- Compute the observation residual

$$\mathbf{r} = \mathbf{z} - \mathbf{\Psi}\xi_i. \tag{7.33}$$

- Choose α_i, and compute ξ_i^+ in a separable approximation problem

$$\xi_i^+ := \arg\min_{\mathbf{u}} (\xi_i - \mathbf{u})^{\mathrm{H}}\mathbf{\Psi}^{\mathrm{H}}\mathbf{r} + \frac{1}{2}\alpha_i\|\xi_i - \mathbf{u}\|_2^2 + \zeta\|\mathbf{u}\|_1. \tag{7.34}$$

For complex valued $\mathbf{z}$, $\mathbf{\Psi}$ and ξ, the problem in (7.34) can be rewritten as

$$\xi_i^+ := \arg\min_{\mathbf{u}} \frac{1}{2}(\boldsymbol{\eta}_i - \mathbf{u})^{\mathrm{H}}(\boldsymbol{\eta}_i - \mathbf{u}) + \frac{\zeta}{\alpha_i}\|\mathbf{u}\|_1, \tag{7.35}$$

with

$$\boldsymbol{\eta}_i = \xi_i - \frac{1}{\alpha_i}\mathbf{\Psi}^{\mathrm{H}}\mathbf{r}. \tag{7.36}$$

For (7.35), an elementwise minimizer [435] is given by

$$\xi_{i,p}^+ = \mathrm{soft}\left(\eta_{i,p}, \frac{\zeta}{\alpha_i}\right) \tag{7.37}$$

where $\xi_{i,p}^+$ denotes the pth element of ξ_i^+, and the soft-threshold function as shown in Figure 7.4, is defined as

$$\mathrm{soft}(a, b) = \frac{\max\{|a| - b, 0\}}{\max\{|a| - b, 0\} + b}a. \tag{7.38}$$

If ξ_i^+ cannot satisfy certain acceptance criterion, such as yielding a decrease in the objective function, increase α_i by a factor $\eta > 1$, i.e., $\alpha_i = \eta\alpha_i$, and repeat this step; see [435] for discussions on the acceptance criterion.

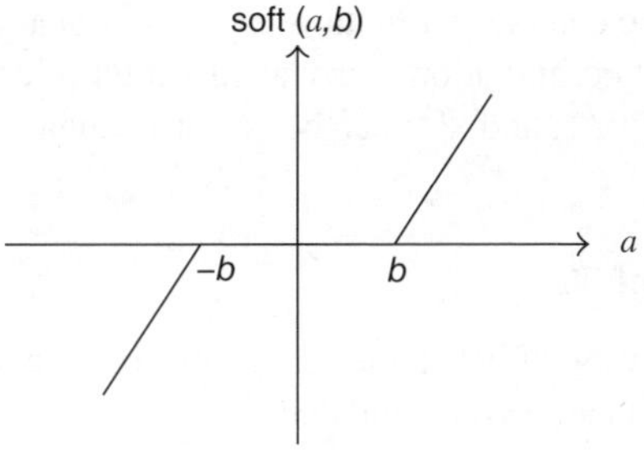

Figure 7.4 Illustration of the soft-threshold function.

(3) *Update estimate.* The estimate in the $(i+1)$th iteration is obtained via

$$\xi_{i+1} = \xi_i + \gamma_i(\xi_i^+ - \xi_i) \tag{7.39}$$

where $\gamma_i \in (0, 1]$ specifies the step length.

(4) *Repeat.* Set $i = i + 1$, and repeat steps (2)-(3) until the stopping criterion holds.

The convergence rate of the ℓ_1-norm minimization algorithms highly depends on the initial estimate ξ_0. A *warm-start* method is investigated in [435]. For channel estimation in OFDM block transmissions, channel estimates in preceding blocks can serve as initial estimate of the channel in the current block. We defer our discussion on the block-to-block receiver processing to Chapter 10.

Due to the ℓ_1-norm relaxation, the solution will not be exactly sparse, but will exhibit numerous small values that do not contribute significantly to the estimation error. If an exactly sparse solution is sought, an additional thresholding or debiasing stage [435] can remove the small components. Specifically in the debiasing stage, components of small amplitude in $\hat{\xi}$ are set as zero, and indices of large components are taken as locations of effective channel paths. With this information, the coefficient matrix in (7.7) or (7.23) can then be reconstructed. The debiased solution is obtained as the LS solution.

Multiple variants under the above framework exist; see Section A.2. In this book, we take the SpaRSA algorithm developed in [435] as an example to illustrate the application of compressive sensing for underwater acoustic channel estimation. Other sparse recovery algorithms can be applied with slight modifications.

7.4.3 Matrix-Vector Multiplication via FFT

To estimate the sparse vectors in (7.8) and (7.23), the sparse recovery algorithms require frequent computation of the multiplication between the dictionary matrix $\mathbf{\Psi}$ and a tentative solution $\hat{\xi}$, and of the multiplication between $\mathbf{\Psi}^{\mathrm{H}}$ and the measurement residual vector $\mathbf{r}$ (see, e.g., the OMP algorithm). Efficient implementation methods are vital in the practical receiver design.

For the ICI-ignorant sparse channel estimation, as the dictionary matrix $\mathbf{\Psi}$ in (7.8) itself is a partial FFT matrix, efficient computation is trivial–the multiplication $\mathbf{\Psi}\hat{\xi}$ can be simply performed via a $\lambda_b K$-point FFT of $\hat{\xi}$, and $\mathbf{\Psi}^T \mathbf{r}$ can be performed via a $\lambda_b K$-point inverse FFT of $\hat{\xi}$.

7.4.3.1 Implementation of $\mathbf{\Psi}\hat{\xi}$

To seek an efficient computation of $\mathbf{\Psi}\hat{\xi}$ in the ICI-aware sparse channel estimation, the dictionary representation in (7.23) can be reformulated as

$$\Psi\hat{\xi} = \sum_{i=1}^{N_{\rm Dop}} \left(\mathbf{\Gamma}(b^{(i)}, \hat{\epsilon})\mathbf{s}_{\rm P}\right) \circ \left(\mathbf{\Phi}_{\lambda_b}\xi^{(i)}\right) \tag{7.40}$$

where $\circ$ denotes an element-wise product between two matrices, which is called *Hadamard product*, and $\mathbf{\Phi}_{\lambda_b}$ is a $K \times N_{\rm de}$ partial FFT matrix defined as

$$\mathbf{\Phi}_{\lambda_b} = \begin{bmatrix} 1 & \cdots & e^{j2\pi\frac{(p-1)}{2\lambda_b}} & \cdots & e^{j2\pi\frac{N_{\rm de}-1}{2\lambda_b}} \\ \vdots & \ddots & \vdots & \ddots & \vdots \\ 1 & \cdots & e^{-j2\pi\frac{m(p-1)}{\lambda_b K}} & \cdots & e^{-j2\pi\frac{m(N_{\rm de}-1)}{\lambda_b K}} \\ \vdots & \ddots & \vdots & \ddots & \vdots \\ 1 & \cdots & e^{-j2\pi\left(\frac{K}{2}-1\right)\frac{(p-1)}{\lambda_b K}} & \cdots & e^{-j2\pi\left(\frac{K}{2}-1\right)\frac{(N_{\rm de}-1)}{\lambda_b K}} \end{bmatrix}. \tag{7.41}$$

The computational load therefore can be saved by performing the multiplication between $\mathbf{\Phi}_{\lambda_b}$ and a tentative solution of $\hat{\xi}^{(i)}$ with a $\lambda_b K$-point FFT. Meanwhile, noting that the multiplication $\mathbf{\Gamma}(b^{(i)}, \hat{\epsilon})\mathbf{s}_P$ is independent of the tentative delay $\overline{\tau}_p$, the product matrix can be pre-computed prior to executing the solution-finding algorithm. To further reduce the computational load, one can retain only D off-diagonals on the template $\mathbf{\Gamma}(b^{(i)}, \hat{\epsilon})$, by assuming a limited ICI leakage from D directly neighboring subcarriers on each side.

7.4.3.2 Implementation of $\mathbf{\Psi}^{\rm H}\mathbf{r}$

Based on the dictionary representation in (7.23), the multiplication can be reformulated as

$$\mathbf{\Psi}^{\rm H}\mathbf{r} = \left[\mathbf{\Psi}^{(1)} \ \cdots \ \mathbf{\Psi}^{(N_b)}\right]^{\rm H}\mathbf{r} \tag{7.42}$$

where the submatrix $\mathbf{\Psi}^{(i)}$ of size $K \times N_{\rm de}$ is defined as

$$\mathbf{\Psi}^{(i)} = \left[\mathbf{\Lambda}(\overline{\tau}_1)\mathbf{\Gamma}(b^{(i)}, \hat{\epsilon})\mathbf{s}_{\rm P} \ \cdots \ \mathbf{\Lambda}(\overline{\tau}_{N_{\rm de}})\mathbf{\Gamma}(b^{(i)}, \hat{\epsilon})\mathbf{s}_{\rm P}\right]. \tag{7.43}$$

The multiplication $(\mathbf{\Psi}^{(i)})^{\rm H}\mathbf{r}$ can be effectively implemented as

$$\left(\mathbf{\Psi}^{(i)}\right)^{\rm H}\mathbf{r} = \mathbf{\Phi}_{\lambda_b}^{\rm H}\left[\left(\mathbf{\Lambda}(b^{(i)}, \hat{\epsilon})\mathbf{s}_{\rm P}\right)^* \circ \mathbf{r}\right] \tag{7.44}$$

where $(\cdot)^*$ denote the conjugate operation. Based on (7.44), the matrix-vector multiplication can be performed via a $\lambda_b K$-point inverse FFT after an element-wise multiplication between the pre-computed vector $(\mathbf{\Lambda}(b^{(i)}, \hat{\epsilon})\mathbf{s}_{\rm P})^*$ and the residual vector $\mathbf{r}$. Given $N_{\rm Dop}$ tentative Doppler values, $N_{\rm Dop}$ inverse FFTs are necessary to obtain the multiplication $\mathbf{\Psi}^{\rm H}\mathbf{r}$.

7.4.4 Computational Complexity

From (7.40), one can see that the channel estimation complexity is roughly linear with the number of representative Doppler values $N_{\rm Dop}$. The study in [171] shows that the average runtime to solve the estimation problem in (7.40) with $N_{\rm Dop} = 15$ tentative Doppler values is about 30 times larger than that with $N_{\rm Dop} = 1$, indicating that even being efficiently implemented, the ICI-aware channel estimation still requires a much higher computational load that its ICI-ignorant counterpart. In Chapter 9, we will look into a practical receiver design with a progressive ICI coefficient estimation, which has a robust performance in time-varying channels while maintaining a relative low computational complexity.

7.5 Extension to Multi-Input Channels

For a multi-input channel with $N_{\rm t}$ transmitters, the channel at each receiving element consists of $N_{\rm t}$ individual subchannels. Channel estimation at each receiving element hence involves recovering $N_{\rm t}$ sets of channel parameters. In spite of the difference in the problem size, techniques for the multi-input channel estimation are essentially the same as those used for the single-input channel estimation.

Focusing on one particular receiving element, a brief discussion will be provided next on how to extend the sparse estimation method developed for the single-input channel to the multi-input channel.

7.5.1 ICI-Ignorant Sparse Channel Sensing

Consider a multi-input multi-output (MIMO) channel with $N_{\rm t}$ transmitters and $N_{\rm r}$ receivers. The system model at the νth receiving element derived in (5.60) is repeated as

$$z_\nu[m] = \sum_{\mu=1}^{N_{\rm t}} H_{\nu,\mu}[m] s_\mu[m] + \eta_\nu[m] \tag{7.45}$$

for $m = -K/2, \cdots, K/2 - 1$, where the channel coefficient is represented by path parameters as

$$H_{\nu,\mu}[m] = \sum_{p=1}^{N_{{\rm pa},\nu,\mu}} \xi_{\nu,\mu,p} e^{-j2\pi \frac{m}{T} \bar{\tau}_{\nu,\mu,p}}. \tag{7.46}$$

Following the same spirit of sparse channel estimation in the ICI-ignorant receiver for single-input channels, the path delays between the μth transmitter and the νth receiving element are quantized into a set of representative values,

$$\bar{\tau}_{\nu,\mu} \in \left\{ 0, \frac{1}{\lambda_{\rm b} B}, \frac{2}{\lambda_{\rm b} B}, \cdots, \frac{N_{\rm de} - 1}{\lambda_{\rm b} B} \right\} \tag{7.47}$$

for $\mu = 1, \cdots, N_{\rm t}$.

The dictionary representation of the frequency measurement vector on pilot subcarriers at the νth receiving element is formulated as

$$\mathbf{z}_{\nu,{\rm P}} = \sum_{\mu=1}^{N_{\rm t}} \mathbf{\Psi}_{\nu,\mu} \boldsymbol{\xi}_{\nu,\mu} + \boldsymbol{\eta}_{\nu,{\rm P}}, \tag{7.48}$$

where $\mathbf{\Psi}_{\nu,\mu}$ and $\boldsymbol{\xi}_{\nu,\mu}$ are similarly defined as $\mathbf{\Psi}$ and $\boldsymbol{\xi}$ in (7.7), respectively.

Define a dictionary matrix $\boldsymbol{\Psi}_\nu$ and a vector ξ_ν as

$$\boldsymbol{\Psi}_\nu = \left[\boldsymbol{\Psi}_{\nu,1} \cdots \boldsymbol{\Psi}_{\nu,N_t}\right], \qquad \xi_\nu = \left[\xi_{\nu,1}^T \cdots \xi_{\nu,N_t}^T\right]^T. \tag{7.49}$$

We have (7.48) rewritten as

$$\mathbf{z}_{\nu,P} = \boldsymbol{\Psi}_\nu \xi_\nu + \boldsymbol{\eta}_\nu, \tag{7.50}$$

which shares an identical form as (7.8) except that the size of ξ_ν is N_t times of that in (7.8). The sparse recovery algorithms discussed in Section 7.4 therefore can be applied.

With the estimate $\hat{\xi}_\nu$, the channel coefficient $H_{\nu,\mu}[m]$ can be reconstructed as

$$\hat{H}_{\nu,\mu}[m] = \sum_{p=0}^{N_{de}-1} \hat{\xi}_{\nu,\mu,p} e^{-j2\pi \frac{mp}{\lambda_b K}} \tag{7.51}$$

where $\hat{\xi}_{\nu,\mu,p}$ is the $(p+1)$th element in $\hat{\xi}_{\nu,\mu}$.

Relative to the single-input channel, more frequency measurements are necessary to estimate the N_t unknown subchannels simultaneously. In practical systems, one design example is to divide the total pilot subcarriers into N_t nonoverlapping sets of equal size, each set corresponding to one transmitter. The numerical results in Figure 7.3 shows the average mean square channel estimation error with different sizes of the pilot set for each transmitter and receiver pair.

7.5.2 ICI-Aware Sparse Channel Sensing

The input–output relationship in the MIMO channel derived in (5.41) is repeated as

$$\mathbf{z}_\nu = \sum_{\mu=1}^{N_t} \mathbf{H}_{\nu,\mu} \mathbf{s}_\mu + \mathbf{w}_\nu, \tag{7.52}$$

with

$$\mathbf{H}_{\nu,\mu} = \sum_{p=1}^{N_{pa,\nu,\mu}} \xi_{\nu,\mu,p} \boldsymbol{\Lambda}(\overline{\tau}_{\nu,\mu,p}) \boldsymbol{\Gamma}(b_{\nu,\mu,p}, \hat{\epsilon}_\nu). \tag{7.53}$$

Similarly to the two symbol vectors $\mathbf{s}_P$ and $\mathbf{s}_D$ defined in (7.15) for the single-input channel, two symbol vectors $\mathbf{s}_{\mu,P}$ and $\mathbf{s}_{\mu,D}$ are defined for the symbols from the μth transmitter. The input–output relationship in (7.52) is rewritten as

$$\mathbf{z}_\nu = \sum_{\mu=1}^{N_t} \mathbf{H}_{\nu,\mu} \mathbf{s}_{\mu,P} + \sum_{\mu=1}^{N_t} \mathbf{H}_{\nu,\mu} \mathbf{s}_{\mu,D} + \mathbf{w}_\nu, \tag{7.54}$$

$$= \sum_{\mu=1}^{N_t} \mathbf{H}_{\nu,\mu} \mathbf{s}_{\mu,P} + \boldsymbol{\eta}_{\nu,D}, \tag{7.55}$$

where $\boldsymbol{\eta}_{\nu,D}$ is the equivalent noise including frequency components from unknown data symbols and ambient noise. Substituting (7.53) into (7.55) yields

$$\mathbf{z}_\nu = \sum_{\mu=1}^{N_t} \sum_{p=1}^{N_{pa,\nu,\mu}} \xi_{\nu,\mu,p} \boldsymbol{\Lambda}(\overline{\tau}_{\nu,\mu,p}) \boldsymbol{\Gamma}(b_{\nu,\mu,p}, \hat{\epsilon}_\nu) \mathbf{s}_{\mu,P} + \boldsymbol{\eta}_{\nu,D}. \tag{7.56}$$

A sparse estimation of path parameters in the N_t-th subchannel can be carried out based on (7.56).

To linearize the channel estimation input–output relationship in (7.56), define a representative set of path parameters for the μth subchannel,

$$\overline{\tau}_{\nu,\mu} \in \left\{0, \frac{1}{\lambda_b B}, \frac{2}{\lambda_b B}, \cdots, \frac{N_{de}-1}{\lambda_b B}\right\}, \tag{7.57}$$

$$b_{\nu,\mu} \in \left\{-b_{\nu,\mu,\max}, -b_{\nu,\mu,\max} + \Delta b_{\nu,\mu}, \cdots, b_{\nu,\mu,\max}\right\}, \tag{7.58}$$

where $b_{\nu,\mu,\max}$ and $\Delta b_{\nu,\mu}$ specify the Doppler searching range and resolution of the μth subchannel, respectively, leading to a total of $N_{\nu,\mu,\text{Dop}}$ tentative Doppler scale values. For convenience, here an identical representative set of delays is assumed for all subchannels; one can easily extend our discussion to the settings that different subchannels use different path delay searching ranges and resolutions.

Corresponding to the $\left(\sum_{\mu=1}^{N_t} N_{de} N_{\nu,\mu,\text{Dop}}\right)$ representative paths, define two coefficient vectors in order

$$\xi_{\nu,\mu}^{(i)} = \left[\xi_{\nu,\mu,1}^{(i)} \cdots \xi_{\nu,\mu,N_{de}}^{(i)}\right]^{T} \tag{7.59}$$

$$\xi_{\nu,\mu} = \left[(\xi_{\nu,\mu}^{(1)})^{T} \cdots \left(\xi_{\mu}^{(N_{\nu,\mu,\text{Dop}})}\right)^{T}\right]^{T} \tag{7.60}$$

for $i = 1, \cdots, N_{\nu,\mu,\text{Dop}}$ and $\mu = 1, \cdots, N_t$, where $\xi_{\nu,\mu}$ consists of combination coefficients of the μth subchannel over $N_{de} N_{\nu,\mu,\text{Dop}}$ entries. Stacking all the $\xi_{\nu,\mu}$ into a long vector yields

$$\xi_{\nu} = \left[\xi_{\nu,1}^{T} \cdots \xi_{\nu,N_t}^{T}\right]^{T}. \tag{7.61}$$

A dictionary representation of the frequency measurement vector in (7.56) follows as

$$\mathbf{z}_{\nu} = \sum_{\mu=1}^{N_t} \sum_{i=1}^{N_{\nu,\mu,\text{Dop}}} \sum_{p=1}^{N_{de}} \xi_{\nu,\mu,p}^{(i)} \mathbf{\Lambda}(\overline{\tau}_{\nu,\mu,p}) \mathbf{\Gamma}(b_{\nu,\mu}^{(i)}, \hat{\epsilon}_{\nu}) \mathbf{s}_{\mu,\text{P}} + \boldsymbol{\eta}_{\nu,\text{D}} \tag{7.62}$$

$$:= \mathbf{\Psi}\xi + \boldsymbol{\eta}_{\text{D}} \tag{7.63}$$

where the dictionary matrix $\mathbf{\Psi}$ has $\left(\sum_{\mu=1}^{N_t} N_{de} N_{\nu,\mu,\text{Dop}}\right)$ columns in total. Techniques for sparse channel estimation discussed in Section 7.3.2 can be directly applied.

With the estimate $\hat{\xi}_{\nu}$, the channel matrix of the μth subchannel can be reconstructed as

$$\hat{H}_{\nu,\mu}[m,k] = \sum_{i=1}^{N_{\nu,\mu,\text{Dop}}} \sum_{p=1}^{N_{de}} \hat{\xi}_{\nu,\mu,p}^{(i)} e^{-j2\pi \frac{m}{T} \overline{\tau}_p} G\left(\frac{f_m + \hat{\epsilon}_{\nu}}{1 + b_{\nu,\mu}^{(i)}} - f_k\right) \tag{7.64}$$

For an efficient implementation, (7.62) can be reformulated as

$$\mathbf{z}_{\nu} = \sum_{\mu=1}^{N_t} \sum_{i=1}^{N_{\nu,\mu,\text{Dop}}} \left(\mathbf{\Gamma}(b_{\nu,\mu}^{(i)}, \hat{\epsilon}_{\nu}) \mathbf{s}_{\nu,\mu,\text{P}}\right) \circ \left(\mathbf{\Phi}_{\lambda_b} \xi_{\nu,\mu}^{(i)}\right) + \boldsymbol{\eta}_{\nu,\text{D}} \tag{7.65}$$

which shows that $\left(\sum_{\mu=1}^{N_t} N_{\nu,\mu,\text{Dop}}\right)$ FFTs of $\lambda_b K$ points are required for each multiplication between the dictionary matrix $\mathbf{\Psi}_{\nu}$ and a tentative estimate $\hat{\xi}_{\nu}$.

7.6 Noise Variance Estimation

Following the OFDM signal design in Section 2.3, the frequency measurements at null subcarriers can be used for noise variance estimation. In the ICI-ignorant receiver processing, the ICI is taken as the additive noise, hence the equivalent noise includes both ICI and ambient noise. The noise variance can be estimated as

$$\hat{\sigma}_{\eta}^{2} = \frac{1}{|S_{\mathrm{N}}|} \sum_{m \in S_N} |z[m]|^2, \tag{7.66}$$

where $|S_{\mathrm{N}}|$ is the number of null subcarriers. The SNR in the frequency domain can be estimated as

$$\hat{\gamma} = \frac{\underset{m \in S_{\mathrm{A}}}{\mathbb{E}} \left[|z[m]|^2 \right]}{\underset{m \in S_{\mathrm{N}}}{\mathbb{E}} \left[|z[m]|^2 \right]} - 1. \tag{7.67}$$

In the ICI-aware receiver processing with a limited ICI-leakage assumption of depth D, the equivalent noise consists of both residual ICI and ambient noise. Hence, based on the frequency measurements at null subcarriers, the noise variance can be estimated as

$$\hat{\sigma}_{\eta}^{2} = \mathbb{E} \sum_{m \in \overline{S}_{\mathrm{N}}} \left[\left| z[m] - \sum_{k=m-D}^{m+D} \hat{H}[m,k]\hat{s}[k] \right|^2 \right], \tag{7.68}$$

where $\overline{S}_{\mathrm{N}}$ denotes the set of null subcarriers within the signal band, and $\hat{H}[m,k]$ and $\hat{s}[k]$ denote the estimated ICI coefficient and transmitted symbol, respectively. The SNR in the frequency domain can be estimated as

$$\hat{\gamma} = \frac{\underset{m \in S_{\mathrm{A}}}{\mathbb{E}} \left[\left| \sum_{k=m-D}^{m+D} \hat{H}[m,k]\hat{s}[k] \right|^2 \right]}{\underset{m \in \overline{S}_{\mathrm{N}}}{\mathbb{E}} \left[\left| z[m] - \sum_{k=m-D}^{m+D} \hat{H}[m,k]\hat{s}[k] \right|^2 \right]} - 1. \tag{7.69}$$

In the MIMO channel with N_{t} transmitters and N_{r} receivers, the noise variance at the νth receiver is estimated as

$$\hat{\sigma}_{\eta,\nu}^{2} = \underset{m \in \overline{S}_{\mathrm{N}}}{\mathbb{E}} \left[| z_{\nu}[m] - \sum_{\mu=1}^{N_t} \sum_{k=m-D}^{k=m+D} \hat{H}_{\nu,\mu}[m,k]\hat{s}_{\mu}[k]|^2 \right], \tag{7.70}$$

where $\hat{H}_{\nu,\mu}[m,k]$ and $\hat{s}_{\mu}[k]$ denote the estimated ICI coefficient and transmitted symbol, respectively.

7.7 Noise Prewhitening

In the ICI-ignorant/aware receiver processing, the unmodelled ICI and the ambient noise are taken as one equivalent noise, with the equivalent noise at the mth subcarrier denoted as $\eta[m]$

in (5.47) for the ICI-ignorant processing and in (5.56) for the ICI-aware processing with a limited ICI-leakage assumption. As for the signal itself, the variance of the residual ICI is frequency-dependent as (i) the transmitter often has a nonideal transmit voltage response (TVR), and (ii) underwater acoustic propagation introduces frequency dependent attenuation [266]. Furthermore, the ambient noise in underwater acoustic environments makes the equivalent noise more colored [266], especially in the presence of ambient interference, such as shrimp noise. Accordingly noise pre-whitening procedure is beneficial to facilitate receiver processing [34].

7.7.1 Noise Spectrum Estimation

Consider a simple method to estimate the variance $\sigma_\eta^2[m] = \mathbb{E}\{|\eta[m]|^2\}$ of the ICI-plus-noise across all subcarriers. It assumes that the spectrum is generally smooth, and can be approximated by a Qth-order polynomial $\mathcal{P}_Q(m) = \sum_{q=0}^{Q} p_q m^q$, either in the linear domain

$$\sigma_\eta^2[m] = P_Q(m), \tag{7.71}$$

or in the log domain

$$\sigma_\eta^2[m] = 10^{\mathcal{P}_Q(m)/10}, \tag{7.72}$$

where $m = -K/2, \ldots, K/2 - 1$. The parameter p_0 represents the noise variance at the center frequency. The white noise model is included as a special case with $Q = 0$.

The measurements on the null subcarriers are used to estimate the model parameters. Two methods are proposed next based on the log-domain approximation in (7.72).

- *Linear regression in log-domain*: A simple linear regression model in the log-domain can be formulated as

$$\{\mathcal{P}_Q\}_{\rm LR} = \arg\min_{\{p_q\}} \sum_{m\in\mathcal{S}_{\rm N}} \left| 10\log_{10}|z[m]|^2 - \sum_{q=0}^{Q} p_q m^q \right|^2. \tag{7.73}$$

  This method is of very low complexity. However, fitting in the log-domain tends to lead to a negative bias on p_0 (i.e., underestimating the noise variance), as small values are amplified in the log-domain. A simple remedy is to apply some smoothing on the observations $|z_m|^2$ before transforming to the log domain.
- *Maximum likelihood based variance estimator*: By the central limit theorem, $\eta[m]$ can be viewed to have a Gaussian distribution. Hence, $|\eta[m]|^2$ is exponentially distributed. The maximum likelihood (ML) solution for the model parameters can be formulated as

$$\{\mathcal{P}_Q\}_{\rm ML} = \arg\max_{\{p_q\}} \sum_{m\in\mathcal{S}_{\rm N}} - \left[\ln\left(10^{\mathcal{P}_Q(m)/10}\right) + \frac{|z[m]|^2}{10^{\mathcal{P}_Q(m)/10}}\right]. \tag{7.74}$$

  For any large Q the complexity of an exhaustive search quickly becomes prohibitive. The ML approach is therefore mainly used for the two dimensional problem, i.e., $Q = 1$.

Furthermore to keep the complexity low, a multi-grid search can be applied or a final solution can be improved via simple interpolation techniques.

7.7.2 *Whitening in the Frequency Domain*

Once the variance of the equivalent noise has been estimated, the data can be easily whitened as

$$z_{\mathrm{w}}[m] = \frac{z[m]}{\hat{\sigma}_{\eta}[m]}, \qquad m = -\frac{K}{2}, \cdots, \frac{K}{2} - 1 \tag{7.75}$$

based on which the receiver processing algorithms, such as the equalization algorithms in Chapter 8, and the sum-product algorithm for LDPC decoding, which requires the statistical description of system noise, can be applied.

7.8 Bibliographical Notes

Estimation of the linear time-invariant frequency selective radio channel has been extensively based on, e.g., subspace fitting [436], model order fitting using a generalized Akaike information criterion [326], zero-tap detection [60], or Monte Carlo Markov Chain methods [322]. More recently, advances in the field of compressive sensing [24, 53, 55, 100] have led to extensive investigations on sparse channel estimation, e.g., sparseness in delay only [18, 59, 88, 130, 204, 303, 339, 437] and very recently for sparseness in delay and Doppler [19, 378]. Specifically on UWA channels, the matching pursuit (MP) algorithm and its variants have been used both in [188, 240] for a single carrier system and in [201] for a multicarrier system. A geometric mixed-norm approach has been developed in [152] for sparse channel estimation and tracking, and an iterative adaptive approach (IAA) have been proposed in [251] for sparse channel estimation in single-carrier transmissions where the Bayesian information criterion (BIC) is adopted to determine the number of dominant channel taps.

In ICI-ignorant OFDM receiver processing, the channel estimation problem can also be linked to the direction finding problem from the array processing literature [403]. Particularly, applications of the direction finding algorithms, such as root-MUSIC and ESPRIT, have been explored in [38]. In the time-varying scenario, besides the sparse channel estimator presented in this chapter, another known approach to estimating this class of channel is the use of a basis expansion model (BEM) to reflect the time-varying nature of the UWA channel, see e.g., [185, 232, 320]. Estimation of the sparse channel with high dimensionality has been studied in [442].

8

Data Detection

This chapter presents the data detection (a.k.a channel equalization) module of the OFDM receivers to be examined in later chapters. Due to the existence of channel coding in practical systems, the data detection module will compute the soft information as needed by the channel decoder. In a noniterative receiver as shown in Figure 8.1(a), the data detection module computes soft information based on the collected measurements only, and passes it to the channel decoder. In an iterative receiver as shown in Figure 8.1(b), the data detection module is coupled with the channel decoder with information exchange between the two modules, where the soft information from the channel decoder is used as prior information of data symbols in data detection. Such an iterative data detection and channel decoding is refereed to as *turbo equalization*.

In the context of OFDM receiver design for underwater acoustic channels, data detection is often carried out on each OFDM block. With a generic notation, denote **s** as the vector containing K information symbols transmitted in one OFDM block, and **z** as the vector collecting all relevant measurements at the receiver. The exact relationship between **z** and **s** will be specified later on based on different assumptions on the underlying channel. Aligned with the nonbinary LDPC codes as discussed in Chapter 3, the soft information will be in the form of the log-likelihood-ratios (LLRs) of each symbol. Assume that each symbol $s[k]$ is taken from a size-M constellation $\{\alpha_0, \ldots, \alpha_{M-1}\}$; see Figures 3.2 and 3.3 for widely used symbol constellations. The channel decoder provides prior knowledge about the symbol $s[k]$ based on the reliability information of other coded symbols exploiting the code structure. The logarithm of an *a priori* probability ratio is defined as

$$L_i^{\mathrm{apr}}[k] := \ln \frac{\Pr(s[k] = \alpha_i)}{\Pr(s[k] = \alpha_0)}, \tag{8.1}$$

and a corresponding vector is defined as

$$\mathbf{L}^{\mathrm{apr}}[k] = \left[L_0^{\mathrm{apr}}[k]\ L_1^{\mathrm{apr}}[k]\ \ldots\ L_{M-1}^{\mathrm{apr}}[k]\right]^{\mathrm{T}}. \tag{8.2}$$

OFDM for Underwater Acoustic Communications, First Edition. Shengli Zhou and Zhaohui Wang.
© 2014 John Wiley & Sons, Ltd. Published 2014 by John Wiley & Sons, Ltd.

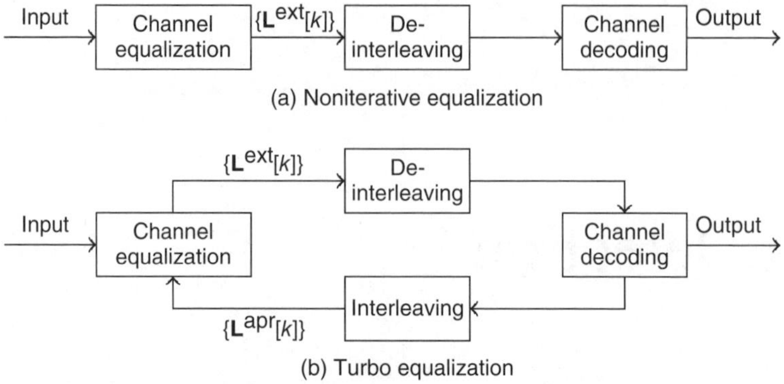

Figure 8.1 Illustration of noniterative and turbo equalization, $\{\mathbf{L}^{\text{ext}}[k]\}$ denotes the extrinsic information from the channel equalizer. $\{\mathbf{L}^{\text{apr}}[k]\}$ represents the extrinsic information from channel decoder, which is used as the *a priori* information for channel equalization.

Now define the *a posteriori* probability (APP) of the symbol $s[k]$ being α_i, given all the measurements and the prior information of all the symbols, as $\Pr(s[k] = \alpha_i | \mathbf{z}, \{\mathbf{L}^{\text{apr}}[k]\})$. The logarithm of an *a posteriori* probability ratio is

$$L_i^{\text{app}}[k] := \ln \frac{\Pr(s[k] = \alpha_i | \mathbf{z}, \{\mathbf{L}^{\text{apr}}[k]\})}{\Pr(s[k] = \alpha_0 | \mathbf{z}, \{\mathbf{L}^{\text{apr}}[k]\})} \tag{8.3}$$

and a corresponding vector is

$$\mathbf{L}^{\text{app}}[k] = \left[L_0^{\text{app}}[k] \ L_1^{\text{app}}[k] \ \ldots \ L_{M-1}^{\text{app}}[k]\right]^{\text{T}}. \tag{8.4}$$

When the data detector passes soft information back to the channel decoder, the contribution from the channel decoder should be deducted. The so called extrinsic LLR, as needed by the channel decoder, is defined as

$$L_i^{\text{ext}}[k] := \ln \frac{\Pr(s[k] = \alpha_i | \mathbf{z}, \{\mathbf{L}^{\text{apr}}[m]\}_{m \neq k})}{\Pr(s[k] = \alpha_0 | \mathbf{z}, \{\mathbf{L}^{\text{apr}}[m]\}_{m \neq k})} \tag{8.5}$$

The corresponding vector

$$\mathbf{L}^{\text{ext}}[k] = \left[L_0^{\text{ext}}[k] \ L_1^{\text{ext}}[k] \ \ldots \ L_{M-1}^{\text{ext}}[k]\right]^{\text{T}} \tag{8.6}$$

is related to $\mathbf{L}^{\text{app}}[k]$ and $\mathbf{L}^{\text{apr}}[k]$ through

$$\mathbf{L}^{\text{ext}}[k] = \mathbf{L}^{\text{app}}[k] - \mathbf{L}^{\text{apr}}[k]. \tag{8.7}$$

The collection of $\{\mathbf{L}^{\text{ext}}[k]\}_{\forall k}$ is passed to the channel decoder for decoding. In the case of noniterative receiver, or, in the initial round of the turbo equalization, there is no prior information, hence $L_i^{\text{apr}}[k] = 0$, and the extrinsic information $L_i^{\text{ext}}[k]$ is equivalent to the *a posteriori* information $L_i^{\text{app}}[k]$.

How difficult it is to compute the extrinsic LLRs depends on the system model which links the measurements in $\mathbf{z}$ with the information symbols in $\mathbf{s}$.

- In Section 8.1, we consider the detection algorithms for ICI-ignorant OFDM receivers with a single data stream. Since ICI is ignored, data detection is performed on each OFDM subcarrier individually. The soft information of each symbol can be easily computed and fed into the channel decoder.
- In Section 8.2, we consider the ICI-aware receiver with full ICI among OFDM subcarriers. We present two block detectors, the maximum *a posteriori* (MAP) detector and the soft-input soft-output linear minimum mean squared error (MMSE) detector.
- In Section 8.3, we consider the ICI-aware receiver where the ICI is limited to only close neighbors. The ICI mitigation problem is then viewed as equivalent to the intersymbol interference (ISI) equalization problem in the presence of time-varying channels. A trellis-structure based equalization method and a factor-graph based equalization method are presented respectively.
- In Section 8.4, we consider data detection in an OFDM system with multiple transmitters. Depending on the assumptions on the ICI, the receivers in Sections 8.2 and 8.3 are extended to the multiple-transmitter case.
- In Section 8.5, the Markov Chain Monte Carlo (MCMC) method is adopted for data detection for the OFDM system with multiple transmitters. Both ICI-ignorant and ICI-aware systems are considered.

Data detection has been intensively investigated in the last several decades, and the tradeoff between data detection performance and computational complexity is well understood. The detection algorithms in this chapter are taken from the literature directly and presented in the context of coded OFDM. Other algorithms, not covered here, can be also applied to reach a desired tradeoff between performance and complexity.

8.1 Symbol-by-Symbol Detection in ICI-Ignorant OFDM Systems

8.1.1 Single-Input Single-Output Channel

For the system with single receiving element, the input–output relationship is copied from (5.47) as

$$z[m] = H[m]s[m] + \eta[m], \quad m = -\frac{K}{2}, \cdots, \frac{K}{2} - 1, \tag{8.8}$$

where the data symbol $s[m]$ takes value from a finite constellation set $\mathcal{M} = \{\alpha_0, \alpha_1, \cdots, \alpha_{M-1}\}$.

Assume that constellation points are equally probable. A hard decision on the symbol $s[m]$ can be made through an exhaustive search

$$\hat{s}[m] = \arg \min_{\alpha_i \in \mathcal{M}} |z[m] - H[m]\alpha_i|^2 \tag{8.9}$$

$$= \arg \min_{\alpha_i \in \mathcal{M}} |\hat{s}_{\text{LS}}[m] - \alpha_i|^2 \tag{8.10}$$

where $\hat{s}_{\text{LS}}[m]$ is the least squares (LS) estimate of $s[m]$:

$$\hat{s}_{\text{LS}}[m] = \frac{z[m]}{H[m]}. \tag{8.11}$$

The soft information as needed by the channel decoder is

$$L_i^{\text{ext}}[m] = \ln \frac{\Pr(s[m] = \alpha_i | z[m])}{\Pr(s[m] = \alpha_0 | z[m])} \tag{8.12}$$

$$= \ln \frac{f(z[m] | s[m] = \alpha_i)}{f(z[m] | s[m] = \alpha_0)}. \tag{8.13}$$

where $f(z[m]|s[m])$ is the likelihood function. Assuming that the noise follows a complex Gaussian distribution $\eta[m] \sim \mathcal{CN}(0, \sigma_\eta^2)$, the likelihood function is

$$f(z[m]|s[m]) = \frac{1}{\pi\sigma_\eta^2} \exp\left(\frac{|z[m] - H[m]s[m]|^2}{\sigma_\eta^2}\right). \tag{8.14}$$

Thus the extrinsic LLR can be computed as

$$L_i^{\text{ext}}[m] = \frac{1}{\sigma_\eta^2}\left[\left|z[m] - H[m]\alpha_0\right|^2 - \left|z[m] - H[m]\alpha_i\right|^2\right] \tag{8.15}$$

$$= \frac{1}{\sigma_\eta^2}[2\Re\{z^*[m]H[m](\alpha_i - \alpha_0)\} - |H[m]|^2(|\alpha_i|^2 - |\alpha_0|^2)]. \tag{8.16}$$

8.1.2 Single-Input Multi-Output Channel

For a single-input multi-output (SIMO) channel with N_r receiving elements, the frequency measurement on the mth subcarrier at the νth element is

$$z_\nu[m] = H_\nu[m]s[m] + \eta_\nu[m] \tag{8.17}$$

for $\nu = 1, \cdots, N_\text{r}$. We assume that noises at all receiving elements are independently and identically distributed (i.i.d), with $\eta_\nu[m] \sim \mathcal{CN}(0, \sigma_\eta^2)$, $\forall \nu$.

8.1.2.1 Maximum Ratio Combining

After the maximum ratio combining (MRC) of frequency measurements at all receiving elements, one can obtain a sufficient statistic of the frequency measurements

$$z_{\text{mrc}}[m] = \frac{\sum_{\nu=1}^{N_\text{r}} z_\nu[m]H_\nu^*[m]}{\sqrt{\sum_{\nu=1}^{N_\text{r}} |H_\nu[m]|^2}}. \tag{8.18}$$

Define an equivalent channel coefficient

$$H_{\text{mrc}}[m] := \left[\sum_{\nu=1}^{N_\text{r}} |H_\nu[m]|^2\right]^{1/2}. \tag{8.19}$$

A simplified expression of (8.18) is

$$z_{\text{mrc}}[m] = H_{\text{mrc}}[m]s[m] + \eta_{\text{mrc}}[m], \tag{8.20}$$

where the noise after MRC is

$$\eta_{\text{mrc}}[m] := \frac{1}{H_{\text{mrc}}[m]} \sum_{v=1}^{N_r} H_v^*[m]\eta_v[m], \tag{8.21}$$

which can be verified to follow a complex Gaussian distribution $\eta_{\text{mrc}}[m] \sim \mathcal{CN}(0, \sigma_\eta^2)$.

8.1.2.2 Extrinsic LLR Computation

Notice that the input–output relationship after MRC in (8.20) has an identical form as (8.8), the LS estimate of the information symbol is similarly obtained as

$$\hat{s}_{\text{LS}}[m] = \frac{z_{\text{mrc}}[m]}{H_{\text{mrc}}[m]}. \tag{8.22}$$

The extrinsic LLR in the SIMO channel is formulated as

$$L_i^{\text{ext}}[m] = \frac{2}{\sigma_\eta^2}\Re\{z_{\text{mrc}}^*[m]H_{\text{mrc}}[m](\alpha_i - \alpha_0)\} - \frac{1}{\sigma_\eta^2}|H_{\text{mrc}}[m]|^2(|\alpha_i|^2 - |\alpha_0|^2) \tag{8.23}$$

$$= \frac{1}{\sigma_\eta^2}\left[2\Re\left\{\sum_{v=1}^{N_r} z_v^*[m]H_v[m]\left(\alpha_i - \alpha_0\right)\right\} - \sum_{v=1}^{N_r} |H_v[m]|^2(|\alpha_i|^2 - |\alpha_0|^2)\right]. \tag{8.24}$$

8.2 Block-Based Data Detection in ICI-Aware OFDM Systems

In an OFDM receiver with intercarrier interference, we copy the frequency measurement at the mth subcarrier in (5.24) as

$$z[m] = \sum_{k=-K/2}^{K/2-1} H[m,k]s[k] + w[m], \quad m = -\frac{K}{2}, \cdots, \frac{K}{2} - 1 \tag{8.25}$$

Stacking frequency measurements at all subcarriers into a vector yields

$$\underbrace{\begin{bmatrix} z[\frac{K}{2}-1] \\ \vdots \\ z[-\frac{K}{2}] \end{bmatrix}}_{:=\mathbf{z}} = \underbrace{\begin{bmatrix} H[\frac{K}{2}-1,\frac{K}{2}-1] & \cdots & H[\frac{K}{2}-1,-\frac{K}{2}] \\ \vdots & \ddots & \vdots \\ H[-\frac{K}{2},\frac{K}{2}-1] & \cdots & H[-\frac{K}{2},-\frac{K}{2}] \end{bmatrix}}_{:=\mathbf{H}} \underbrace{\begin{bmatrix} s[\frac{K}{2}-1] \\ \vdots \\ s[-\frac{K}{2}] \end{bmatrix}}_{:=\mathbf{s}} + \underbrace{\begin{bmatrix} w[\frac{K}{2}-1] \\ \vdots \\ w[-\frac{K}{2}] \end{bmatrix}}_{:=\mathbf{w}} \tag{8.26}$$

where $\mathbf{z}$, $\mathbf{s}$ and $\mathbf{w}$ are vectors of size $K \times 1$, and $\mathbf{H}$ is the channel matrix of size $K \times K$. Define $\mathbf{h}_m$ is the column of $\mathbf{H}$ corresponding to the symbol $s[m]$, $\mathbf{s}_m^-$ as the vector $\mathbf{s}$ with $s[m]$ removed, and $\mathbf{H}_m^-$ as the matrix $\mathbf{H}$ with $\mathbf{h}_m$ removed. We rewrite (8.26) as

$$\mathbf{z} = \mathbf{H}\mathbf{s} + \mathbf{w} \tag{8.27}$$

$$= \mathbf{h}_m s[m] + \mathbf{H}_m^- \mathbf{s}_m^- + \mathbf{w}. \tag{8.28}$$

The ambient noise samples are assumed following independent and identical distribution (i.i.d) with $w[m] \sim \mathcal{CN}(0, \sigma_w^2), \forall m$.

8.2.1 MAP Equalizer

Assuming the independence of the information symbols, the prior probability of $\mathbf{s}^-$ is expressed as

$$\Pr(\mathbf{s}^-) = \prod_{k=-K/2,\, k\neq m}^{K/2-1} \Pr(s[k]) \tag{8.29}$$

where $\Pr(s[k])$ is computed from the LLRV $\mathbf{L}^{\mathrm{apr}}[k]$ from the decoder; see e.g., (3.41) and (3.46) on how to convert LLRs to probabilities. With the Gaussian noise, the likelihood function is

$$f(\mathbf{z}\,|\,\mathbf{s}^-, s[m] = \alpha_i) \propto \exp\left(-\frac{1}{\sigma_{\mathrm{w}}^2}||\mathbf{z} - \mathbf{h}\alpha_i - \mathbf{H}^-\mathbf{s}^-||^2\right) \tag{8.30}$$

The extrinsic LLR to be passed to the decoder is then:

$$L_i^{\mathrm{ext}}[m] = \ln \frac{\sum_{\mathbf{s}^-} f(\mathbf{z}\,|\,\mathbf{s}^-, s[m] = \alpha_i)\Pr(\mathbf{s}^-)}{\sum_{\mathbf{s}^-} f(\mathbf{z}\,|\,\mathbf{s}^-, s[m] = \alpha_0)\Pr(\mathbf{s}^-)}. \tag{8.31}$$

The computational complexity of (8.31) is on the order of $\mathcal{O}(M^K)$. Hence, the MAP detector is not practical for block detection of an OFDM block with a reasonable K. Nevertheless the general presentation of the MAP detector is useful, and its applicability is justified in the scenario considered in Section 8.4.1.

8.2.2 Linear MMSE Equalizer with A Priori Information

8.2.2.1 MMSE Formulation

With the constraint that only linear operation can be applied on the vector $\mathbf{z}$ to obtain an estimate of $s[m]$, the estimate of $s[m]$ that minimizes the cost function $E[|\hat{s}[m] - s[m]|^2]$ is [393]:

$$\hat{s}_{\mathrm{MMSE}}^{\mathrm{app}}[m] = \bar{s}[m] + \mathrm{Cov}(s[m], \mathbf{z})\mathrm{Cov}^{-1}(\mathbf{z}, \mathbf{z})(\mathbf{z} - \mathbb{E}[\mathbf{z}]). \tag{8.32}$$

Based on the prior information, the mean and variance of symbol $s[m]$ are obtained as

$$\bar{s}[m] = \sum_{\alpha_i \in \mathcal{M}} \alpha_i \cdot \Pr(s[m] = \alpha_i), \tag{8.33}$$

$$\bar{\sigma}_s^2[m] = \sum_{\alpha_i \in \mathcal{M}} (\alpha_i - \bar{s}[m])^2 \cdot \Pr(s[m] = \alpha_i). \tag{8.34}$$

Stack the mean of all data symbols into a vector $\hat{\mathbf{s}}$, and the variances of all data symbols into a diagonal matrix $\boldsymbol{\Sigma}_s$,

$$\bar{\mathbf{s}} = \left[\bar{s}\left[-\frac{K}{2}\right] \cdots \bar{s}\left[\frac{K}{2} - 1\right]\right]^{\mathrm{T}} \tag{8.35}$$

$$\boldsymbol{\Sigma}_s = \mathrm{diag}\left(\sigma_s^2\left[-\frac{K}{2}\right], \cdots, \quad \sigma_s^2\left[\frac{K}{2} - 1\right]\right). \tag{8.36}$$

The mean of the frequency measurement vector is obtained as

$$\mathbb{E}[\mathbf{z}] = \mathbf{H}\bar{\mathbf{s}}. \tag{8.37}$$

The covariance matrix of $\mathbf{z}$ and the cross-covariance matrix between $s[m]$ and $\mathbf{z}$ can be computed as:

$$\mathrm{Cov}(\mathbf{z}, \mathbf{z}) = \mathbf{H}\boldsymbol{\Sigma}_s\mathbf{H}^{\mathrm{H}} + \sigma_{\mathrm{w}}^2\mathbf{I}, \tag{8.38}$$

$$\mathrm{Cov}(s[m], \mathbf{z}) = \sigma_s^2[m]\mathbf{h}_m^{\mathrm{H}}. \tag{8.39}$$

The MMSE estimate in (8.32) can be expanded as

$$\hat{s}_{\mathrm{MMSE}}^{\mathrm{app}}[m] = \bar{s}[m] + \sigma_s^2[m] \cdot \mathbf{f}_m^{\mathrm{H}}(\mathbf{z} - \mathbf{H}\bar{\mathbf{s}}), \tag{8.40}$$

where the linear filter $\mathbf{f}_m$ is defined as

$$\mathbf{f}_m := (\mathbf{H}\boldsymbol{\Sigma}_s\mathbf{H}^{\mathrm{H}} + \sigma_{\mathrm{w}}^2\mathbf{I})^{-1}\mathbf{h}_m. \tag{8.41}$$

The variance of the MMSE estimate is

$$\tilde{\sigma}_s^2[m] = \mathrm{Cov}(\hat{s}_{\mathrm{MMSE}}^{\mathrm{app}}[m], \hat{s}_{\mathrm{MMSE}}^{\mathrm{app}}[m]) = \sigma_s^4[m]\mathbf{f}_m^{\mathrm{H}}\mathbf{h}_m. \tag{8.42}$$

8.2.2.2 Extrinsic LLR Computation

Note that the MMSE estimate in (8.40) depends on the prior information on $s[m]$ via $\bar{s}[m]$ and $\sigma_s^2[m]$. To compute the extrinsic information needed by the channel decoder, it is required that the MMSE estimate $\hat{s}[m]$ shall be independent from the prior information on $s[m]$. To achieve this goal, the receiver shall calculate the MMSE estimate of $\hat{s}[m]$ by ignoring the prior information on $s[m]$ while unitizing the prior information on all other symbols.

Let σ_s^2 be the average power of symbols in the constellation. Enforcing $\bar{s}[m] = 0$ and $\sigma_s^2[m] = E_s$ when computing the MMSE estimate $\hat{s}[m]$, one obtains:

$$\hat{s}_{\mathrm{MMSE}}^{\mathrm{ext}}[m] = 0 + \sigma_s^2 \cdot \mathbf{f}_m'^{\mathrm{H}}(\mathbf{z} - \mathbf{H}\bar{\mathbf{s}} + \bar{s}[m]\mathbf{h}_m) \tag{8.43}$$

with

$$\mathbf{f}_m' := (\mathbf{H}\boldsymbol{\Sigma}_s\mathbf{H}^{\mathrm{H}} + \sigma_{\mathrm{w}}^2\mathbf{I} + (\sigma_s^2 - \hat{\sigma}_s^2[m])\mathbf{h}_m\mathbf{h}_m^{\mathrm{H}})^{-1}\mathbf{h}_m. \tag{8.44}$$

Using the matrix inversion lemma, $\mathbf{f}_m'$ is related to $\mathbf{f}_m$ as

$$\mathbf{f}_m' = (1 + (\sigma_s^2 - \bar{\sigma}_s^2[m])\mathbf{f}_m^{\mathrm{H}}\mathbf{h}_m)^{-1}\mathbf{f}_m. \tag{8.45}$$

The MMSE estimate can be reformulated as

$$\hat{s}_{\mathrm{MMSE}}^{\mathrm{ext}}[m] = K_m \cdot \mathbf{f}_m^{\mathrm{H}}(\mathbf{z} - \mathbf{H}\bar{\mathbf{s}} + \bar{s}[m]\mathbf{h}_m) \tag{8.46}$$

with

$$K_m := \frac{\sigma_s^2}{1 + (\sigma_s^2 - \bar{\sigma}_s^2[m])\mathbf{f}_m^{\mathrm{H}}\mathbf{h}_m}. \tag{8.47}$$

As suggested in [393], the key to simplify the extrinsic LLR computation is to assume that $f(\hat{s}^{\text{ext}}_{\text{MMSE}}[m]|s[m] = \alpha_i)$ follows a Gaussian distribution with mean $\mu_{m,i}$ and variance $\sigma^2_{m,i}$,

$$f(\hat{s}^{\text{ext}}_{\text{MMSE}}[m]|s[m] = \alpha_i) = \frac{1}{\pi\sigma^2_{m,i}} \exp\left(-\frac{|\hat{s}^{\text{ext}}_{\text{MMSE}}[m] - \mu_{m,i}|^2}{\sigma^2_{m,i}}\right). \tag{8.48}$$

The mean and variance are obtained as

$$\begin{aligned} \mu_{m,i} &= K_m \cdot \mathbf{f}^{\text{H}}_m(\mathbb{E}[\mathbf{z}|s[m] = \alpha_i] - \mathbf{H}\bar{\mathbf{s}} + \bar{s}[m]\mathbf{h}_m) \\ &= K_m \cdot \alpha_i \cdot \mathbf{f}^{\text{H}}_m\mathbf{h}_m \end{aligned} \tag{8.49}$$

$$\begin{aligned} \sigma^2_{m,i} &= K^2_m \cdot \mathbf{f}^{\text{H}}_m\text{Cov}(\mathbf{z}, \mathbf{z}|s[m] = \alpha_i)\mathbf{f}_m \\ &= K^2_m \cdot \mathbf{f}^{\text{H}}_m(\mathbf{H}\boldsymbol{\Sigma}_s\mathbf{H}^{\text{H}} + \sigma^2_{\text{w}}\mathbf{I} - \sigma^2_s[m]\mathbf{h}_m\mathbf{h}^{\text{H}}_m)\mathbf{f}_m \\ &= K^2_m \cdot (\mathbf{f}^{\text{H}}_m\mathbf{h}_m - \sigma^2_s[m]\mathbf{f}^{\text{H}}_m\mathbf{h}_m\mathbf{h}^{\text{H}}_m\mathbf{f}_m). \end{aligned} \tag{8.50}$$

The extrinsic LLR can be computed as

$$L^{\text{ext}}_i[m] = \ln\frac{f(\hat{s}^{\text{ext}}_{\text{MMSE}}[m]|s[m] = \alpha_i)}{f(\hat{s}^{\text{ext}}_{\text{MMSE}}[m]|s[m] = \alpha_0)} \tag{8.51}$$

$$= \frac{2\Re\{(\alpha^*_i - \alpha^*_0)\mathbf{f}^{\text{H}}_m(\mathbf{z} - \mathbf{H}\bar{\mathbf{s}} + \bar{s}[m]\mathbf{h}_m)\} - (|\alpha_i|^2 - |\alpha_0|^2)\mathbf{f}^{\text{H}}_m\mathbf{h}_m}{1 - \sigma^2_s[m]\mathbf{f}^{\text{H}}_m\mathbf{h}_m}, \tag{8.52}$$

which requires the computation of $\mathbf{f}^{\text{H}}_m(\mathbf{z} - \mathbf{H}\hat{\mathbf{s}})$ and $\mathbf{f}^{\text{H}}_m\mathbf{h}_m$.

One can relate the extrinsic LLR computation with the MMSE estimate $\hat{s}^{\text{app}}_{\text{MMSE}}$ and its variance $\tilde{\sigma}^2_s[m]$. Based on (8.40) and (8.42), we have

$$\mathbf{f}^{\text{H}}_m(\mathbf{z} - \mathbf{H}\bar{\mathbf{s}}) = \frac{\hat{s}^{\text{app}}_{\text{MMSE}}[m] - \bar{s}[m]}{\sigma^2_s[m]}, \quad \mathbf{f}^{\text{H}}_m\mathbf{h}_m = \frac{\tilde{\sigma}^2_s[m]}{\sigma^4_s[m]}. \tag{8.53}$$

Hence, the extrinsic LLR in (8.52) can be expressed as

$$L^{\text{ext}}_i[m] = \frac{2\Re\{(\alpha^*_i - \alpha^*_0)[\sigma^2_s[m](\hat{s}^{\text{app}}_{\text{MMSE}}[m] - \bar{s}[m]) + \tilde{\sigma}^2_s[m]\bar{s}[m]]\} - (|\alpha_i|^2 - |\alpha_0|^2)\tilde{\sigma}^2_s[m]}{\sigma^2_s[m](\sigma^2_s[m] - \tilde{\sigma}^2_s[m])}. \tag{8.54}$$

The relationship of (8.54) will be used in Section 8.3.2.

8.2.2.3 Complexity

For each symbol $s[m]$, the linear filter $\mathbf{f}_m$ as defined in (8.41) needs to be computed. Note that only one matrix conversion in the form of $(\mathbf{H}\boldsymbol{\Sigma}_s\mathbf{H}^{\text{H}} + \sigma^2_{\text{w}}\mathbf{I})^{-1}$ is needed to calculate the $\mathbf{f}_m$'s for all symbols. The linear MMSE equalizer has a computational complexity on the order of $\mathcal{O}(K^3)$, which is affordable for a moderate K.

8.2.3 Extension to the Single-Input Multi-Output Channel

Consider a SIMO channel with N_r receiving elements. For each element $v = 1, \cdots, N_r$, the frequency measurement vector is expressed as

$$\mathbf{z}_v = \mathbf{H}_v \mathbf{s} + \mathbf{w}_v \tag{8.55}$$

where the matrices and vectors are defined similar to those in (8.27). Stacking frequency measurements at all receiving elements into a vector yields

$$\underbrace{\begin{bmatrix} \mathbf{z}_1 \\ \vdots \\ \mathbf{z}_{N_r} \end{bmatrix}}_{:=\mathbf{z}_{\text{simo}}} = \underbrace{\begin{bmatrix} \mathbf{H}_1 \\ \vdots \\ \mathbf{H}_{N_r} \end{bmatrix}}_{:=\mathbf{H}_{\text{simo}}} \mathbf{s} + \underbrace{\begin{bmatrix} \mathbf{w}_1 \\ \vdots \\ \mathbf{w}_{N_r} \end{bmatrix}}_{:=\mathbf{w}_{\text{simo}}}. \tag{8.56}$$

The system model in (8.56) shares an identical structure as that in (8.26), except that the matrices and vectors have different sizes. The MAP equalizer in Section 8.2.1 and the MMSE equalizer in Section 8.2.2 presented for the SISO channel can be directly applied to the SIMO channel.

8.3 Data Detection for OFDM Systems with Banded ICI

To reduce the computational complexity involved in (8.25), a banded assumption of the channel matrix is usually adopted. With the band-limited ICI assumption, we repeat the input–output relationship of the frequency measurement on the mth subcarrier in (5.56) as

$$z[m] = \sum_{k=m-D}^{m+D} H[m,k]s[k] + \eta[m], \tag{8.57}$$

where $\eta[m]$ represents additive noise and the residual ICI beyond the band. For simplification, we assume that $\eta[m]$ are i.i.d. Gaussian distributed $\eta[m] \sim \mathcal{CN}(0, \sigma_\eta^2)$. The likelihood function is then

$$f(z[m]|\{s[k]\}_{k=m-D}^{m+D}) = \frac{1}{\pi\sigma_\eta^2} \exp\left(-\frac{1}{\sigma_\eta^2}\left|z[m] - \sum_{k=m-D}^{m+D} H[m,k]s[k]\right|^2\right). \tag{8.58}$$

8.3.1 BCJR Algorithm and Log-MAP Implementation

The Bahl-Cocke-Jelinek-Raviv (BCJR) algorithm is a computationally efficient algorithm to obtain the *exact* posterior probability over a trellis [17]. Here we choose to follow the presentation in [408], which facilitates a direct log-main implementation. Such an implementation is refereed to as? the Log-MAP detector.

First, we need to define the trellis structure. With the variable change $\ell = m + D - k$, (8.57) is rewritten as

$$z[m] = \sum_{\ell=0}^{2D} H[m, m + D - \ell]s[m + D - \ell] + \eta[m]. \tag{8.59}$$

To simplify the notation, let us define the sequences

$$\tilde{z}[m] = z[m - D], \quad \tilde{\eta}[m] = \eta[m - D] \tag{8.60}$$

and a set of time-varying coefficients

$$\tilde{h}[m; \ell] = H[m - D, m - \ell], \quad \ell = 0, \ldots, 2D. \tag{8.61}$$

Eq. (8.59) then simplifies to

$$\tilde{z}[m] = \sum_{\ell=0}^{2D} \tilde{h}[m; \ell]s[m - \ell] + \tilde{\eta}[m], \tag{8.62}$$

which shows the convolution between the sequence $s[m]$ and the sequence $\tilde{h}[m; \ell]$. The ICI mitigation problem is then related to a channel equalization problem in the presence of intersymbol interference, where the finite-impulse-response channel, $\tilde{h}[m; \ell], \ell = 0, \ldots, 2D$, has time-varying tap weights.

Due to the memory length $2D$, define the state at index m as

$$(s[m - 1], \cdots, s[m - 2D]), \tag{8.63}$$

which leads to a total of M^{2D} states. As shown in Figure 8.2, let us use ζ' and ζ to represent the states in transition at index m. The joint probability $\Pr(\zeta', \zeta, \{\tilde{z}[k]\}_{\forall k})$ can be partitioned as:

$$\begin{aligned}\Pr(\zeta', \zeta, \{\tilde{z}[k]\}_{\forall k}) &= \Pr(\zeta', \zeta, \{\tilde{z}[k]\}_{k<m}, \tilde{z}[m], \{\tilde{z}[n]\}_{k>m}) \\ &= \Pr(\zeta', \{\tilde{z}[k]\}_{k<m}) \Pr(\tilde{z}[m], \zeta|\zeta') \Pr(\{\tilde{z}[k]\}_{k>m})|\zeta),\end{aligned} \tag{8.64}$$

where the second equality is based on the Markov chain property. Define three quantities associated with the state transitions at index m as

$$\alpha_m(\zeta') = \ln \ \Pr(\zeta', \{\tilde{z}[n]\}_{n<m}) \tag{8.65}$$

$$\beta_m(\zeta) = \ln \ \Pr(\{\tilde{z}[n]\}_{n>m}|\zeta) \tag{8.66}$$

$$\gamma_m(\zeta', \zeta) = \ln \ \Pr(\tilde{z}[m], \zeta|\zeta'), \tag{8.67}$$

where $\gamma_m(\zeta', \zeta)$ is the branch metric associated with the state transition from ζ' to ζ. Rewriting the likelihood function in (8.58) as $f(\tilde{z}[m]||\zeta', \zeta)$ and with the prior information on $s[m]$, the branch metric is

$$\gamma_m(\zeta', \zeta) = \ln \ f(\tilde{z}[m]||\zeta', \zeta) + \ln \ \Pr(\zeta|\zeta'). \tag{8.68}$$

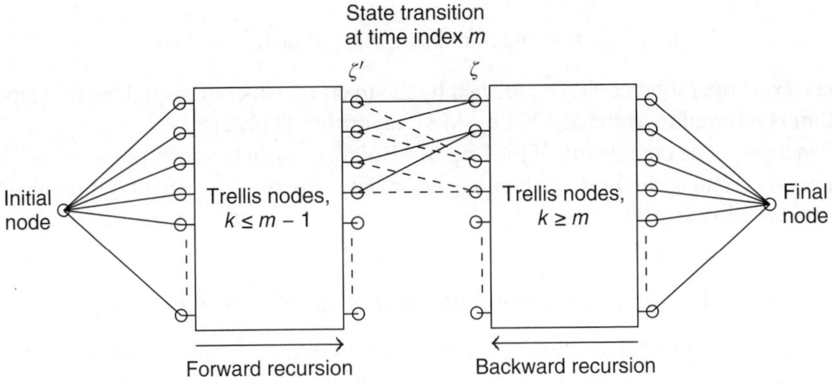

Figure 8.2 The illustration of the trellis structure. (Source: Viterbi 1998 [408], figure 1, p. 261. Reproduced with permission of IEEE.).

The Log-MAP algorithm runs as follows.

- **Forward recursion**: Starting from $m = -K/2$ to $K/2 - 1$, compute $\alpha_m(\zeta)$ recursively for each state ζ:

$$\alpha_m(\zeta) = \ln \sum_{\zeta'} \exp\ (\alpha_{m-1}(\zeta') + \gamma_m(\zeta', \zeta)) \tag{8.69}$$

- **Backward recursion**: Starting from $m = K/2 - 1$ to $-K/2$, compute $\beta_m(\zeta')$ recursively for each state ζ':

$$\beta_{m-1}(\zeta') = \ln \sum_{\zeta} \exp\ (\gamma_m(\zeta', \zeta) + \beta_m(\zeta)) \tag{8.70}$$

- Compute the a posteriori probability ratio in the log-domain for each symbol $s[m]$ as

$$\begin{aligned} L_i^{\text{app}}[m] = \ln & \sum_{\zeta', \zeta:\ s[m]=\alpha_i} \exp\ (\alpha_{m-1}(\zeta) + \gamma_m(\zeta', \zeta) + \beta_m(\zeta)) \\ - \ln & \sum_{\zeta', \zeta:\ s[m]=\alpha_0} \exp\ (\alpha_{m-1}(\zeta') + \gamma_m(\zeta', \zeta) + \beta_m(\zeta)) \end{aligned} \tag{8.71}$$

  The extrinsic LLR $L_i^{\text{ext}}[m]$ can be obtained through (8.7).

The Log domain representation allows for efficient implementation. Equations (8.69), (8.70) and (8.71) can be evaluated using the Jacobian logarithm [111, 215, 440]

$$\max{}^*(x, y) = \max(x, y) + \ln\ (1 + e^{-|x-y|}), \tag{8.72}$$

where the second term can be implemented using a look-up table. It follows from the definition that

$$\max{}^*(x, y) = \ln\ (e^x + e^y) \tag{8.73}$$

and

$$\max{}^*(x, y, z) = \max{}^*[\max{}^*(x, y), z] = \ln\ (e^x + e^y + e^z) \tag{8.74}$$

The $\max^*(x, y)$ operation could be replaced by the $\max(x, y)$ operation, and the corresponding algorithm is referred to as the MAX-Log-MAP algorithm [111, 215].

As D increases, the complexity of the Log-MAP algorithm on the order of $\mathcal{O}(M^{(2D+1)})$ grows *exponentially*. Hence, the Log-MAP equalizer is only suitable for a small ICI depth D and a small constellation size M.

8.3.2 Factor-Graph Algorithm with Gaussian Message Passing

Factor graph and related algorithms, such as the sum-product algorithm (SPA) and the Gaussian message passing (GMP) principle, have been under extensive investigation in recent years [223, 258, 259]. The application of the factor graph algorithm for ISI equalization has been thoroughly documented in [104]. In the following, we will sketch the main idea on the use of the factor graph algorithm for the mitigation of the banded ICI. Please refer to [104] for detailed algorithm presentations, including fast and numerically stable implementations, of the factor-graph based algorithm presented in the context of ISI equalization.

8.3.2.1 Message Passing in the Factor Graph

With the Gaussian message passing principle, all the messages are constrained to be in the forms of Gaussian probability density function, denoted by $\hat{f}(\cdot)$ in the following.

Now define the variable nodes and check function nodes associated with a factor graph. Based on the channel input and output model in (8.57), the variable nodes are defined through $(2D + 1) \times 1$ vectors

$$\mathbf{x}_m = [s[m - D], \ \ldots\ , s[m], \ \ldots\ s[m + D]]^{\mathrm{T}} \tag{8.75}$$

for $m = -K/2, \ \ldots\ , K/2 - 1$. Surrounding each variable node, there are four check function nodes as shown in Figure 8.3.

- The check node corresponding to the prior knowledge from the channel decoder on each symbol $s[m]$ will pass a message
$$\hat{f}(s[m]) = \frac{1}{\pi\bar{\sigma}_s^2[m]} \exp\left\{ -\frac{1}{\bar{\sigma}_s^2[m]} |s[m] - \bar{s}[m]|^2 \right\}, \tag{8.76}$$
where the mean and variance is computed from $\mathbf{L}^{\mathrm{apr}}[m]$.
- The check node corresponding to the observation $z[m]$ will pass the message in the form of the likelihood function $f(z[m]|\mathbf{x}_m)$, as shown in (8.58).
- The check node $C_{m-1,m}$ is introduced to specify the constraint that $\mathbf{x}_m$ and $\mathbf{x}_{m-1}$ share $2D$ entries $\{s[m - D], \ \ldots\ , s[m + D - 1]\}$. Based on the incoming message from the left $\hat{f}(\mathbf{x}_{m-1} \to C_{m-1,m})$, the outgoing message to the right $\hat{f}(C_{m-1,m} \to \mathbf{x}_m)$ is generated as
$$\hat{f}(C_{m-1,m} \to \mathbf{x}_m) = \int \hat{f}(\mathbf{x}_{m-1} \to C_{m-1,m})\, ds[m - D - 1], \tag{8.77}$$
where the marginalization is over the symbol $s[m - D - 1]$.

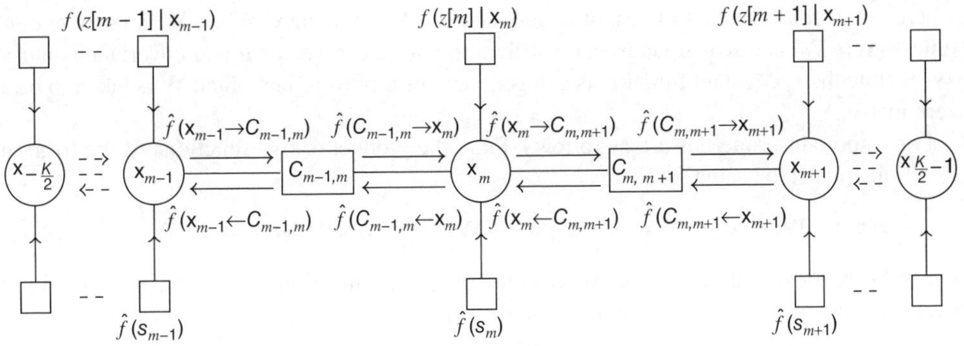

Figure 8.3 The factor graph based detector for mitigation of banded ICI, where all the messages illustrated on the graph are in the form of Gaussian probability distribution functions (pdf).

- The check node $C_{m,m+1}$ is introduced to specify the constraint that $\mathbf{x}_m$ and $\mathbf{x}_{m-1}$ share $2D$ entries $\{s[m-D+1], \ldots, s[m+D]\}$. Based on the incoming message from the right $\hat{f}(C_{m,m+1} \leftarrow \mathbf{x}_{m+1})$, the outgoing message to the left $\hat{f}(\mathbf{x}_m \leftarrow C_{m,m+1})$ is generated as

$$\hat{f}(\mathbf{x}_m \leftarrow C_{m,m+1}) = \int \hat{f}(C_{m,m+1} \leftarrow \mathbf{x}_{m+1})\, ds[m+D+1], \tag{8.78}$$

where the marginalization is over the the symbol $s[m+D+1]$.

The message passing over the factor graph in Figure 8.3 is carried out as follows.

- Forward message passing. Starting from the node $\mathbf{x}_{-K/2}$ to the node $\mathbf{x}_{K/2-1}$, generate the message at each variable node to be passed to its right check function as:

$$\hat{f}(\mathbf{x}_m \to C_{m,m+1}) = \hat{f}(C_{m-1,m} \to \mathbf{x}_m)\hat{f}(s[m])f(z[m]|\mathbf{x}_m) \tag{8.79}$$

- Backward message passing. Starting from the node $\mathbf{x}_{K/2-1}$ to the node $\mathbf{x}_{-K/2}$, generate the message at each variable node to be passed to its left check function as:

$$\hat{f}(C_{m-1,m} \leftarrow \mathbf{x}_m) = \hat{f}(\mathbf{x}_m \leftarrow C_{m,m+1})\hat{f}(s[m])f(z[m]|\mathbf{x}_m) \tag{8.80}$$

- After the forward and backward message passing, the a posteriori probability of the state vector $\mathbf{x}_m$ can be computed as

$$\hat{f}^{\text{app}}(\mathbf{x}_m) = \hat{f}(C_{m-1,m} \to \mathbf{x}_m)\hat{f}(s[m])f(z[m]|\mathbf{x}_m)\hat{f}(\mathbf{x}_m \leftarrow C_{m,m+1}) \tag{8.81}$$

The marginalization of $\hat{f}^{\text{app}}(\mathbf{x}_m)$ over $2D$ variables $\{s[m-D], s[m-1], s[m+1], \ldots, s[k+D]\}$ will lead to $\hat{f}^{\text{app}}(s[m])$.

8.3.3 Computations related to Gaussian Messages

Following the presentations in [104], introduce the function

$$\gamma(\mathbf{x}, \boldsymbol{\mu}, \mathbf{W}) \propto \exp\{-(\mathbf{x}-\boldsymbol{\mu})^{\mathrm{H}}\mathbf{W}(\mathbf{x}-\boldsymbol{\mu})\} \tag{8.82}$$

where $\mathbf{x}$ and $\boldsymbol{\mu}$ are vectors of length N and $\mathbf{W}$ is an $N \times N$ matrix. When $\mathbf{W}$ is positive definite, $\gamma(\mathbf{x}, \boldsymbol{\mu}, \mathbf{W})$ is a scaled Gaussian distribution with mean vector $\boldsymbol{\mu}$ and covariance matrix $\mathbf{W}^{-1}$. Note that a constant function is a degenerate form of (8.82) in which $\mathbf{W}$ is taken to be a zero matrix.

Two important properties are as follows. First, the product of two functions of the form in (8.82) maintains the same form:

$$\gamma(\mathbf{x}, \boldsymbol{\mu}_1, \mathbf{W}_1)\gamma(\mathbf{x}, \boldsymbol{\mu}_2, \mathbf{W}_2) \propto \gamma(\mathbf{x}, (\mathbf{W}_1 + \mathbf{W}_2)^\dagger(\mathbf{W}_1\boldsymbol{\mu}_1 + \mathbf{W}_2\boldsymbol{\mu}_2), \mathbf{W}_1 + \mathbf{W}_2) \tag{8.83}$$

where $\mathbf{W}^\dagger$ is the pseudo-inverse of $\mathbf{W}$. Second, the marginalization of the function as in (8.82) over some set of variables maintains the same form

$$\int_{-\infty}^{\infty} \gamma\left(\begin{bmatrix}\mathbf{x}\\ \mathbf{y}\end{bmatrix}, \begin{bmatrix}\boldsymbol{\mu}_x\\ \boldsymbol{\mu}_y\end{bmatrix}, \begin{bmatrix}\mathbf{W}_x & \mathbf{W}_{xy}\\ \mathbf{W}_{yx} & \mathbf{W}_y\end{bmatrix}\right) d\mathbf{x} \propto \gamma(\mathbf{y}, \boldsymbol{\mu}_y, \mathbf{W}_y - \mathbf{W}_{yx}\mathbf{W}_x^\dagger\mathbf{W}_{xy}) \tag{8.84}$$

Hence, when a factor graph represents a Gaussian distribution or a degenerate form, all messages will also be Gaussian or in degenerate forms. As a result, only mean vectors and the (pseudo) inverse of the covariance matrices need to be populated in the factor graph. With the complexity on the order of $\mathcal{O}((2D+1)^3)$, the factor graph based equalizer is often appealing for a computational point of view, since D is usually a small number and the complexity does not depend on the constellation size M.

8.3.3.1 Extrinsic LLR Computation

The message $\hat{f}^{\text{app}}(s[m])$ specifies the mean and covariance of $s[m]$ conditioned on all the measurements and the prior information, from which the MMSE estimate $\hat{s}_{\text{MMSE}}^{\text{app}}[m]$ as the conditional mean and the variance $\tilde{\sigma}_s^2[m]$ can be extracted. When computing the LLRs needed by the channel decoder, the prior information on the symbol itself needs to be excluded. Given the availability of the MMSE estimates based on a posteriori and a priori probabilities, and the corresponding variances, the expression in (8.54) can be used to compute the extrinsic LLRs directly.

8.3.4 Extension to SIMO Channel

For the single-input multi-output channel, the channel input–output relationship is extended from (8.57) to

$$z_\nu[m] = \sum_{k=m-D}^{m+D} H_\nu[m,k]s[k] + \eta_\nu[m], \tag{8.85}$$

where $\nu = 1, \ldots, N_r$. For simplicity, assume that the noises are i.i.d. Gaussian distributed, $\eta_\nu[m] \sim \mathcal{CN}(0, \sigma_\eta^2)$. The likelihood function is

$$f(z_\nu[m] | \{s[k]\}_{k=m-D}^{m+D}) = \frac{1}{\pi\sigma_\eta^2} \exp\left(-\frac{1}{\sigma_\eta^2}\left|z_\nu[m] - \sum_{k=m-D}^{m+D} H_\nu[m,k]s[k]\right|^2\right). \tag{8.86}$$

ICI equalization methods developed for the SISO channel can directly be applied to the SIMO channel, simply having more measurements available due to the receive diversity.

- For the Log-MAP algorithm, change $f(\tilde{z}[m]|\zeta',\zeta)$ in (8.68) for the branch metric computation by $\prod_{\nu=1}^{N_r} f(\tilde{z}_\nu[m]|\zeta',\zeta)$.
- For the factor-graph based equalization, change the likelihood function $f(z[m]|\mathbf{x}_m)$ to $\prod_{\nu=1}^{N_r} f(z_\nu[m]|\mathbf{x}_m)$.

8.4 Symbol Detectors for MIMO OFDM

8.4.1 ICI-Ignorant MIMO OFDM

For a multi-input multi-output (MIMO) channel with N_t inputs and N_r receiving elements, we repeat the frequency measurement on the mth subcarrier at the νth receiving element in (5.51) as

$$z_\nu[m] = \sum_{\mu=1}^{N_t} H_{\nu,\mu}[m]s_\mu[m] + \eta_\nu[m], \tag{8.87}$$

for $\nu = 1, \cdots, N_r$, where $s_\mu[m]$ denotes the transmitted symbol on the mth subcarrier from the μth transmitter. Stacking the frequency measurements in (8.87) at all receiving elements yields,

$$\underbrace{\begin{bmatrix} z_1[m] \\ \vdots \\ z_{N_r}[m] \end{bmatrix}}_{:=\mathbf{z}[m]} = \underbrace{\begin{bmatrix} H_{1,1}[m] & \cdots & H_{1,N_t}[m] \\ \vdots & \ddots & \vdots \\ H_{N_r,1}[m] & \cdots & H_{N_r,N_t}[m] \end{bmatrix}}_{:=\mathbf{H}[m]} \underbrace{\begin{bmatrix} s_1[m] \\ \vdots \\ s_{N_t}[m] \end{bmatrix}}_{:=\mathbf{s}[m]} + \underbrace{\begin{bmatrix} \eta_1[m] \\ \vdots \\ \eta_{N_r}[m] \end{bmatrix}}_{:=\boldsymbol{\eta}[m]} \tag{8.88}$$

On each subcarrier m, the matrix-vector formulation is

$$\mathbf{z}[m] = \mathbf{H}[m]\mathbf{s}[m] + \boldsymbol{\eta}[m]. \tag{8.89}$$

Denote $\mathbf{s}_\mu^-[m]$ as $\mathbf{s}[m]$ with $s_\mu[m]$ removed. Assuming that $\eta[m] \sim \mathcal{CN}(0,\sigma_\eta^2)$, $\forall\,\mu$, the likelihood function of $s_\mu[m]$ can be expanded as

$$f(\mathbf{z}[m]|\mathbf{s}_\mu^-[m], s_\mu[m] = \alpha_i) \propto \exp\left(-\frac{1}{\sigma_\eta^2}\left\|\mathbf{z}[m] - \mathbf{h}_\mu[m]\alpha_i - \mathbf{H}_\mu^-[m]\mathbf{s}_\mu^-[m]\right\|^2\right) \tag{8.90}$$

with $\mathbf{h}_\mu[m]$ denoting the μth column of $\mathbf{H}[m]$, and $\mathbf{H}_\mu^-[m]$ as $\mathbf{h}_\mu[m]$ removed. Following the same derivation in Section 8.2.1, the extrinsic LLR to be passed to the decoder is

$$L_{i,\mu}^{\text{ext}}[m] = \ln \frac{\sum_{\mathbf{s}_\mu^-[m]} f(\mathbf{z}[m]|\mathbf{s}_\mu^-[m], s_\mu[m] = \alpha_i)\Pr(\mathbf{s}_\mu^-[m])}{\sum_{\mathbf{s}_\mu^-[m]} f(\mathbf{z}[m]|\mathbf{s}_\mu^-[m], s_\mu[m] = \alpha_0)\Pr(\mathbf{s}_\mu^-[m])} \tag{8.91}$$

The MAP detector has the complexity of $\mathcal{O}(M^{N_t})$ per subcarrier.

Similarly, the linear MMSE detectors can be applied with complexity $\mathcal{O}(N_t^3)$ per subcarrier; Note that the formulations in (8.89) and (8.26) are the same except different notations on the noise.

Due to the small problem size on each OFDM subcarriers, a variety of detectors from the open literature can be applied as well, such as the sphere decoding algorithm, the MCMC detector, and the lattice reduction based linear equalizers.

8.4.2 Full-ICI Equalization

For a MIMO channel with full ICI modeled at receiver side, we repeat the frequency measurement vector at the νth receiving element in (5.35) as

$$\mathbf{z}_\nu = \sum_{\mu=1}^{N_t} \mathbf{H}_{\nu,\mu}\mathbf{s}_\mu + \mathbf{w}_\nu, \qquad \nu = 1, \cdots, N_r. \tag{8.92}$$

Stacking frequency measurement vectors at all receiving elements, we have

$$\underbrace{\begin{bmatrix} \mathbf{z}_1 \\ \vdots \\ \mathbf{z}_{N_r} \end{bmatrix}}_{:=\mathbf{z}_{\text{mimo}}} = \underbrace{\begin{bmatrix} \mathbf{H}_{1,1} & \cdots & \mathbf{H}_{1,N_t} \\ \vdots & \ddots & \vdots \\ \mathbf{H}_{N_r,1} & \cdots & \mathbf{H}_{N_r,N_t} \end{bmatrix}}_{:=\mathbf{H}_{\text{mimo}}} \underbrace{\begin{bmatrix} \mathbf{s}_1 \\ \vdots \\ \mathbf{s}_{N_t} \end{bmatrix}}_{:=\mathbf{s}_{\text{mimo}}} + \underbrace{\begin{bmatrix} \mathbf{w}_1 \\ \vdots \\ \mathbf{w}_{N_r} \end{bmatrix}}_{:=\mathbf{w}_{\text{mimo}}} \tag{8.93}$$

The block-based MAP and MMSE detectors can be applied to the model

$$\mathbf{z}_{\text{mimo}} = \mathbf{H}_{\text{mimo}}\mathbf{s}_{\text{mimo}} + \mathbf{w}_{\text{mimo}}. \tag{8.94}$$

However, the matrices and vectors have large size, and computation complexity is high even for the linear MMSE equalizer.

8.4.3 Banded-ICI Equalization

For a MIMO channel with a banded-ICI assumption at receiver side, the frequency measurement on the mth subcarrier at the νth receiving element is copied from (5.60) as

$$z_\nu[m] = \sum_{\mu=1}^{N_t} \sum_{k=m-D}^{m+D} H_{\nu,\mu}[m,k] s_\mu[k] + \eta_\nu[m]. \tag{8.95}$$

Assume that the noise is Gaussian distributed $\eta_\nu[m] \sim \mathcal{CN}(0, \sigma_\eta^2)$. Define a likelihood function as

$$f(z_\nu[m] | \{s_\mu[k]\}_{\forall \mu, \forall k}) = \frac{1}{\pi \sigma_\eta^2} \exp\left(-\frac{1}{\sigma_\eta^2} \left| z_\nu[m] - \sum_{\mu=1}^{N_t} \sum_{k=m-D}^{m+D} H_{\nu,\mu}[m,k] s_\mu[k] \right|^2 \right). \tag{8.96}$$

8.4.3.1 The Log-MAP Detector

Following the same steps to reach (8.62), we rewrite (8.95) into a convolutional form

$$\tilde{z}_\nu[m] = \sum_{\mu=1}^{N_t} \sum_{\ell=0}^{2D} \tilde{h}_{\nu,\mu}[m;\ell] s_\mu[m-\ell] + \tilde{\eta}_\nu[m], \tag{8.97}$$

with proper definitions on $\tilde{z}_\nu[m]$, $\tilde{\eta}_\nu[m]$ and the time-varying coefficients $\tilde{h}_{\nu,\mu}[m;\ell]$. The Log-MAP algorithm itself directly applies with the following two changes.

- The state in (8.63) is now changed to

$$(s_1[m-1], \ldots, s_{N_t}[m-1], \ldots\ s_1[m-2D], \ldots,, \cdots, s_{N_t}[m-2D]). \tag{8.98}$$

Hence, the number of states increases to M^{2DN_t}.
- The likelihood function $f(\tilde{z}[m]|\zeta', \zeta)$ used in (8.68) for the branch metric computation needs to be updated as $\prod_{\nu=1}^{N_r} f(\tilde{z}_\nu[m]|\zeta', \zeta)$, where $f(\tilde{z}_\nu[m]|\zeta', \zeta)$ is redefined along the expression in (8.96).

8.4.3.2 The Factor-Graph based Detector

The factor-graph based equalization algorithm as described in Section 8.3.2 can be directly applied with slight changes.

- The variable node in the factor graph is redefined as

$$\begin{aligned}\mathbf{x}_m = [&s_1[m-D], \ldots, s_{N_t}[m-D], \ldots, s_1[m], \ldots, s_{N_t}[m], \ldots, \\ &s_1[m+D], \ldots, s_{N_t}[m+D]]^{\mathrm{T}}\end{aligned} \tag{8.99}$$

which has length $N_t(2D+1)$.
- The check function $f(z[m]|\mathbf{x}_m)$ due to observations needs to be updated as $\prod_{\nu=1}^{N_r} f(z_\nu[m]|\mathbf{x}_m)$.
- The check function $C_{m,m-1}$ need to be updated as $\mathbf{x}_{m-1}$ and $\mathbf{x}_m$ share $2N_tD$ symbols. The marginalization operations in (8.77) are over N_t symbols $\{s_\mu[m-D-1]\}_{\mu=1}^{N_t}$.
- The check function $C_{m,m+1}$ need to be updated as $\mathbf{x}_m$ and $\mathbf{x}_{m+1}$ share $2N_tD$ symbols. The marginalization operations in (8.78) are over N_t symbols $\{s_\mu[m+D+1]\}_{\mu=1}^{N_t}$.

With these modifications, the message passing as described in (8.79) – (8.81) remains unchanged. The computational complexity increases to be the order of $\mathcal{O}(N_t^3(2D+1)^3)$.

8.5 MCMC Method for Data Detection in MIMO OFDM

The Markov-Chain Monte-Carlo (MCMC) method has been successfully applied for multiuser detection and MIMO detection [72, 117]. It has been applied for ISI equalization for single-carrier transmissions [309, 411] as well. In this section, we follow the presentation in [180, 181] on the MCMC method for ICI and co-channel interference (CCI) equalization in MIMO OFDM systems.

8.5.1 MCMC Method for ICI-Ignorant MIMO Detection

Note that the high complexity of the MAP equalizer lies in the exponential complexity to compute (8.91), where all the possible combinations are involved. In fact, only a handful of combinations, the *importance set*, contribute significantly to the summation in the numerator and denominator of (8.91). The MCMC method tries to find the *importance set* by browsing the possible choices of postulated data sequences in an efficient manner.

Table 8.1 Procedure of the Gibbs sampler

Initialize $\mathbf{s}^{(-N_{\rm bu})}[m]$ based on $\Pr(\mathbf{s}[m])$
For $n = -N_{\rm bu} + 1 : N_{\rm sa}$ draw $s_1^{(n)}[m]$ from $\Pr(s_1[m] \| s_2^{(n-1)}[m], \cdots, s_{N_{\rm t}}^{(n-1)}[m], \mathbf{z}[m], \mathbf{L}_1^{\rm apr}[m])$, ⋮ draw $s_\mu^{(n)}[m]$ from $\Pr(s_\mu[m] \| s_1^{(n)}[m], \cdots, s_{\mu-1}^{(n)}[m], s_{\mu+1}^{(n-1)}[m], \cdots, s_{N_{\rm t}}^{(n-1)}[m], \mathbf{z}[m], \mathbf{L}_\mu^{\rm apr}[m])$, ⋮ draw $s_{N_{\rm t}}^{(n)}[m]$ from $\Pr(s_{N_{\rm t}}[m] \| s_1^{(n)}[m], \cdots, s_{N_{\rm t}-1}^{(n)}[m], \mathbf{z}[m], \mathbf{L}_{N_{\rm t}}^{\rm apr}[m])$ End

The MCMC equalizer consists of two steps. The first step is termed as *burn-in* period, in which $N_{\rm bu}$ iterations of the Gibbs sampling are performed to let the Markov Chain converge to its nearest distribution [301]. In the second step, an important sample set Ω is generated with $N_{\rm sa}$ loops of Gibbs sampling after removing the redundant samples, based on which the extrinsic LLRs will be computed. The initializing samples $\mathbf{s}[m]^{(-N_{\rm bu})}$ can be drawn based on the prior information $\Pr(\mathbf{s}[m])$. The sampling procedure is summarized in Table 8.1.

The marginal probability used in the sampling process is computed as follows:

$$\Pr(s_\mu[m] | s_1^{(n)}[m], \cdots, s_{\mu-1}^{(n)}[m], s_{\mu+1}^{(n-1)}[m], \cdots, s_{N_{\rm t}}^{(n-1)}[m], \mathbf{z}[m], \mathbf{L}_\mu^{\rm apr}[m])$$
$$\propto f(\mathbf{z}[m] | s_1^{(n)}[m], \cdots, s_{\mu-1}^{(n)}[m], s_\mu[m], s_{\mu+1}^{(n-1)}[m], \cdots, s_{N_{\rm t}}^{(n-1)}[m]) \Pr(s_\mu[m]) \quad (8.100)$$

which can be readily evaluated based on the likelihood function expression in (8.90).

After the sampling process, there are multiple samples available. Denote $|\Omega|$ as the cardinality of the importance set, and $\mathbf{s}_\mu^{-(n)}[m]$ as $\mathbf{s}^{(n)}[m]$ with $s_\mu^{(n)}[m]$ removed. With these important samples, the LLR as needed by the channel decoder is computed as

$$L_{i,\mu}^{\rm ext}[m] \simeq \ln \frac{\sum_{n=1}^{|\Omega|} f(\mathbf{z}[m] | \mathbf{s}_\mu^{-(n)}[m], s_\mu[m] = \alpha_i) \Pr(\mathbf{s}_\mu^{-(n)}[m])}{\sum_{n=1}^{|\Omega|} f(\mathbf{z}[m] | \mathbf{s}_\mu^{-(n)}[m], s_\mu[m] = \alpha_0) \Pr(\mathbf{s}_\mu^{-(n)}[m])}. \quad (8.101)$$

Hence, the MCMC equalizer only uses the importance set to approximate the optimal solution.

The complexity of the MCMC equalizer, on the order of $\mathcal{O}(|\Omega| M N_{\rm t})$, depends on the size of the importance set Ω, the constellation size M, and the number of transmitters $N_{\rm t}$.

8.5.2 MCMC Method for Banded-ICI MIMO Detection

With slight changes, the MCMC method descried in Section 8.5.1 can be applied to the MIMO OFDM detection with banded ICI. Compared to the ICI-ignorant MIMO detection, there are $N_{\rm t}$ parallel data streams $\{s_\mu[k]\}_{\mu=1}^{N_{\rm t}}$. There are two different ways to define a mixed sequence from which the samples are drawn sequentially. Denote ℓ as the symbol index in the mixed sequence. One way is to draw all the symbols from one transmitter continuously before moving on to another transmitter, as

$$\{\tilde{s}[\ell]\}_{\ell=1}^{N_t K} = \left\{ s_1 \left[-\frac{K}{2} \right], \cdots, s_1 \left[\frac{K}{2} - 1 \right], \cdots, s_{N_t} \left[-\frac{K}{2} \right], \cdots, s_{N_t} \left[\frac{K}{2} - 1 \right] \right\} \quad (8.102)$$

The other is to group symbols in the same subcarriers next to each other, with a sequence defined as

$$\{\tilde{s}[\ell]\}_{\ell=1}^{N_t K} = \{\cdots, s_1[k], \cdots, s_{N_t}[k], s_1[k+1], \cdots, s_{N_t}[k+1], \cdots\} \quad (8.103)$$

The sampling procedure in Table 8.1 can be applied. Assume that symbol $\tilde{s}[\ell]$ corresponds to $s_\mu[m]$. The marginal probability used in the sampling process is

$$\begin{aligned} &\Pr(\tilde{s}[\ell] | \tilde{s}^{(n)}[1], \cdots, \tilde{s}^{(n)}[\ell-1], \tilde{s}^{(n-1)}[\ell+1], \cdots, \tilde{s}^{(n-1)}[KN_t], \{z_\nu[k]\}_{\forall\nu,\forall m}, \mathbf{L}_\mu^{\text{apr}}[m]) \\ &\propto f(\{z_\nu[k]\}_{\forall\nu,\forall m} | \tilde{s}^{(n)}[1], \cdots, \tilde{s}^{(n)}[\ell-1], \tilde{s}[\ell], \tilde{s}^{(n-1)}[\ell+1], \cdots, \tilde{s}^{(n-1)}[KN_t]) \Pr(s_\mu[m]) \\ &= \prod_{\forall\nu} \prod_{\forall k} f(z_\nu[k] | \tilde{s}^{(n)}[1], \cdots, \tilde{s}^{(n)}[\ell-1], \tilde{s}[\ell], \tilde{s}^{(n-1)}[\ell+1], \cdots, \tilde{s}^{(n-1)}[KN_t]) \Pr(s_\mu[m]) \end{aligned} \quad (8.104)$$

where the likelihood function can be evaluated based on (8.96).

Denote $\{\tilde{s}^{(n)}[\ell]\}_{\mu,m}^{-}$ as $\{\tilde{s}^{(n)}[\ell]\}$ with $s_\mu^{(n)}[m]$ removed. Similar to (8.101), the LLR is computed as

$$L_{i,\mu}^{\text{ext}}[m] \simeq \ln \frac{\sum_{n=1}^{|\Omega|} \prod_{\forall\nu} \prod_{\forall k} f(z_\nu[k] | \{\tilde{s}^{(n)}[\ell]\}_{\mu,m}^{-}, s_\mu[m] = \alpha_i) \Pr(\{\tilde{s}^{(n)}[\ell]\}_{\mu,m}^{-})}{\sum_{n=1}^{|\Omega|} \prod_{\forall\nu} \prod_{\forall k} f(z_\nu[k] | \{\tilde{s}^{(n)}[\ell]\}_{\mu,m}^{-}, s_\mu[m] = \alpha_0) \Pr(\{\tilde{s}^{(n)}[\ell]\}_{\mu,m}^{-})}. \quad (8.105)$$

Note that $s_\mu[m]$ only contributes to the mth and its $2D$ neighboring subcarriers. For further computational complexity reduction, the LLR in (8.105) can be approximated as

$$L_{i,\mu}^{\text{ext}}[m] \approx \ln \frac{\sum_{n=1}^{|\Omega|} \prod_{\forall\nu} \prod_{k=m-D}^{m+D} f(z_\nu[k] | \{\tilde{s}^{(n)}[\ell]\}_{\mu,m}^{-}, s_\mu[m] = \alpha_i) \Pr(\{\tilde{s}^{(n)}[\ell]\}_{\mu,m}^{-})}{\sum_{n=1}^{|\Omega|} \prod_{\forall\nu} \prod_{k=m-D}^{m+D} f(z_\nu[k] | \{\tilde{s}^{(n)}[\ell]\}_{\mu,m}^{-}, s_\mu[m] = \alpha_0) \Pr(\{\tilde{s}^{(n)}[\ell]\}_{\mu,m}^{-})}. \quad (8.106)$$

8.6 Bibliographical Notes

Symbol by symbol detection in additive white Gaussian channels and flat fading channels are well documented in textbooks on digital communications. Block based data detection has been extensively studied in the past 20 years in the multiuser detection and MIMO communication context. An optimal detector based on the maximum likelihood criterion is often computationally prohibitive. Linear equalizers, such as the zero-forcing (ZF) and MMSE detectors have low complexity but may suffer from severe performance loss. There are a variety of suboptimal or near-optimal detectors being developed in the literature such as: decision-feedback equalizers [318, 366], lattice reduction aided linear equalizers [464], sphere decoding algorithm,

MCMC-based equalizer. These equalization methods strike a balance in equalization performance and computational complexity between (near-)MAP equalizers and linear equalizers. The books [166, 406] are good references dedicated to data detection algorithms.

Channel equalization in the presence of intersymbol interference channels has been well treated in textbooks on digital communications. The use of Viterbi algorithm for maximum likelihood sequence estimation dates back to 1973 [122], while turbo equalization was first studied in 1995 [103]. The use of MCMC detectors for ISI channels was reported in [309], and the use of linear MMSE equalizers and a factor-graph based representation are introduced in [393] and [104] for ISI channels, respectively.

9

OFDM Receivers with Block-by-Block Processing

Section 5.4 provided a brief categorization of OFDM receivers based on the ICI level and the block-level processing procedure in burst transmissions. Chapter 9 to Chapter 12 will present OFDM receiver designs for different communication scenarios over single-input channels; while Chapter 13 to Chapter 15 will focus on various communication scenarios over multi-input channels. These chapters illustrate how the individual modules developed in Chapters 6–8 can be integrated to form complete receivers for specific applications.

For single-input channels, the focuses of the four chapters are briefly summarized as follows.

- *Chapter 9: Block-by-block receivers.* This chapter is centered on the OFDM receiver design to process each block individually. Several receiver processing schemes will be developed to accommodate the channel temporal variations within each block.
- *Chapter 10: Block-to-block adaptive receivers.* For the underwater acoustic channel with coherence time larger than a block duration, block-to-block receivers will be developed in this chapter to exploit the channel coherence property across OFDM blocks, which can achieve similar performance to the block-by-block processing with less pilot overhead.
- *Chapter 11: Receiver design for deep water acoustic communications.* As a counterpart to shallow water acoustic communications which has been a major focus of underwater communication literature, this chapter looks into the receiver processing techniques for deep water acoustic communications, with special focus on the *deep water horizontal channel* which has widely separated multipath clusters.
- *Chapter 12: Receiver design with external interference cancellation.* The underwater acoustic channel is known prone to interference from various sources, whereas scant attention has been paid to the external interference mitigation. Chapter 12 discusses a general interference cancellation method in the OFDM receiver processing.

OFDM for Underwater Acoustic Communications, First Edition. Shengli Zhou and Zhaohui Wang.
© 2014 John Wiley & Sons, Ltd. Published 2014 by John Wiley & Sons, Ltd.

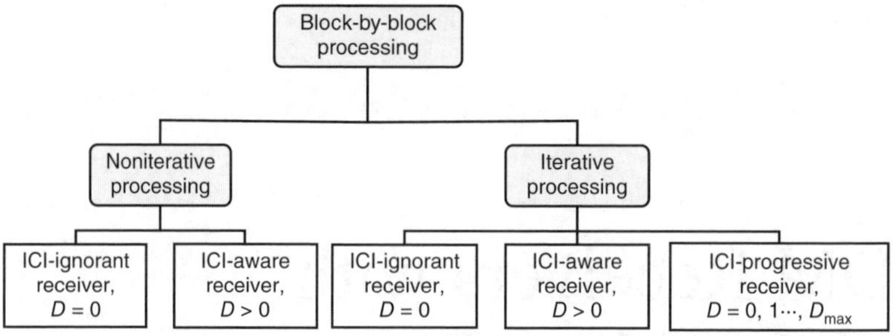

Figure 9.1 Illustration of block-by-block receiver categories, D : ICI depth.

The layout of this chapter on block-by-block OFDM receivers, as categorized in Figure 9.1, is as follows.

- Section 9.1 focuses on a noniterative ICI-ignorant receiver, where the ICI is treated as additive noise.
- Section 9.2 focuses on a noniterative ICI-aware receiver for communications over channels with large temporal variations, where the ICI is addressed explicitly in the receiver design.
- Section 9.3 briefly discusses iterative processing of the ICI-ignorant and ICI-aware receivers.
- Section 9.4 presents an ICI-progressive receiver that adapts the ICI-level to channel variations. Sections 9.5 and 9.6 contain extensive performance results of the ICI-progressive receiver via simulation and recorded data from field experiments.

9.1 Noniterative ICI-Ignorant Receiver

For a system with N_r receive elements, the input–output relationship for the ICI-ignorant receiver processing is copied from (5.47) as

$$z_\nu[m] = H_\nu[m]s[m] + \eta_\nu[m], \tag{9.1}$$

for $m = -K/2, \cdots, K/2 - 1$, and $\nu = 1, \cdots, N_r$.

9.1.1 Noniterative ICI-Ignorant Receiver Structure

The goal of receiver processing is to estimate the channel coefficients and recover transmitted information symbols. To this end, a noniterative ICI-ignorant receiver structure is depicted in Figure 9.2.

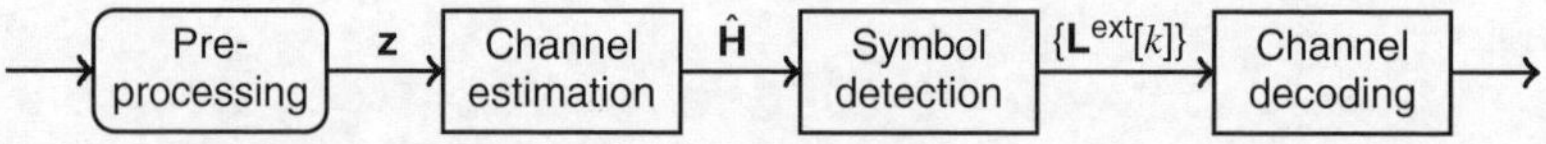

Figure 9.2 The noniterative ICI-ignorant receiver structure.

The receiver modules are briefly described in the following.

(i) *ICI-ignorant channel and noise variance estimation:* Taking the frequency measurements at pilot subcarriers as input, the ICI-ignorant channel estimator developed in Section 7.2 can be used to reconstruct the channel coefficients at all subcarriers. The frequency measurements at null subcarriers can be used to estimate the noise variance, as discussed in Section 7.6.

(ii) *Symbol detection and channel decoding:* An ICI-ignorant symbol detector in Section 8.1 can be adopted to recover the transmitted information symbols. The maximum ratio combination of frequency measurements at all the receiving elements is used to improve the detection performance. The calculated extrinsic information of data symbols is then fed into the channel decoder to decipher the information bits, as discussed in Chapter 3.

9.1.2 Simulation Results: ICI-Ignorant Receiver

The underwater acoustic channel is simulated according to the specifications in Section 5.5.1. The channel parameters are $N_{\rm pa} = 15$, $\Delta\tau = 1$ ms, $\Delta P_{\rm pa} = 20$ dB, $T_{\rm g} = 24.6$ ms, and $v_0 = 0$ m/s. The OFDM parameters are identical to the parameters used in the SPACE experiment, which are specified in Table B.1.

Figure 9.3 compares the ICI-ignorant receiver performance in the time-invariant channel and the time-varying channel with mild Doppler spread. One can find that all receivers can still achieve a low BLER, but at different levels of SNR. This implies that the level of ICI is below the necessary SNR for the LDPC code to decode successfully. The performance loss in Figure 9.3(b) is about 1.5 dB compared to the ICI-free case in Figure 9.3(a). It is postulated that the performance loss is due to the unaddressed ICI, and that channel estimation is not significantly affected by the model mismatch of the linear time invariant channel assumption. Between the considered channel estimators, the compressed sensing based algorithm outperforms the least squares estimator.

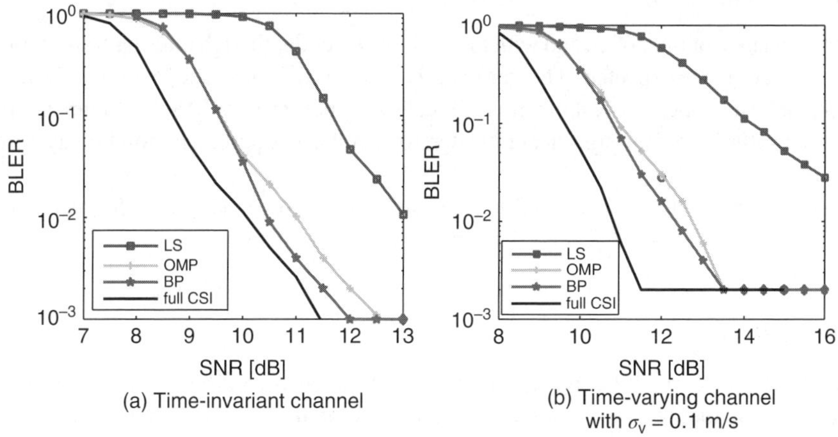

Figure 9.3 Performance comparisons for ICI-ignorant receivers with different channel estimation methods: least squares (LS), orthogonal matching pursuit (OMP), and basis pursuit (BP).

9.1.3 Experimental Results: ICI-Ignorant Receiver

The data sets collected in the SPACE08 experiment are used for the receiver performance evaluation. The experiment setup is described in Appendix B. The recorded data files in two different days, Julian Dates 297 and 300, are considered, where one day has rather calm sea and one day has severe wind activity, respectively. On Julian Date 300, the five files recorded during the afternoon were severely distorted and therefore unusable; only the remaining seven files recorded during the morning and evening are decoded. The received OFDM blocks with a 16-QAM constellation and a rate-$1/2$ nonbinary LDPC code are used, leading to a spectral efficiency 1.1 bits/s/Hz and a data rate 10.4 kb/s.

To test the performance of receiving algorithms, one common practice with the recorded data is to investigate the performance as a function of the number of phones combined. With more phones, the SNR after combining increases, hence performance improvement is due to both diversity effect and SNR increase. In this experimental data processing throughout this chapter, the phones are selected sequentially across the array, from top to the bottom. Channel estimation is performed for each phone individually. Channel equalization is performed based on signals received at multiple receiving elements.

The block-error-rate (BLER) is adopted as the performance measure, which is the average number of erroneous OFDM blocks after LDPC decoding. This is a reasonable performance criterion, since on unreliable channels such as UWA, it can be expected that there is a mechanism in place to recover lost blocks, e.g., automatic repeat-request (ARQ) or a higher layer block erasure code. In this context it has been recently shown that BLER's around 10^{-1} to 10^{-2} achieve optimal overall spectral efficiency [41], when combined with a higher layer erasure code.

The sample channel responses based on the LS estimators at different receiver locations are shown in Figure 9.4. The BLER performance combining an increasing number of phones is shown next.

9.1.3.1 S1 Data (60 m)

At a short distance of only 60 m and with the shallow water depth, rich multipath and significant Doppler variation are expected. This makes this receiver the most challenging in terms of its channel response, but the easiest in terms of received signal strength or SNR. Figure 9.4 reveals that there are three to four significant clusters of similar strength. The total delay spread is around 10 ms.

The BLER performance for Julian Dates 297 and 300 is shown in Figure 9.5, where the order of compressed sensing and LS stays the same as in the simulation.

9.1.3.2 S3 Data (200 m)

The middle distance might be the best tradeoff between channel difficulty and received SNR. The example channel responses in Figure 9.4 seem to be more contained, with a more dominating first cluster. The BLER performance in Figure 9.5 is generally better compared to the S1 receiver, where the LS performance improves relative to the sparse estimators on Julian date 297.

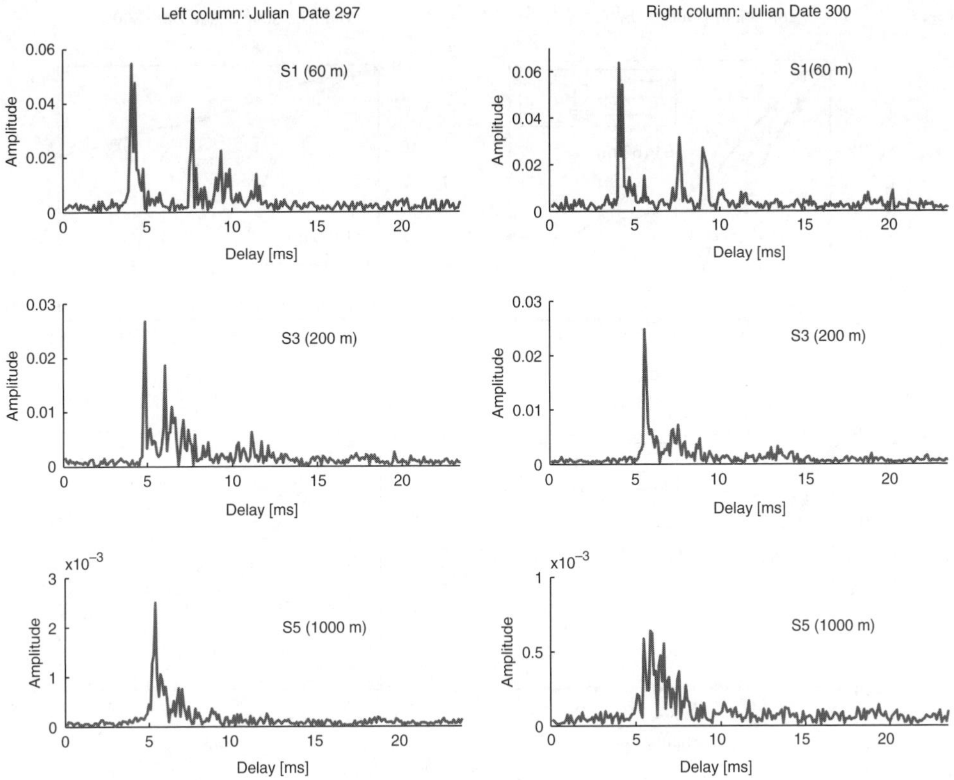

Figure 9.4 Examples of channel impulse responses from the SPACE08 experiment, as estimated using the Least Squares method.

9.1.3.3 S5 Data (1000 m)

At the 1 km distance only one significant cluster can be spotted in the channel estimates, and at the day with a large wave height (Julian Date 300) the received energy seems to be vanishingly small, c.f. Figure 9.4. Accordingly the trend of the LS channel estimator closing in on the compressed sensing algorithms continues. On the day with a large wave height the performance is generally not good, with even the CS algorithms successfully recovering only about 80% of the OFDM blocks.

9.2 Noniterative ICI-Aware Receiver

With a banded assumption on the ICI depth, the input–output relationship for the ICI-aware receiver is copied from (5.56) as

$$z_\nu[m] = \sum_{k=m-D}^{m+D} H_\nu[m,k]s[k] + \eta_\nu[m], \tag{9.2}$$

for $m = -K/2, \cdots, K/2-1$, and $\nu = 1, \cdots, N_r$.

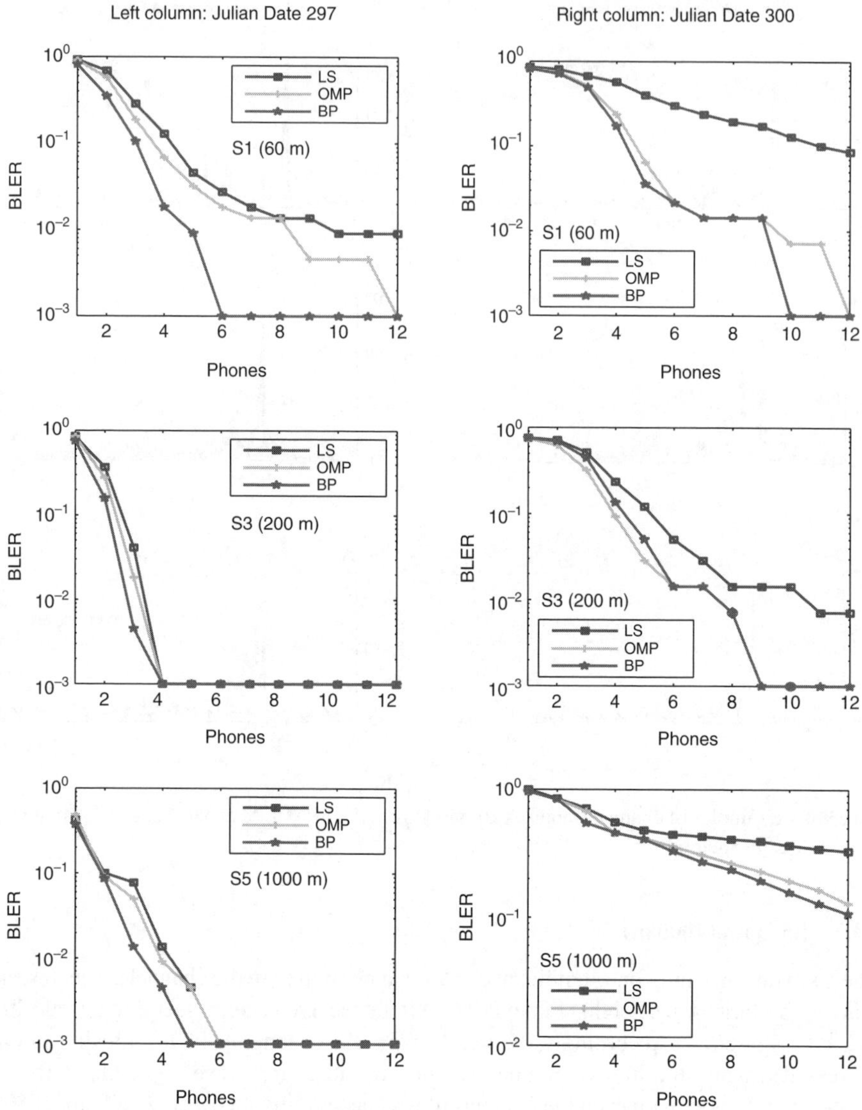

Figure 9.5 Performance results from the SPACE08 experiment using ICI-ignorant receivers at three different locations (S1, S3, and S5), first on a day with calm weather, then a day with large wave heights.

9.2.1 Noniterative ICI-Aware Receiver Structure

Similar to the ICI-ignorant receiver, the goal of receiver processing is to estimate the channel coefficients and recover transmitted information symbols. A noniterative ICI-aware receiver structure is depicted in Figure 9.6.

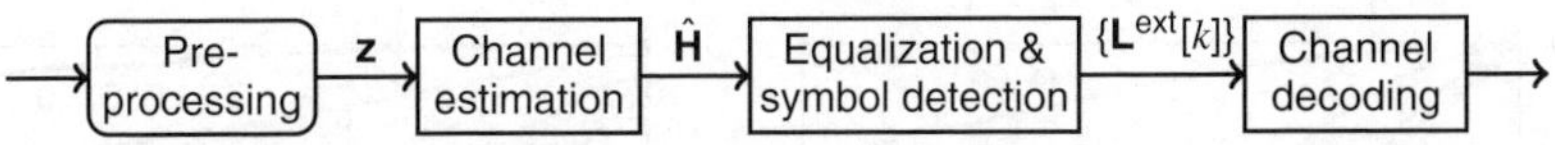

Figure 9.6 The noniterative ICI-aware receiver structure.

The receiver modules are briefly sketched in the following.

(i) *ICI-aware channel and noise variance estimation:* Taking the frequency measurements at the pilot and null subcarriers as input, the channel coefficients at all subcarriers and the noise variance can be reconstructed at each receiving element using the ICI-aware channel estimation and noise variance estimation methods discussed in Sections 7.3 and 7.6, respectively.

(ii) *ICI equalization and channel decoding:* Based on the frequency measurements at the data subcarriers, a banded-ICI equalizer in Section 8.3 can be adopted to recover the transmitted information symbols. The calculated extrinsic information of data symbols is then fed into the channel decoder to decipher the information bits, as discussed in Chapter 3.

9.2.2 *Simulation Results: ICI-Aware Receiver*

During the simulation, single receive element is assumed at the receiver side. A sampling rate being twice the bandwidth is used as the baseband sampling rate. The time-varying underwater acoustic channel is simulated as specified in Section 5.5.1. The OFDM signalling format is identical to that in the SPACE08 experiment as illustrated in Section B.1, but different from the regular pilot subcarrier distribution in the SPACE08 experiment, irregularly distributed pilot subcarriers are introduced: besides the 256 regularly distributed pilot subcarriers as in the SPACE08 experiment, 96 subcarriers from the rest active subcarriers are selected to form an irregular pilot subcarrier distribution pattern, leading to 352 pilot subcarriers and 576 data subcarriers in total. The newly added pilots are grouped in clusters between zero subcarriers and the regularly distributed pilots, creating groups of five consecutive known subcarriers. Adjacent observations are needed to the ICI coefficients in time-varying channels.

Since 96 coded symbols are assumed known while the same LDPC code structure is used (code truncation), this leads to an equivalent coding rate of $r_c = (336 - 96)/(672 - 96) \approx 0.4$. With a 16-QAM the spectral efficiency and the data rate are updated as

$$\alpha = \frac{T}{T+T_g} \cdot \frac{576}{1024} \cdot r_\mathrm{c} \cdot \log_2 16 = 0.76\,\mathrm{bits/s/Hz}, \tag{9.3}$$

$$R = \alpha B = 7.4\,\mathrm{kb/s}. \tag{9.4}$$

To assess the need for explicit ICI equalization, Figure 9.7 shows the BLER performance with different values of ICI-depth in the presence and absence of the channel state information (CSI) respectively. A linear MMSE equalizer discussed in Section 8.2.2 is used for ICI equalization. One can observe that more ICI can be removed by choosing a larger D, whereas the receiver has to accept higher computational complexity. Due to the change in coding rate, 1 dB gap can be observed in the scenarios with and without full CSI when sufficient levels of ICI is removed.

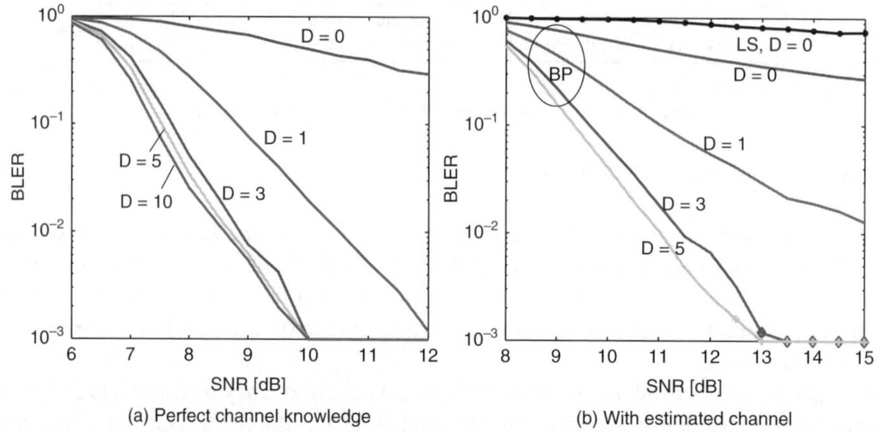

Figure 9.7 BLER performance, only D off-diagonals from each side are kept in the channel matrix for data demodulation. The simulated channel has a severe Doppler spread with $\sigma_v = 0.25$ m/s.

9.2.3 *Experimental Results: ICI-Aware Receiver*

During the SPACE08 experiment, the decoding performance in Figure 9.5 indicates that the performance on Julian Date 300 was limited, most likely due to the ICI caused by significant Doppler spread which degrades the effective SNR of the ICI-ignorant receivers. The effectiveness of the ICI-aware receiver is now tested using the data sets collected on Julian Date 300.

To improve the channel estimation performance in the presence of severe ICI, similar to the irregular pilot distribution in simulations for the ICI-aware receiver evaluation, 96 data subcarriers are converted into additional pilots by assuming that 96 data symbols are known a priori. For a 16-QAM constellation, the spectral efficiency and the data rate are computed in (9.3) and (9.4), respectively.

The performance improvement for ICI-aware receivers can be seen in Figure 9.8. As a comparison the LS and sparse channel estimators operating ICI-ignorant ($D = 0$) are included, as they also benefit from the additional pilots and reduced coding rate. These plots clearly highlight that on channels with severe Doppler spread, adopting ICI-aware channel estimation and equalization yields significant performance gain, and that using sparse channel estimation another significant performance gain can be achieved compared to a conventional LS channel estimator.

9.3 Iterative Receiver Processing

Iterative processing has been widely used to improve the receiver decoding performance. In this section, a brief overview on the iterative operations in the ICI-ignorant and ICI-aware receivers is provided. In these two iterative receivers the system model keeps unchanged during iterations. In Section 9.4, we will look into another type of iterative receiver where the system model keeps updated according to the channel condition.

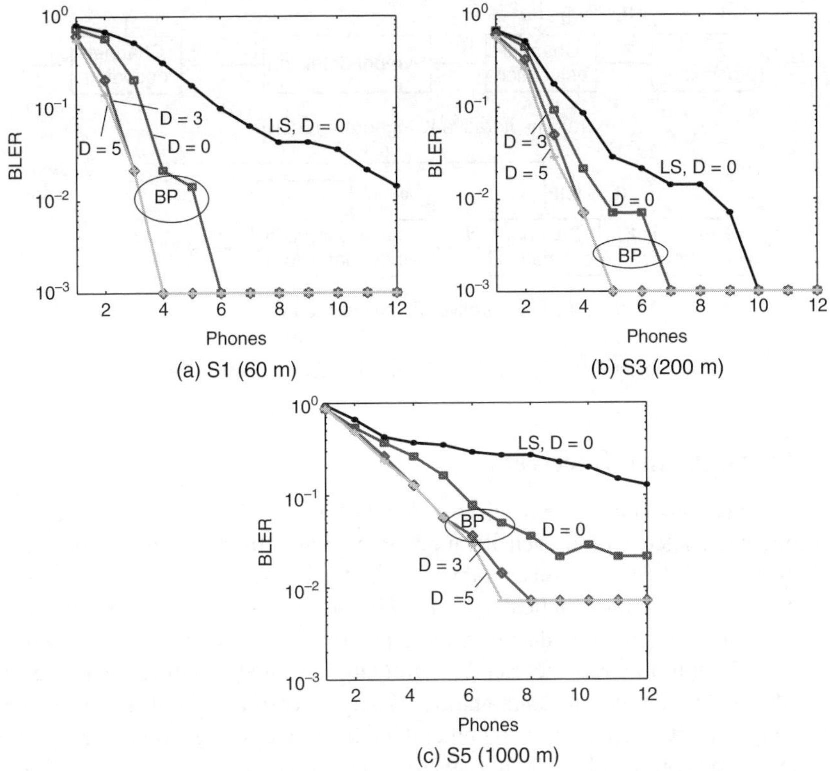

Figure 9.8 Performance results from the SPACE08 experiment using the ICI-aware receiver. The results are averaged over data files collected in Julian date 300 and OFDM blocks with a 16-QAM constellation at a data rate of 7.4 kb/s.

9.3.1 *Iterative ICI-Ignorant Receiver*

One typical iterative receiver structure for ICI-ignorant is shown in Figure 9.9, where the channel estimation is refined based on the estimated data symbols from the channel decoder. Due to the absence of ICI, the feedback from the channel decoder will not be used for symbol detection. Based on the *a posteriori* probability from the decoding, several feedback strategies from the channel decoder for channel estimation have been be described in Section 3.4. Please refer to [201, 203] for more discussion on the iterative ICI-ignorant receiver.

9.3.2 *Iterative ICI-Aware Receiver*

The iterative ICI-aware receiver is shown in Figure 9.9(b). Similar to the iterative ICI-ignorant receiver, the channel estimation is included into the iteration loop for channel estimate refinement. The interaction between the channel equalization and decoding is identical to that in the turbo equalization as discussed in Chapter 8, in which the extrinsic information is exchanged between equalizer and channel decoder.

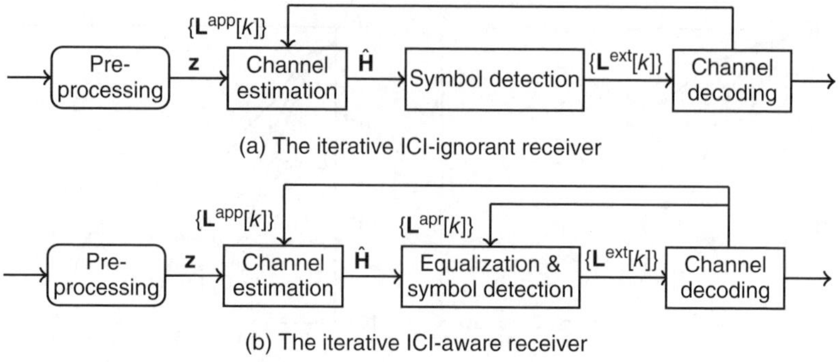

Figure 9.9 Illustration of iterative receiver structures.

9.4 ICI-Progressive Receiver

Note that in the canonical iterative receiver design in Figure 9.9, the receiver processing procedure keeps unchanged within each iteration. Particularly for UWA communications, the assumed channel ICI depth is a constant throughout iterations. This design philosophy might raise one critical issue to the practical system which has no access to the information about channel variations. For a reliable data transfer, a practical receiver prefers a *safe* hence a relatively large ICI depth to cover the scenario with large channel variations. However, for the channel with small variations, the computational complexity of the ICI-aware receiver assuming a relatively large ICI depth, is much higher than that of the ICI-ignorant receiver, meaning more power will be consumed by the ICI-aware processing unit.

Since the computational efficient receiving algorithm is critical to underwater networks, it is necessary to design a receiver to adapt the ICI model to the unknown channel condition during iterations. The ICI-progressive receiver is essentially an iterative receiver in nature, which follows the turbo principle. However, in contrast to the canonical iterative receiver structure (e.g. [217]), the system model used for channel identification and data demodulation changes at each iteration. It starts with a simple channel model that allows for ICI-ignorant processing, and then proceeds to ICI-aware processing where the severity of the assumed ICI increases as the iteration goes on. The soft information obtained from the previous iteration contributes to channel estimation and data demodulation for the current iteration. This way, the receiver can self adapt to the *unknown* degree of channel variation progressively. The proposed receiver keeps the complexity low when the channel conditions are good, while still maintaining excellent performance when the channel conditions deteriorate.

The effective system model for the ICI-progressive processing is

$$
\begin{aligned}
\mathrm{z} &= \mathbf{H}_D \mathrm{S} + \underbrace{(\mathbf{H} - \mathbf{H}_D)\mathrm{S} + \mathrm{w}}_{:=\eta} \\
&= \mathbf{H}_D S + \eta \qquad (9.5)
\end{aligned}
$$

where η is the effective noise. In the proposed progressive receiver, the parameter D increases as the iteration goes on, and hence more severe ICI can be addressed as the receiver processing proceeds to deal with channels with large Doppler spread.

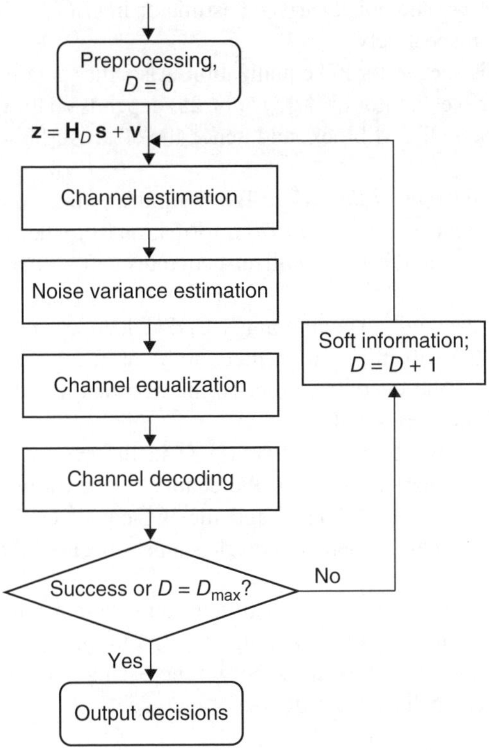

Figure 9.10 Progressive receiver structure.

As depicted in Figure 9.10, the ICI-progressive receiver consists of the following steps.

(i) *Channel estimation:* Estimate the channel matrix $\mathbf{H}_D$ based on the assumed channel model given in (9.5). The sparse channel estimators developed in Sections 7.2 and 7.3 are adopted.

During the first iteration, only the measurements on the pilot subcarriers are used for channel estimation. In later iterations, the inputs to the channel estimator include (i) the measurement vector $\mathbf{z}$, (ii) the pilot symbols, and (iii) the *a posteriori* probabilities (APP) of the information symbols from the channel decoder. Here, either soft decisions or hard decisions on information symbols are used. Denote the decision on the information symbol by $\bar{s}[k]$. Denote $\mathcal{S}_{\rm P}$, $\mathcal{S}_{\rm N}$ and $\mathcal{S}_{\rm D}$ as the sets of pilot subcarriers, null subcarriers, and data subcarriers, respectively. Hence, an estimate of $\mathbf{s}$ (denoted by $\hat{\mathbf{s}}$) used for channel estimation can be formed as

$$\hat{s}[k] = \begin{cases} s[k], & k \in \mathcal{S}_{\rm P} \\ 0, & k \in \mathcal{S}_{\rm N} \\ \bar{s}[k], & k \in \mathcal{S}_{\rm D}. \end{cases} \tag{9.6}$$

(ii) *Noise variance estimation:* After channel estimation, the variance of the effective noise η is updated, as more ICI will be modeled as opposed to being treated as additive noise with an increasing D. The noise variance estimator in (7.66) and (7.70) are adopted for $D = 0$ and $D > 1$, respectively.

Noise variance is needed for ICI equalization. Also, the sparsity factor ζ in the ℓ_1-norm based sparse channel estimator in (7.13) or (7.25) depends on the effective signal-to-noise ratio (SNR) for each OFDM block, and hence the effective SNR will be updated as the iteration goes on.

(iii) *ICI equalization and symbol detection:* By using the estimated channel matrix $\mathbf{H}_D$, the equivalent noise variance, and the *a priori* information from the nonbinary LDPC decoder in the previous iteration, the ICI equalizer generates soft output on the reliability of the data symbols.

(iv) *Nonbinary LDPC decoding:* The nonbinary LDPC decoder yields the decoded information symbols and the soft information that can be used for channel estimation and ICI equalization. During the decoding process, the decoder will declare success if all the parity check conditions are satisfied.

(v) *Iterations among steps (i) to (iv):* Increase D in the system model and increase the assumed maximum Doppler spread of the channel to be estimated. Feed back the soft information to the channel estimator and the ICI equalizer. Iteration stops when the decoder declares a success, or when D reaches a pre-specified number $D_{\max}$.

In the receiver depicted in Figure 9.10, each iteration is associated with a different D. The receiver can also iterate multiple times among step (ii) to step (iv) on channel estimation, equalization and decoding for each fixed D, before increasing D to update the system model. For exposition simplicity, such a possibility is not included.

9.5 Simulation Results: ICI-Progressive Receiver

With an identical simulation setup described in Section 9.2.2, Figures 9.11 and 9.12 show the performance of the progressive receiver with QPSK and 16-QAM constellations, respectively, where only one receive phone is used. The BLER after LDPC decoding is used as the

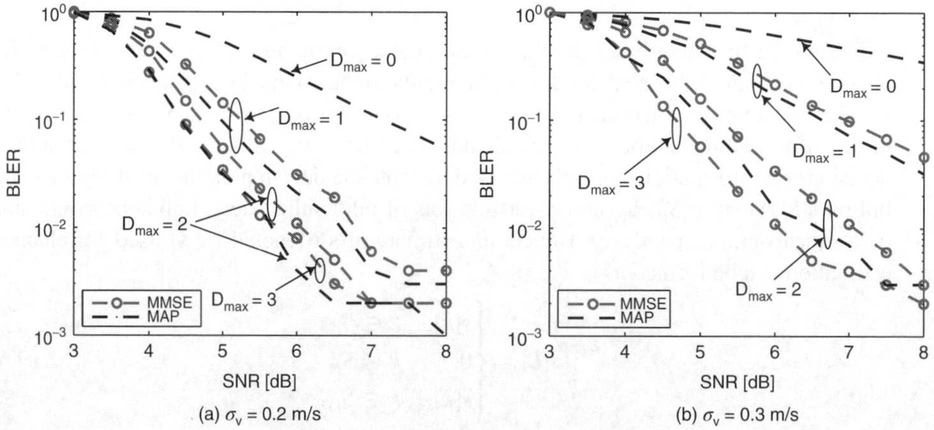

Figure 9.11 Simulated performance for the progressive receiver with different $D_{\max}$, QPSK.

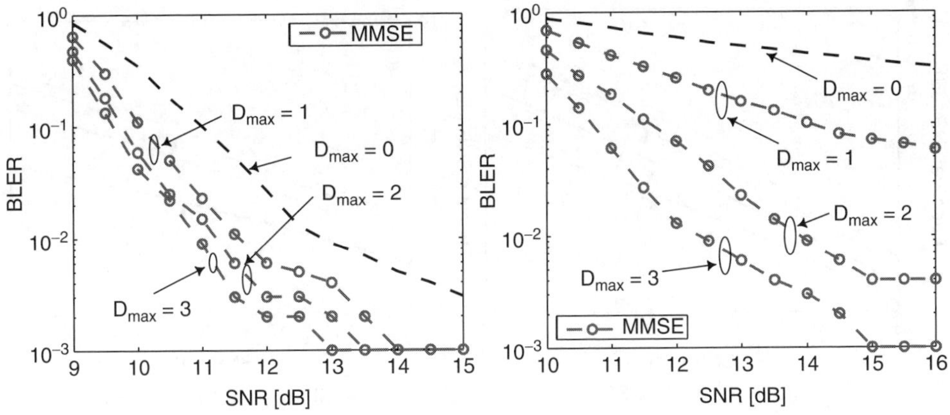

(a) $\sigma_v = 0.1$ m/s (b) $\sigma_v = 0.2$ m/s

Figure 9.12 Simulated performance for the progressive receiver with different D_{max}, 16-QAM.

performance metric. The results are averaged over at least 1000 channel realizations or when 50 block errors are detected. The related parameters are $D_{max} = 3$, and for sparse channel estimation in (7.13) or (7.25), the delay resolution $\lambda_b = 2$, the Doppler resolution $\Delta b = 4 \times 10^{-5}$. As $\Delta b = \Delta v/c$, the corresponding velocity step size is $\Delta v = 0.06$ m/s. During the iteration process, $N_b^D = 7, 11$ and 15 for $D = 1$, 2 and 3, respectively. Due to the high complexity, only the MAP equalization results for QPSK up to $D_{max} = 2$ are reported. It can be observed that the MAP equalization outperforms the linear MMSE equalizer slightly, while both of them achieve significant performance improvement relative to the ICI-ignorant receiver.

For 16-QAM with good channel conditions (see Figure 9.12(a)), at an operating SNR of 11 dB, more than 90% OFDM symbols can be decoded in the first round, i.e., using the ICI-ignorant receiver. In more adverse channel conditions (see Figure 9.12(b)), the ICI-ignorant receiver has very poor performance, decoding barely half of the OFDM symbols at 13 dB, while with $D_{max} = 1$ about 80% of the OFDM symbols can be decoded, almost 97% at $D_{max} = 2$, and more than 99% for $D_{max} = 3$. This also means that only 20% of the time $D = 2$ has to be used and less than 3% of the time the algorithm runs to $D = 3$.

In the progressive receiver, the effective noise variance is re-estimated during each iteration, as shown in (7.68). Define the effective SNR as the energy ratio of the signal portion to the effective noise as in (7.69). Figure 9.13 illustrates how the effective SNR changes during the progressive process across a certain range of SNR, where $\sigma_v = 0.3$ m/s and $\sigma_v = 0.2$ m/s for QPSK and 16-QAM, respectively. As more ICI is addressed, rather than being regarded as additive noise, the effective SNR increases as the iteration goes on.

The complexity issue is briefly explored next. Because both the sparse channel estimation and the LDPC decoding are iterative processes, which can stop at any time once the stopping criteria are met, the FLOPs of individual algorithms will not be counted. Instead, the average receiver processing time per block for the proposed receiver is adopted. The numerical results were carried on under MATLAB 2007b, on a personal computer with an Intel(R) Core(TM)2 CPU 6600@2.4 GHz and 3GB of memory. A total of 10^4 OFDM blocks were tested for each SNR point. Figure 9.14 shows the overall complexity of the progressive receiver

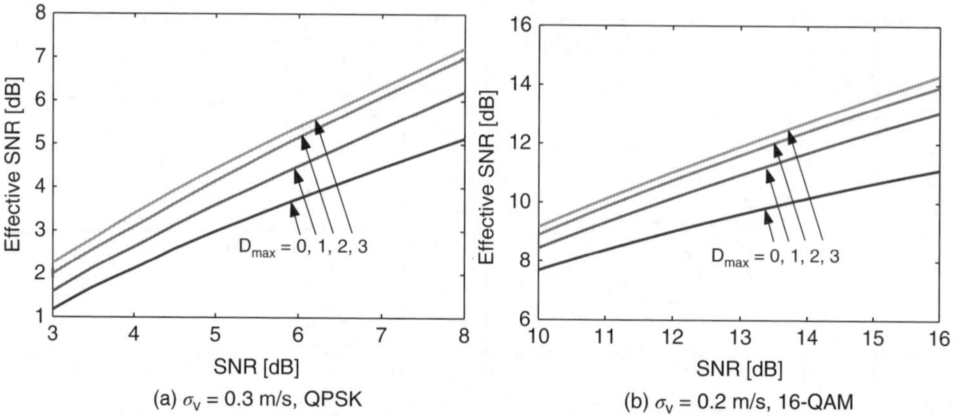

Figure 9.13 Estimated effective SNR during the progressive process. The linear MMSE equalization.

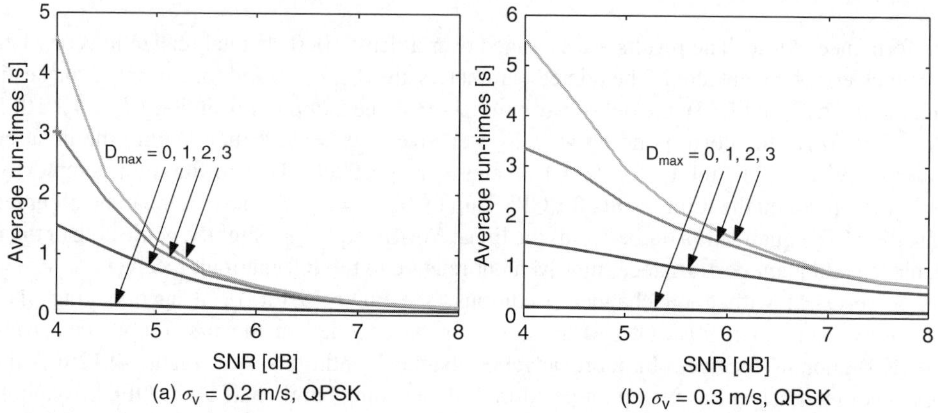

Figure 9.14 Average CPU run-times per OFDM block with the progressive receiver for different $D_{\max}$. The linear MMSE equalization.

for the setting of Figure 9.11. In Figure 9.14(a) when the channel conditions are good, the average total run-times for different $D_{\max}$ are close to the ICI-ignorant receiver at the medium to high SNR region. This verifies that the proposed receiver structure keeps the complexity low automatically when the channel conditions are good. In Figure 9.14(b) corresponding to more challenging channels, the trend is similar that the average run-times decrease as the SNR improves. However, the complexity is larger than the ICI-ignorant receiver, as a large portion of OFDM blocks can only be recovered after explicit ICI mitigation. In this setting, the receiver complexity with $D_{\max} = 2, 3$ converges to that of $D_{\max} = 1$ at high SNR, suggesting that the $D = 2$ and $D = 3$ iterations are used infrequently.

The progressive receiver needs to implement all the functions of different D values. However, the progressive receiver will likely be run on software-defined modems [445], where storage is not a concern. Rather, the processing speed is the main focus in order to meet real-time data processing requirements.

9.6 Experimental Results: ICI-Progressive Receiver

The ICI-progressive receiver is evaluated using the data files recorded in the SPACE08 experiment on Julian dates 295–302 at three receiving element arrays, labeled as S1, S3, and S5, which were 60 m, 200 m, and 1000 m from the transmitter, respectively. The Doppler resolution and the dictionary size are the same as used in the simulation. The typical channel responses can be found in Figure 9.4.

9.6.1 BLER Performance

Table 9.1 reports the number of OFDM blocks that *have not been decoded correctly* as D increases in the progressive receiver, using the MMSE equalizer, with different number of phones combined. The data across eight days (Julian dates 295–302) is used. Since some recorded files are corrupted, there are a total of 1560, 1640 and 1600 blocks processed for S1, S3 and S5, respectively. Combining 12 receiving elements, *all blocks* in S1 and S3 are decoded correctly using the progressive receiver when it reaches $D = 3$. There are 9 blocks that cannot be decoded in S5.

Figure 9.15 shows the block success rate averaged over the eight consecutive days using the proposed progressive receiver with the MMSE equalizer. Due to the geometry, rich multipath and significant Doppler variation are expected at short (S1) to medium (S3) ranges. When the number of receiving elements is small, the performance of the ICI-ignorant receiver ($D = 0$) is limited, and many more OFDM symbols can be decoded by applying the progressive procedure, with a larger D. When the number of receiving elements is large, the ICI-ignorant receiver already achieves excellent results for all the blocks. Checking the results using four receiving elements, about 90% OFDM blocks can be decoded at the $D = 0$ stage, and the success rate increases to 95% when $D_{\text{max}} = 1$, and up to 98.8% when $D_{\text{max}} = 3$.

For S5, similar trends as S1 and S3 can be observed, but the gap between the ICI-ignorant and progressive receivers gets smaller. When four receiving elements are combined, over 93% blocks can be decoded by ignoring the ICI, and the success rate increases to 96% when the progressive receiver reaches $D = 3$.

9.6.2 Environmental Impact

Using four receiving elements for combining, Table 9.1 shows that there are 19 out of 1560 blocks with decoding errors in S1, 57 out of 1640 blocks with decoding errors in S3, and 66 out of 1600 blocks with decoding errors in S5, for the progressive receiver with $D_{\text{max}} = 3$. Figure 9.16 illustrates the success level of each transmission of 20 OFDM blocks across the 8-day period. Each day, there are about 12 files recorded (a few files are corrupted). "All success" means that all 20 blocks in that file, of duration $20(T + T_g) = 2.59$ s, can be decoded, while "With errors" means that some blocks cannot be decoded out of 20 blocks in the file.

Table 9.1 The number of undecoded OFDM blocks for different values of D_{max} in the progressive receiver. Julian dates 295-302. 16-QAM, rate 1/2 coding, linear MMSE based ICI equalization

System	S1 (60 m); 1560 blocks				S3 (200 m); 1640 blocks				S5 (1000 m); 1600 blocks			
# of	D_{max}				D_{max}				D_{max}			
Phones	0	1	2	3	0	1	2	3	0	1	2	3
1	1178	1048	951	877	1229	1141	1090	1046	809	758	743	732
2	731	565	431	298	775	679	630	583	395	337	296	277
3	350	215	123	73	470	368	299	259	179	136	119	104
4	152	77	38	19	213	126	80	57	109	86	76	66
5	70	32	13	5	88	47	24	15	82	65	51	37
6	36	19	8	4	44	17	9	8	68	46	30	23
7	24	12	6	2	16	9	4	0	53	34	21	18
8	19	10	4	1	11	3	0	0	45	22	14	12
9	16	7	3	0	4	2	0	0	37	19	14	13
10	13	6	2	0	2	0	0	0	26	16	14	13
11	11	5	2	0	0	0	0	0	25	17	11	11
12	9	5	2	0	0	0	0	0	23	12	10	9

Source: Huang 2011, Table I, p. 1533. Reproduced with permission of IEEE.

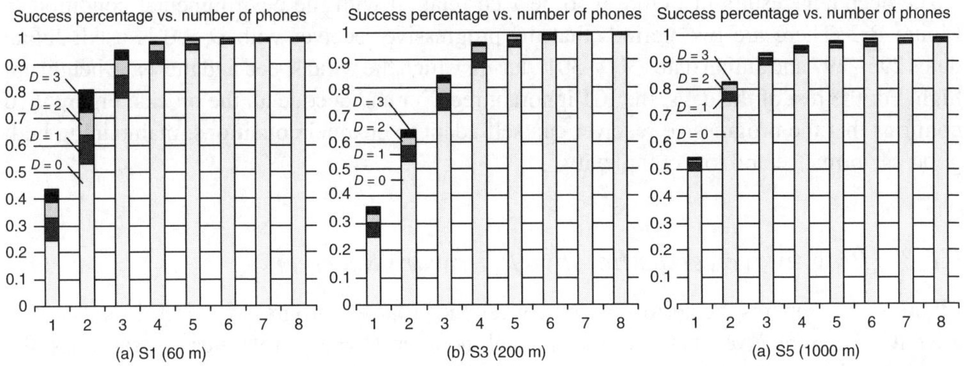

Figure 9.15 The block success percentage averaged over Julian dates 295–302, SPACE08, linear MMSE based ICI equalization. (Source: Huang 2011, figure 9, p. 1534. Reproduced with permission of IEEE.).

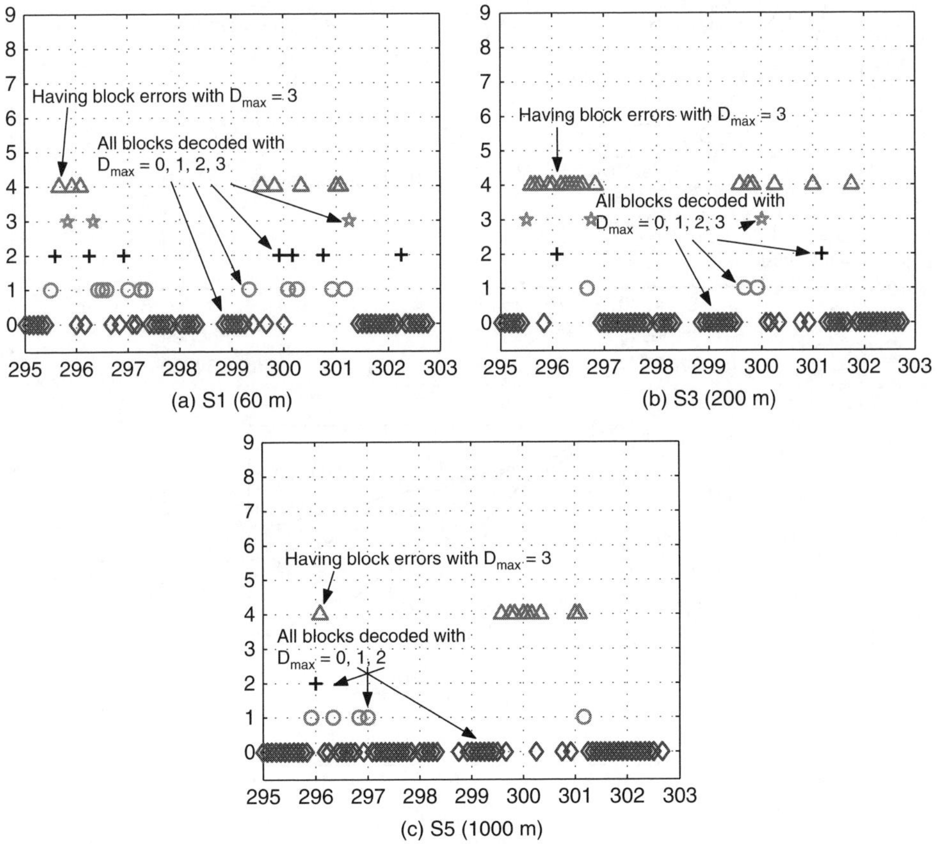

Figure 9.16 Success level for each transmission of 20 OFDM blocks; four receiving elements. Markers are placed at convenient heights for illustration purposes only.

The decoding results in Figure 9.16 bear correlation with the environmental condition in Figure B.2. There are two periods that the progressive receiver with $D > 0$ is used: Julian dates 296–297 and Julian dates 300–301, during which the wind speed and the wave height are high. For the rest of the days, the ICI-ignorant receiver can decode all the blocks. Figure 9.16 confirms that the progressive receiver can self adapt to channel conditions, maintaining both good performance and low complexity.

9.6.3 Progressive versus Iterative ICI-Aware Receivers

Figure 9.17 compares the performance between the proposed progressive receiver and an iterative ICI-aware receiver that fixes the channel model at $D = 3$, but iterates several times; the channel with large Doppler spread (Julian date 300) is considered. Obviously, the latter receiver

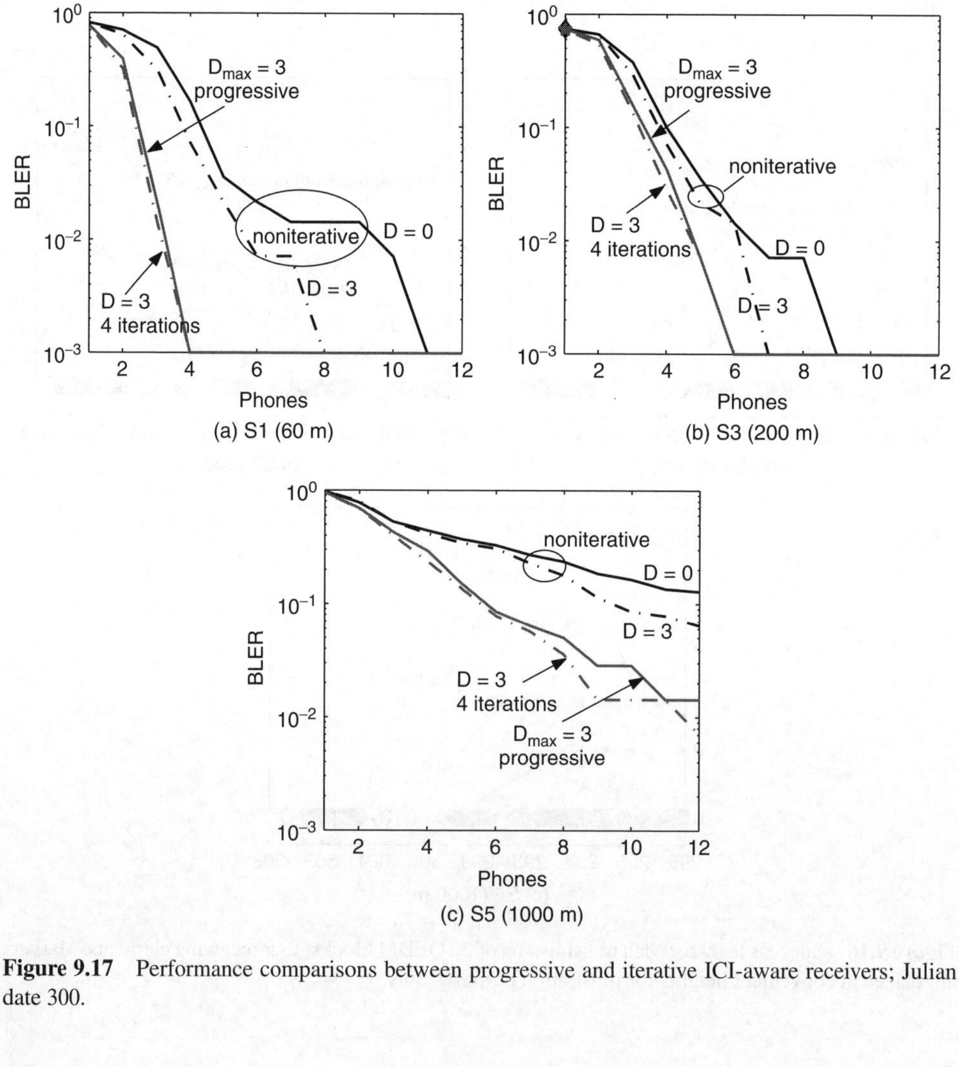

Figure 9.17 Performance comparisons between progressive and iterative ICI-aware receivers; Julian date 300.

has much higher complexity, and is not well-motivated in good channel conditions. After the first iteration, the ICI-aware receiver outperforms the ICI-ignorant receiver. As the iteration continues, the progressive receiver catches up with the iterative ICI-aware receiver, and the performance difference is negligible. Hence, the progressive receiver collects the performance benefits as the iterative ICI-aware receiver, but enjoys much lower complexity in various channel conditions.

9.7 Discussion

Different receivers have different characteristics. In practical systems, there are various factors in selecting a suitable receiver structure.

- ICI-ignorant receivers assume that all the paths have a similar Doppler rate, and hence the ICI can be ignored after the main Doppler scale effect compensation. As such, the channel mixing matrix is diagonal, leading to low-complexity channel estimation and data demodulation. The ICI-ignorant receiver works well in good channel conditions, but its performance degrades considerably in adverse channel conditions.
- ICI-aware receivers assume that all the paths have different Doppler rates, and hence ICI exists and should be explicitly dealt with by the receiver. The ICI-aware receiver achieves excellent performance even in adverse channel conditions. However, a large number of pilots is needed for channel estimation, leading to a spectral efficiency reduction. Moreover, the complexity is much higher than that of ICI-ignorant receivers.
- The ICI-progressive receiver adapts themselves to channel conditions, without any prior knowledge on the channel time-varying properties. It will have as low complexity as the ICI-ignorant receiver in good channel conditions, while achieving as excellent performance as the ICI-aware receiver in adverse channel conditions. Furthermore, compared to the non-iterative receiver, the progressive receiver does not require any extra pilot overhead.

9.8 Bibliographical Notes

Studies of underwater acoustic OFDM prior to 2005 can be found in e.g., [29, 48, 77, 86], [211, 225, 458]. The research activities have been increased significantly since 2006.

This chapter focuses on the block by block receivers for a single-transmitter system. The receivers in [201–203, 235, 457] assume no ICI after Doppler compensation, while the receivers in [38] and [392] addressing the ICI explicitly. The ICI progressive receiver was developed in [180]. Other block by block receiver processing approaches include the multiband OFDM approach [232] and combination of OFDM with the time-reversal approach [149, 257]. The block-to-block receivers that utilize the channel coherence across blocks will be covered in Chapter 10, and the extension to MIMO OFDM will be covered in Chapter 13.

10

OFDM Receiver with Clustered Channel Adaptation

Chapter 9 focuses on the block-by-block receiver processing, where channel paths are estimated solely based on the measurements from one individual block. This chapter will present an OFDM receiver leveraging the measurements from both the current and past blocks. This allows the system to reduce the number of pilots while maintaining robust operation over fast-varying channels.

This chapter is organized as follows.

- Section 10.1 illustrates channel dynamics, notably different variations across multipath clusters. Section 10.2 presents a cluster-based channel variation model to parameterize the channel variation from block to block.
- Section 10.3 describes the design of an adaptive OFDM receiver, developing key receiver modules incorporating channel adaptation.
- Sections 10.4 and 10.5 contain performance results of the block-to-block adaptive receiver using data sets collected in two field experiments.

10.1 Illustration of Channel Dynamics

In general, underwater acoustic channels have large temporal variations. From the receiver point of view, whether the channels vary fast or slow depending on the relationship between the channel coherence time and the transmission block length. Fig. 10.1 presents the channel estimate samples in two experiments. Two important observations are as follows

- Channel paths tend to be clustered; according to the ray-tracing model [141], there are many small paths centering around the eigen-paths associated with the medium refractions and surface/bottom bounces.
- The channel coherence time varies across clusters, as different clusters are usually generated according to different mechanisms. For example, the clusters formed by medium refraction usually have larger coherence time than those formed by surface/bottom bounces.

OFDM for Underwater Acoustic Communications, First Edition. Shengli Zhou and Zhaohui Wang.
© 2014 John Wiley & Sons, Ltd. Published 2014 by John Wiley & Sons, Ltd.

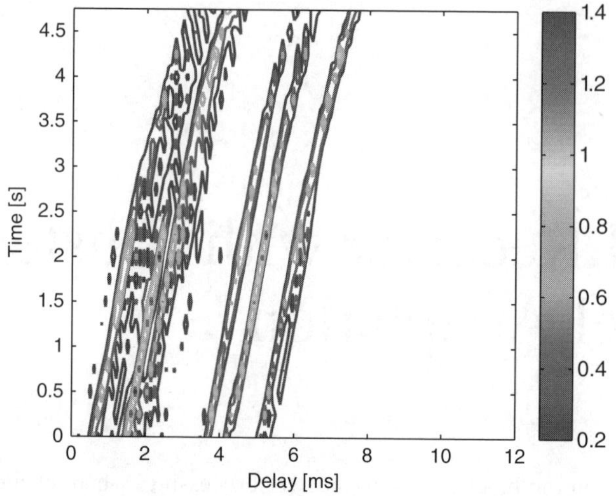

(a) Mobile experiment: MACE10

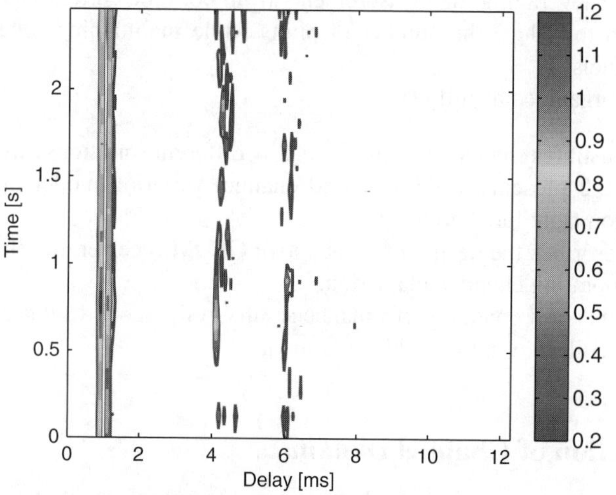

(b) Stationary experiment: SPACE08

Figure 10.1 Time variations of block-by-block channel estimates over 20 consecutive OFDM blocks. A constant-time (horizontal) slice across delay represents the magnitude of the channel impulse response. No resampling is performed.

A model that describes the channel variation accurately is important for the adaptive receiver design.

10.2 Modeling Cluster-Based Block-to-Block Channel Variation

Assume that the data burst is divided into multiple blocks, with each block of duration T_{bl}, and let n denote the block index. Denote $A_p[n]$, $\tau_p[n]$ and $a_p[n]$ as the amplitude, initial delay and the Doppler rate of the pth path during the nth block transmission, respectively. A path-based channel model with path-specific Doppler scales yields the channel impulse response

$$h(t;\tau,n) = \sum_{p=1}^{N_{\rm pa}} A_p[n]\delta(\tau - (\tau_p[n] - a_p[n]t)). \tag{10.1}$$

Based on the channel model in (10.1), one may want to model the variation of each path individually. However, the effort of characterizing the variation of each path could be similar to that of directly estimating the parameters of each path. Here, utilizing the clustering property of channel paths, a cluster-based model for channel variations across blocks was proposed in [426] via two steps:

- *Step 1*: Split the channel paths into $N_{\rm cl}$ clusters as

$$\left\{\left(A_p[n],\tau_p[n],a_p[n]\right)\right\}_{p=1}^{N_{\rm pa}} = \bigcup_{i=1}^{N_{\rm cl}} \left\{\left(A_p[n],\tau_p[n],a_p[n]\right); \forall p \in \Omega_i\right\} \tag{10.2}$$

  where Ω_i is a collection of paths within the ith cluster.
- *Step 2*: Assume that all paths within one cluster share the same variations $\{\overline{\gamma}_i, \Delta\tau_i, \Delta b_i\}$ on the complex amplitude, the delay, and the Doppler scale between adjacent blocks; however, variations on different clusters are independent. The triplets of paths within the ith cluster of the nth block after the cluster-offset compensation are

$$\begin{cases} A_p[n|n-1] = A_p[n-1]\cdot\overline{\gamma}_i, \\ \tau_p[n|n-1] = \tau_p[n-1] + \Delta\tau_i, \\ a_p[n|n-1] = a_p[n-1] + \Delta a_i, \end{cases} \qquad \forall p \in \Omega_i \tag{10.3}$$

  where $i = 1,\cdots,N_{\rm cl}$.

Consider the OFDM modulation with K subcarriers. Denote f_k as the kth subcarrier frequency. Define $\hat{a}[n]$ and $\hat{\epsilon}[n]$ as the estimate of the main Doppler scale factor and the residual Doppler frequency shift at the nth block, respectively. The channel impulse response in (10.1) translates into a matrix which is similarly formulated as in (5.29),

$$\mathbf{H}[n] = \sum_{p=1}^{N_{\rm pa}} \xi_p[n]\mathbf{\Lambda}(\overline{\tau}_p[n])\mathbf{\Gamma}(b_p[n],\hat{\epsilon}[n]), \tag{10.4}$$

where the two generic $K \times K$ matrices $\mathbf{\Lambda}(\tau)$ and $\mathbf{\Gamma}(b,\epsilon)$ are defined as

$$[\mathbf{\Lambda}(\tau)]_{m,m} := e^{-j2\pi\frac{m}{T}\tau}, \qquad [\mathbf{\Gamma}(b,\epsilon)]_{m,k} := G\left(\frac{f_m+\epsilon}{1+b} - f_k\right) \tag{10.5}$$

with $G(f)$ being the Fourier transform of the pulse shaping filter at transmitter, and

$$1 + b_p[n] := \frac{1+a_p[n]}{1+\hat{a}[n]}, \qquad \xi_p[n] := \frac{A_p[n]}{1+b_p[n]}e^{-j2\pi(f_c+\hat{\epsilon}[n])\overline{\tau}_p[n]}, \qquad \overline{\tau}_p[n] := \frac{\tau_p[n]}{1+b_p[n]}. \tag{10.6}$$

It is suggested from (10.4) that the channel can be uniquely characterized by $N_{\rm pa}$ triplets $\{\xi_p[n], \tau_p[n], b_p[n]\}$. Based on the relationship in (10.6), prediction of path parameters in (10.3) can be approximated as

$$\begin{cases} \xi_p[n|n-1] = \xi_p[n-1] \cdot \gamma_i, \\ \overline{\tau}_p[n|n-1] = \overline{\tau}_p[n-1] + \Delta\overline{\tau}_i, \\ b_p[n|n-1] = b_p[n-1] + \Delta b_i, \end{cases} \qquad \forall p \in \Omega_i. \tag{10.7}$$

After the receiver estimates the offsets $\{\gamma_i, \Delta b_i, \Delta\overline{\tau}_i\}_{i=1}^{N_{\rm cl}}$, the path parameters after cluster-offset compensation in (10.7) will be used to assist channel estimation of the nth block.

10.3 Cluster-Adaptation Based Block-to-Block Receiver

Denote $\mathbf{s}[n]$ as the transmitted symbol vector which has pilot symbols multiplexed with information symbols. Denote $\mathbf{z}[n]$ as the frequency measurement vector at the nth block. The input–output relationship is expressed as

$$\mathbf{z}[n] = \mathbf{H}[n]\mathbf{s}[n] + \mathbf{w}[n] \tag{10.8}$$

where $\mathbf{w}[n]$ denotes the ambient noise vector. The goal of the receiver processing is to estimate the channel matrix $\mathbf{H}[n]$ and to recover the information symbols in $\mathbf{s}[n]$.

The receiver incorporating the cluster-adaptation based channel estimator consists of four steps as shown in Figure 10.2, with each step briefly described in the following.

(i) *Cluster offset parameter estimation/compensation*. Estimate the offset parameters of each cluster based on the estimated path parameters $\{\hat{\xi}_p[n-1], \hat{\overline{\tau}}_p[n-1], \hat{b}_p[n-1]\}_{p=1}^{N_{\rm pa}}$ of the $(n-1)$th block and the frequency observation at pilot subcarriers of the nth block. Pass the offset-compensated path parameters $\{\hat{\xi}_p[n|n-1], \hat{\overline{\tau}}_p[n|n-1], \hat{b}_p[n|n-1]\}$ to the hybrid sparse channel estimation.

(ii) *Hybrid sparse channel estimation*. Perform hybrid sparse channel estimation based on the measurement vector $\mathbf{z}$[n] of the nth block.the compensated channel parameters of each cluster, and the estimated cluster variances $\{\hat{\sigma}_i^2[n-1]\}_{i=1}^{N_{\rm cl}}$. Pass the estimated channel matrix $\hat{\mathbf{H}}[n]$ to the symbol detection module.

(iii) *Symbol detection and channel decoding*. Symbol detection can be performed according to the maximum *a posterior* (MAP) or the minimum mean square error (MMSE) criteria. A linear MMSE equalizer described in Section 8.2.2 is adopted here for receiver development. After inputting the linear MMSE estimate of information symbols to a channel decoder, the *a posteriori* probabilities (APP) of information symbols can be obtained at the channel decoder output, which are used to calculate both soft and hard estimates of information symbols. In this chapter, the soft decisions of information symbols are used to refine the channel estimate.

(iv) *Refined channel estimation and cluster variation estimation*. Perform the sparse channel estimation based on the estimate of information symbols and the measurement vector $\mathbf{z}[n]$ within the nth block to refine the channel estimate $\{\hat{\xi}_p[n], \hat{\overline{\tau}}_p[n], \hat{b}_p[n]\}_{p=1}^{N_{\rm pa}}$, and update cluster variances $\{\hat{\sigma}_i^2[n]\}_{i=1}^{N_{\rm cl}}$; both of them are passed to the next block.

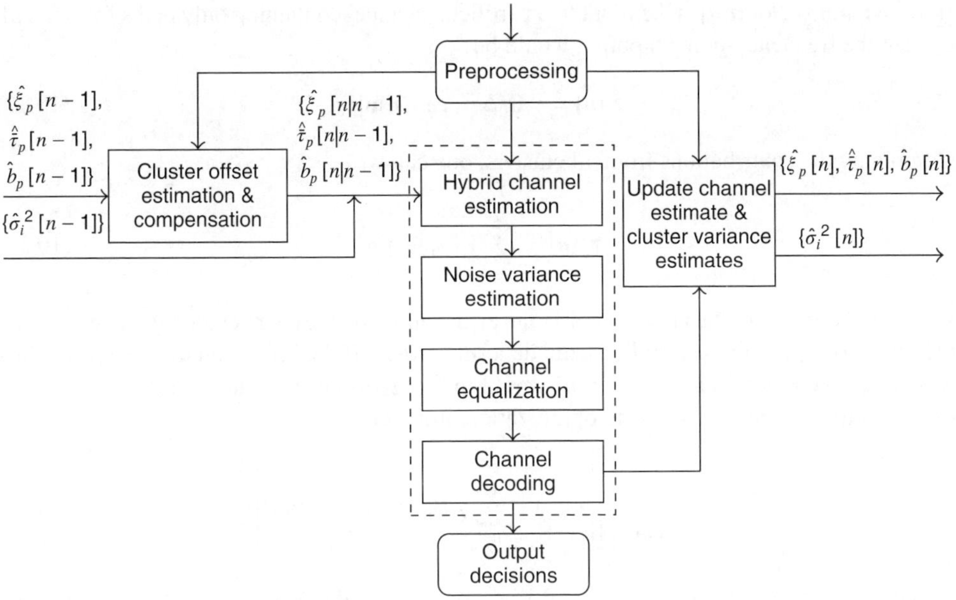

Figure 10.2 Flow chart of the processing for the nth block in the block-to-block receiver design.

For computational efficiency, a limited ICI-leakage assumption is usually adopted,

$$H[m,k;n] = 0, \qquad \forall |m-k| > D \tag{10.9}$$

where $H[m,k;n]$ denotes the (m,k)th element of $\mathbf{H}[n]$, and D is termed as ICI depth. When $D = 0$, ICI is taken as ambient noise; when $D > 0$, ICI is explicitly considered in the receiver processing.

Symbol detection and channel decoding have been presented in Chapters 8 and 3, respectively. In the following, we will focus on the other three key receiver modules.

10.3.1 Cluster Offset Estimation and Compensation

For the nth block, the receiver knows symbols on the pilot and null subcarriers. We define a $K \times K$ selector matrix $\mathbf{\Phi}$ which is a diagonal matrix with unit entry at pilot and null subcarriers, and zeros elsewhere. Define

$$\bar{\mathbf{z}}_{\rm P}[n] = \mathbf{\Phi}\mathbf{z}[n], \qquad \mathbf{u}[n] = \mathbf{\Phi}\mathbf{s}[n]. \tag{10.10}$$

These two vectors, which are known to the receiver, will be used to estimate the offset triplets $\{\gamma_i, \Delta\bar{\tau}_i, \Delta b_i\}_{i=1}^{N_{\rm cl}}$.

Based on the channel variation model in (10.7), define a $K \times K$ generic matrix

$$\mathbf{B}\left(\Delta\bar{\tau}_i, \Delta b_i; n\right) = \sum_{p\in\Omega_i} \hat{\xi}_p[n-1]\mathbf{\Lambda}(\bar{\tau}_p[n|n-1])\mathbf{\Gamma}(b_p[n|n-1], \hat{\epsilon}[n]). \tag{10.11}$$

If the symbol vector $\mathbf{u}[n]$ is the input to an artificial channel containing only paths from the ith cluster, the frequency-domain output would be

$$\mathbf{z}_i[n] = \gamma_i \mathbf{B}(\Delta\overline{\tau}_i, \Delta b_i; n)\mathbf{u}[n]. \tag{10.12}$$

Combining the contributions from all clusters, one can write

$$\overline{\mathbf{z}}_P[n] = \sum_{i=1}^{N_{cl}} \mathbf{z}_i[n] + \boldsymbol{\eta}[n], \tag{10.13}$$

where $\boldsymbol{\eta}(n)$ consists of the ambient noise, the error caused by the channel model mismatch, and the intercarrier interference spilled from data subcarriers. If the offset parameters for all clusters are correctly estimated, $\sum_{i=1}^{N_{cl}} \mathbf{z}_i[n]$ would match $\overline{\mathbf{z}}_P[n]$ closely. Hence, offset parameters can be estimated via the following optimization problem

$$\min_{\{\gamma_i, \Delta\overline{\tau}_i, \Delta b_i\}_{i=1}^{N_{cl}}} \left\| \overline{\mathbf{z}}_P[n] - \sum_{i=1}^{N_{cl}} \gamma_i \mathbf{B}(\Delta\tau_i, \Delta b_i; n)\mathbf{u}[n] \right\|_2^2. \tag{10.14}$$

Once the offset triplets $\{\hat{\gamma}_i, \Delta\hat{\overline{\tau}}_i, \Delta\hat{b}_i\}_{i=1}^{N_{cl}}$ are estimated, estimates of path parameters $\{\hat{\xi}_p[n|n-1], \hat{\overline{\tau}}_p[n|n-1], \hat{b}_p[n|n-1]\}$ are passed to the channel estimation module.

Two methods to solve the problem in (10.14) are presented in the sequel.

- *Exhaustive search*: The tentative measurement vector $\mathbf{z}_i[n]$ in (10.12) is linear in γ_i, but nonlinear with respect to $\Delta\overline{\tau}_i$ and Δb_i. Note that for each setting of $\{\Delta\overline{\tau}_i, \Delta b_i\}_{i=1}^{N_{cl}}$, the complex amplitudes $\{\gamma_i\}_{i=1}^{N_{cl}}$ can be easily obtained via the least-squares approach. Hence, a brute force method is to try all the possible values $\{\Delta b_i, \Delta\overline{\tau}_i\}_{i=1}^{N_{cl}}$ and choose the optimal one.

  Specifically, define a two-dimensional grid where $\{\Delta b_i, \Delta\overline{\tau}_i\}$ falls on

  $$\Delta\overline{\tau}_i \in \{-\Delta\tau_{max}, \Delta\tau_{max} + \delta_\tau, \cdots, \Delta\tau_{max}\} \tag{10.15}$$

  $$\Delta b_i \in \{-\Delta b_{max}, -\Delta b_{max} + \delta_b, \cdots, \Delta b_{max}\} \tag{10.16}$$

  for $i = 1, \cdots, N_{cl}$, where δ_τ and δ_b are the step sizes in the two-dimensional delay and Doppler plane, which depend on the channel variation across blocks. Typically, one can refer the variability of estimated channels in preceding blocks to decide a reasonable search window size. Although here all clusters are assumed to share a common set of representative offset values, extension to the scenario defining different representative sets for different clusters is straightforward, and will not be presented here.

  Assume that there are N_1 grid points on the delay dimension, and N_2 grid points on the Doppler scale dimension. The exhaustive search has a complexity proportional to $(N_1 N_2)^{N_{cl}}$, which becomes prohibitive when the number of clusters is large.
- *Orthogonal matching pursuit*: A suboptimal greedy algorithm, the orthogonal matching pursuit (OMP) algorithm discussed in Section 7.4, can be used to solve (10.14) at low computational complexity. On the delay-Doppler plane as defined in (10.15) and (10.16), let $\gamma_{i,\ell,j}$ denote the complex amplitude corresponding to the grid point $(\Delta\overline{\tau}[\ell], \Delta b[j])$ for the

ith cluster. The frequency observation vector $\bar{\mathbf{z}}_{\mathrm{P}}[n]$ can be formulated as

$$\bar{\mathbf{z}}_{\mathrm{P}}[n] = \sum_{i=1}^{N_{\mathrm{cl}}} \sum_{\ell=1}^{N_1} \sum_{j=1}^{N_2} \gamma_{i,j,\ell} \mathbf{B}(\Delta\bar{\tau}[\ell], \Delta b[j]; n)\mathbf{u}[n] + \boldsymbol{\eta}[n]. \tag{10.17}$$

Define a vector of size $K \times 1$,

$$\mathbf{p}_{i,j,\ell} = \mathbf{B}(\Delta\bar{\tau}[\ell], \Delta b[j]; n)\mathbf{u}[n]. \tag{10.18}$$

The observation vector in (10.17) can be rewritten as

$$\bar{\mathbf{z}}_{\mathrm{P}}[n] = \left[\mathbf{p}_{1,1,1}, \cdots, \mathbf{p}_{N_{\mathrm{cl}},N_1,N_2}\right] \begin{bmatrix} \gamma_{1,1,1} \\ \vdots \\ \gamma_{N_{\mathrm{cl}},N_1,N_2} \end{bmatrix} + \boldsymbol{\eta}[n]. \tag{10.19}$$

When the OMP algorithm is used to find $\hat{\gamma}_{i,\ell,j}$, it enforces *only one delay and Doppler offset pair with nonzero amplitude for each cluster*. Hence, the OMP algorithm will stop after N_{cl} steps, and its complexity is linear in the number of clusters.

10.3.1.1 A Special Case with Only Amplitude and Delay Variations

Consider an important special case which has zero Doppler scale variation (i.e., $\Delta b_i = 0$), and only amplitude and delay variations

$$\begin{cases} \xi_p[n|n-1] = \hat{\xi}_p[n-1] \cdot \gamma_i, \\ \bar{\tau}_p[n|n-1] = \hat{\bar{\tau}}_p[n-1] + \Delta\bar{\tau}_i. \end{cases} \quad \forall p \in \Omega_i \tag{10.20}$$

It follows that

$$\mathbf{B}(\Delta\bar{\tau}_i, 0; n)\mathbf{u}[n] = \sum_{p\in\Omega_i} \hat{\xi}_p[n-1]\boldsymbol{\Lambda}(\bar{\tau}_p[n|n-1])\boldsymbol{\Gamma}(\hat{b}_p[n-1], \hat{\epsilon}[n])\mathbf{u}[n] \tag{10.21}$$

Note that

$$\boldsymbol{\Lambda}(\bar{\tau}_p[n|n-1]) = \boldsymbol{\Lambda}(\Delta\bar{\tau}_i)\boldsymbol{\Lambda}(\hat{\bar{\tau}}_p[n-1]), \tag{10.22}$$

and define a vector,

$$\hat{\mathbf{p}}_i[n] := \sum_{p\in\Omega_i} \hat{\xi}_p[n-1]\boldsymbol{\Lambda}(\hat{\bar{\tau}}_p[n-1])\boldsymbol{\Gamma}(\hat{b}_p[n-1], \hat{\epsilon}[n])\mathbf{u}[n] \tag{10.23}$$

which can be precomputed. The optimization problem in (10.14) is simplified to

$$\min_{\{\gamma_i, \Delta\bar{\tau}_i\}_{i=1}^{N_{\mathrm{cl}}}} \left\| \bar{\mathbf{z}}_{\mathrm{P}}[n] - \sum_{i=1}^{N_{\mathrm{cl}}} \gamma_i \boldsymbol{\Lambda}(\Delta\bar{\tau}_i)\hat{\mathbf{p}}_i[n] \right\|_2^2. \tag{10.24}$$

Similarly, the optimization problem can be solved by exhaustive search and the OMP algorithm. The exhaustive search along the delay grid in (10.15) has a complexity $N_1^{N_{\mathrm{cl}}}$. The OMP algorithm still needs N_{cl} steps; however, within each step, there are much less templates to correlate. This special case with $\Delta b_i = 0$ hence has a major complexity advantage relative to the general case with $\Delta b_i \neq 0$.

10.3.2 Cluster-Adaptation Based Sparse Channel Estimation

The resources available to estimate the channel of the nth block include:

(i) the offset-compensated triplets $\{\hat{\xi}_p[n|n-1], \hat{\bar{\tau}}_p[n|n-1], \hat{b}_p[n|n-1]\}$;
(ii) the frequency observation vector $\bar{\mathbf{z}}_{\rm P}[n]$ and the corresponding symbol vector $\mathbf{u}[n]$.

To utilize the $\{(\hat{\xi}_p[n|n-1], \hat{\bar{\tau}}_p[n|n-1], \hat{b}_p[n|n-1]), \forall p \in \Omega_i\}_{i=1}^{N_{\rm cl}}$ in a way similar to the pilot-based channel estimation approach, a set of artificial measurements $\{\check{\mathbf{z}}_i[n]\}_{i=1}^{N_{\rm cl}}$ can be constructed by passing a known symbol vector $\check{\mathbf{s}}[n]$ through $N_{\rm cl}$ channels formed by paths from $N_{\rm cl}$ clusters, respectively. The transmitted symbol vector $\check{\mathbf{s}}(n)$ is arbitrary but known, and is often constructed by drawing each symbol from a QPSK constellation.

Define two channel matrices corresponding to the ith cluster as

$$\hat{\mathbf{H}}_i[n|n-1] = \sum_{p\in\Omega_i} \hat{\xi}_p[n|n-1]\mathbf{\Lambda}(\hat{\bar{\tau}}_p[n|n-1])\mathbf{\Gamma}(\hat{b}_p[n|n-1], \hat{\epsilon}[n]), \tag{10.25}$$

$$\mathbf{H}_i[n] = \sum_{p\in\Omega_i} \xi_p[n]\mathbf{\Lambda}(\bar{\tau}_p[n])\mathbf{\Gamma}(b_p[n], \hat{\epsilon}[n]). \tag{10.26}$$

The artificial measurements are constructed as

$$\check{\mathbf{z}}_i[n] = \hat{\mathbf{H}}_i[n|n-1]\check{\mathbf{s}}(n) = \mathbf{H}_i[n]\check{\mathbf{s}}[n] + \check{\mathbf{w}}_i[n], \tag{10.27}$$

where $\check{\mathbf{w}}_i(n)$ denotes the error caused by the channel prediction inaccuracy,

$$\check{\mathbf{w}}_i(n) := \left(\hat{\mathbf{H}}_i[n|n-1] - \mathbf{H}_i[n]\right)\check{\mathbf{s}}[n]. \tag{10.28}$$

The receiver thus has measurements from different sources

$$\begin{cases} \bar{\mathbf{z}}_{\rm P}[n] = \mathbf{H}[n]\mathbf{u}[n] + \mathbf{w}[n], \\ \check{\mathbf{z}}_i[n] = \mathbf{H}_i[n]\check{\mathbf{s}}[n] + \check{\mathbf{w}}_i[n], \qquad i = 1, \cdots, N_{\rm cl}. \end{cases} \tag{10.29}$$

These measurements have different reliabilities. For simplicity, $\mathbf{w}[n]$ can be assumed following a zero-mean Gaussian distribution with covariance matrix $\hat{\sigma}_0^2[n]\mathbf{I}_K$, and similarly, $\check{\mathbf{w}}_i[n]$ follows a zero-mean Gaussian distribution with covariance matrix $\hat{\sigma}_i^2[n-1]\mathbf{I}_K$. The noise variance $\sigma_0^2[n]$ can be estimated based on frequency measurements at null subcarriers as in (7.66). Estimates of $\sigma_i^2[n-1]$, $i = 1, \ldots, N_{\rm cl}$ are passed from the previous block; see the estimation procedure in Section 10.3.3.

For sparse channel estimation, define two sets formed by all the possible values of the path delay and the Doppler scale factor, respectively,

$$\bar{\tau} \in \left\{0, \frac{T}{\beta K}, \frac{2T}{\beta K}, \cdots, \frac{N_{\rm de}-1}{\beta K}\right\} \tag{10.30}$$

$$b \in \{-b_{\max}, -b_{\max} + \Delta b, \cdots, b_{\max}\} \tag{10.31}$$

where $T/(\beta K)$ and Δb are the uniform sampling steps on the delay and the Doppler scale, respectively.

Denote $\xi_{j,\ell}$ as the complex amplitude of the path at the $(\overline{\tau}_\ell, b_j)$ grid, and define

$$\mathbf{a}_{j,\ell} := \mathbf{\Lambda}(\overline{\tau}_\ell)\mathbf{\Gamma}(b_j, \hat{\epsilon}[n])\mathbf{u}[n], \qquad \mathbf{b}_{j,\ell} := \mathbf{\Lambda}(\overline{\tau}_\ell)\mathbf{\Gamma}(b_j, \hat{\epsilon}[n])\check{\mathbf{s}}[n]. \tag{10.32}$$

The frequency measurements at the pilot subcarriers of the nth block can be expressed as

$$\overline{\mathbf{z}}_P[n] = \sum_{j=1}^{N_{\rm Dop}} \sum_{\ell=1}^{N_{\rm de}} \xi_{j,\ell}\mathbf{a}_{j,\ell} + \mathbf{w}[n]. \tag{10.33}$$

Define

$$\mathbf{A} := \left[\mathbf{a}_{1,1} \;\cdots\; \mathbf{a}_{N_{\rm Dop},N_{\rm de}}\right], \tag{10.34}$$

$$\xi := \left[\xi_{1,1} \;\cdots\; \xi_{N_{\rm Dop},N_{\rm de}}\right]^{\rm T}. \tag{10.35}$$

The measurement vector in (10.33) can be recast as

$$\overline{\mathbf{z}}_P[n] = \mathbf{A}\xi + \mathbf{w}[n]. \tag{10.36}$$

Define $\mathbf{\Theta}_i$ as a selector of channel paths within the ith cluster, being a diagonal matrix of size $N_{\rm Dop}N_{\rm de} \times N_{\rm Dop}N_{\rm de}$, and with unit entry for grids within a zone where the paths of the ith cluster can reside, and zeros elsewhere (see Figures 10.5 and 10.7 for the plots on the zones for different clusters). Similar to (10.33), define

$$\mathbf{B} := \left[\mathbf{b}_{1,1}, \cdots, \mathbf{b}_{N_{\rm Dop},N_{\rm de}}\right], \tag{10.37}$$

which leads to a similar expression for the artificial measurements

$$\check{\mathbf{z}}_i[n] = \mathbf{B}\mathbf{\Theta}_i\xi + \check{\mathbf{w}}[n]. \tag{10.38}$$

Now we are ready to present two sparse channel estimators, where one does not enforce paths to fall into the specified zones for all clusters and the other one does.

- *Hybrid channel estimation without zone information*: For the hybrid channel estimation without zone information, the solution is obtained as

$$\hat{\xi} = \arg\min_{\xi} \frac{1}{\hat{\sigma}_0^2[n]}\|\overline{\mathbf{z}}_P[n] - \mathbf{A}\xi\|^2 + \sum_{i=1}^{N_{\rm cl}} \frac{1}{\hat{\sigma}_i^2[n-1]}\|\check{\mathbf{z}}_i[n] - \mathbf{B}\mathbf{\Theta}_i\xi\|^2 + \zeta\|\xi\|_1 \tag{10.39}$$

  where $|\xi|_1$ denotes the ℓ_1 norm of vector ξ, and ζ controls the sparsity of the solution.
- *Hybrid channel estimation with zone information*: For the hybrid channel estimation with zone information, the solution is obtained as

$$\hat{\xi} = \arg\min_{\xi} \frac{1}{\hat{\sigma}_0^2[n]}\left\|\overline{\mathbf{z}}_P[n] - \mathbf{A}\sum_{i=1}^{N_{\rm cl}} \mathbf{\Theta}_i\xi\right\|^2 + \sum_{i=1}^{N_{\rm cl}} \frac{1}{\hat{\sigma}_i^2[n-1]}\|\check{\mathbf{z}}_i[n] - \mathbf{B}\mathbf{\Theta}_i\xi\|^2 + \zeta\|\xi\|_1. \tag{10.40}$$

  Clearly, in the solution of (10.40), the entries outside the specified zones of $N_{\rm cl}$ clusters are zero.

Compressive sensing techniques can be used to solve (10.39) and (10.40). In this chapter, the SpaRSA algorithm from [435] is used. Once $\hat{\xi}$ is obtained, $\hat{\mathbf{H}}[n]$ can be computed from (10.4), which will be used for symbol detection.

10.3.3 Channel Re-estimation and Cluster Variance Update

Denote $\bar{\mathbf{s}}[n]$ as the soft decision on the symbol vector with the channel decoding. The input-output relationship for channel estimation becomes

$$\mathbf{z}[n] = \mathbf{H}[n]\bar{\mathbf{s}}[n] + \boldsymbol{\eta}[n] \tag{10.41}$$

where $\boldsymbol{\eta}[n]$ includes the ambient noise and the error caused by the symbol estimation inaccuracy. The block-by-block sparse channel estimator in Chapter 7 applies directly. The refined estimates on the path parameters are passed to the next block.

Based on the updated channel estimate $\hat{\xi}$, variance of the artificial measurements corresponding to the ith cluster can be updated as

$$\hat{\sigma}_i^2[n] = \frac{1}{K}\left\|\check{\mathbf{z}}_i[n] - \mathbf{B}\boldsymbol{\Theta}_i\hat{\xi}\right\|^2 \tag{10.42}$$

The variance $\hat{\sigma}_i^2[n]$ will be be used for the hybrid channel estimation of the $(n+1)$th block. A smoothing operation can be adopted as well:

$$\hat{\sigma}_i^2[n] = \nu\hat{\sigma}_i^2[n-1] + (1-\nu)\frac{1}{K}\|\check{\mathbf{z}}_i[n] - \mathbf{B}\boldsymbol{\Theta}_i\hat{\xi}\|^2 \tag{10.43}$$

where the forgetting factor ν needs to be tuned.

10.4 Experimental Results: MACE10

The MACE10 experiment setup is described in Appendix B, with OFDM parameters summarized in Table B.2. Despite 256 pilots in the original signal design, only a subset of pilot subcarriers, which is denoted as $\overline{S}_{\rm P}$, is used to investigate performance of the cluster-adaptation based channel estimator. For a constellation size M, the spectral efficiency and the data rate are formulated as

$$\alpha = \frac{1}{2}\cdot\frac{T}{T+T_{\rm g}}\cdot\frac{|S_{\rm D}|}{|S_{\rm D}|+|\overline{S}_{\rm P}|+|S_{\rm N}|}\cdot\log_2 M \text{ bits/s/Hz}, \tag{10.44}$$

$$R = \alpha B \text{ bit/s}. \tag{10.45}$$

To remove the Doppler effect caused by the mobility of the source array, an overall resampling operation is performed for each transmission. The overall resampling factor is estimated based on a CP-OFDM preamble prior to each transmission as discussed in Section 6.2.

For multipath clustering, three algorithms have been tested in [426].

(1) From a cluster detection point of view, a Page-test along the delay axis, can be used to determine the starting and ending points of the delay of each cluster at each possible Doppler rate value [2];

(2) From a clustering point of view, the traditional k-means algorithm [311] can be employed after a thresholding operation to eliminate noise samples in the channel estimate;
(3) In scenarios with a stable clustering structure, cluster locations can also be taken as priors.

It was observed in [426] that there is not much difference of the three strategies in terms of receiver decoding performance. The experimental results presented in this chapter are based on the assumption that the clustering structure does not change within each transmission and the fact that the Page-test is adopted for clustering using the channel estimate of the preamble.

For the MACE10 data, the ICI-ignorant receiver with $D = 0$ is adopted. The step size for the delay offset estimation in (10.15) is $\delta_{\overline{\tau}} = 1/(2B)$ with the maximum delay offset $\Delta\overline{\tau}_{\max} = 8\delta_{\overline{\tau}}$, and the step size for the Doppler offset estimation in (10.16) is $\delta_b = \Delta b_{\max}/7$ where the maximum Doppler offset is set as $\Delta b_{\max} = 5 \times 10^{-4}$ which corresponds to a maximum speed offset $\Delta v_{\max} = \Delta b_{\max} c = 0.75$ m/s where $c = 1500$ m/s is the sound speed in water. Hence, the numbers of searching grids on the delay and Doppler domains are $N_1 = 17$ and $N_2 = 15$, respectively. For the sparse channel estimation, the maximum Doppler scale in (10.30) is taken as $b_{\max} = 5 \times 10^{-4}$ with a step size $\Delta b = b_{\max}/7$.

10.4.1 BLER Performance with an Overall Resampling

To investigate the performance of the cluster-adaptation based channel estimator, $|\overline{\mathcal{S}}_{\mathrm{P}}| = 32$ pilots are used for channel estimation and 24 data subcarriers are introduced as extra pilots to estimate the Doppler scale offset, which leads to a spectral efficiency and a data rate

$$\alpha = \frac{T}{T + T_g} \cdot \frac{336 - 24}{672 + 32 + 96} \cdot \log_2 16 = 1.31 \text{ bits/s/Hz}, \tag{10.46}$$

$$R = \alpha B = 6.39 \text{ kb/s}. \tag{10.47}$$

By treating all paths as in one cluster, the block-error-rate (BLER) performance of the receiver corresponding to several channel estimation schemes is shown in Figure 10.3, with the setting of each scheme listed in the following.

(1) *Pilot-based block-by-block processing*: Each block is processed individually with the ICI-ignorant receiver developed in Chapter 9.
(2) *Block-to-block processing without offset compensation*: The channel is assumed stationary over blocks, setting $\Delta\overline{\tau} = 0, \Delta b = 0$, and $\gamma = 1$. The hybrid channel estimation is performed without using the zone information of clusters;
(3) *Block-to-block processing with $\Delta b = 0$ and without zone information*: The channel variation is only modeled with the delay and amplitude offsets, and set $\Delta b = 0$. The hybrid channel estimation is performed without using the zone information of clusters;
(4) *Block-to-block processing with $\Delta b = 0$ and zone information*: The channel variation is only modeled with the delay and amplitude offsets, and set $\Delta b = 0$. The hybrid channel estimation is performed with the zone information of clusters;
(5) *Block-to-block processing with $\Delta b \neq 0$ and zone information*: the channel variation is modeled with the delay, amplitude and Doppler scale offsets. The hybrid channel estimation is performed with the zone information of clusters.

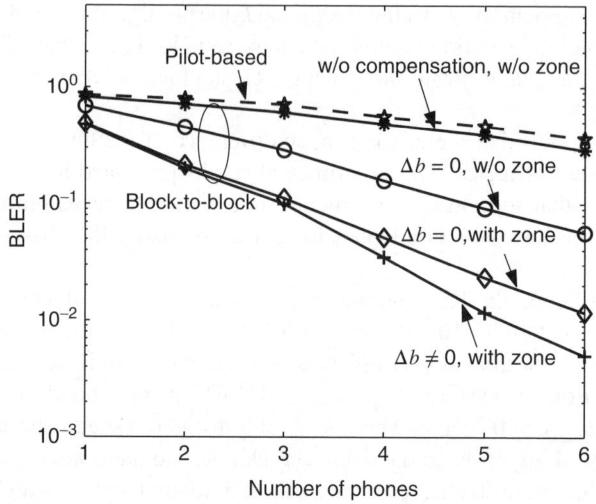

Figure 10.3 BLER performance of the ICI-ignorant receiver, 16-QAM, with an overall resampling. All the channel paths are treated as in one cluster.

Several observations from Figure 10.3 are in the following.

- Exploiting the time-coherence of UWA channels boosts the decoding performance significantly, as can be seen from the performance of the pilot-based block-by-block processing and that of schemes corresponding to the block-to-block processing;
- Compensation of channel variation over blocks is necessary, as can be seen from the performance of the block-to-block processing without the offset compensation and that of the methods with the offset compensation;
- Utilizing the zone information is beneficial, as can be seen from the performance of the block-to-block processing without zone information and that of the methods with the zone information;
- Introducing the Doppler scale offset improves the decoding performance, as can be seen from the performance of the block-to-block processing with $\Delta b = 0$ and that with $\Delta b \neq 0$.

Figure 10.4 shows the estimated Doppler speed of each block, i.e. the accumulation of Δb estimates up to the nth block, using the offset estimation method in Section 10.3.1. One can find that the estimate matches well with the Doppler speed estimate using the null-subcarrier based method. Hence, even after an overall resampling operation performed on each data burst, the residual Doppler scaling effect within each block is not negligible. The block-to-block processing is able to estimate the Doppler scale variation across blocks, which leads to performance improvement.

10.4.2 BLER Performance with Refined Resampling

In this subsection, the cluster-adaptation based channel estimation after a refined resampling operation on each individual block is evaluated. The resampling factor in each block is

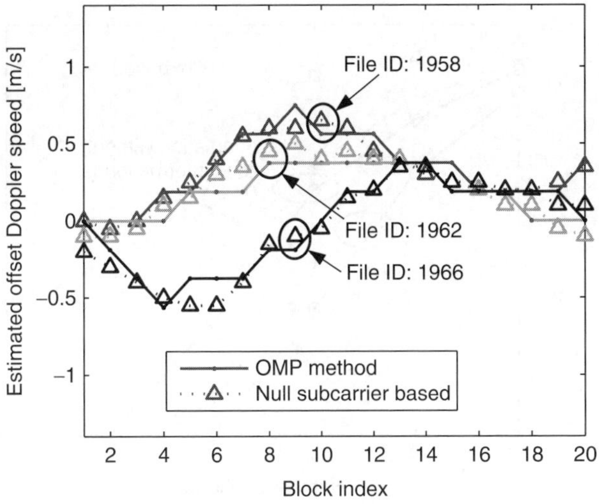

Figure 10.4 Samples of estimated offset Doppler speed over 20 OFDM blocks after an overall resampling operation. All the channel paths are treated as in one cluster.

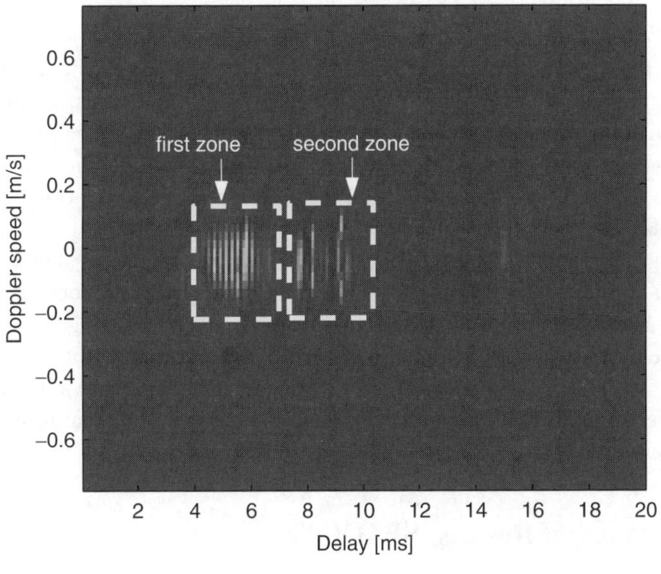

Figure 10.5 MACE10: Sample of channel estimates with an OFDM preamble. A Page-test based clustering algorithm is used.

estimated using the null-subcarrier based method. As shown in Figure 10.5, the paths after the refined resampling are quite centered around the zero Doppler scale. It hence suffices to assume $\Delta b_i = 0$ and test the performance of the cluster-adaptation based channel estimation using different numbers of clusters.

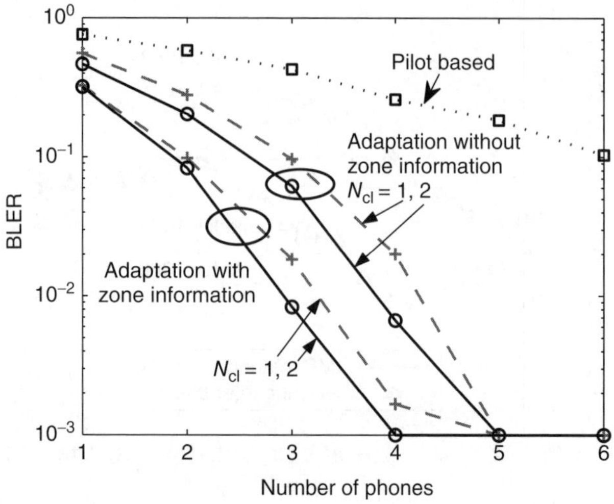

Figure 10.6 BLER performance of the ICI-ignorant receiver with different number of phones combined, 16-QAM, with refined resampling for each block, 64 pilots. The curves marked with "without zone information" correspond to the formulation in (10.39), and those marked with "with zone information" correspond to the formulation in (10.40).

To compare the channel estimation performance with different number of clusters, two clustering schemes are considered: (i) treating the whole channel as a single cluster; and (ii) dividing channel paths into two clusters as shown in Figure 10.5. $|\overline{S}_P| = 64$ pilots are used for channel estimation, which corresponds to a spectral efficiency $\alpha = 1.36$ bits/s/Hz and a data rate $R = 6.62$ kb/s, resulting a 23% increase relative to the original signal design with 256 pilots.

The decoding performance of several channel estimation schemes is shown in Figure 10.6. One can see a significant performance gap between the block-by-block channel estimator and the cluster-adaptation based channel estimators. Moreover, utilizing the zone information and treating the multipath clusters individually bring considerable performance improvement. Although paths in the second zone in Figure 10.5 can be further divided into different clusters, the decoding performance with more than two clusters has been found almost identical to that with two clusters.

10.5 Experimental Results: SPACE08

The SPACE08 experimental setup and OFDM parameter settings have been described in Chapter 9. Due to the interesting clustered channel structure, only the data sets collected by the receiver labeled as S1 which was 60 m away from the transmitter are considered. Despite 256 pilots in the original signal design, only $|\overline{S}_P| = 64$ pilot subcarriers are used for channel estimation. Formulation of the data rate is identical to (10.45). The ICI bandwidth $D = 1$ is adopted during the receiver processing. To perform the ICI-aware channel estimation, 24 data subcarriers are introduced as extra pilots for ICI coefficients estimation, leading to a spectral efficiency $\alpha = 1.22$ bits/s/Hz and a data rate $R = 11.9$ kb/s.

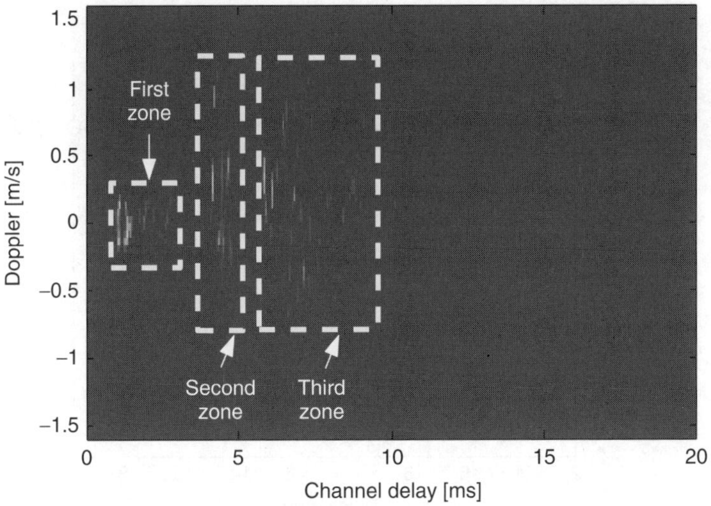

Figure 10.7 SPACE08: Sample of channel estimates with an OFDM preamble. Paths in the first zone correspond to direct transmission, while paths in the second and the third zones correspond to the surface/bottom reflections. A Page-test based clustering algorithm is used.

During the experiment, since both transmitter and receiver were stationary, no resampling operation is applied to the received signal, and it is assumed that there is no Doppler scale offset over consecutive blocks, i.e., $\{\Delta b_i = 0\}$ for all clusters. The maximum delay offset and grid size are identical to the values described in Section 10.4, so as the parameter values for the hybrid channel estimation.

Sample of one particular estimated channel impulse response is demonstrated in Figure 10.7, where channel paths are grouped into three clusters: the first cluster corresponds to direct transmissions, the second and the third clusters correspond to the surface and bottom reflections, respectively.

As can be seen from Figure 10.1, paths formed by the direct transmissions are quite stable, while the paths constituted by the reflected paths tend to scatter around with a very large fluctuation. To compare the channel estimation performance with different number of clusters, three clustering schemes are tested: (i) treating all the paths as in one single cluster; (ii) dividing channel paths into two clusters, i.e. one cluster formed by refractions and the other cluster formed by reflections; and (iii) dividing channel paths into three clusters, i.e. one cluster by refractions and two clusters by reflections, respectively, as shown in Figure 10.7.

Figure 10.8 shows the estimated variances of three clusters normalized by the power of each cluster averaged over twelve phones of data files collected in Julian date 297, which confirms that the first cluster is much more stable than the other two clusters. It hence is advantageous to model different variations for different clusters.

Figure 10.9 demonstrates the averaged BLER performance of files collected in Julian date 297 and Julian date 300 using the cluster-adaptation based channel estimator and the pilot-based block-by-block channel estimator. One can find that the block-to-block receiver outperforms the pilot-based block-by-block receiver considerably. By modeling the variations

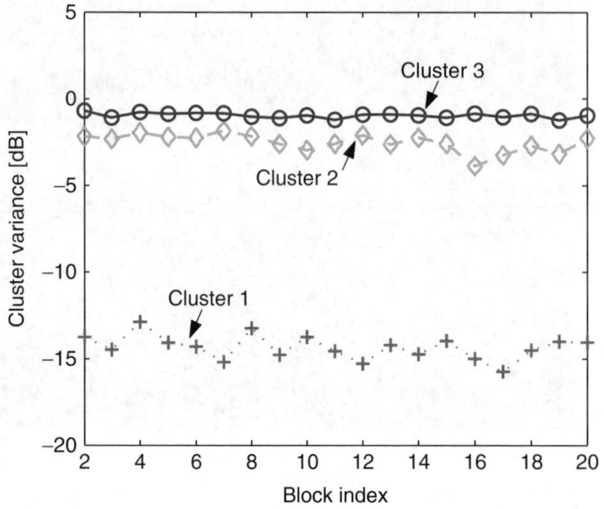

Figure 10.8 Estimated variances of three clusters normalized by the power of each cluster, averaged over twelve phones of data files collected in Julian date 297.

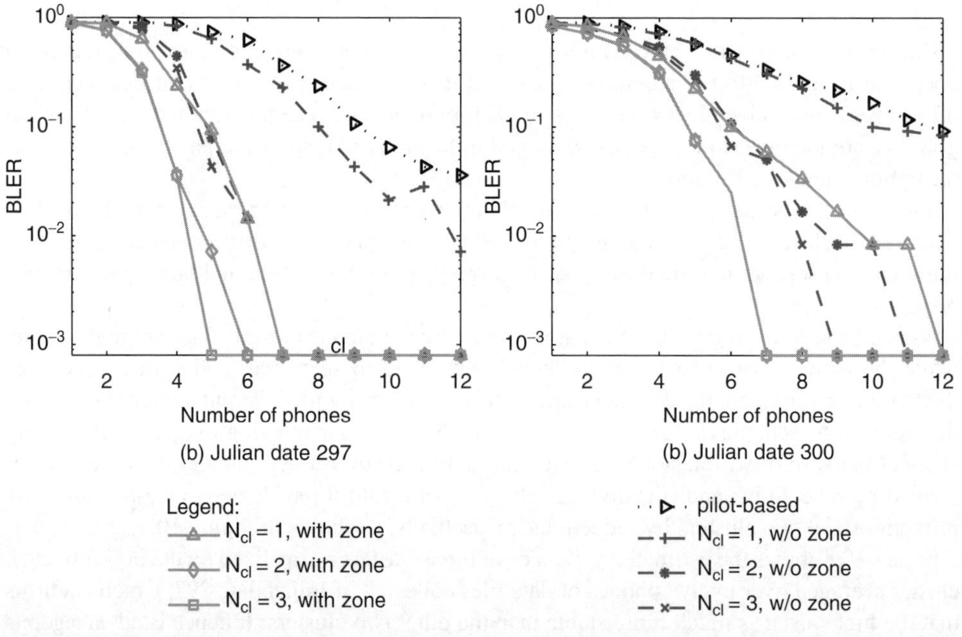

Figure 10.9 BLER performance of the ICI-aware receiver ($D = 1$) with different channel estimation schemes, 16-QAM. N_{cl}: the number of clusters.

of multipath clusters independently and utilizing the cluster support information for hybrid channel estimation, the system performance has been significantly improved.

10.6 Discussion

Chapter 9 presents receivers based on block-by-block processing, while Chapter 10 presents an adaptive receiver based on block-to-block processing. The concepts can be combined in practical systems.

- The first possible extension is that the adaptive receiver in Figure 10.2 can be combined with the iterative receivers developed in Chapter 9 to improve the block decoding performance. With the estimates of information symbols from the channel decoder, all the frequency measurements of the current block can be used for the hybrid channel estimation.
- The second possible combination is that the adaptive receiver can be used to initialize the ICI-progressive receiver. Here, the channel prediction from the preceding block is based on the ICI depth $D = 0$. The hybrid channel estimation yields an initial ICI-ignorant channel estimate which is used to start the receiver iteration. As iteration goes on, the ICI depth D will increase gradually. Assuming that all paths are within one cluster, the above initialization method has been tested in [182] for both single-input and multi-input channels, confirming that a considerable reduction on the pilot overhead can be achieved.

10.7 Bibliographical Notes

Adaptive receiver has played an important role in the advancement of underwater acoustic communications. Symbol-level adaptation was adopted in the seminal work for single-carrier transmissions in [366, 367], where a phase-locked loop tracks the channel phase variations. Block-level adaptation was utilized for single-carrier block transmissions with frequency domain equalization in [376, 468]. For OFDM transmissions, a block-adaptive receiver was proposed in [361, 363] for the single-input channel and later extended in [67] for the multiple-input channel, where a single parameter referred to as the Doppler scaling factor is introduced to model the time variability of the channel across blocks. The clustered-adaption approach presented in this chapter is based on the work in [425, 426].

11

OFDM in Deep Water Horizontal Communications

For the OFDM receiver designs in Chapters 9 and 10, one underlying assumption is that the channel delay spread is less than the guard time and hence there is no interblock interference (IBI). This is suitable for shallow water acoustic channels or deep water vertical channels, where the channel delay spread is often around several tens of milliseconds. This chapter focuses on two scenarios where the channel consists of long separately clusters, leading to an extremely large delay spread, e.g., on the order of seconds. The first scenario is in deep-water *horizontal* channel, as illustrated in Figure 11.1, where the transmitter and receiver are horizontally separated, and could also be at different water depths. The second scenario is an underwater broadcasting networking, where multiple surface nodes broadcast the same information to underwater nodes.

This line of research is motivated by the strong need of forming acoustic local area networks (ALAN) in deep oceans, as illustrated in Figure 11.2. One example is the cellular network in the Atlantic Undersea Test and Evaluation Center (AUTEC) located around Andros Island near the Tongue of the Ocean, Bahamas, where 96 fixed nodes are deployed in an area of size $30 \times 50\,\mathrm{km}^2$. The water depth is about 1.5 km to 2 km, and the distance of nodes is larger than 4 km [159, 428]. Acoustic communications are in extensive daily use between mobile users and fixed network nodes. Figure 11.1 illustrates the horizontal channel structure in the AUTEC network, where the transmitter and the receiver are both anchored closely to the sea floor. The first cluster as shown consists of both direct transmission paths and paths arising from bottom reflections. Given the short distance of both transmitter and receiver to the sea floor, the first cluster has a very small delay spread. The paths associated with the first surface reflection and possible bottom refections constitute the second cluster, which has a relatively large delay spread and a severe Doppler spread due to the dispersion caused by reflections. The third cluster and beyond are formed by the paths with more than one surface reflections. The energy of the third cluster and beyond has been observed much smaller than the first two clusters, thus can be neglected.

The widely separated clusters in the deep sea horizontal channels incur severe inter-block interference (IBI) in the received signal. This chapter is devoted to designing receiver

OFDM for Underwater Acoustic Communications, First Edition. Shengli Zhou and Zhaohui Wang.
© 2014 John Wiley & Sons, Ltd. Published 2014 by John Wiley & Sons, Ltd.

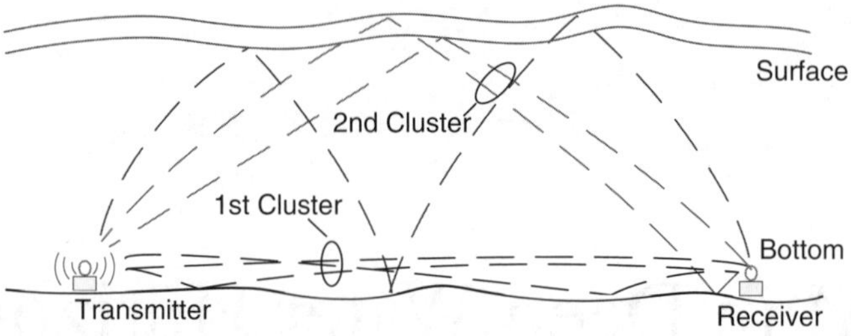

Figure 11.1 Illustration of deep water acoustic communication channels.

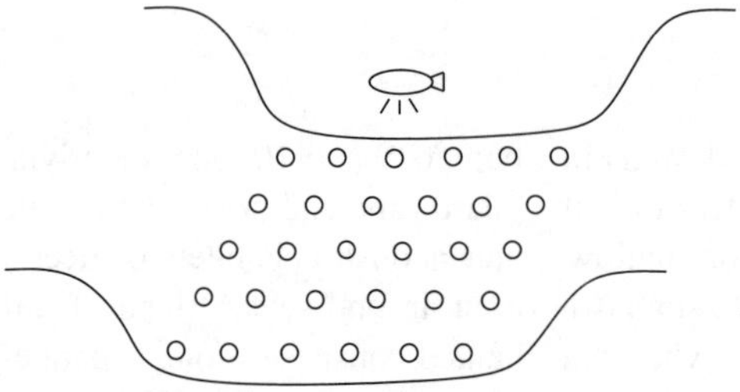

Figure 11.2 Illustration of a deep-water bottom network.

processing algorithms to address both IBI and possible ICI in the received signal. This chapter is organized as follows.

- Section 11.1 introduces the system model where clustered physical channel is converted into a summation of two quasi-synchronous channels.
- Sections 11.2 and 11.3 describe a decision-feedback based equalizer and a factor-graph based equalizer, respectively, for joint IBI and ICI equalization.
- Section 11.4 presents an iterative block-to-block receiver, where all the received blocks within one batch of transmission are decoded jointly.
- Sections 11.5 and 11.6 contain the performance results using both simulation and experimental data sets collected from the AUTEC network.
- Section 11.7 applies the receiver developed for the deep water horizontal channel to an underwater broadcasting network, where emulated experimental results are provided.

11.1 System Model for Deep Water Horizontal Communications

This section presents an approach to modeling the multipath channel with widely-separated clusters, upon which an input–output relationship will be built. Throughout this chapter, we limit our discussion to the channel with two clusters – a typical situation observed from field experiments.

11.1.1 Transmitted Signal

Consider an OFDM system with K subcarriers. Denote f_k as the kth subcarrier frequency,

$$f_k = f_c + \frac{k}{T}, \qquad k = -\frac{K}{2}, \cdots, \frac{K}{2} - 1 \tag{11.1}$$

where f_c is the center frequency, and T is the OFDM symbol duration. Let $s[k;n]$ denote the symbol on the kth subcarrier of the nth block. The passband signal of the nth transmitted block can be expressed as

$$\tilde{s}(t;n) = 2\Re\left(\sum_{k=-K/2}^{K/2-1} s[k;n]e^{j2\pi f_k t} g(t)\right), \tag{11.2}$$

where $g(t)$ is the rectangular pulse shaping window, defined in (2.9). To avoid the inter-symbol interference caused by channel dispersion, a guard interval is usually inserted between consecutive OFDM symbols. Denote T_g as the length of the guard interval. The total OFDM block duration is thus $T_{bl} := T + T_g$.

For a transmission burst with N_{bl} blocks, the transmitted signal is

$$\tilde{x}(t) = \sum_{n=1}^{N_{bl}} \tilde{s}(t - nT_{bl}; n), \qquad t \in [0, N_{bl}T_{bl}]. \tag{11.3}$$

One difference exists concerning the signal design for the channels with widely-separated clusters and that for the shallow water channels. Shallow water acoustic channels usually have small or moderate delay spreads, and hence the guard interval is usually larger than the maximum channel delay spread, so that the IBI is avoided at the receiver [38, 201, 232, 235, 361, 392]. For channels with widely-separated clusters, it is recommend to have the guard interval larger than the maximum channel delay spread, as that would lead to a significant data rate reduction. With T_g smaller than the channel delay spread, one has to address the IBI in the received signal.

11.1.2 Modeling Clustered Multipath Channel

For OFDM block transmissions over a channel with the intercluster delay $\tau_{1,2}$ much larger than the OFDM block length T_{bl}, decompose the intercluster delay as

$$\tau_{1,2} = \Delta T_{bl} + \varsigma, \tag{11.4}$$

where Δ is an integer and ς is the residual within $[-T_{bl}/2, T_{bl}/2]$. To model the two-cluster channel, denote $h^{(1)}(t, \tau; n)$ as the channel impulse response of the first cluster for the nth transmitted block,

$$h^{(1)}(t, \tau; n) = \sum_{p=1}^{N_{pa}^{(1)}} A_p^{(1)}[n]\delta\left(\tau - (\tau_p^{(1)}[n] - a_p^{(1)}[n]t)\right) \tag{11.5}$$

where $N_{\rm pa}^{(1)}$ denotes the number of paths, $A_p^{(1)}[n]$, $\tau_p^{(1)}[n]$ and $a_p^{(1)}[n]$ denote the amplitude, the initial delay and the Doppler rate of the pth path within the first cluster, respectively. The impulse response of the second cluster $h^{(2)}(t,\tau;n)$ is similarly defined but with an offset of $\Delta T_{\rm bl}$ in path delays; see Figure 11.3. The overall channel impulse response corresponding to the nth transmitted block can be formulated as

$$h(t,\tau;n) = h^{(1)}(t,\tau;n) + h^{(2)}(t,\tau-\Delta T_{\rm bl};n). \tag{11.6}$$

The decomposition in (11.4) reveals that the nth transmitted block propagating along the first cluster will be superimposed with the $(n-\Delta)$th transmitted block propagating along the second cluster, which implies that the received signal can be taken as the summation of the signals from two *virtual users*, while the two *virtual* channels, $h^{(1)}(t,\tau;n)$ and $h^{(2)}(t,\tau;n-\Delta)$, are quasi-synchronous with an offset of ς. Let χ_i denote the delay spread of the ith cluster. When $\varsigma \geq 0$, $h^{(2)}(t,\tau;n-\Delta)$ lags behind $h^{(1)}(t,\tau;n)$; as illustrated in Figure 11.3. The impulse responses of $h^{(1)}(t,\tau;n)$ and $h^{(2)}(t,\tau;n-\Delta)$ are within the interval of $[0,\max\{\chi_1,\varsigma+\chi_2\}]$. On the other hand, when $\varsigma < 0$, $h^{(1)}(t,\tau;n)$ lags behind $h^{(2)}(t,\tau;n-\Delta)$, and their impulse responses are within the interval of $[\varsigma,\max\{\chi_1,\chi_2+\varsigma\}]$. The combined delay spread of the two channels is

$$T_{\rm ch} := \begin{cases} \max\{\chi_1,\varsigma+\chi_2\}, & \varsigma \geq 0 \\ \max\{\chi_1+|\varsigma|,\chi_2\}, & \varsigma < 0. \end{cases} \tag{11.7}$$

11.1.3 Received Signal

Design a guard interval larger than the effective delay spread of the quasi-synchronous channel in (11.7), i.e., $T_g \geq T_{\rm ch}$. The receiver can partition the received signals into blocks of duration $T_{\rm bl}$, without interblock interference due to the channel spreading within each cluster; see Figure 11.4. The nth received block is expressed as

$$\begin{aligned}\tilde{y}(t;n) &= \sum_{p=1}^{N_{\rm pa}^{(1)}} A_p^{(1)}[n]\tilde{s}\left((1+a_p^{(1)}[n])t-\tau_p^{(1)}[n];n\right) \\ &\quad + \sum_{p=1}^{N_{\rm pa}^{(2)}} A_p^{(2)}[n-\Delta]\tilde{s}\left((1+a_p^{(2)}[n-\Delta])t-\tau_p^{(2)}[n-\Delta];n-\Delta\right)+\tilde{n}(t;n),\end{aligned} \tag{11.8}$$

where $\tilde{n}(t;n)$ is the ambient noise.

After receiver preprocessing, the input–output relationship in the frequency domain is expressed as

$$\mathbf{z}[n] = \mathbf{H}^{(1)}[n]\mathbf{s}[n] + \mathbf{H}^{(2)}[n-\Delta]\mathbf{s}[n-\Delta] + \mathbf{w}[n] \tag{11.9}$$

which can be rewritten as

$$\mathbf{z}[n] = \left[\mathbf{H}^{(1)}[n]\ \mathbf{H}^{(2)}[n-\Delta]\right]\begin{bmatrix}\mathbf{s}[n] \\ \mathbf{s}[n-\Delta]\end{bmatrix} + \mathbf{w}[n], \tag{11.10}$$

where $\mathbf{z}[n]$ is frequency measurement vector of the nth received block at all the subcarriers, $\mathbf{s}[n]$ is the transmitted symbol vector of the nth transmitted block, and the channel matrix is

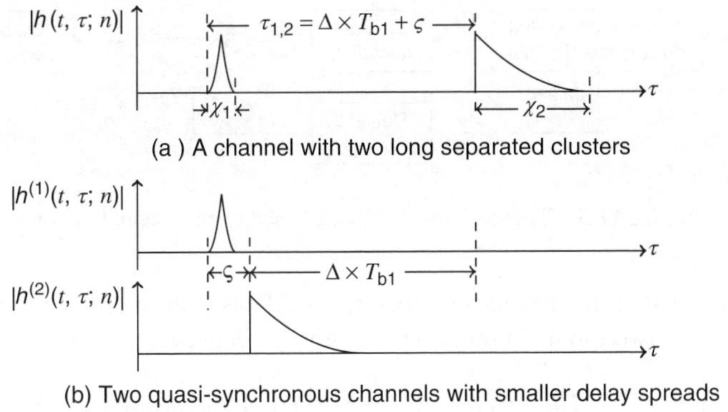

Figure 11.3 Illustration of a deep water horizontal channel with two clusters. Δ: an integer; T_{bl} : time duration of each transmitted block.

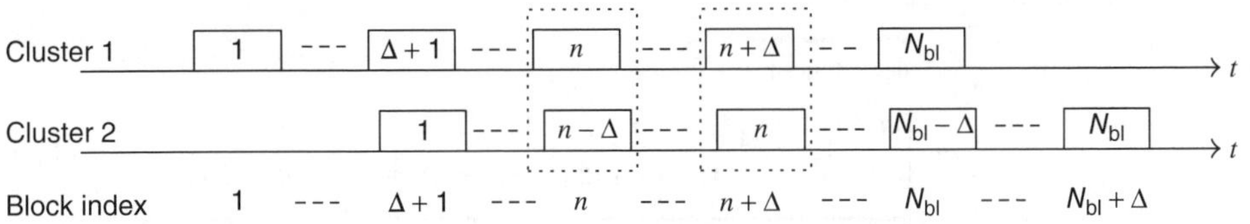

Figure 11.4 Illustration of received signal; $\varsigma > 0$ in this example.

formulated as

$$\mathbf{H}^{(i)}[n] = \sum_{p=1}^{N_{\mathrm{pa}}^{(i)}} \xi_p^{(i)}[n]\mathbf{\Lambda}(\overline{\tau}_p^{(i)}[n])\mathbf{\Gamma}(b_p^{(i)}[n], \hat{\epsilon}[n]), \tag{11.11}$$

with the two generic $K \times K$ matrices $\mathbf{\Lambda}(\tau)$ and $\mathbf{\Gamma}(b, \epsilon)$ defined in (5.30) and (5.31), respectively, $\hat{\epsilon}[n]$ being the estimated Doppler shift, and the triplet $(\xi_p^{(i)}[n], \overline{\tau}_p^{(i)}[n], b_p^{(i)}[n])$ is associated to $(A_p^{(i)}[n], \tau_p^{(i)}[n], a_p^{(i)}[n])$ as in (5.26).

11.2 Decision-Feedback Based Receiver Design

As shown in (11.10), the received signal is the block-level convolution of the transmitted blocks and the channel clusters. Given the convolutional structure of the system model, a block-by-block equalization with decision feedback as depicted in Figure 11.5 is developed.

Since one transmitted block is included in two received blocks, it is wise to incorporate the two received blocks for symbol detection. Eq. (11.10) suggests that $\mathbf{s}[n]$ shows up in both of the nth and $(n+\Delta)$th received blocks. Stacking two received blocks yields

$$\begin{bmatrix} \mathbf{z}[n] \\ \mathbf{z}[n+\Delta] \end{bmatrix} = \begin{bmatrix} \mathbf{H}^{(2)}[n-\Delta] & \mathbf{H}^{(1)}[n] & \mathbf{0} \\ \mathbf{0} & \mathbf{H}^{(2)}[n] & \mathbf{H}^{(1)}[n+\Delta] \end{bmatrix} \begin{bmatrix} \mathbf{s}[n-\Delta] \\ \mathbf{s}[n] \\ \mathbf{s}[n+\Delta] \end{bmatrix} + \begin{bmatrix} \mathbf{w}[n] \\ \mathbf{w}[n+\Delta] \end{bmatrix}. \tag{11.12}$$

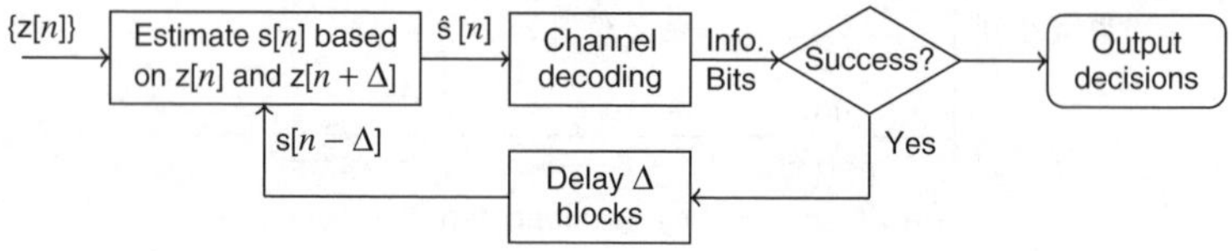

Figure 11.5 The decision-feedback based channel equalization.

Depending on whether the interfering block $\mathbf{s}[n-\Delta]$ has been successfully recovered upon detecting block $\mathbf{s}[n]$, two input–output relationships for data detection are possible.

- *Equalization without decision feedback*: From (11.12), the desired symbol vector $\mathbf{s}[n]$ can be estimated by treating the interference from symbol vectors $\mathbf{s}[n-\Delta]$ and $\mathbf{s}[n+\Delta]$ as part of an equivalent noise $\mathbf{w}$, as illustrated in

$$\begin{bmatrix}\mathbf{z}[n]\\ \mathbf{z}[n+\Delta]\end{bmatrix} = \underbrace{\begin{bmatrix}\mathbf{H}^{(1)}[n]\\ \mathbf{H}^{(2)}[n]\end{bmatrix}\mathbf{s}[n]}_{\text{desired signal}} + \underbrace{\begin{bmatrix}\mathbf{H}^{(2)}[n-\Delta] & \mathbf{0}\\ \mathbf{0} & \mathbf{H}^{(1)}[n+\Delta]\end{bmatrix}\begin{bmatrix}\mathbf{s}[n-\Delta]\\ \mathbf{s}[n+\Delta]\end{bmatrix}}_{\text{IBI from the preceding and succeeding blocks}} + \underbrace{\begin{bmatrix}\mathbf{w}[n]\\ \mathbf{w}[n+\Delta]\end{bmatrix}}_{\text{ambient noise}}. \tag{11.13}$$

- *Equalization with decision feedback*: If the block $\mathbf{s}[n-\Delta]$ transmitted prior to the current block $\mathbf{s}[n]$ has been successfully detected, it can be used to facilitate the estimation of $\mathbf{s}[n]$. After subtracting the contribution of $\mathbf{s}[n-\Delta]$ from the received blocks, the input–output relationship can be put as

$$\begin{bmatrix}\mathbf{z}[n] - \mathbf{H}^{(2)}[n-\Delta]\hat{\mathbf{s}}[n-\Delta]\\ \mathbf{z}[n+\Delta]\end{bmatrix} = \underbrace{\begin{bmatrix}\mathbf{H}^{(1)}[n]\\ \mathbf{H}^{(2)}[n]\end{bmatrix}\mathbf{s}[n]}_{\text{desired signal}} + \underbrace{\begin{bmatrix}\mathbf{0}\\ \mathbf{H}^{(1)}[n+\Delta]\end{bmatrix}\mathbf{s}[n+\Delta]}_{\text{IBI from the succeeding block}} + \underbrace{\begin{bmatrix}\mathbf{w}[n]\\ \mathbf{w}[n+\Delta]\end{bmatrix}}_{\text{ambient noise}}. \tag{11.14}$$

Both (11.13) and (11.14) define a detection problem with colored noise. A linear MMSE equalizer developed in Section 8.2.2 can be applied.

11.3 Factor-Graph Based Joint IBI/ICI Equalization

11.3.1 Probabilistic Problem Formulation

For the sake of computational efficiency, the channel matrix $\mathbf{H}^{(i)}[n]$ is usually assumed having energy concentrated on the main diagonal and several off-diagonals, i.e.,

$$H^{(i)}[m,k;n] \approx 0, \qquad \forall |m-k| > D \tag{11.15}$$

where D is termed as the ICI depth, meaning that only the ICI from D-direct neighboring subcarriers is considered explicitly.

The input–output relationship corresponding to (11.10) at each subcarrier is

$$z[m;n] = \sum_{k=m-D}^{m+D} H^{(1)}[m,k;n]s[k;n] + \sum_{k=m-D}^{m+D} H^{(2)}[m,k;n-\Delta]s[k;n-\Delta] + \eta[m;n], \tag{11.16}$$

where $\eta[m;n]$ denotes an equivalent noise consisting of ambient noise and ignored ICI.

Define two vectors:

$$\begin{aligned}\mathbf{h}_{n,k} &:= [H^{(1)}[k,k-D;n],\cdots,H^{(1)}[k,k+D;n],\\ &\qquad H^{(2)}[k,k-D;n-\Delta],\cdots,H^{(2)}[k,k+D;n-\Delta]]^{\mathrm{T}},\\ \xi_{n,k} &:= \left[s[k-D;n],\cdots,s[k+D;n],\ \ s[k-D;n-\Delta],\cdots,s[k+D;n-\Delta]\right]^{\mathrm{T}}.\end{aligned}$$

The input–output relationship in (11.16) can be rewritten as

$$z[k;n] = \mathbf{h}_{n,k}^{\mathrm{T}}\xi_{n,k} + \eta[k;n]. \tag{11.17}$$

Assume that (i) transmitted symbols are independent, and (ii) frequency noise samples are independent across subcarriers. With the second assumption, the frequency measurements can be shown independent conditional on the transmitted symbols. The *a priori* probability and the likelihood function of transmitted symbols are expressed, respectively, as

$$\Pr(\{\mathbf{s}[n]\}_{n=1}^{N_{\mathrm{bl}}}) = \prod_{n=1}^{N_{\mathrm{bl}}}\prod_{k=-K/2}^{K/2-1}\Pr(s[k;n]), \tag{11.18}$$

$$f\left(\{\mathbf{z}[n]\}_{n=1}^{N_{\mathrm{bl}}+\Delta} \mid \{\mathbf{s}_n\}_{b=1}^{N_{\mathrm{bl}}}\right) = \prod_{n=1}^{N_{\mathrm{bl}}+\Delta}\prod_{k=-K/2}^{K/2-1} f(z[k;n] \mid \xi_{n,k}). \tag{11.19}$$

The *a posteriori* probability can be obtained as

$$\begin{aligned}&\Pr\left(\{\mathbf{s}[n]\}_{n=1}^{N_{\mathrm{bl}}} \mid \{\mathbf{z}[n]\}_{n=1}^{N_{\mathrm{bl}}+\Delta}\right)\\ &\qquad\propto \left[\prod_{n=1}^{N_{\mathrm{bl}}+\Delta}\prod_{k=-K/2}^{K/2-1} f(z[k;n] \mid \xi_{n,k})\right] \times \left[\prod_{n=1}^{N_{\mathrm{bl}}}\prod_{k=-K/2}^{K/2-1}\Pr(s[k;n])\right].\end{aligned} \tag{11.20}$$

Hence, the optimal estimate of the information symbols can be obtained via

$$\{\hat{\mathbf{s}}[n]\}_{n=1}^{N_{\mathrm{bl}}} = \arg\max \Pr\left(\{\mathbf{s}[n]\}_{n=1}^{N_{\mathrm{bl}}} \mid \{\mathbf{z}[n]\}_{n=1}^{N_{\mathrm{bl}}+\Delta}\right). \tag{11.21}$$

Solving (11.21) requires a very high computational complexity, especially when the number of OFDM blocks per data burst and the number of subcarriers are large. To make the problem tractable, one can exploit the fact that each symbol only shows up in several frequency measurements of two blocks. The posterior probability of each symbol thus can be obtained by performing the probability marginalization over a factor graph.

11.3.2 Factor-Graph Based Equalization

Gaussian approximation is adopted for the prior probability function and the likelihood function of transmitted symbols. Let $\hat{f}(s[k;n])$ represent the Gaussian approximation of $\Pr(s[k;n])$, with the mean and variance denoted by $\bar{s}[k;n]$ and $\sigma^2_{s,k,n}$, respectively, and $\hat{f}(z[k;n] \mid \xi_{n,k})$ represents the Gaussian approximation of $f(z[k;n] \mid \xi_{n,k})$ where the variance of equivalent noise $\eta[k;n]$ in (11.17) is denoted by $\sigma^2_{\eta,k,n}$. The probability density functions can be put as

$$\hat{f}(s[k;n]) \propto \exp\left(-\frac{1}{\sigma^2_{s,k,n}}|s[k;n]-\bar{s}[k;n]|^2\right), \tag{11.22}$$

$$\hat{f}(z[k;n] \mid \xi_{n,k}) \propto \exp\left(-\frac{1}{\sigma^2_{\eta,k,n}}\left|z[k;n]-\mathbf{h}^{\mathrm{T}}_{n,k}\xi_{n,k}\right|^2\right), \tag{11.23}$$

where the mean $\bar{s}[k;n]$ and variance $\sigma^2_{s,k,n}$ of $\hat{f}(s[k;n])$ are computed based on the prior probability $\Pr(s[k;n])$, and estimation of $\sigma^2_{\eta,k,n}$ will be discussed in the later section.

Taking the symbol vector $\xi_{n,k}$ as the variable node, the factor graph representation of (11.20) is shown in Figure 11.6, where the function nodes are formed by the prior probability density function, the likelihood function, and two delta functions $\delta_1(\xi_{n,k}, \xi_{n,k+1})$ and $\delta_2(\xi_{n,k}, \xi_{n+\Delta,k})$ which are introduced to ensure the consistency of identical symbols in adjacent variables. The messages $m_1 \sim m_{12}$ in the graph represent either the prior probability function or the marginal probability density function related to the function nodes. The posterior probability of each individual symbol vector $\xi_{n,k}$ can be found by passing messages over the graph according to the sum-product algorithm [223] and the Gaussian message passing principle [259].

According to the sum-product algorithm and the Gaussian message passing principle, the outgoing message of each variable node is the product of incoming messages from all the other edges. Take the variable node $\xi_{n,k}$ in Figure 11.6 as an example. The outgoing message m_8 can be updated as

$$m_8 = m_1 m_2 m_3 m_5 m_7. \tag{11.24}$$

Meanwhile, the outgoing message from the delta functions corresponds to the extraction of the probability distribution of common symbols among consecutive variable nodes from the incoming message.

To update all the messages in the graph in Figure 11.6, all the messages are initialized as one. Then the messages are passed from the left node to the right node row by row. Once the last row has been updated, the messages are updated in an inverse direction, i.e., from the right node to the left node, and from the last row to the first row.

The posterior probability of the variable node is obtained as

$$\hat{f}(\xi_{n,k} \mid \{\mathbf{z}[n]\}_{n=1}^{N_{\mathrm{bl}}+\Delta}) = m_1 m_2 m_3 m_5 m_7 m_9, \tag{11.25}$$

from which the posterior probability of each individual symbol $\hat{f}(s[k;n] \mid \{\mathbf{z}[n]\}_{n=1}^{N_{\mathrm{bl}}+\Delta})$ can be directly obtained. With the Gaussian approximation, computation in (11.24) and (11.25) can be simplified by operating only over the mean and covariance matrices, as illustrated in Section 8.3.2.

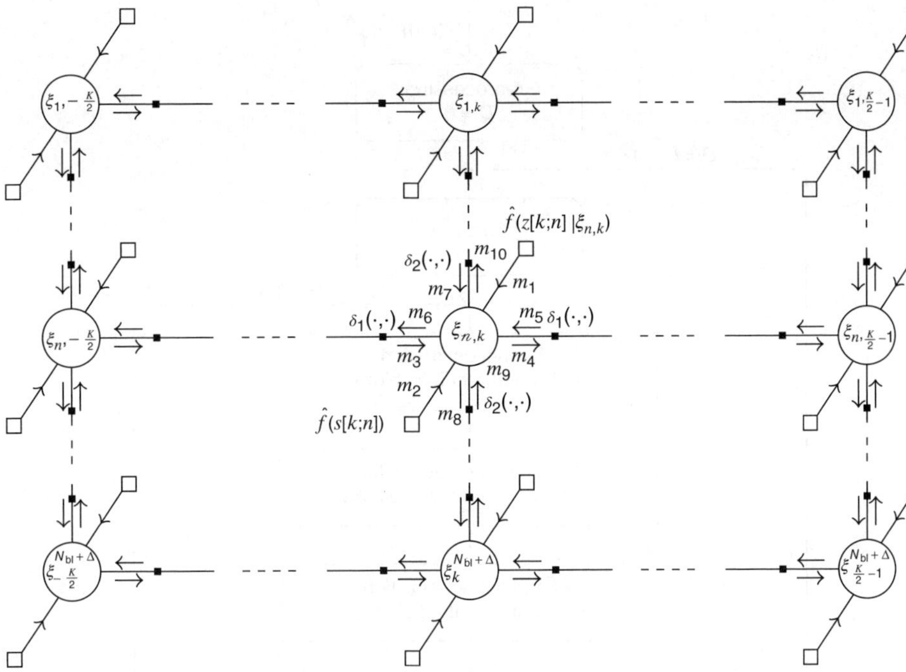

Figure 11.6 Factor-graph based joint IBI/ICI equalization, empty boxes represent the prior probability density function nodes; filled boxes represent the likelihood function nodes; messages $m_1 \sim m_{12}$ represent either the prior probability density function or the marginal probability density function over the factor nodes.

Three remarks on the factor-graph based equalization are in the following.

- For the channel with *either* IBI or ICI, the factor graph corresponding to one column or one row of the graph in Figure 11.6 does not have cycles. However, for the channel under consideration with both IBI and ICI, the factor-graph in Figure 11.6 is not cycle-free. Nonetheless, message-passing algorithms can achieve excellent performance over graphs with cycles, e.g., decoding of the low-density-parity-check codes.
- For the receiver with turbo equalization, it is the extrinsic information of each data symbol that should be computed and sent to the channel decoder. Calculation of the extrinsic information based on the obtained *a posteriori* probability has been discussed in Section 8.3.2.
- To ensure the well-conditioned property of covariance matrices of messages in both horizontal and vertical message propagation, the prior information $\hat{f}(s[k;n])$ can be factorized into several components during the implementation. For the practical implementation of the factor graph in Figure 11.6, please refer to [421].

11.4 Iterative Block-to-Block Receiver Processing

Due to the block-level convolution shown in (11.10), it is necessary to recover the transmitted symbols based on all the received blocks of one data burst. An iterative block-to-block receiver

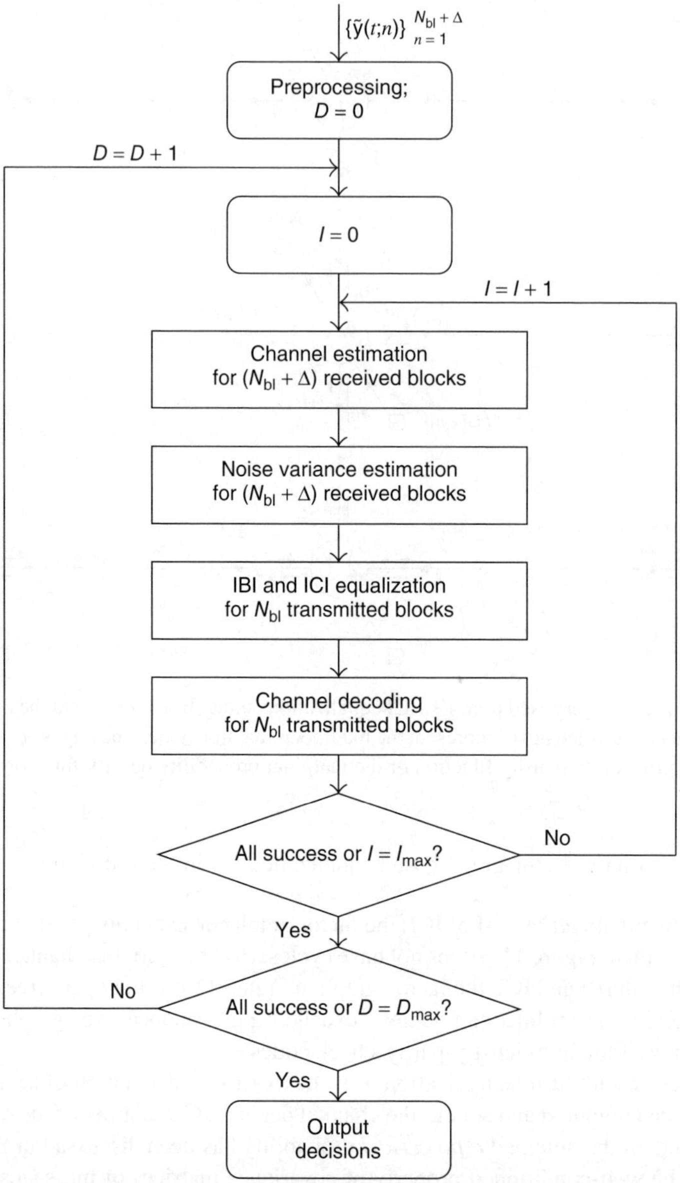

Figure 11.7 Flow chart of the iterative batched-processing receiver with ICI-progressive processing.

can be developed, as shown in Figure 11.7, where blocks within one burst are decoded as one batch. To accommodate the unknown ICI in the received signal, after I_{max} iterative operations, the channel ICI depth D is updated to include more ICI into the system model. Once the parity check conditions of all blocks are satisfied during the channel decoding, or the ICI depth reaches a predetermined threshold D_{max}, the iterative process stops.

The receiver modules are briefly sketched in the sequel.

- *Clustered Channel Estimation*: The channel estimation is performed for each received block individually. Note that the sub-channels corresponding to $\mathbf{H}^{(1)}[n]$ and $\mathbf{H}^{(2)}[n-\Delta]$ in (11.10) are synchronized on the block level with a small offset on the channel support, as shown in Figure 11.3. As such, estimation of a channel with two widely-separated clusters is converted to the estimation of two quasi-synchronous virtual channels. Hence, based on the knowledge of the delay offset, the channel estimation methods in multiuser systems developed in Section 7.5 can be applied to the current channel setting.

  In the iterative receiver structure, the input of the channel estimator at each received block includes the frequency observation vector $\mathbf{z}[n]$, the pilot symbols and the *a posteriori* probabilities of information symbols fed back from the decoder. In this chapter, the soft decisions of information symbols are used for channel estimation; please refer back to Chapter 3 on the soft decision computation based on the *a posteriori* probabilities from the decoder.
- *Noise Variance Estimation*: With the estimated channel matrices and the soft decisions of information symbols, the noise variance estimate at each received block is updated as

$$\hat{\sigma}_\eta^2[n] = \mathbb{E}_{m\in\mathcal{S}_{\rm N}}\left[\left|z[m;n] - \sum_{k=m-D}^{m+D}\left(\hat{H}^{(1)}[m,k;n]\hat{s}[k;n] + \hat{H}^{(2)}[m,k;n-\Delta]\hat{s}[k;n-\Delta]\right)\right|^2\right] \tag{11.26}$$

  which is used as the estimate of $\{\sigma^2_{\eta,n,k}, \forall k\}$ in (11.22) for the factor-graph based channel equalization.
- *Channel Equalization and Decoding*: Based on the extrinsic and *a posterior* probabilities from the channel decoder, the channel equalizer developed in Sections 11.2 or 11.3 can be applied. Specifically,
  - For the decision-feedback based equalizer, only the decision on $\mathbf{s}[n-\Delta]$ is used to mitigate IBI in (11.12), as shown in (11.12). The linear MMSE equalizer is adopted for symbol detection, in which the extrinsic information fed from the channel decoder is used as *a priori* information.
  - The extrinsic probabilities will be translated to the *a priori* probabilities of transmitted symbols for the factor-graph based equalization.

After obtaining the extrinsic probabilities of transmitted symbols from the channel equalizer, they will be fed into the channel decoder, and both *a posteriori* probabilities and extrinsic probabilities of information symbols will be updated, these then being fed back for channel estimation and equalization in the next loop.

11.5 Simulation Results

Each cluster of the time-varying horizontal acoustic channel is simulated according to the specifications in Section 5.5.1. The channel parameters are $N_{\rm pa} = 10$, $\Delta\tau = 1$ ms, $\Delta P_{\rm pa} = 20$ dB, $T_{\rm g} = 24.6$ ms, $\upsilon_0 = 0$ m/s and $\sigma_{\rm v} = 0.10$ m/s. The interarrival time of the two clusters follows a uniform distribution with $\Delta \sim \mathcal{U}[0, 2]$ and $\varsigma \sim \mathcal{U}[-0.2T_{\text{bl}}, 0.2T_{\text{bl}}]$. The OFDM parameters are identical to the parameters used in an experiment, which are specified in Table B.1, except that there are $N_{\text{bl}} = 10$ blocks in each transmission.

In the following, three receiver configurations will be compared in three cases.

- *Configuration 1*: Receiver design which treats signal propagating along the second cluster as ambient noise; the linear MMSE equalizer is adopted;
- *Configuration 2*: Receiver design with decision-feedback based equalization;
- *Configuration 3*: Receiver design with factor-graph based equalization.

In the sequel, the three configurations are referred to as *the receiver without IBI mitigation*, *the DFE based receiver* and *the factor-graph based receiver*, respectively.

Test Case 11.5.1 In this test, the ICI-depth is fixed as $D = 1$, and compare the performance of the factor-graph based receiver and the DFE based receiver. The average powers of both clusters are simulated as identical, and only one receiving element is used. To estimate the ICI coefficients, 96 data subcarriers are converted to pilot subcarriers. With a 16-QAM constellation, the data rate is $R = 7.4$ kb/s.

For all the simulation results, at least 100 block errors are collected for each signal-to-noise-ratio (SNR) level. Figure 11.8 depicts the block-error-rate (BLER) performance of the two receivers. For the factor-graph based receiver, one can observe significant performance improvements with iterations between the equalizer and the decoder. Moreover, relative to the DFE based receiver, the factor-graph based receiver has more than 1 dB gain when the iteration time reaches three, since the latter can utilize the decoding results of all the other blocks to facilitate the decoding of a particular block, while only the knowledge of the previous blocks can be utilized in the former.

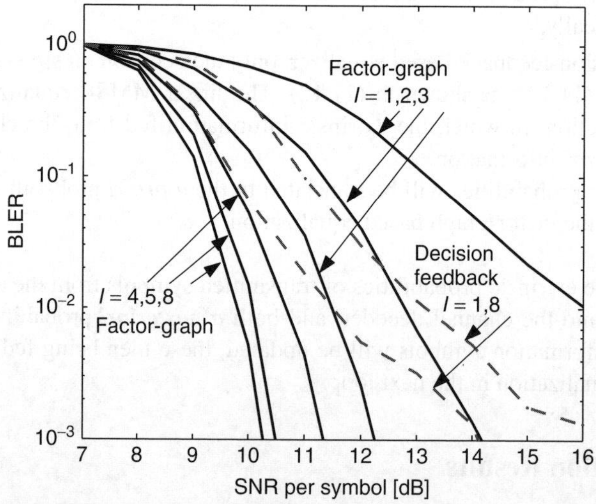

Figure 11.8 BLER performance of the factor-graph based receiver versus the DFE based receiver with mild Doppler spreads $\sigma_v = 0.10$ m/s. ICI depth is fixed at $D = 1$.

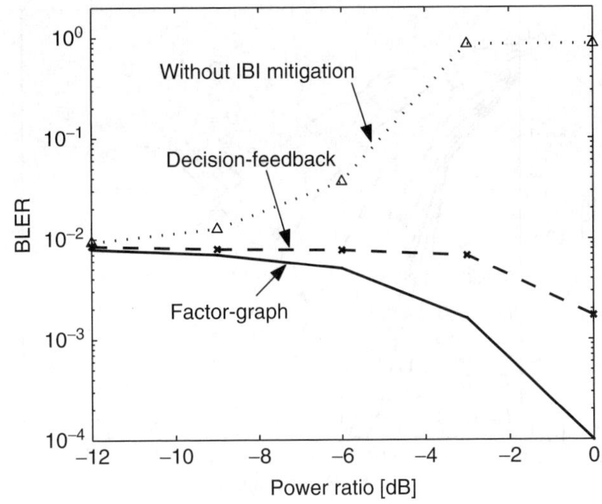

Figure 11.9 BLER performance of three receivers with mild Doppler spreads $\sigma_v = 0.10$ m/s. SNR of the first cluster is fixed at 11 dB; ICI depth is fixed at $D = 1$. Five iterations are performed.

Test Case 11.5.2 To get insights on how the performance of both the factor-graph based receiver and the DFE based receiver with interference cancellation change at different inter-cluster interference levels, fix the SNR of the signal arriving along the first cluster to be 11 dB, and vary the power ratio of the signal arriving along the first cluster to that arriving along the second cluster from −12 dB to 0 dB. The same OFDM parameter set in test case 11.5.1 is used.

Figure 11.9 shows the performance curves of the factor-graph based IBI/ICI-aware receiver, the DFE based receiver, and the receiver without IBI mitigation, with different power ratios between the second and the first cluster.

- As the signal arriving along the second cluster becomes stronger, performance of both the factor-graph based receiver and the DFE based receiver increases, while the performance of the receiver without IBI mitigation decreases. This result shows that signal arriving along the second cluster can be utilized by both the factor-graph based receiver and the DFE based receiver to improve decoding performance, while the decoding performance of the receiver without IBI mitigation deteriorates due to the intercluster interference.
- The performance gap between the factor-graph based receiver and the DFE based receiver gets pronounced as the power of the second cluster increases. This is due to the fact that the DFE based receiver can only partially mitigate the intercluster interference while the factor-graph based receiver can fully exploit all the available information from both clusters.

Test Case 11.5.3 In this test, the performance of the factor-graph based receiver is evaluated in ICI-progressive framework, where the receiver starts with $D = 0$ and 256 regularly distributed pilot subcarriers are used for channel estimation. With a 16-QAM constellation, the data rate is $R = 10.4$ kb/s. Figure 11.10 plots the BLER performance of the progressive receiver, which shows that progressively updating the system model improves the performance quickly.

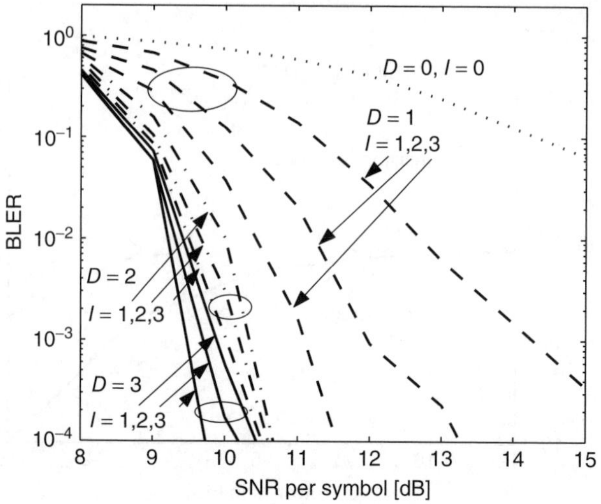

Figure 11.10 BLER performance of the factor-graph based ICI-progressive receiver with mild Doppler spreads $\sigma_v = 0.10$ m/s.

11.6 Experimental Results in the AUTEC Environment

In this section, the performance of the iterative receiver is evaluated using the data sets collected from an experiment conducted in the AUTEC network in March 2010, which is abbreviated as the AUTEC10 experiment. Detailed descriptions on the AUTEC network are presented in this chapter. Among 96 network nodes, node 45 is used as the transmitter, and the remaining nodes as receivers.

The OFDM parameters in the AUTEC10 experiment are specified in Table 11.1. In the signal design, 860 subcarriers are formed by the random distribution of three pilot subcarrier patterns and three data subcarrier patterns,

- pilot patterns: $\{[\text{N P N}]_{(50)}, [\text{P P}]_{(16)}, [\text{P P P}]_{(114)}\}$
- data patterns: $\{[\text{D}]_{(48)}, [\text{D D}]_{(105)}, [\text{D D D}]_{(26)}\}$,

Table 11.1 OFDM parameters in AUTEC10 experiment

Carrier frequency	f_c	11 kHz
Bandwidth	B	5 kHz
No. subcarriers	K	860
Symbol duration	T	170.7 ms
Subcarrier spacing	Δf	5.86 Hz
Guard interval	T_g	250 ms
# of null subcarriers	$\lvert S_N \rvert$	0
# of pilot subcarriers	$\lvert S_P \rvert$	424
# of data subcarriers	$\lvert S_D \rvert$	336
# of blocks in each transmission	N_{bl}	10

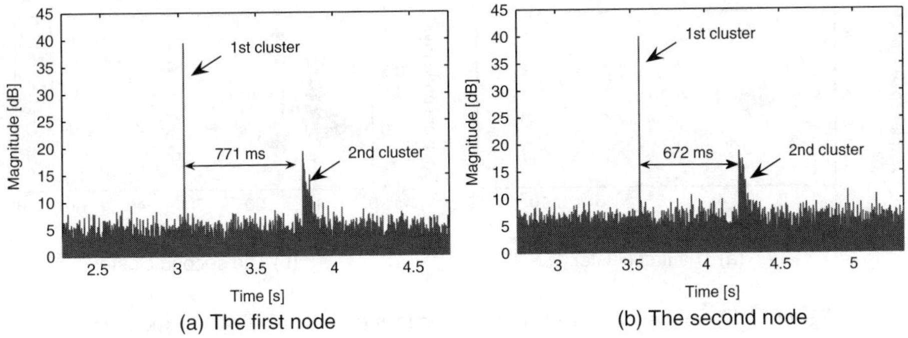

Figure 11.11 Samples of HFM correlation results at two receiving nodes in the AUTEC10 experiment.

where N, P and D denote null subcarrier, pilot subcarrier and data subcarrier, respectively, and the sub-index denotes the number of different patterns, which results in 424 pilot subcarriers and 336 data subcarriers in total. With a rate-1/2 nonbinary LDPC code and a QPSK constellation, the data rate is computed as

$$R = \frac{1}{2} \cdot \frac{336}{0.1707 + 0.25} \cdot \log_2 4 = 798.7\,\text{bits/s} \tag{11.27}$$

or $\frac{3}{8}R = 299.5$ Bytes/(3s). There are 22 transmissions in total, with 10 blocks in each transmission.

In this experiment, a hyperbolic frequency modulated (HFM) waveform [222] was used as the preamble to probe the channel. The matched filtering operation at the receiver side provides the multipath profile of the two-cluster channel. As a detection scheme for Sonar applications, the modified Page test [2] is employed at the matched filter output to detect the rising and falling edges of clusters; the intercluster delay $\tau_{1,2}$ and the delay spreads χ_1 and χ_2 thus become available (c.f. Page test in Section 6.1).

Two receiving nodes are used for receiver decoding. Distances between these two receiving nodes and the transmitter are about 3.7 km and 4.4 km, respectively. The average input SNR of received signals is about 17.9 dB at the first node, and about 14.9 dB at the second node. Samples of correlation results of the received signal with an HFM local replica at nodes with two cluster-structures are shown in Figure 11.11, which shows that power of the signal arriving along the second cluster is about 20 dB lower relative to that of the signal arriving along the first cluster. The intercluster delay of these two nodes can be rounded into two OFDM blocks, with the residual term being $\varsigma = -70$ ms and $\varsigma = -169$ ms, respectively.

Samples of channel estimates at one receiving node are shown in Figure 11.12. Again, the channel estimates agree well with the HFM correlations results in Figure 11.11. Combining the received signals from these two receiving nodes, the decoding results of 22 transmissions are shown in Figure 11.13. It can be observed that:

- As the number of iterations increases, more and more blocks become decodable.
- With 5 progressive operations, all blocks can be decoded with the factor-graph based ICI-progressive receiver, while only one block error occurs for the DFE based

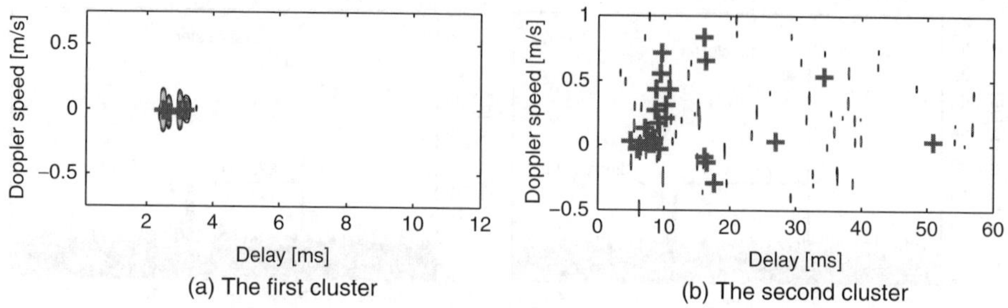

Figure 11.12 Sample of channel estimation results in the AUTEC10 experiment.

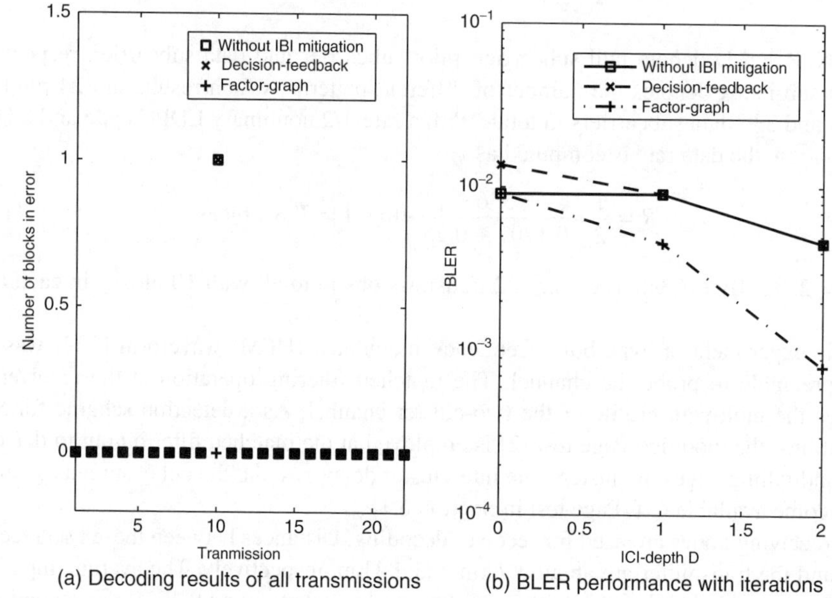

Figure 11.13 Decoding results of three receivers in the AUTEC10 experiment; 22 transmissions are involved. $I_{\max} = 2$ for each value of D.

ICI-progressive receiver and the receiver without IBI mitigation. Due to the significant power difference between the two clusters, the advantage of the factor-graph based receiver over the other two receivers is small, which agrees with the results in Figure 11.9.

The above receiver processing schemes have also been compared in [420, 421] using another field experimental data sets collected over the AUTEC network, which the second cluster has much larger power than the first cluster. The advantage of the factor-graph based receiver over the DFE based receiver is pronounced in that experiment [421].

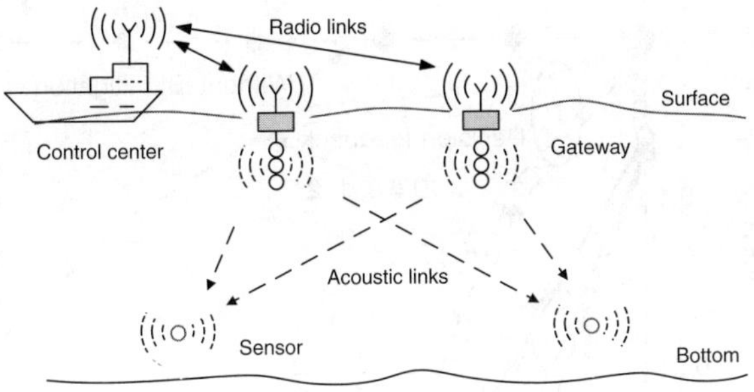

Figure 11.14 Illustration of an underwater broadcasting network with multiple gateways.

11.7 Extension to Underwater Broadcasting Networks

11.7.1 Underwater Broadcasting Networks

The receiver for deep-water horizontal acoustic communications applies to other communication scenarios with similar channel characteristics. One application example is the underwater broadcasting network [5, 295, 310] as shown in Figure 11.14, in which the gateways communicate with a control center using radio links, then broadcast the information they received from the control center to underwater sensors via acoustic links using an identical transmission waveform. Related research on this broadcasting network with multiple gateways can be found, e.g., [90, 186, 187]. The broadcasting network illustrated in Figure 11.14 falls into a large category called single frequency networks (SFNs). The concept of SFNs has been widely used in Digital Audio Broadcasting (DAB) and Digital Video Broadcasting (DVB) systems [245].

Note that the time-difference-of-arrivals of signals from different gateways at one particular sensor could be much larger than the typical OFDM block length. For example, if the distances between the sensor to the two transmitters differ by 450 meters, the relative delay is about 300 ms, which is much larger than the typical block duration. The IBI will occur in the received signal at the sensor. To recover the broadcast information, one can regard the received signal as the one from a single source but passing through a channel with widely-separated multipath clusters. The receiver developed for the deep water horizontal acoustic channel with widely separated clusters therefore applies.

We next evaluate the receiver performance using emulated data sets.

11.7.2 Emulated Experimental Results: MACE10

The experimental setting and the OFDM parameters have been described in Appendix B. Out of two tows in the experiment, only the data sets collected in the first tow are considered. There are 31 transmissions in total, with 20 blocks in each transmission. One transmission file recorded during the turn of the source is excluded, where the SNR of the received signal is quite low. With a rate-1/2 nonbinary LDPC code and a 16-QAM constellation, the data rate is $R = 5.4$ kb/s.

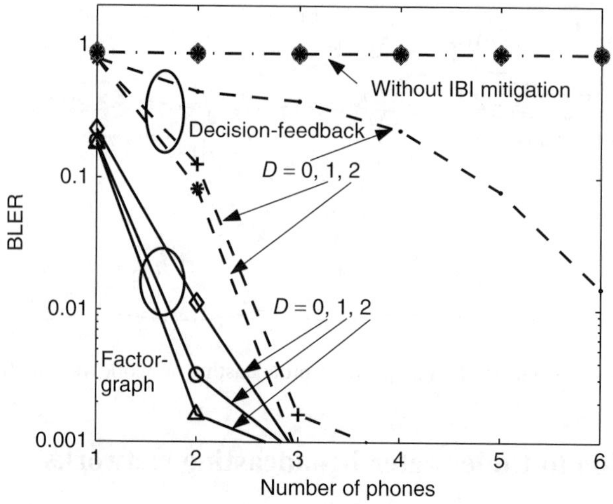

Figure 11.15 BLER performance of three receivers, averaged over 30 transmissions. $I_{max} = 2$ for each value of D.

Note that this experiment was carried out in shallow water with a single source. To create a scenario that multiple gateways transmit the same information to underwater sensors, the received signal of each transmission is shifted two OFDM blocks behind, and take the shifted signal as the signal from the second source. Then, by adding up the shifted signal with the originally received signal, a set of semi-experimental signal is obtained, which can be regarded as the received signal at one sensor in the underwater broadcasting network with two gateways. The intercluster block delay of the broadcasting channel is $\Delta = 2$, and the fractional delay ς follows a uniform distribution $\varsigma \sim \mathcal{U}[-0.1T_{bl}, 0.1T_{bl}]$.

The receiver without IBI mitigation, the DFE based receiver and the factor-graph based receiver will be tested in three cases.

Test Case 11.7.1 During the preprocessing, the received signal is resampled to compensate the mobility of the source array. With five iterations, the decoding results of the receiver without IBI mitigation, the DFE based receiver and the factor-graph based receiver are shown in Figure 11.15. One can see that the receiver without IBI mitigation almost cannot work, and there is a considerable performance gap between the DFE based receiver and the factor-graph based receiver. Meanwhile, one can find that the decoding performance converges after three to four iterations for the DFE based receiver and the factor-graph based receiver.

Test Case 11.7.2 Based on the estimated ambient noise variance $\hat{\sigma}_\eta^2$, the white Gaussian noise is added to the received signal to create a semi-experimental data set to test three receivers under different SNR levels. Decoding results of the receiver without IBI mitigation, the DFE based receiver and the factor-graph based receiver are shown in Figure 11.16. One can see that the decoding performance keeps decreasing as the added noise variance level increases, and the factor-graph based receiver has the best performance.

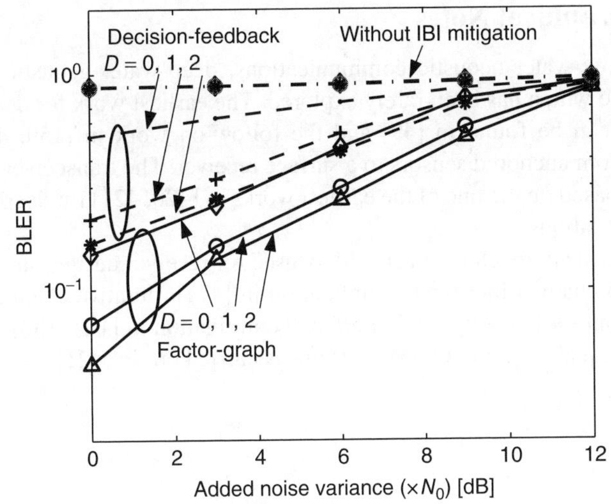

Figure 11.16 BLER performance of three receivers by adding white Gaussian noise with different variances, averaged over 30 transmissions. Two phones are combined. $I_{max} = 2$ for each value of D.

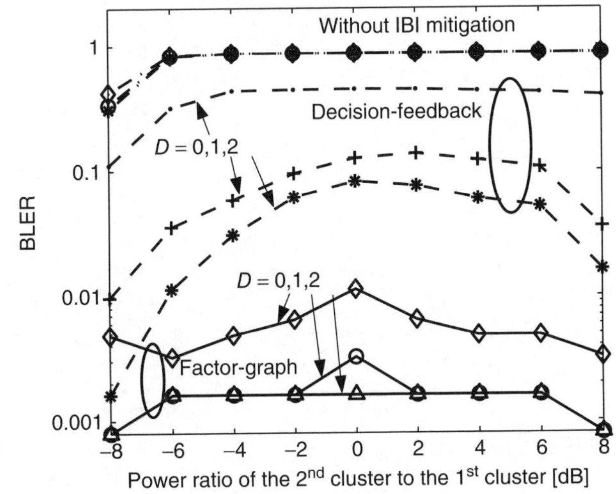

Figure 11.17 BLER performance of three receivers with different power ratios between two clusters, averaged over 30 transmissions. Two phones are combined. Three curves for each receiver correspond to $D = 0, 1, 2$, respectively, and $I_{max} = 2$ for each value of D.

Test Case 11.7.3 To test the three receivers with different power ratios between the second cluster and the first cluster, the weighting coefficients are varied when adding the delay-shifted signal with the originally received signal. Decoding results of the receiver without IBI mitigation, the DFE based receiver and the factor-graph based receiver are shown in Figure 11.17. One can see that the DFE based receiver is sensitive to the relative power levels of the two clusters, while the decoding performance of the factor-graph based receiver is very stable.

11.8 Bibliographical Notes

Relative to shallow water acoustic communications, deep-water acoustic communications belong to a regime which has been rarely explored. The earliest work for deep-water acoustic communications can be found in [49] and the follow-on work in [159, 472] for multiple access of sea-bottom anchored sensors to a surface receiver. The transceiver design presented in this chapter is based on the one of the earliest works in [420, 421] for deep-water horizontal acoustic communications.

Factor graph based approach has been widely used for receiver design, such as joint channel estimation and co-channel interference mitigation in [471], iterative channel estimation and LDPC decoding in [296], intersymbol interference mitigation in [104, 150]. The extension to joint IBI and ICI equalization in OFDM systems is carried out in [421].

12

OFDM Receiver with Parameterized External Interference Cancellation

Underwater acoustic communication channels are prone to external interference from various sources, such as the interference from marine animals, human activities and sonar operations. Malicious jammers could also inject interfering signals into the communication channel to destroy the communication link. External interference could incur significant performance degradation if not accounted for.

This chapter focuses on explicit mitigation of an external interference with a certain time duration and bandwidth. One example of the type of interference under consideration is the waveform from sonar users, as depicted in Figure 12.1. Exploiting the prior information on the time duration and frequency band of the interference, a *parameterized interference cancellation approach* will model the interference explicitly via a number of parameters, where the number of parameters is equal to the time-frequency-product of the interference. With the parameterized interference model, an iterative OFDM receiver with interference cancellation is designed to perform data processing and interference estimation/cancellation iteratively. The chapter is organized as follows.

- Section 12.1 introduces the parameterization method that represents an external interference via a number of parameters.
- Sections 12.2 and 12.2.2 present an OFDM receiver, where OFDM receiver decoding is performed with interference estimation and cancellation iteratively.
- Sections 12.3, 12.4 and 12.5 contain simulation and experimental results that validate the performance of the iterative receiver.

12.1 Interference Parameterization

Interference is different from the ambient noise. The interference considered in this chapter overlaps with the OFDM signals partially in the frequency band and partially in the block duration, as illustrated in Figure 12.1.

OFDM for Underwater Acoustic Communications, First Edition. Shengli Zhou and Zhaohui Wang.
© 2014 John Wiley & Sons, Ltd. Published 2014 by John Wiley & Sons, Ltd.

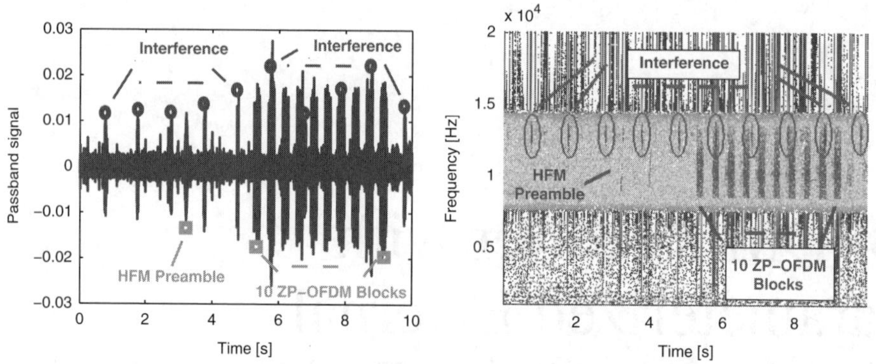

(a) Time-domain waveform (b) Time-frequency spectrum

Figure 12.1 Samples of the time domain waveform and the time-frequency spectrum of the received signal at one hydrophone after bandpass filtering. The transmitted signal consists of a hyperbolic frequency modulated (HFM) preamble followed by 10 zero-padded OFDM blocks. The circle and the square in subfigure (a) denote the locations of interference and useful signal, respectively. (Source: Wang 2012 [422], figure 1, p. 1783. Reproduced with permission of IEEE.).

Denote the center frequency, bandwidth and time duration of the interference as $f_{\rm Ic}$, $B_{\rm I}$ and $T_{\rm I}$, respectively. The bandwidth $B_{\rm I}$ is taken as the 3-dB bandwidth of the interference, but does not need to be very accurate. As will be seen later, a loose upper bound on $B_{\rm I}$ is sufficient. Let $\tilde{I}(t)$ denote the passband waveform of the interference, and $I(t)$ the baseband waveform. Since $I(t)$ is time limited, it adopts a Fourier-series representation as

$$I(t) = \sum_{\ell=-\infty}^{\infty} c_\ell e^{j2\pi \frac{\ell}{T_{\rm I}} t}, \qquad t \in [0,\, T_{\rm I}] \tag{12.1}$$

where c_ℓ is the coefficient on the basis $e^{j2\pi \frac{\ell}{T_{\rm I}} t}$. Without loss of generality, assume that $N_{\rm I} = \lceil B_{\rm I} T_{\rm I} \rceil$ is an even number. Since $I(t)$ is bandwidth-limited to $[-B_{\rm I}/2, B_{\rm I}/2)$, the coefficient c_ℓ is approximately zero for $\ell < -N_{\rm I}/2$ or $\ell \geq N_{\rm I}/2$. Hence, the baseband signal in (12.1) can be rewritten as

$$I(t) \approx \sum_{\ell=-N_{\rm I}/2}^{N_{\rm I}/2-1} c_\ell e^{j2\pi \frac{\ell}{T_{\rm I}} t}, \qquad t \in [0,\, T_{\rm I}]. \tag{12.2}$$

The corresponding passband signal is

$$\tilde{I}(t) = 2\Re\left(\sum_{\ell=-N_{\rm I}/2}^{N_{\rm I}/2-1} c_\ell e^{j2\pi \bar{f}_\ell t} \right), \qquad t \in [0,\, T_{\rm I}] \tag{12.3}$$

where $\bar{f}_\ell := f_{\rm Ic} + \ell/T_{\rm I}$. The Fourier transform of $\tilde{I}(t)$ in the frequency band $\mathcal{B}_{\rm I} := [f_{\rm Ic} - B_{\rm I}/2, f_{\rm Ic} + B_{\rm I}/2)$ can be expressed as

$$\tilde{\mathcal{I}}(f) = \sum_{\ell=-N_{\rm I}/2}^{N_{\rm I}/2-1} c_\ell \frac{\sin(\pi(f-\bar{f}_\ell)T_{\rm I})}{\pi(f-\bar{f}_\ell)} e^{-j\pi(f-\bar{f}_\ell)T_{\rm I}}, \qquad \forall f \in \mathcal{B}_{\rm I}. \tag{12.4}$$

Note that the interference overlaps the received OFDM signal with an unknown delay. Define $\tau_{\rm I}$ as the delay of the interference relative to the starting point of the OFDM block in which it resides. After the preprocessing of the OFDM receiver, the interference component at the mth subcarrier is

$$\Upsilon[m] = \int_0^{T+T_{\rm g}} \tilde{I}\left(\frac{t-\tau_{\rm I}}{1+\hat{a}}\right) e^{-j2\pi(f_m+\hat{\epsilon})t} dt. \tag{12.5}$$

Following the derivation to (12.4), $\Upsilon[m]$ can be formulated as

$$\Upsilon[m] = e^{-j2\pi\frac{m}{T}\bar{\tau}_{\rm I}} \sum_{\ell=-N_{\rm I}/2}^{N_{\rm I}/2-1} \rho_{m,\ell} u_\ell \tag{12.6}$$

where

$$u_\ell := (1+\hat{a})T_{\rm I} e^{-j2\pi(f_{\rm c}+\hat{\epsilon})\bar{\tau}_{\rm I}} c_\ell, \qquad \bar{\tau}_{\rm I} := \frac{\tau_{\rm I}}{1+\hat{a}},$$

$$\rho_{m,\ell} := \frac{\sin(\pi((1+\hat{a})(f_m+\hat{\epsilon}) - \bar{f}_\ell)T_{\rm I})}{\pi((1+\hat{a})(f_m+\hat{\epsilon}) - \bar{f}_\ell)T_{\rm I}} e^{-j\pi((1+\hat{a})(f_m+\hat{\epsilon})-\bar{f}_\ell)T_{\rm I}}.$$

Stacking interference components at all subcarriers into a vector Υ yields

$$\Upsilon = \Lambda(\bar{\tau}_{\rm I})\Gamma_{\rm I}\mathbf{u} \tag{12.7}$$

where $\Lambda(\bar{\tau}_{\rm I})$ is a $K \times K$ diagonal matrix, $\Gamma_{\rm I}$ is a $K \times N_{\rm I}$ matrix, and $\mathbf{u}$ is an $N_{\rm I} \times 1$ vector,

$$[\Lambda(\bar{\tau}_{\rm I})]_{m,m} = e^{-j2\pi\frac{m}{T}\bar{\tau}_{\rm I}}, \qquad [\Gamma_{\rm I}]_{m,l} = \rho_{m,\ell}, \qquad \mathbf{u} = \left[u_{-N_{\rm I}/2} \;\cdots\; u_{N_{\rm I}/2-1}\right]^{\rm T}. \tag{12.8}$$

12.2 An Iterative OFDM Receiver with Interference Cancellation

For easy exposition, we focus on an OFDM system in single-input single-output channels. Denote K as the total number of subcarriers. The received signal in the presence of interference is then formulated as

$$\mathbf{z} = \mathbf{H}\mathbf{s} + \Lambda(\bar{\tau}_{\rm I})\Gamma_{\rm I}\mathbf{u} + \mathbf{w}, \tag{12.9}$$

where $\mathbf{z}$ is the frequency measurements at all the subcarriers, $\mathbf{s}$ is the transmitted symbol vector, $\mathbf{H}$ is a $K \times K$ channel matrix, and $\mathbf{w}$ is the ambient noise vector.

The received signal model in (12.9) shows that four sets of parameters: (i) the channel matrix $\mathbf{H}$; (ii) the information symbol vector $\mathbf{s}$; (iii) the interference vector $\mathbf{u}$; and (iv) the delay $\tau_{\rm I}$, need to be estimated from the same measurements $\mathbf{z}$. Although pilots are usually multiplexed with information symbols to separate channel estimation and information symbol detection in the conventional OFDM receiver, frequency measurements at both pilot and data subcarriers within the interference band are contaminated.

To estimate these four sets of parameters, an iterative receiver for interference detection/estimation, channel estimation and information symbol detection was proposed in [422], as shown in Figure 12.2. Relative to the iterative OFDM receivers in Figures 9.9 and 9.10 for the scenario without interference, the receiver structure includes several new modules for interference detection and estimation. It is thus a more general framework than the one in which only the self-interference in the form of ICI is addressed.

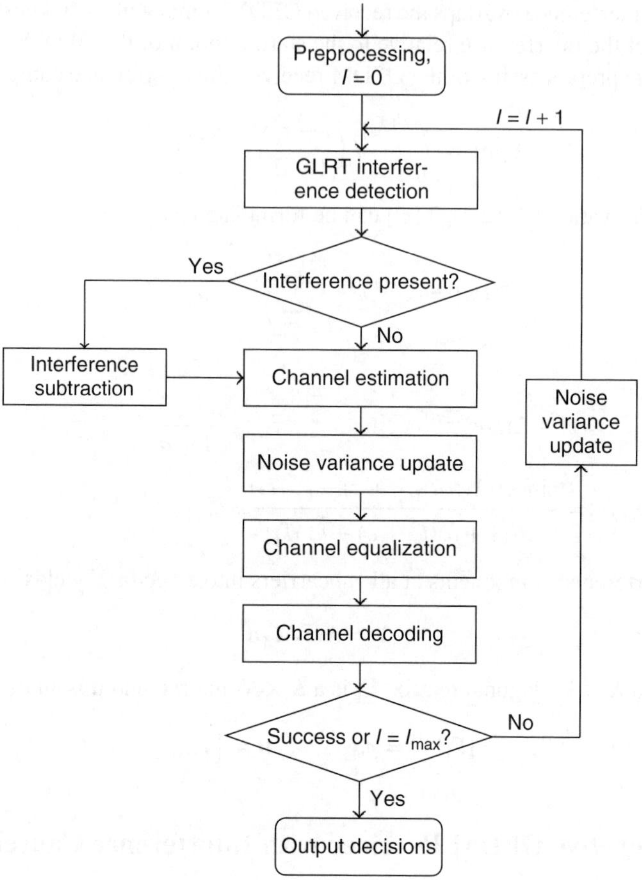

Figure 12.2 Receiver structure for interference mitigation and data detection.

After a preprocessing to remove the main Doppler effect and residual Doppler shift via the resampling operation and the Doppler compensation (see Chapter 5), and setting the initial value of the iteration index $I = 0$, the iterative OFDM receiver operates according to the following steps.

(i) *GLRT interference estimation/detection.* A generalized log-likelihood ratio test (GLRT) detector is used to detect the interference. During this process, maximum likelihood (ML) estimates of the interference $\mathbf{u}$ and the delay τ_I are obtained. Comparison of the GLRT statistics with a predetermined threshold yields a decision about the presence of the interference.

(ii) *Interference subtraction.* If interference exists, the contributions of interference will be estimated and subtracted from the measurements.

(iii) *Channel estimation.* Based on both pilot symbols and the soft decisions on information symbols fed back from the decoder, the ICI-ignorant or ICI-aware channel estimation methods in Sections 7.2 and 7.3 can be applied.

(iv) *Channel equalization and decoding.* The linear MMSE channel equalizer in Section 8.2.2 and channel decoding schemes in Chapter 3 can be used. Interactions between the two models are identical to the iterative receivers developed in Chapters 9 - 11. The decoded information symbols and soft decisions about the information symbols from the channel decoder are used for both channel estimation and linear MMSE equalization in the next iteration.

(v) *Noise variance update.* Due to the iterative estimation of the interference, the channel matrix and information symbols, the noise variance is kept updated within each iteration.

Once the information symbols have been successfully decoded, i.e., the parity check conditions of the nonbinary LDPC decoder are satisfied, or the number of iterations reaches a predetermined threshold $I_{\max}$, the iterative loop will stop.

Initialization of the proposed iterative receiver is discussed in Section 12.2.1. A brief description on sparse channel estimation, channel equalization and decoding, and noise variance estimation will be discussed in Sections 12.2.3 and 12.2.4. A detailed description of GLRT interference detection will be presented in Section 12.2.2.

12.2.1 *Initialization*

As discussed in Chapter 2, null subcarriers are usually inserted into the signal band to estimate the carrier frequency offset and the variance of the ambient noise. To initialize the iterative interference cancellation in Figure 12.2, the following operations are carried out.

(1) Based on the frequency measurements at null subcarriers, a linear interpolation is performed to estimate the variance of interference within the signal band.

(2) Assume that the interference components at all the frequency subcarriers are independent. The pre-whitening operation in Section 7.7 is performed for the frequency measurements.

(3) With the pre-whitened frequency measurements, the channel estimation is carried out based on the frequency measurements at pilot subcarriers, which is followed by the ICI equalization and nonbinary LDPC decoder to estimate the information symbols and to recover the information bits.

(4) If the parity check conditions of the nonbinary LDPC decoder cannot be satisfied, the estimated information symbols are fed into the GLRT interference detector to initialize the iterative interference mitigation.

12.2.2 *Interference Detection and Estimation*

Let us first assume that both channel and symbol estimates ($\hat{\mathbf{H}}$ and $\hat{\mathbf{s}}$) are available. Within the interference band $\mathcal{B}_{\rm I}$, there are $M_{\rm I} = \lceil B_{\rm I} T \rceil$ subcarriers contaminated. Denote the set of subcarriers within the interference band as $\{i_1, \cdots i_{M_{\rm I}}\}$. Define a selector matrix $\boldsymbol{\Theta}$ of size $M_{\rm I} \times K$ with unit entry at the (m, i_m)th position $(m = 1, \cdots, M_{\rm I})$ and zeros elsewhere. The selection matrix $\boldsymbol{\Theta}$ is determined based on the prior knowledge of the interference frequency band.

The relevant measurements within the interference frequency band are contained in

$$\bar{\mathbf{z}} = \boldsymbol{\Theta}(\mathbf{z} - \hat{\mathbf{H}}\hat{\mathbf{s}}) = \mathbf{B}(\bar{\tau}_{\rm I})\mathbf{u} + \bar{\mathbf{w}}, \tag{12.10}$$

where

$$\mathbf{B}(\overline{\tau}_{\rm I}) := \boldsymbol{\Theta}\boldsymbol{\Lambda}(\overline{\tau}_{\rm I})\boldsymbol{\Gamma}_I, \qquad \overline{\mathbf{w}} = \boldsymbol{\Theta}\mathbf{w} + \boldsymbol{\Theta}(\hat{\mathbf{H}}\hat{\mathbf{s}} - \mathbf{H}\mathbf{s}). \tag{12.11}$$

Here, $\overline{\mathbf{w}}$ denotes the equivalent additive noise within the frequency band, which consists of the ambient noise and the residual noise due to imperfect channel and information symbol estimates.

Assume that the noise samples are independent and follow a complex Gaussian distribution $\mathcal{CN}(0, \sigma^2_{B_{\rm I}}\mathbf{I}_{M_{\rm I}})$, where $\sigma^2_{B_{\rm I}}$ denotes the noise variance. The likelihood function of the interference component $\overline{\mathbf{z}}$ in the presence of interference is

$$f(\overline{\mathbf{z}} \mid \overline{\tau}_{\rm I}, \mathbf{u}) \propto \exp\left[-\frac{1}{\sigma^2_{B_{\rm I}}}\|\overline{\mathbf{z}} - \mathbf{B}(\overline{\tau}_{\rm I})\mathbf{u}\|^2\right]. \tag{12.12}$$

Let $\mathcal{H}_0$ and $\mathcal{H}_1$ denote the absence and presence of interference, respectively. To detect the presence of interference in one particular OFDM block, define the generalized log-likelihood ratio test (GLRT) statistic

$$\begin{aligned}
L(\overline{\mathbf{z}}) &= \max_{\{\overline{\tau}_{\rm I},\mathbf{u}\}} \ln \frac{f(\overline{\mathbf{z}} \mid \overline{\tau}_{\rm I}, \mathbf{u}, \mathcal{H}_1)}{f(\overline{\mathbf{z}}|\mathcal{H}_0)} \\
&= \max_{\{\overline{\tau}_{\rm I},\mathbf{u}\}} \ln \frac{\exp\left[-\|\overline{\mathbf{z}} - \mathbf{B}(\overline{\tau}_{\rm I})\mathbf{u}\|^2/\sigma^2_{B_{\rm I}}\right]}{\exp\left[-\|\overline{\mathbf{z}}\|^2/\sigma^2_{B_{\rm I}}\right]} \\
&= \max_{\{\overline{\tau}_{\rm I},\mathbf{u}\}} \frac{1}{\sigma^2_{B_{\rm I}}}\left[\overline{\mathbf{z}}^{\rm H}\mathbf{B}(\overline{\tau}_{\rm I})\mathbf{u} + \mathbf{u}^{\rm H}\mathbf{B}^{\rm H}(\overline{\tau}_{\rm I})\overline{\mathbf{z}} - \mathbf{u}^{\rm H}\mathbf{B}^{\rm H}(\overline{\tau}_{\rm I})\mathbf{B}(\overline{\tau}_{\rm I})\mathbf{u}\right] \gtrless \Gamma_{\rm th}
\end{aligned} \tag{12.13}$$

where $\Gamma_{\rm th}$ is a predetermined threshold.

Define an objective function to be maximized over $\{\overline{\tau}_{\rm I}, \mathbf{u}\}$,

$$J := \overline{\mathbf{z}}^{\rm H}\mathbf{B}(\overline{\tau}_{\rm I})\mathbf{u} + \mathbf{u}^{\rm H}\mathbf{B}^{\rm H}(\overline{\tau}_{\rm I})\overline{\mathbf{z}} - \mathbf{u}^{\rm H}\mathbf{B}^{\rm H}(\overline{\tau}_{\rm I})\mathbf{B}(\overline{\tau}_{\rm I})\mathbf{u}. \tag{12.14}$$

Setting the derivative $\nabla J_{\mathbf{u}}$ to zero yields the optimal estimate of $\mathbf{u}$ as

$$\hat{\mathbf{u}} = \left[\mathbf{B}^{\rm H}(\overline{\tau}_{\rm I})\mathbf{B}(\overline{\tau}_{\rm I})\right]^{-1}\mathbf{B}^{\rm H}(\overline{\tau}_{\rm I})\overline{\mathbf{z}}, \tag{12.15}$$

Substituting $\hat{\mathbf{u}}$ into (12.14), the estimate of $\overline{\tau}_{\rm I}$ is obtained as

$$\hat{\overline{\tau}}_{\rm I} = \arg\max_{\overline{\tau}_{\rm I}} \overline{\mathbf{z}}^{\rm H}\mathbf{B}(\overline{\tau}_{\rm I})\left[\mathbf{B}^{\rm H}(\overline{\tau}_{\rm I})\mathbf{B}(\overline{\tau}_{\rm I})\right]^{-1}\mathbf{B}(\overline{\tau}_{\rm I})^{\rm H}\hat{\overline{\mathbf{z}}}, \tag{12.16}$$

which can be solved using the grid search.

Based on the estimated parameters $\{\hat{\mathbf{u}}, \hat{\overline{\tau}}_{\rm I}\}$, the test statistic is evaluated as

$$L(\overline{\mathbf{z}}) = \frac{1}{\hat{\sigma}^2_{B_{\rm I}}}\hat{\mathbf{u}}^{\rm H}\mathbf{B}^{\rm H}(\hat{\overline{\tau}}_{\rm I})\mathbf{B}(\hat{\overline{\tau}}_{\rm I})\hat{\mathbf{u}} \gtrless \Gamma_{\rm th}, \tag{12.17}$$

where the estimate $\hat{\sigma}^2_{B_{\rm I}}$ is obtained following the procedure to be discussed in Section 12.2.4. The test statistic $L(\overline{\mathbf{z}})$ is actually the ratio of the energy of the estimated interference to the energy of the equivalent noise. See [422] for discussions on the selection of the detection threshold and efficient implementations.

12.2.3 Channel Estimation, Equalization and Channel Decoding

If the presence of interference is declared from the GLRT detector (either true detection or false alarm), the desired OFDM component can be obtained by subtracting the estimated interference from the received signal

$$\check{\mathbf{z}} = \mathbf{z} - \mathbf{\Lambda}(\hat{\bar{\tau}}_{\mathrm{I}})\mathbf{\Gamma}_{\mathrm{I}}\hat{\mathbf{u}} = \mathbf{H}\mathbf{s} + \check{\mathbf{w}}, \tag{12.18}$$

where $\check{\mathbf{w}}$ denotes the equivalent noise which consists of the ambient noise and the residual interference

$$\check{\mathbf{w}} = \mathbf{w} + [\mathbf{\Lambda}(\bar{\tau}_{\mathrm{I}})\mathbf{\Gamma}_{\mathrm{I}}\mathbf{u} - \mathbf{\Lambda}(\hat{\bar{\tau}}_{\mathrm{I}})\mathbf{\Gamma}_{\mathrm{I}}\hat{\mathbf{u}}]. \tag{12.19}$$

If no interference is detected (either absence of interference or missed detection), simply set $\hat{\mathbf{u}} = 0$ in (12.18). The following processing will be carried out based on $\check{\mathbf{z}}$.

Based on the observation vector $\check{\mathbf{z}}$ and the symbol vector $\hat{\mathbf{s}}$, the sparse channel estimator in Section 7.2 can be used to estimate the channel matrix $\mathbf{H}$, where the banded assumption can be used to reduce computational complexity. Here, the linear MMSE equalizer is used as the performance benchmark. After inputting the linear MMSE estimate of information symbols into the channel decoder, both hard and soft decisions on the information symbols can be obtained, these being fed back for interference detection, channel estimation and equalization in the next iteration. A detailed description of the channel decoder and linear MMSE equalizer can be found in Chapters 3 and 8.

12.2.4 Noise Variance Estimation

Due to the partial-band property of the interference, the noise variance should be estimated individually for noise within and outside of the interference band $\mathcal{B}_{\mathrm{I}}$ based on the frequency measurements at null subcarriers. Based on the estimates of the channel matrix and transmitted symbols, the variance of the equivalent noise outside the interference band, which consists of the ambient noise and the residual ICI due to the banded assumption of the channel matrix, can be estimated as

$$\hat{\sigma}^2_{\overline{\mathcal{B}}_{\mathrm{I}}} = \mathbb{E}_{\{m \in \mathcal{S}_{\mathrm{N}}, f_m \notin \mathcal{B}_{\mathrm{I}}\}}\left[\left|z[m] - \sum_{k=m-D}^{m+D} \hat{H}[m,k]\hat{s}[k]\right|^2\right]. \tag{12.20}$$

For the equivalent noise within the interference band, which consists of the ambient noise, the residual ICI and the residual interference, the noise variance can be estimated as

$$\hat{\sigma}^2_{\mathcal{B}_{\mathrm{I}}} = \mathbb{E}_{\{m \in \mathcal{S}_{\mathrm{N}}, f_m \in \mathcal{B}_{\mathrm{I}}\}}\left[\left|z[m] - \sum_{k=m-D}^{m+D} \hat{H}[m,k]\hat{s}[k] - \sum_{\ell=-N_{\mathrm{I}}/2}^{N_{\mathrm{I}}/2-1} [\mathbf{\Lambda}(\hat{\bar{\tau}}_{\mathrm{I}})]_{m,m}[\mathbf{\Gamma}_{\mathrm{I}}]_{m,l}\hat{u}[\ell]\right|^2\right]. \tag{12.21}$$

The estimated variance is then used for interference detection and information symbol estimation.

12.3 Simulation Results

The underwater acoustic channel is simulated according to the specifications in Section 5.5.1. The channel parameters are $N_{\mathrm{pa}} = 10$, $\Delta\tau = 1\,\mathrm{ms}$, $\Delta P_{\mathrm{pa}} = 20\,\mathrm{dB}$, $T_{\mathrm{g}} = 24.6\,\mathrm{ms}$, and

$\upsilon_0 = 0$ m/s. The interference is generated by passing white Gaussian noise of a given duration through a bandpass filter according to the following parameters: center frequency $f_{\mathrm{Ic}} = 15$ kHz, bandwidth $B_{\mathrm{I}} = 2.4$ kHz and time duration $T_{\mathrm{I}} = 26.2$ ms. The delay of the interference relative to start of each OFDM block is uniformly distributed within $[0, T + T_{\mathrm{g}} - T_{\mathrm{I}}]$.

Throughout this chapter, the signal-to-noise ratio (SNR) and signal-to-interference ratio (SIR) are defined as

$$\mathrm{SNR} = P_{\mathrm{s}}/P_{\mathrm{n}}, \qquad \mathrm{SIR} = P_{\mathrm{s}}/P_{\mathrm{I}}, \tag{12.22}$$

where P_{s} denotes the average power of OFDM frequency measurements within the useful signal band, P_{I} denotes the average power of interference frequency components within the interference frequency band, and P_{n} denotes the variance of the additive noise in frequency domain.

Four configurations are introduced to compare the performance of the proposed receiver with the receiver which does not perform interference cancellation.

- *Configuration 1:* the receiver without interference cancellation in the interference-free environment;
- *Configuration 2:* the proposed receiver in the interference-free environment;
- *Configuration 3:* the receiver without interference cancellation in the presence of interference;
- *Configuration 4:* the proposed receiver in the presence of interference.

The block-error-rate (BLER) performance will be used as the performance metric. When applying the proposed receiver in configurations 2 and 4, it is assumed that the interference has been detected in each OFDM block such that interference cancellation is always performed.

12.3.1 Time-Invariant Channels

To examine the performance of the proposed receiver in a time-invariant channel, the Doppler rate of each path is set zero. The ZP-OFDM parameters and subcarrier distributions are identical to the simulation setup in Section 9.1.2. With a rate-1/2 nonbinary LDPC code and a 16-QAM constellation, the data rate is $R = 10.4$ kb/s.

Averaged over 2000 Monte Carlo runs, Figure 12.3 demonstrates the BLER performance of the four configurations with SIR = 0 dB. In the presence of interference, the receiver without interference cancellation does not work, while the proposed receiver can approach the performance of the conventional receiver in the absence of interference. Meanwhile, in the scenario without interference, there is only slight performance degradation of the proposed receiver relative to the conventional receiver.

Figure 12.4 shows the performance improvement of the proposed receiver with different number of iterations. One can observe a large performance gap between the receiver without iteration, i.e., the receiver which performs interference suppression via noise pre-whitening as described in Section 12.2.1, and the proposed receiver which iteratively mitigates the interference through interference estimation and subtraction.

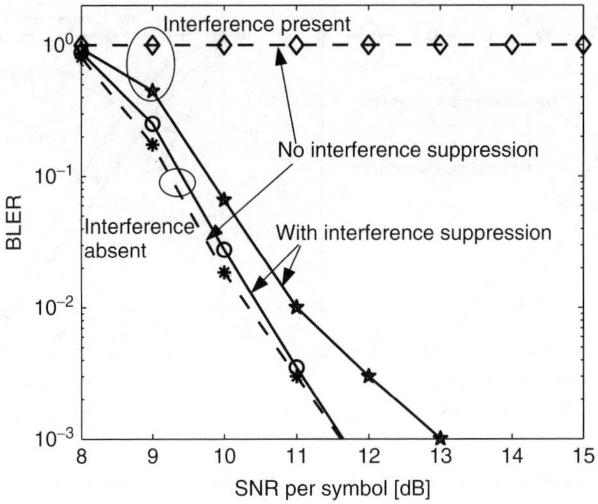

Figure 12.3 BLER performance of several receivers in the time-invariant scenario, SIR = 0 dB.

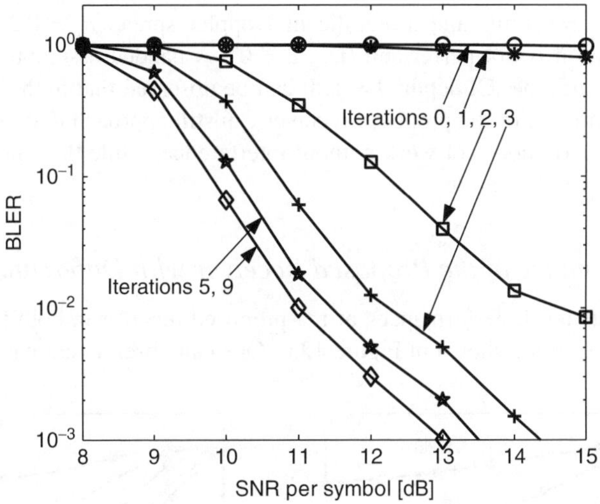

Figure 12.4 BLER performance improvement against iterations of the proposed receiver in the time-invariant scenario, SIR = 0 dB.

12.3.2 Time-Varying Channels

To explore the receiver performance in a time-varying channel, the Doppler rate of each path is drawn *independently* from a zero-mean uniform distribution with the standard deviation σ_v m/s. The ZP-OFDM parameters and subcarrier distributions are identical to the simulation

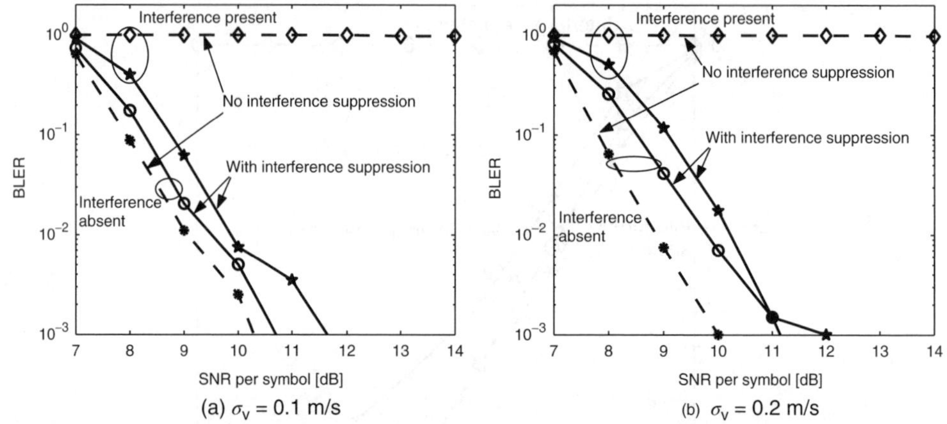

Figure 12.5 BLER performance of receivers in the time-varying channels, $D = 1$, SIR $= 0$ dB.

setup in Section 9.2.2. With a nonbinary LDPC code of rate 0.4 and a 16-QAM constellation, the data rate is $R = 7.4$ kb/s.

Figure 12.5 shows the performance of the four configurations in the scenario with a mild Doppler spread $\sigma_v = 0.1$ m/s and a significant Doppler spread $\sigma_v = 0.2$ m/s, respectively. Resampling operation is not performed (i.e., $\hat{a} = 0$). A banded assumption of the channel matrix is adopted with the ICI depth $D = 1$. It can be observed that in the presence of interference, the performance of the proposed receiver can still approach that of the conventional receiver in the scenario does not work without interference, while the conventional receiver without interference cancellation.

12.3.3 Performance of the Proposed Receiver with Different SIRs

By varying the SIR level, performances of the proposed receiver in both time-invariant and time-varying scenarios are shown in Figure 12.6. One can observe that when the SIR is very

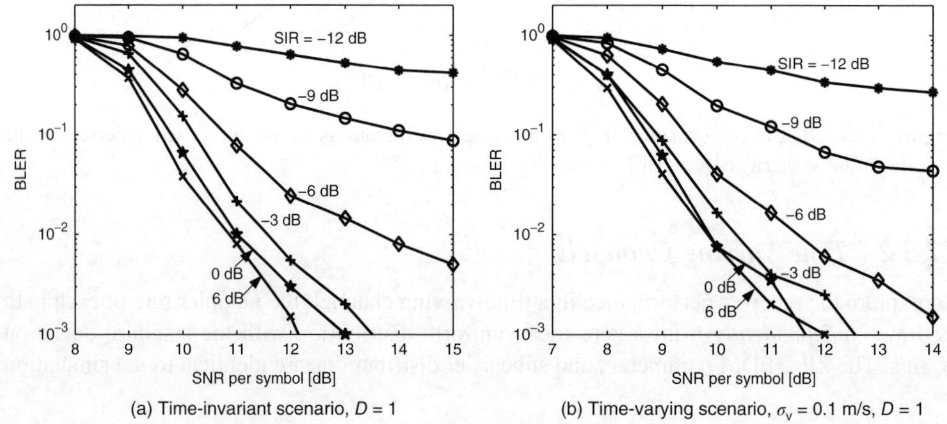

Figure 12.6 BLER performance of the proposed receiver at different SIR levels.

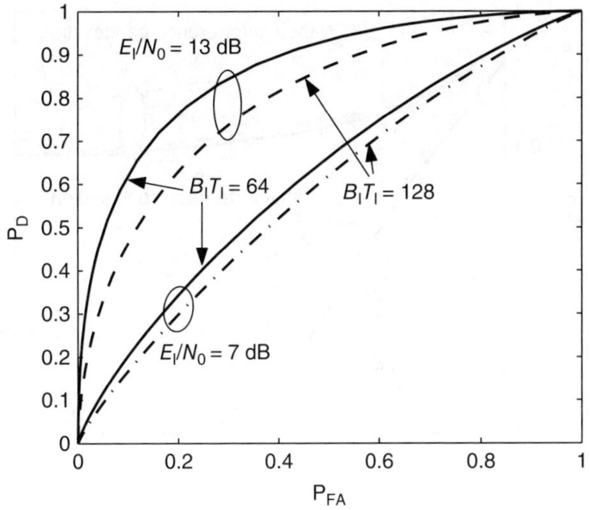

Figure 12.7 ROC of the GLRT interference detector.

small, a BLER error floor shows up, and that as the SIR increases, the decoding performance of the proposed receiver gradually converges to the decoding performance in the scenario without interference.

12.3.4 Interference Detection and Estimation

Define E_I as the interference energy, and N_0 as the power of ambient noise. The receiver operating characteristic (ROC) curve of the GLRT detector with different E_I/N_0 and time-bandwidth product of interference is shown in Figure 12.7. One can observe that with an identical E_I/N_0, the larger time-bandwidth product of the interference, the more difficult for it to be detected, because it appears more like the ambient noise.

12.4 Experimental Results: AUTEC10

The AUTEC10 experimental setup and OFDM parameters are described in Section 11.6. With rate-1/2 nonbinary LDPC coding and QPSK constellation, the data rate is $R = 798.7$ bits/s, or $3R/8 = 299.5$ Bytes/(3s) according to the unit used in the current AUTEC system. A hyperbolic frequency modulated (HFM) waveform with time duration 100 ms and bandwidth 5 kHz was used as the preamble to probe the channel.

There are two transmission trials of an identical data set, with the first trial and the second trial including 18 transmissions and 22 transmissions, respectively. During the experiment, sonar operations were carried out by others simultaneously. At sensors close to the sonar operation, the received signal of all the transmissions in the first trial and the first portion of transmissions in the second trial were contaminated. Based on the received signal waveform in Figure 12.1, the receiver can infer that $T_I \approx 45$ ms and the frequency band [11, 15] kHz, and hence sets $f_{Ic} \approx 13$ kHz, the interfering bandwidth $B_I \approx 2$ kHz.

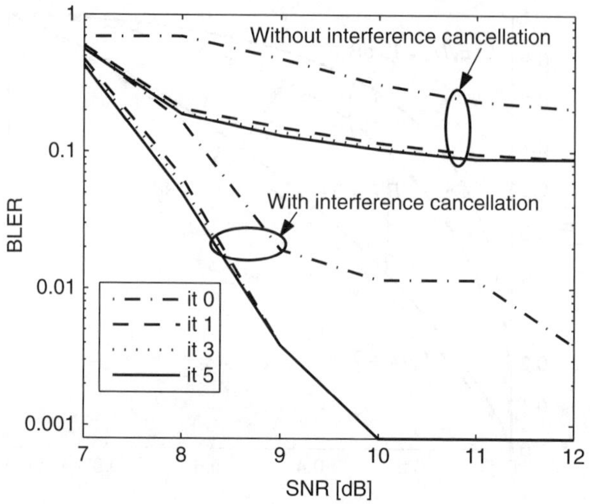

Figure 12.8 Block error rate of 13 transmissions with/without interference cancellation (IC) versus different levels of added noise, $I_{max} = 7$.

In the sequel, the proposed interference cancellation receiver will be tested with the experimental data in two different cases. Note that as opposed to the simulations, only parts of the OFDM blocks contain interference; the interference detection module in the receiver is thus started to decide the necessity of interference cancellation.

Test Case 12.4.1 In this test case, the interference cancellation performance of the receiver at different SNR levels is investigated. Thirteen transmissions with the SIR level ranging from 0 dB to 9 dB from the second trial are used for this test. With the received interference-contaminated waveform shown in Figure 12.1, white Gaussian noise is added to the received signal to generate several semi-experimental data sets with different SNRs.

The decoding performance of the semi-experimental data sets with and without interference mitigation is shown in Figure 12.8. One can observe that the receiver with interference cancellation outperforms the receiver without interference cancellation, and the performance gap between the iterative interference cancellation method and the noise-prewhitening method is large.

Test Case 12.4.2 In this test case, the interference cancellation performance of the receiver at different SIR levels is investigated. By modifying the weighting coefficients, the semi-experimental data sets of different SIRs are constructed through adding the received signal of pure interference with one received OFDM signal which is not contaminated by interference. The received signals at two hydrophones are shown in Figure 12.9, where the signal in Figure 12.9 (a) is interference-free, while the signal in Figure 12.9 (b) is almost purely interference with a very weak useful signal masked by ambient noise.

Figure 12.10 shows the decoding performance of the semi-experimental data sets with and without interference cancellation. Again, one can observe that the receiver with interference

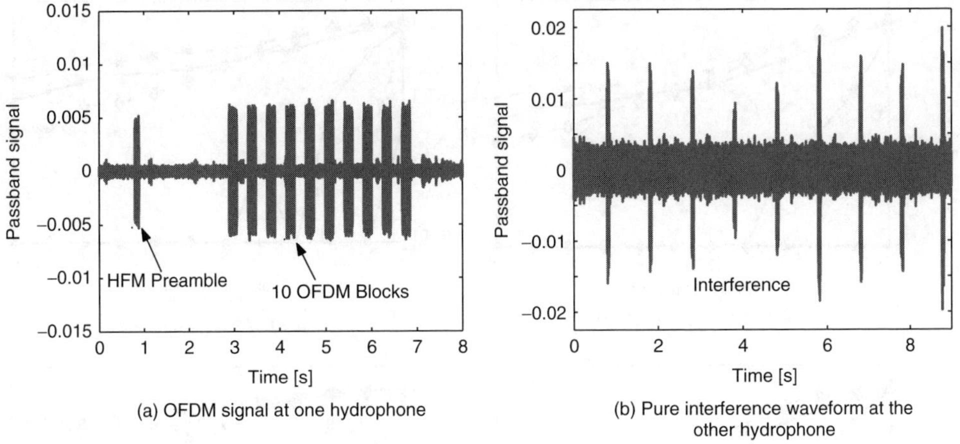

Figure 12.9 Sample of the received signals at two hydrophones after bandpass filtering. The two data sets are added together with different weights to generate semi-experimental data sets at different SIR levels.

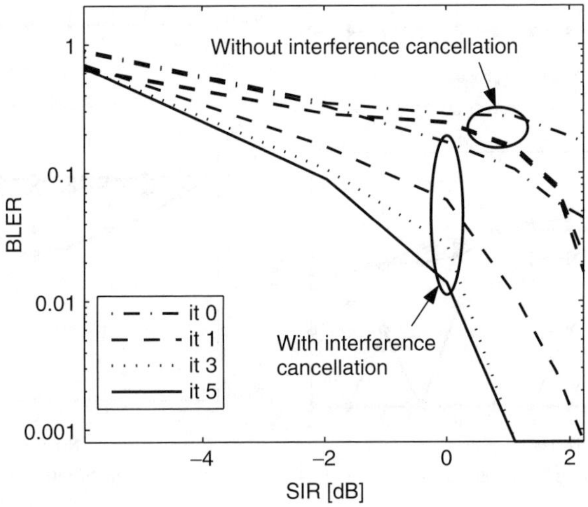

Figure 12.10 Block error rate of 18 transmissions with/without interference cancellation (IC) versus different signal to interference power ratio, $I_{max} = 7$, SNR ≈ 7.9 dB.

cancellation outperforms that without interference cancellation considerably, and the performance improvement brought by the iterative interference cancellation is significant.

12.5 Emulated Results: SPACE08

The experimental setting and OFDM parameter are described in Section B.1. The data recorded on Julian dates 297 and 300 are used to verify performance of the proposed

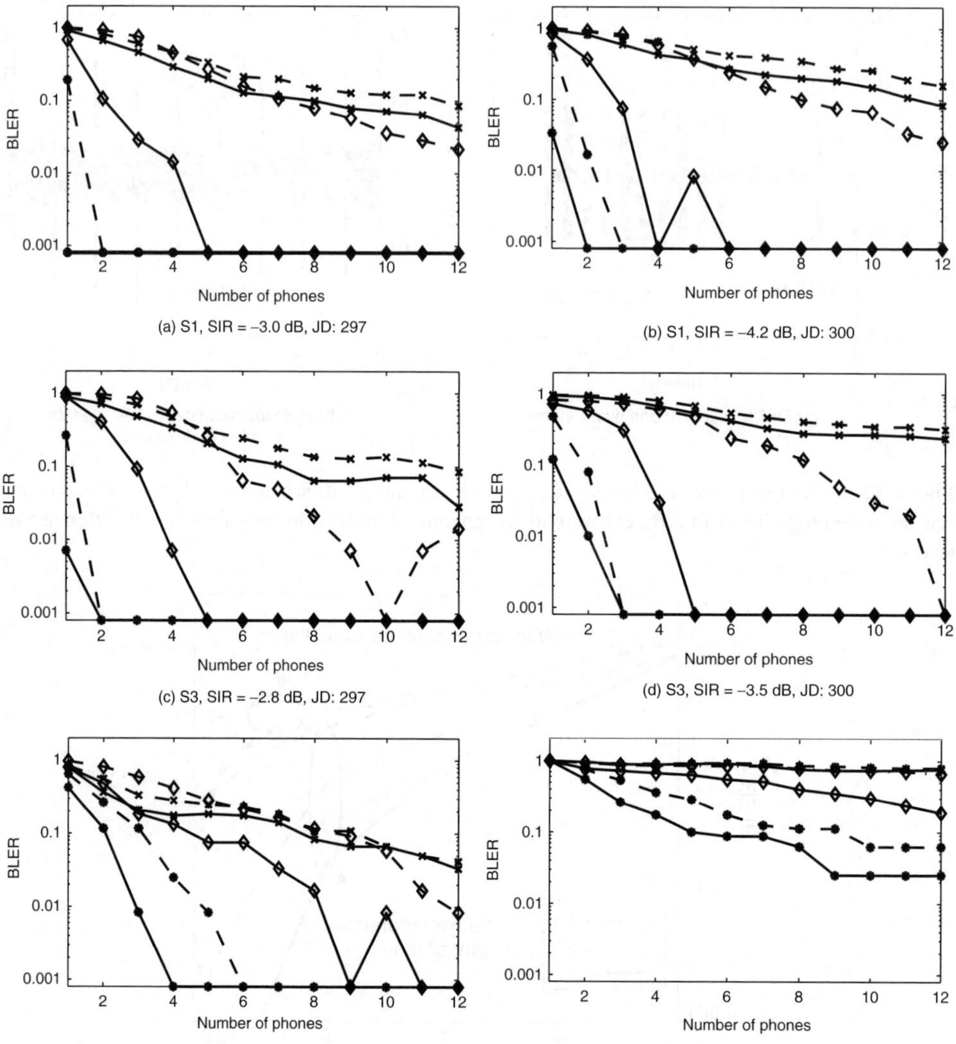

(a) S1, SIR = −3.0 dB, JD: 297

(b) S1, SIR = −4.2 dB, JD: 300

(c) S3, SIR = −2.8 dB, JD: 297

(d) S3, SIR = −3.5 dB, JD: 300

(e) S5, SIR = 0.4 dB, JD: 297

(f) S5, SIR = −2.1 dB, JD: 300

Figure 12.11 BLER performance comparison of receiver with/without interference cancellation, 16-QAM, $D = 3$, JD: Julian date. Marker x: receiver without interference cancellation; marker ⋄: proposed interference cancellation receiver; marker $*$: original BLER without adding interference; dashed line: 0 iteration, solid line: 10th iteration.

method in the time-varying scenario. Note that there was no external interference during this experiment. The interference from the recorded sonar waveform in the AUTEC10 experiment at one hydrophone, as shown in Figure 12.9 (b), is extracted. Each sonar waveform is artificially added to one OFDM block to create a semi-experimental data set with interference contamination. The delay of the added sonar waveform relative to the OFDM block follows a uniform distribution within the interval $[0, T + T_{\mathrm{g}} - T_{\mathrm{I}}]$. To estimate the channel accurately, the setting described in Section 9.2.2 is used, where 352 pilots are used, leading to a data rate $R = 798.7$ bits/s. Again, different from the simulations, the receiver incorporates the interference detection module to determine the necessity of interference cancellation.

The decoding performance of the proposed receiver and the conventional receiver without interference cancellation, and the decoding performance of the signal without adding the interference are depicted in Figure 12.11. One can find that relative to the decoding performance of the original received signal, significant performance degradation of the ICI-aware receiver without interference cancellation is incurred by introducing the interference. Meanwhile, the proposed receiver can effectively eliminate the interference after a number of iterations.

12.6 Discussion

The parameterized interference cancellation method is general in that the number of parameters only depends on the time-bandwidth product of the interference. With a given time-bandwidth product, the time duration or the bandwidth of the interference could be arbitrary. Therefore, it can be used to address other types of interferences, such as the impulsive noise and the narrowband interference. The impulsive noise has very short time duration and large bandwidth which is comparable to system bandwidth. The narrowband interference occupies partial system frequency band and has very large time duration. Both of them have limited number of degrees of freedom and hence allow for a compact parameterization.

Underwater OFDM receiver in the presence of impulsive noise has been investigated in [372]. The impulsive noise is parameterized by a long time-domain vector, which is constrained to have only a few nonzero entries. Hence, a compressive sensing approach is adopted to estimate the impulsive noise in the frequency domain.

One immediate application of the parameterized interference cancellation is in Chapter 15 to mitigate asynchronous multiuser interference, which overlaps the desired OFDM blocks partially in the time domain.

12.7 Bibliographical Notes

Various methods for external interference suppression have been investigated for wireless communications in the last two decades [143, 229]. Mainly, the external interferences are divided into two categories according to the time-frequency characteristics: (i) impulsive interference with short time duration and large bandwidth; and (ii) narrowband interference with small bandwidth and long time duration.

In contrast, rare attention has been paid to interference in underwater acoustic environments. Sporadic progress on the impulsive noise cancellation in water can be found in [78, 307]. The impact of impulsive like noise on OFDM is investigated by [77, 372]. Mitigation of the partial-band partial-block-duration interference in an underwater OFDM system is studied in [422].

13

Co-located MIMO OFDM

Chapters 9–12 presented receiver designs for single-transmitter OFDM systems in different scenarios. From this chapter to Chapter 15, we will look into receiver designs for multi-input multi-output (MIMO) systems, where multiple transmitters could be co-located or spatially distributed.

- *Chapter 13: Co-located MIMO OFDM*. This chapter focuses on the MIMO-OFDM reception for a MIMO system with co-located transmitters, in which, as illustrated in Figure 13.1, multiple parallel data streams are used to increase the bandwidth efficiency. The signals from co-located transmitters are quasi-synchronized at the receiver and experience similar Doppler scaling effects.
- *Chapter 14: Distributed MIMO OFDM*. This chapter considers a particular scenario with multiple geometrically distributed users, where the moving velocity could be drastically different from user to user. The signals from different users are assumed to be quasi-synchronized at the receiver, and the focus is on how to deal with different moving velocities of users.
- *Chapter 15: Asynchronous multiuser OFDM*. This chapter looks into a communication system with multiple geometrically distributed users in which signals from users are misaligned at the receiver but have a similar Doppler effect. It will present an OFDM receiver for asynchronous multiuser reception.

In this chapter, we look into the co-located MIMO-OFDM receiver design in both time-invariant and time-varying channels. Note that with co-located transmitters, various space time coding approaches could be adopted to either improve the system performance or increase the bandwidth efficiency. The spatial multiplexing scheme, where different transmitters send independent data streams, is adopted so that the data rate linearly increases with the number of transmitters. This choice is attractive to underwater acoustic communication systems, which are fundamentally bandwidth-limited.

Compared to the single-input OFDM, challenges in the co-located MIMO-OFDM are two-fold. In terms of channel estimation, the number of frequency-domain channel coefficients increases proportionally to the number of transmitters. In terms of data detection, both co-channel interference (CCI) and potential ICI have to be equalized, leading to a problem of

OFDM for Underwater Acoustic Communications, First Edition. Shengli Zhou and Zhaohui Wang.
© 2014 John Wiley & Sons, Ltd. Published 2014 by John Wiley & Sons, Ltd.

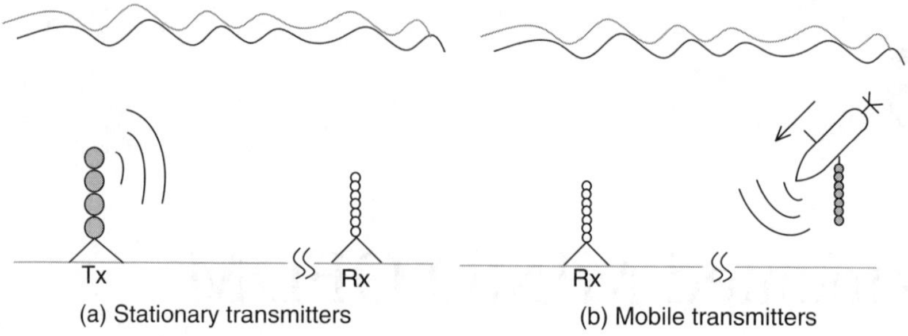

Figure 13.1 Illustration of a co-located underwater MIMO-OFDM system.

size proportional also to the number of transmitters. The rest of this chapter is organized as follows.

- Sections 13.1 to 13.4 present a noniterative and an iterative ICI-ignorant MIMO-OFDM receiver. By ignoring the ICI, data detection is carried on each OFDM subcarrier individually.
- Sections 13.5 to 13.9 describe an ICI-progressive MIMO-OFDM receiver, which is a counterpart of the progressive receiver in Section 9.4. Section 13.10 discusses an initialization method to the ICI-progressive MIMO OFDM receiver.

13.1 ICI-Ignorant MIMO-OFDM System Model

Consider a MIMO-OFDM system with N_t transmitters and N_r receiving elements, and copy the input–output relationship from (5.51) as

$$z_\nu[m] = \sum_{\mu=1}^{N_t} H_{\nu,\mu}[m] s_\mu[m] + \eta_\nu[m], \tag{13.1}$$

where $H_{\nu,\mu}[m]$ is the coefficient that specifies how the symbol transmitted on the mth subcarrier of the μth transmitter contributes to the output on that subcarrier of the νth receiver, and $\eta_\nu[m]$ is the equivalent noise in the frequency domain which consists of the ambient noise and the unmodeled ICI.

A matrix representation of frequency measurements on the mth subcarrier can be easily obtained based on (13.1),

$$\underbrace{\begin{bmatrix} z_1[m] \\ \vdots \\ z_{N_r}[m] \end{bmatrix}}_{:=\mathbf{z}[m]} = \underbrace{\begin{bmatrix} H_{1,1}[m] & \cdots & H_{1,N_t}[m] \\ \vdots & \ddots & \vdots \\ H_{N_r,1}[m] & \cdots & H_{N_r,N_t}[m] \end{bmatrix}}_{:=\mathbf{H}[m]} \underbrace{\begin{bmatrix} s_1[m] \\ \vdots \\ s_{N_t}[m] \end{bmatrix}}_{:=\mathbf{s}[m]} + \underbrace{\begin{bmatrix} \eta_1[m] \\ \vdots \\ \eta_{N_r}[m] \end{bmatrix}}_{:=\boldsymbol{\eta}[m]}. \tag{13.2}$$

For each transmitted symbol of interest $s_\mu[m]$, there exists the co-channel interference from other parallel transmissions on the same subcarrier. We next present an ICI-ignorant MIMO receiver structure to recover the data streams from N_t transmitters.

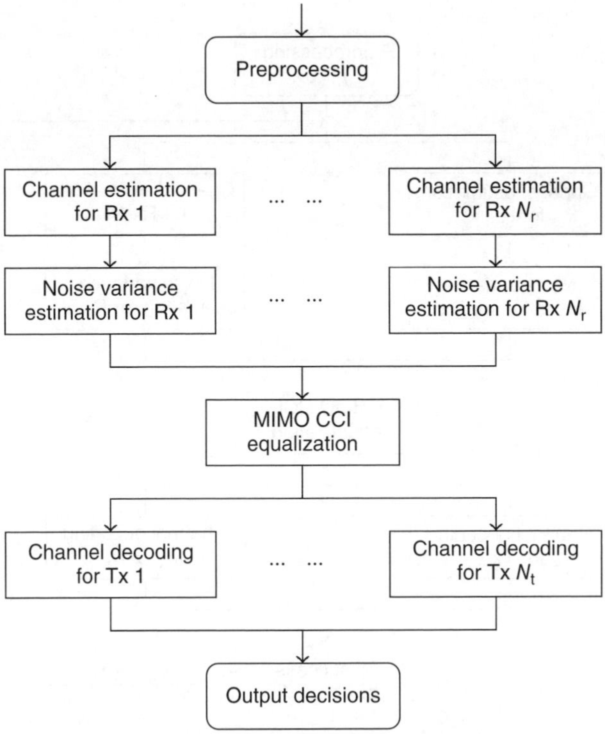

Figure 13.2 The noniterative ICI-ignorant MIMO-OFDM receiver diagram.

13.2 ICI-Ignorant MIMO-OFDM Receiver

13.2.1 Noniterative ICI-Ignorant MIMO-OFDM Receiver

The noniterative ICI-ignorant receiver for a MIMO-OFDM with N_t transmitters and N_r receivers is shown in Figure 13.2.

- *ICI-ignorant channel estimation and noise variance estimation:* If the pilot subcarriers for different transmitters are nonoverlapping, then channel between each transmitter-receiver pair is estimated individually, with the estimation methods described in Sections 7.1 and 7.2. If the pilot subcarriers are overlapping, then the channels from all transmitters to one receiver will be estimated jointly, with the methods described in Section 7.5.1. The noise variance is estimated with (7.66). The estimated channel coefficient $\hat{H}_{\nu,\mu}[m]$ and noise variance are then used for MIMO symbol detection.
- *MIMO demodulation and channel decoding:* Different from the single-user scenario, the signals from N_t data streams are superimposed at the receiver. With the input-output relationship specified by (13.2), the MAP or the MMSE detector developed in Section 8.4.1 can be used on each subcarrier. The estimated LLRs of each data stream are then fed into the channel decoder to recover the information bits from the corresponding transmitter.

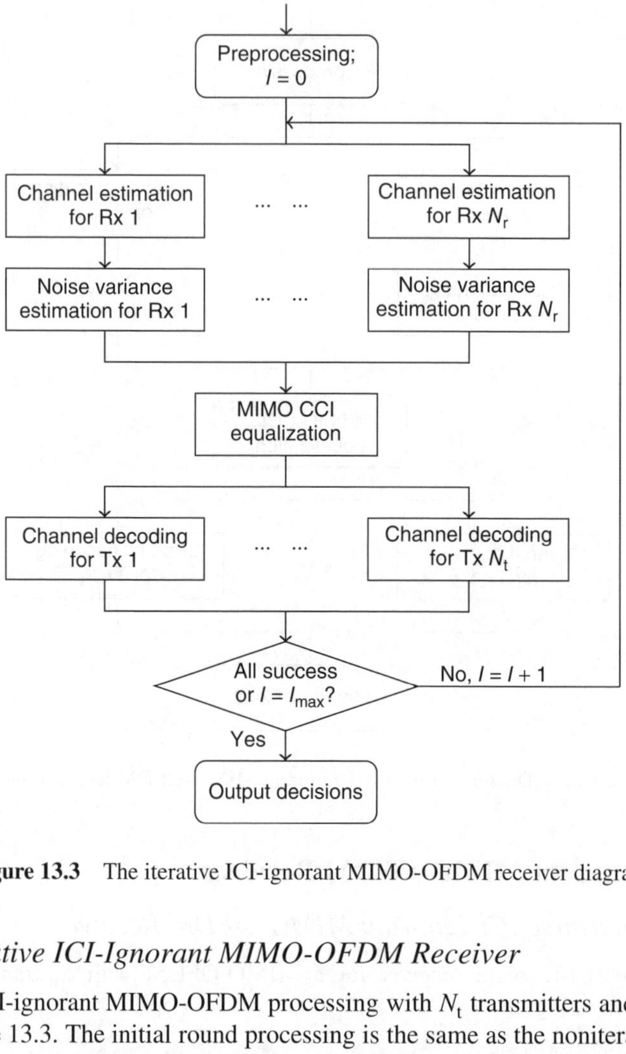

Figure 13.3 The iterative ICI-ignorant MIMO-OFDM receiver diagram.

13.2.2 Iterative ICI-Ignorant MIMO-OFDM Receiver

The iterative ICI-ignorant MIMO-OFDM processing with N_t transmitters and N_r receivers is shown in Figure 13.3. The initial round processing is the same as the noniterative receiver as shown in Figure 13.2. With the outputs from the equalizer, channel decoding is run for each data stream to be decoded. In later iterations, all the estimated information symbols are fed back from the channel decoder and used to refine channel estimation and data detection. During the decoding process, the decoder will declare success if the parity check conditions are satisfied. When one data stream is decoded successfully, its contribution will be substracted from the received signals, and the number of parallel data streams decreases by one. The iterative receiver stops after all N_t streams have been successfully decoded, or after a pre-specified number of runs.

13.3 Simulation Results: ICI-Ignorant MIMO OFDM

Consider MIMO systems with $N_t = 2$ or $N_t = 3$ transmitters. The parameter setting is identical to that in the SPACE08 experiment specified in Table B.1. The 256 pilots are divided into nonoverlapping sets among all transmitters so that each transmitter has roughly the same number of pilots. The pilot patterns are randomly drawn, rendering irregular positioning. This is usually seen as advantageous in compressed sensing theory, as it can guarantee identifiability of active channel taps with high probability [100]. The pilot symbols are drawn from the QPSK constellation whereas the data symbols are drawn from QPSK or 16-QAM constellations. The pilots are scaled to ensure that about one third of the total transmission power is dedicated to channel estimation regardless of the number of transmitters. The data rate after accounting all overheads and rate 1/2 coding is

$$R = \frac{1}{2} \cdot N_t \cdot \frac{672 \cdot \log_2 M}{T + T_g} = \begin{cases} 5.2N_t \text{ kb/s}, & \text{QPSK}, \\ 10.4N_t \text{ kb/s}, & \text{16-QAM}. \end{cases} \tag{13.3}$$

With $N_t = 2$, the data rates are 10.4 kb/s and 20.8 kb/s for QPSK and 16-QAM modulations, respectively. With $N_t = 3$, the data rates are 15.6 kb/s and 31.2 kb/s for QPSK and 16-QAM modulations, respectively.

The underwater acoustic channel between any transmitter and receiver is generated according to the specifications in Section 5.5.1. The channel parameters are $N_{pa} = 15$, $\Delta\tau = 1$ ms, $\Delta P_{pa} = 20$ dB, $T_g = 24.6$ ms, and $v_0 = 0$ m/s. As each OFDM symbol is encoded separately, the block-error-rate (BLER) is used as the figure of merit. In the simulation, each OFDM symbol experiences an independently generated channel. The BLER performance is evaluated at different SNR levels, where SNR is the signal to noise power ratio on the data subcarriers.

The following receivers are considered.

(1) *Noniterative ICI-ignorant receiver:* The sparse channel estimator in Section 7.2 is used.
(2) *The ICI-ignorant receiver with turbo equalization:* The sparse channel estimator is outside the iteration loop. Multiple rounds of iterations occur between the data detection and channel decoding modules.
(3) *The iterative ICI-ignorant receiver (soft decision):* Channel estimation is inside the iteration loop. The soft decisions on information symbols are used for channel estimation:

$$\tilde{s}_\mu[k] = \begin{cases} s_\mu[k], & k \in \mathcal{S}_{P,\mu}, \\ 0, & k \in \mathcal{S}_N \bigcup (\mathcal{S}_P \backslash \mathcal{S}_{P,\mu}), \\ \bar{s}_\mu[k], & k \in \mathcal{S}_D. \end{cases} \tag{13.4}$$

where $\mathcal{S}_{P,\mu}$ is set of pilots for the μth data stream, $\mathcal{S}_P = \bigcup_{\mu=1}^{N_t} \mathcal{S}_{P,\mu}$, and $\bar{s}_\mu[k]$ is the soft symbol estimate, computation of which can be found in (3.43).
(4) *The iterative ICI-ignorant receiver (hard decision):* The hard decisions on information symbols are fed back for channel estimation.

$$\tilde{s}_\mu[k] = \begin{cases} s_\mu[k], & k \in \mathcal{S}_{P,\mu}, \\ 0, & k \in \mathcal{S}_N \bigcup (\mathcal{S}_P \backslash \mathcal{S}_{P,\mu}), \\ \hat{s}_\mu[k], & k \in \mathcal{S}_D. \end{cases} \tag{13.5}$$

where $\hat{s}_\mu[k]$ is the hard decision estimate of $s_\mu[k]$.
(5) *The ICI-ignorant receiver with turbo equalization (full CSI):* The receiver iterates between MIMO detection and LDPC decoding, where the full channel state information (CSI) is assumed at the receiver.

In subsequent figures, *Noniterative*, *Turbo-equalization*, *Soft feedback*, *Hard feedback*, and *Full CSI* are used as legends for the above receivers, respectively. For the iterative receiver, other feedback strategies have been discussed in [172].

Figure 13.4 compares different receivers for two MIMO-OFDM systems under different setups. The maximum number of iterations for performing iterative updating between sparse channel estimation, MIMO detection and nonbinary LDPC decoding is 10, where both the channel estimation and the MIMO detection are updated in each iteration.

Figure 13.4 shows that employing a turbo-equalization receiver gains about 0.5–1 dB over a noniterative receiver. Including channel estimation in the iteration loop leads to gains of about 1 dB for $N_t = 2$ and 1.5 dB for $N_t = 3$. This seems intuitive, as with an increasing number of transmitters there are less pilots available per data stream, making the *additional pilots* from feedback more valuable. The gap between the proposed receivers and the full CSI case is approximately between 0.5 dB and 1 dB.

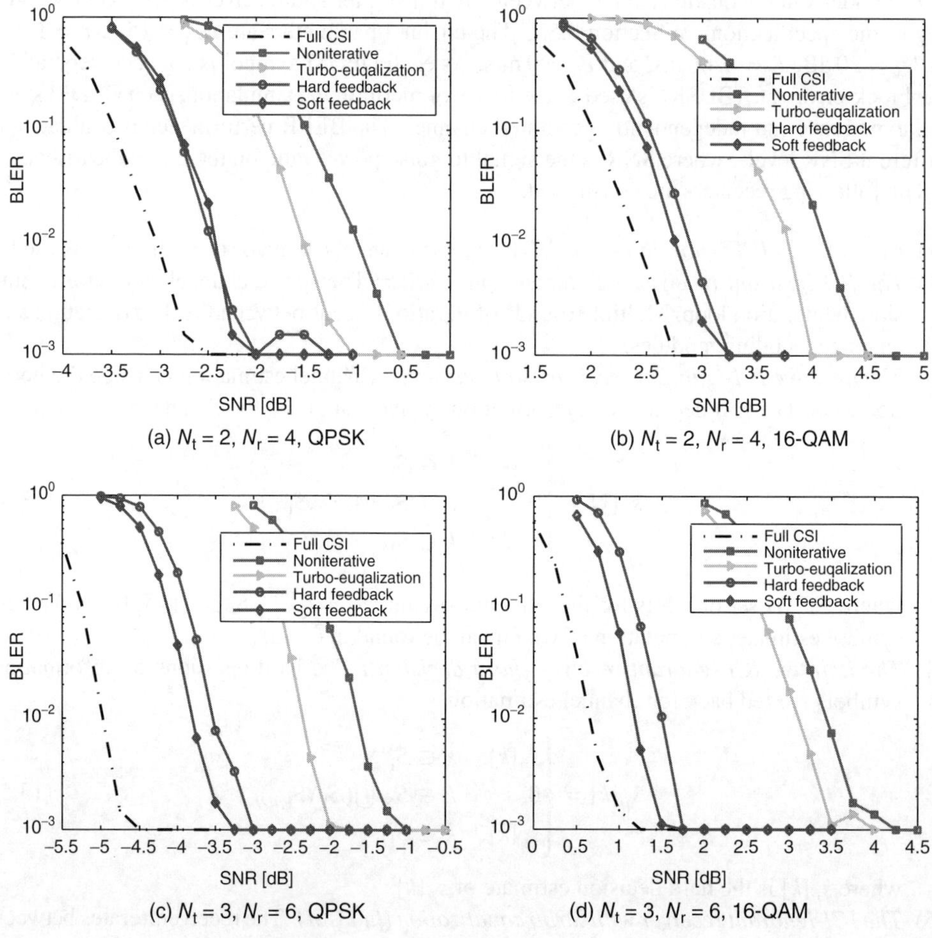

(a) $N_t = 2$, $N_r = 4$, QPSK

(b) $N_t = 2$, $N_r = 4$, 16-QAM

(c) $N_t = 3$, $N_r = 6$, QPSK

(d) $N_t = 3$, $N_r = 6$, 16-QAM

Figure 13.4 Simulation results in different settings.

13.4 SPACE08 Experimental Results: ICI-Ignorant MIMO OFDM

Experimental settings of the SPACE08 experiment can be found in Appendix B. The OFDM parameters are specified in Table B.1. Consider the receivers S3 and S5 located 200 m and 1000 m away from the transmitter. The recorded data files from Julian dates 298 and 299 are used. For each day, there are twelve recorded files consisting of twenty OFDM symbols each. On the Julian date 298, the five files recorded during the afternoon were severely distorted and therefore unusable, and the remaining seven files recorded during the morning and evening are used here. Due to the more challenging environment, only the small-size QPSK constellation is considered. The signaling format for SPACE08 is identical to that in the simulation setup in Section 13.3. The data rates for the MIMO system using QPSK modulation are the same as (13.3).

Performance results are plotted in Figure 13.5 for $N_{\rm t} = 2$ and $N_{\rm t} = 3$, respectively. The maximum number of iterations for performing iterative updating between sparse channel estimation, MIMO detection and nonbinary LDPC decoding is $I_{\rm max} = N_{\rm t}$. A sizable gain using updated channel estimates is observed, while all iterative receivers gain significantly over the noniterative receiver. For the $N_{\rm t} = 3$ setup, the gain of updated channel estimates is more pronounced, matching the observations based on numerical simulation.

13.5 ICI-Aware MIMO-OFDM System Model

For a MIMO-OFDM system with $N_{\rm t}$ transmitters and $N_{\rm r}$ receive elements in a time-varying environment, the frequency measurement on the mth subcarrier at the νth receiver with a band-limited ICI leakage assumption is copied from (5.60) as

$$z_\nu[m] = \sum_{\mu=1}^{N_{\rm t}} \sum_{k=m-D}^{k=m+D} H_{\nu,\mu}[m,k]s_\mu[k] + \eta_\nu[m], \tag{13.6}$$

where D is the ICI depth, $H_{\nu,\mu}[m,k]$ is the coefficient that specifies how the symbol transmitted on the kth subcarrier of the μth transmitter contributes to the output on the mth subcarrier of the νth receiver, and $\eta_\nu[m]$ is the equivalent noise consisting of both residual ICI/CCI and ambient noise.

Define $\mathbf{H}^D_{\nu,\mu}$ as a banded channel matrix with $H_{\nu,\mu}[m,k]$ as its (m,k)th entry, where D specifies the ICI depth on each side. Using the matrix-vector notation, the frequency measurements at the νth receiver can be compactly expressed as

$$\mathbf{z}_\nu = \sum_{\mu=1}^{N_{\rm t}} \mathbf{H}^D_{\nu,\mu}\mathbf{s}_\mu + \boldsymbol{\eta}_\nu, \tag{13.7}$$

where $\mathbf{z}_\nu$ is the frequency domain observation at the νth receiver, $\mathbf{s}_\mu$ is the transmitted data stream at the μth transmitter, and $\boldsymbol{\eta}_\nu$ is the equivalent noise.

13.6 ICI-Progressive MIMO-OFDM Receiver

The ICI-progressive MIMO-OFDM receiver is an extension of the ICI-progressive receiver in the single-user system developed in Chapter 9: it continually updates the ICI model

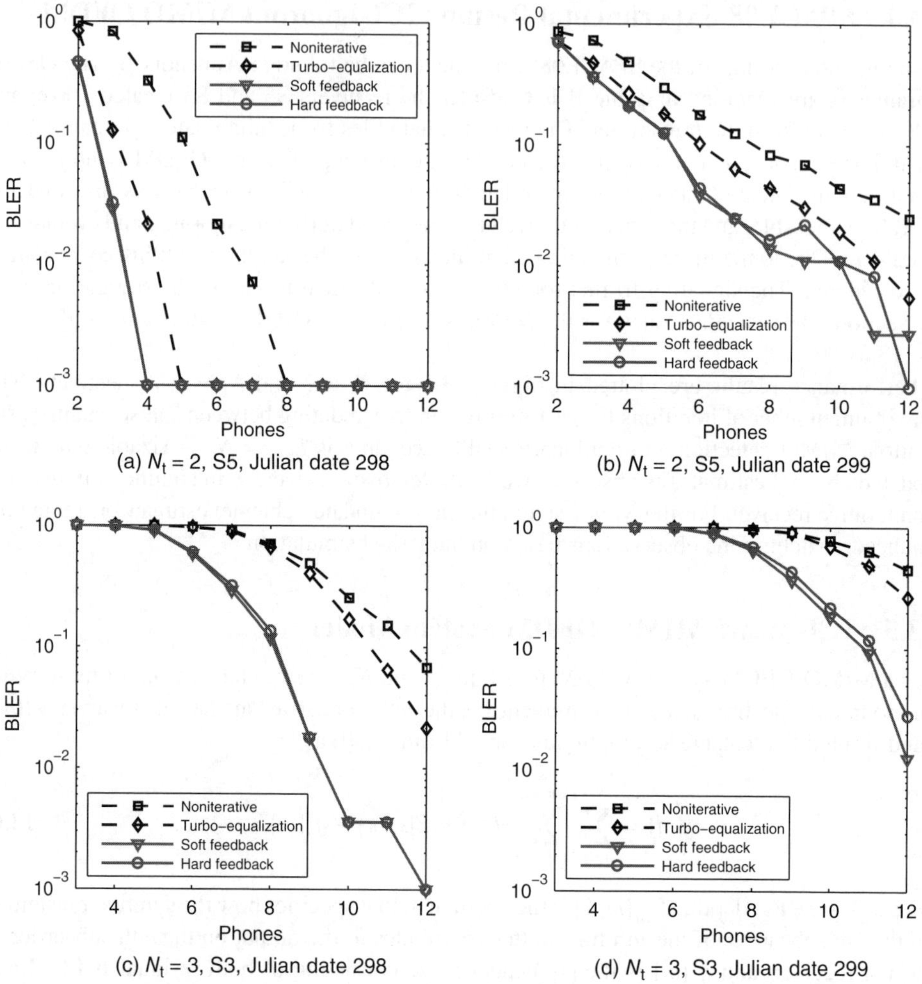

Figure 13.5 Experimental results from the SPACE08 experiment, QPSK, for S3 (200 m) and S5 (1000 m).

during iterations. Compared to the single-user OFDM system, benefits of the ICI-progressive processing in the MIMO-OFDM system are two-fold.

- As in the single-user system, the ICI-progressive MIMO receiver can adapt its system model, thus the processing complexity, towards the unknown severity of the channel conditions.
- The number of pilots per data stream is limited for MIMO-OFDM, which might not be sufficient for channel estimation when using a complicated model at the beginning. The progressive receiver starts with the ICI-ignorant model with a limited number of pilots, while later iterations can afford more detailed ICI modeling as the decoded symbols become more accurate and available for channel estimation.

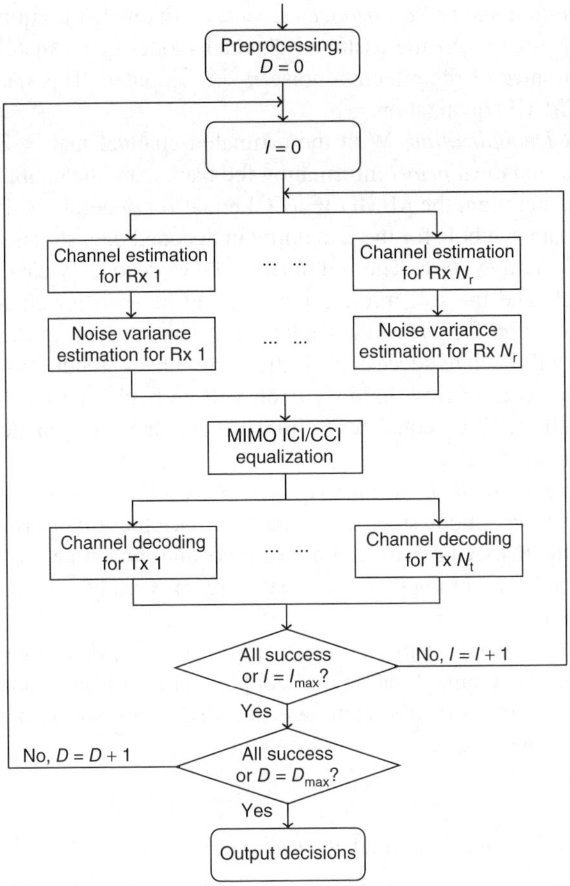

Figure 13.6 The progressive receiver for co-located MIMO-OFDM.

In the following, we first provide an overview of the progressive receiver, then take a brief discussion on the channel and noise variance estimation and joint ICI/CCI equalization.

13.6.1 Receiver Overview

Figure 13.6 shows the structure of an ICI-progressive receiver for MIMO-OFDM systems. The equivalent system model used in the receiver is shown as in (13.7). During the progressive iteration process, the parameter D increases after every few iterations, and hence more severe ICI can be addressed as the receiver processing proceeds to deal with channels with large Doppler spread. The progressive MIMO-OFDM reception structure consists of the following steps.

(i) *Pre-processing.* For each received OFDM block, each receiver applies the preprocessing operation to remove the dominant Doppler effect caused by the platform mobility. Set $D = 0$.

(ii) *Channel estimation and noise variance estimation.* Estimate the equivalent channel mixing matrix $\mathbf{H}_{\nu,\mu}^{D}$ based on the assumed channel model given in (13.7). After channel estimation, variance of the effective noise $\boldsymbol{\eta}_{\nu}$ is computed. This quantity is needed for the MIMO ICI/CCI equalization.

(iii) *MIMO ICI/CCI equalization.* With the estimated channel matrix $\mathbf{H}_{\nu,\mu}^{D}$, the equivalent noise variance, and the *a priori* information fed back from the nonbinary LDPC decoder in the previous iteration, the MIMO ICI/CCI equalizer generates soft output on the reliability of the data symbols for the remaining undecoded data streams.

(iv) *Nonbinary LDPC decoding.* The nonbinary LDPC decoder yields the decoded information symbols and the soft information that can be used for channel estimation and ICI/CCI equalization for each independent data stream. During the decoding process, a data stream will declare success if all the parity check conditions are satisfied. Once one data stream is recovered, hard decision will be made and fed back to the channel estimator and the ICI/CCI equalizer to improve the channel estimates and the detection performance, respectively.

(v) *Iteration among steps (ii) to (iv) for each D.* For each parameter D, it is reasonable to iterate multiple times among steps 2 to 5, and feed back the soft information to the channel estimator and the ICI/CCI equalizer. For example, one can iterate $I_{\max} = N_{\mathrm{t}}$ times on each D for simplicity by assuming that a new data stream is ready to be decoded successfully for each iteration.

(vi) *Further iteration among steps (ii) to (iv) after model update.* Increase D in the system model, and the assumed maximum Doppler spread of the channel to be estimated. Iteration stops when all data streams are decoded successfully, or when D reaches a pre-specified number $D_{\max}$.

13.6.2 Sparse Channel Estimation and Noise Variance Estimation

Since channel estimation is carried out for each receiving element separately, the ICI-progressive channel estimation in Section 9.4 can be directly applied. In each iteration, the soft symbol feedback in (13.4) will be adopted.

During the iteration of $D = 0$, all ICI terms are regarded as additive noise, and variance of the effective noise is measured on the null subcarriers $\mathcal{S}_{\mathrm{N}}$ directly as in the ICI-ignorant MIMO receiver. As D increases, ICI becomes modeled as opposed to being treated as additive noise. The variance of the effective noise hence needs to be updated at each iteration, using (7.70). As the iteration goes on with $D > 0$, the spillover from the neighboring subcarriers to the null subcarriers is extracted as in (7.70). As less ICI and CCI are viewed as additive noise, the effective signal-to-noise ratio (SNR) is expected to increase.

13.6.3 Joint ICI/CCI Equalization

Based on the input–output relationship in (13.7), the joint ICI/CCI equalization methods developed in Sections 8.4.3 and 8.5 can be applied. In this chapter, we mainly focus on the factor-graph based MMSE equalizer and the MCMC equalizer. Particularly for the

MCMC-based joint ICI/CCI equalization there are two possible Gibbs sampling strategies available. One is to draw the samples in a horizontal manner, i.e., symbols in the same data stream are drawn continuously; see (8.102). The other is to draw samples in a vertical manner, i.e., symbols in the same frequency bin are drawn continuously; see (8.103). In the numerical results of this chapter, the former strategy is used.

13.7 Simulation Results: ICI-Progressive MIMO OFDM

Similar to the simulation setting in Section 13.3, the OFDM parameters specified in Table B.1 are used. Consider MIMO systems with $N_t = 2, 3, 4$ transmitters. With $N_t = 2$, QPSK and 16-QAM modulation schemes will be investigated. For $N_t = 3$ and $N_t = 4$, only QPSK results will be presented. The data rate is computed in (13.3).

As Section 13.3, the total 256 pilots are divided into nonoverlapping sets randomly among all transmitters so that each transmitter has roughly the same number of pilots. The pilot symbols are drawn from QPSK constellation and scaled to ensure that one-third of the total transmission power is dedicated to channel estimation regardless of the number of transmitters.

The underwater acoustic channel between any transmitter and receiver pair is generated according to the specifications in Section 5.5.1. The channel parameters are $N_{pa} = 15$, $\Delta\tau = 1$ ms, $\Delta P_{pa} = 20$ dB, $T_g = 24.6$ ms, and $\upsilon_0 = 0$ m/s.

Figure 13.7 shows the MIMO-OFDM performance in mild Doppler spread channels with $\sigma_v = 0.1$ m/s, and in adverse channel conditions with $\sigma_v = 0.2$ m/s. The following receivers are considered.

(1) *A noniterative receiver* with an LS channel estimator.
(2) *An ICI-ignorant iterative receiver* with a sparse channel estimator and a linear MMSE equalizer.
(3) *An ICI-progressive receiver*, denoted as *prog* in the figures, with different D_{max} based on the MMSE equalizer.
(4) *An ICI-progressive receiver* with different D_{max} based on the MCMC equalizer.

From Figure 13.7, one can have the following observations.

- Significant system performance improvement is achieved by the iterative receivers compared to the noniterative receiver, using either MMSE or MCMC equalizer.
- The iterative ICI-ignorant receiver has similar performance to the proposed receiver based on the MMSE equalizer for the mild Doppler spread channels. A sizable performance difference can be observed in the adverse channel conditions.
- The MCMC equalizer outperforms the MMSE equalizer. The performance gap becomes obvious when more transmitters are involved, or as the channel conditions become more adverse. For the mild Doppler spread channels as shown in Figure 13.7, the performance gap increases as N_t increases. One possible reason could be that the MCMC equalizer is more effective for CCI mitigation, while the ICI patterns in OFDM systems are relatively well-behaved [180]. Comparing the same transmitters and modulation setting one can see that the performance gap between the MCMC and MMSE equalizers widens as σ_v increases.

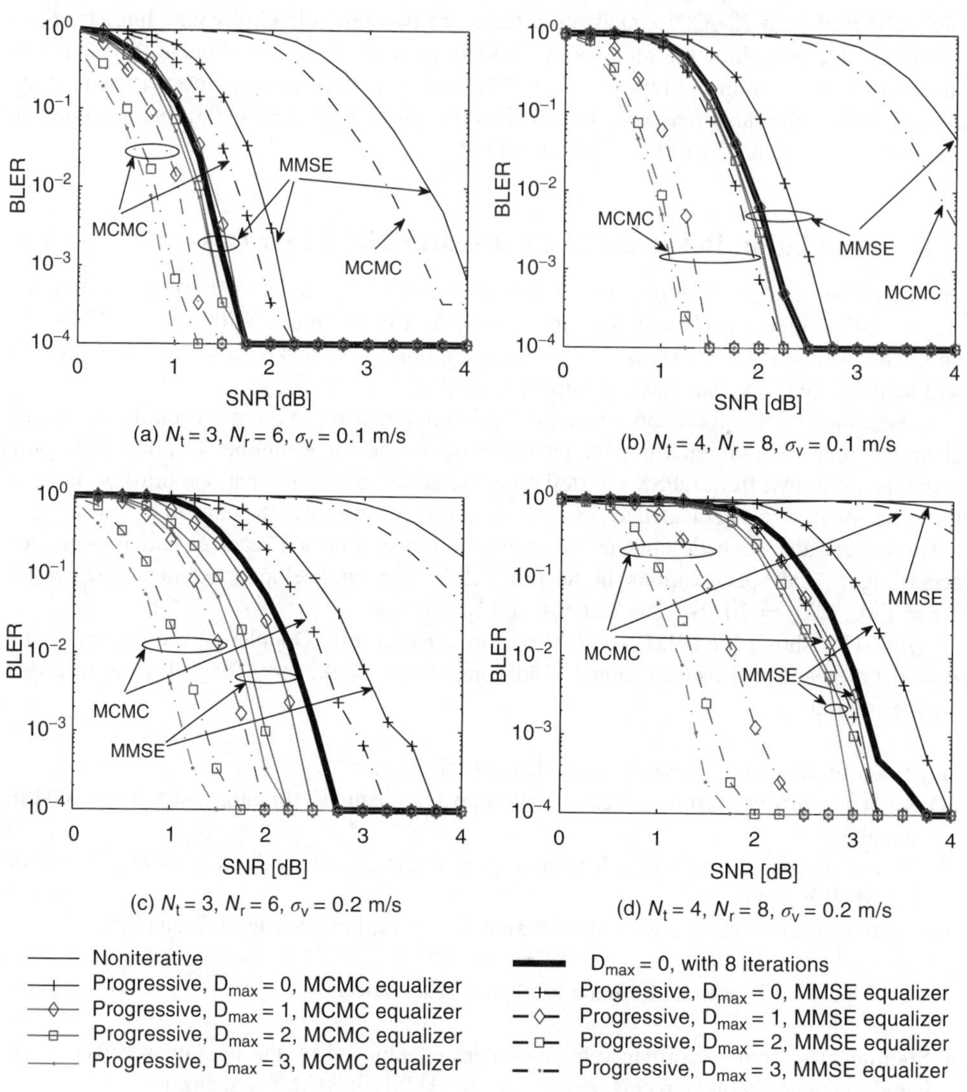

Figure 13.7 Simulated performance for the progressive receiver with different D_{max}. For the progressive receiver, there are $I_{max} = N_t$ iterations in each D.

13.8 SPACE08 Experiment: ICI-Progressive MIMO OFDM

For SPACE08 experiment, the performance results for Julian dates 298–299 (Oct. 25–26, 2008) are reported here at the receivers S3 and S5, which were 200 m and 1000 m from the transmitter, respectively.

Performance results are plotted in Figure 13.8(a)-(d) for $N_t = 3$ with different transmitter-receiver distances. For comparison, the corresponding iterative ICI-ignorant receiver based on the MMSE equalizer with eight iterations is also included. Compared with

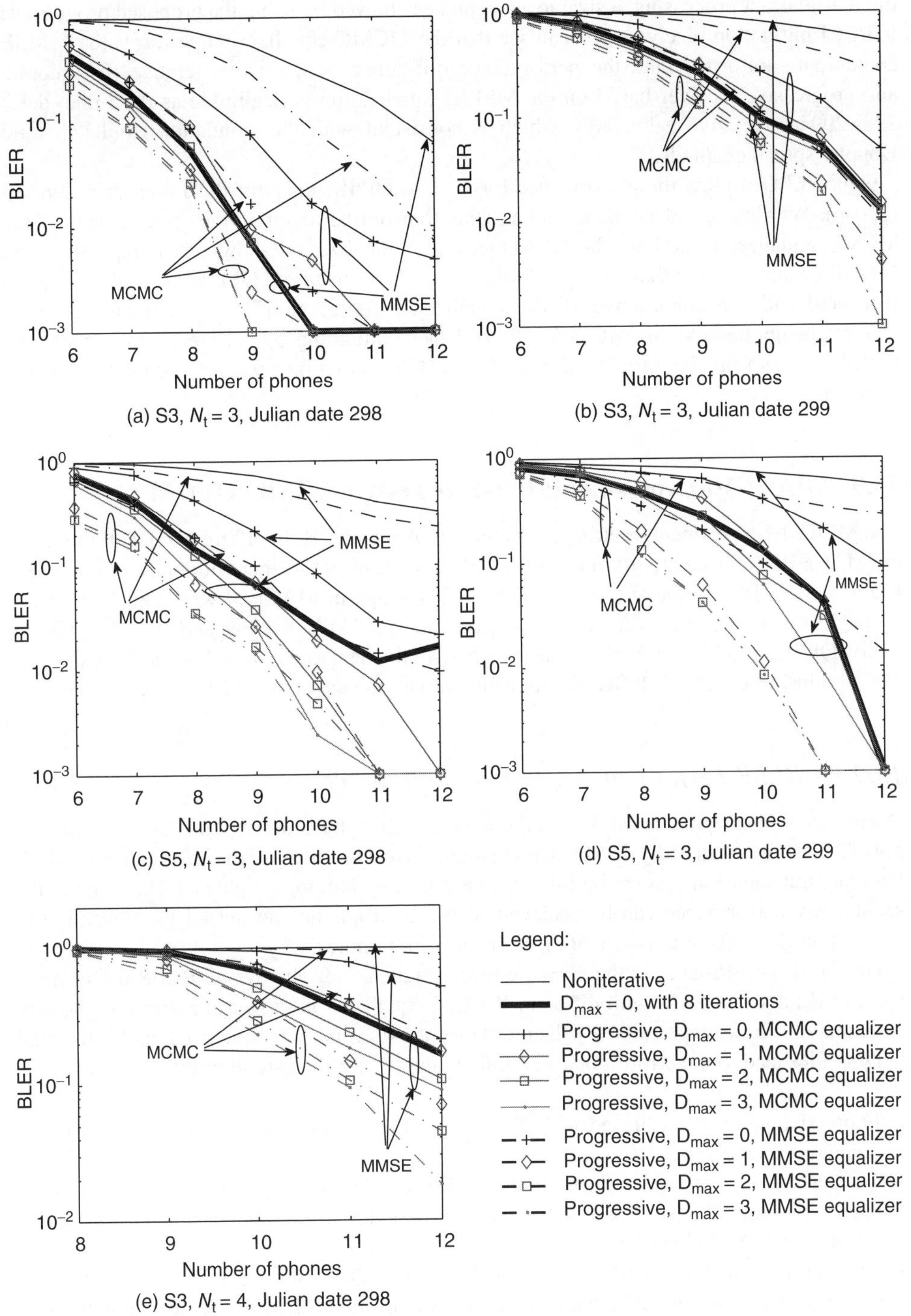

Figure 13.8 SPACE08 experimental results with $N_t = 3, 4$ transmitters, QPSK, S3 (200 m) and S5 (1000 m); For the progressive receiver, there are N_t iterations in each D. The spectral efficiency is 1.60 bits/s/Hz when $N_t = 3$, and is 2.13 bits/s/Hz when $N_t = 4$.

the ICI-ignorant processing, a sizable gain can be achieved by using the proposed progressive ICI/CCI mitigation receiver. One can see that the MCMC equalizer outperforms the MMSE counterpart uniformly, and the performance difference between the iterative ICI-ignorant and progressive receiver based on the MMSE equalization is negligible as the Julian dates 298–299 are relative calm days, which is consistent with the simulation results in mild Doppler spread channels.

Figure 13.8(e) plots the performance for $N_t = 4$, QPSK, only for the Julian date 298, S3 (200 m). With more independent data streams transmitted together, the effectiveness of the MCMC equalizer to address the two-dimensional equalization shows up, compared to the MMSE equalizer. Note that for $N_t = 4$, almost all the transmitted OFDM symbols cannot be recovered with the noniterative MMSE equalizer, while about 15% of them can be recovered by the noniterative MCMC counterpart. When combining $N_r = 12$ receive-elements, the BLER reduces from 10^{-1} to 2×10^{-2} with the MCMC equalizer, when incorporating the ICI explicitly.

13.9 MACE10 Experiment: ICI-Progressive MIMO OFDM

The MACE10 experimental setting is provided in Appendix B. In different MIMO settings of the MACE10 experiment, different constellations were used together with rate $1/2$ nonbinary LDPC codes. The available transmitted MIMO settings in MACE10 were QPSK, 8-QAM, 16-QAM for $N_t = 2$ transmitters with the data rate as 5.4 kb/s, 8.1 kb/s and 10.7 kb/s, respectively; QPSK, 8-QAM for $N_t = 3$ transmitters with the data rate as 8.1 kb/s and 12.1 kb/s. For $N_t = 4$ transmitters, only QPSK was transmitted with the data rate as 10.7 kb/s.

13.9.1 BLER Performance with Two Transmitters

During the tow duration of two hours, there were 31 transmissions in total. One transmission out of 31 data sets was badly distorted, preventing correct decoding of the information symbols; ten other transmissions cannot be fully decoded after combining 12 phones. One can see that satisfactory performance can be achieved with 8 hydrophones during all the transmissions except four data sets at the 44th, 56th, 68th and 72th minutes.

The QPSK performance in the $N_t = 2$ setting is quite good. The overall BLER performance for 8-QAM and 16-QAM with different ICI/CCI equalizers is shown in Figures 13.9(a)-(b). Note that all files except the badly distorted one at 64th minute are included, and some challenging files lead to the error floors. The following observations are in order.

- Impressive gains are achieved with the proposed progressive receiver compared with the noniterative receivers.
- The performance difference between the MMSE and MCMC equalizers is negligible when the receiver uses $D_{max} = 1$ or less than five hydrophones are combined, which is true in both 8-QAM and 16-QAM settings.
- An error floor can be observed in both 8-QAM and 16-QAM settings with the MMSE equalizer. However, the MCMC equalizer can decode all the OFDM blocks after combining 7 phones in 8-QAM; and more 16-QAM OFDM blocks can be decoded by the MCMC equalizer than the MMSE counterpart, leading to a lower error floor.

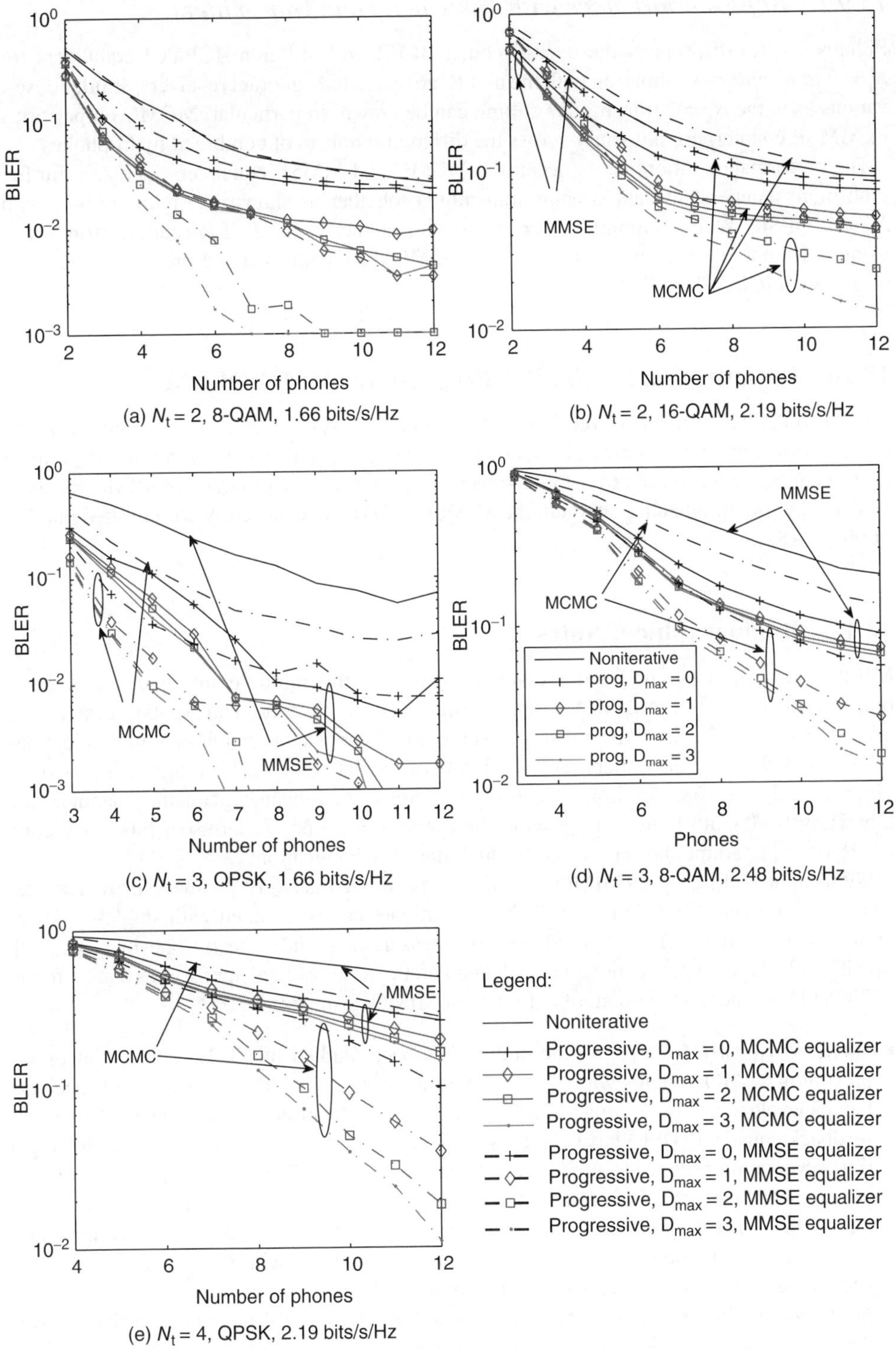

Figure 13.9 Overall BLER results for MACE10 with $N_t = 2, 3, 4$. For the progressive receiver, there are N_t iterations in each D.

13.9.2 BLER Performance with Three and Four Transmitters

Figures 13.9(c)-(d) depicts the overall coded BLER with different ICI/CCI equalizers for $N_t = 3$ transmitters. Comparing with the noniterative and ICI-ignorant receivers, similar observations as in the $N_t = 2$ transmitters setting can be drawn. In particular, MCMC outperforms its MMSE counterpart uniformly across the different numbers of combined hydrophones.

The performance difference between the MCMC and MMSE equalizers increases further with more independent data streams transmitted together as shown in Figure 13.9(e) with $N_t = 4$, QPSK. When combining $N_r = 12$ receive-elements, the BLER reduces from 10^{-1}, using ICI-ignorant receivers, to 10^{-2}, using the MCMC equalizer in the progressive receiver to deal with ICI/CCI explicitly.

13.10 Initialization for the ICI-Progressive MIMO OFDM

Similar to the ICI-progressive receiver in the single-user system, an adaptive block-to-block initialization approach discussed in Section 10.6 can be applied to obtain an initial estimate of the channel between each transmitter-receiver pair for the ICI-progressive MIMO receiver. With the hybrid initialization method, the MIMO-OFDM works decently with limited number of pilots [182].

13.11 Bibliographical Notes

MIMO can help to improve the diversity gain, or, the multiplexing gain, or a combination of both. With MIMO, signal design over both time and spatial dimensions are possible. MIMO have been well covered in textbooks on wireless communications [145, 284, 389], and also there are a plenty of books dedicated to the MIMO topic: [43, 106, 142, 144, 168, 193, 194, 227, 304, 409]. Since underwater acoustic channels are inherently bandwidth-limited, improving the spectral efficiency in terms of bits per second per Hertz is of paramount importance for high data rate applications.

Multi-input multi-output (MIMO) techniques have been actively pursued for underwater acoustic communications. Different MIMO techniques can be coupled with single-carrier or multicarrier systems. Early explorations and measurements have been done in [209, 210]. MIMO with stationary transmitters distributed and the time-reversal approaches can be found in 2006 [357]. There are a boom of activities of MIMO design after 2007.

- *Single-carrier MIMO.* Space time coding has been studied by [331]. For spatial multiplexing in MIMO systems with parallel data transmissions, there are a variety of receivers being developed. One approach is time-domain equalization. A multi-channel decision feedback equalizer (DFE) has been presented in [331], A single-channel DFE following a time reversal preprocessing module has been used in [347, 357]. In [376], iterative block decision-feedback equalizer (BDFE) was proposed with successive interference cancellation (SIC) in each iteration. The other approach is the frequency-domain equalization [461], where a frequency-domain turbo equalizer combined with phase rotation and soft successive interference cancellation was proposed.
- *Multicarrier MIMO.* The majority of work for MIMO OFDM has been on the receiver design with spatial multiplexing. The first category of approaches are the block-by-block

receivers. The block-by-block receivers rely on embedded pilot symbols in each OFDM block for channel estimation [233], which can be further refined by using soft symbol estimates from the channel decoder [172, 181]. The second category of approaches are the adaptive block-to-block processing [67, 364]. The adaptive receiver uses the channels estimated from the previous block for data detection of the current block, after proper phase compensation [67, 364]. The combination of Alamouti space-time block coding with OFDM has also been investigated in [319] and [234].

14

Distributed MIMO OFDM

Chapter 13 was concentrated on co-located MIMO OFDM where data streams arriving at the receiving element experience similar mobilities. This chapter considers an OFDM system with multiple distributed users as depicted in Figure 14.1, where different users could have significantly different mobility patterns. Via some coordination mechanisms, the signals from different users are assumed to be synchronized on the OFDM block level and the major challenge for receiver design lies on how to deal with the significantly different Doppler distortions on different data streams. In such scenarios, a single resampling operation at the receiver does not suffice to remove the main Doppler distortion of all data streams.

- Section 14.2 introduces a front-end processing module that consists of multiple resampling operations, as proposed in [390, 391].
- Section 14.3 presents one receiver termed as *a multiuser detection (MUD) based iterative receiver* [183]. It adopts a frequency-domain oversampling approach to generate discrete samples, on which joint channel estimation and multiuser data detection are carried out in an iterative fashion.
- Section 14.4 presents another receiver termed as *a single-user detection (SUD) based iterative receiver* [183]. Adopting the front end with multiple resampling branches suggested in Section 14.2, it introduces to each branch a module for multiuser interference (MUI) reconstruction/cancellation prior to data decoding, thus leading to a receiver framework with iterative MUI cancellation and data decoding.

The MUD-based receiver is more computationally demanding than the SUD-based receiver. Performance results reveal that the MUD-based receiver outperforms the SUD-based counterparts at the first several iterations, but the performance gap decreases quickly as the iteration goes on. With emulated data sets, it was shown that the SUD-based receiver can handle various distributed MIMO OFDM settings with satisfactory performance [183].

OFDM for Underwater Acoustic Communications, First Edition. Shengli Zhou and Zhaohui Wang.
© 2014 John Wiley & Sons, Ltd. Published 2014 by John Wiley & Sons, Ltd.

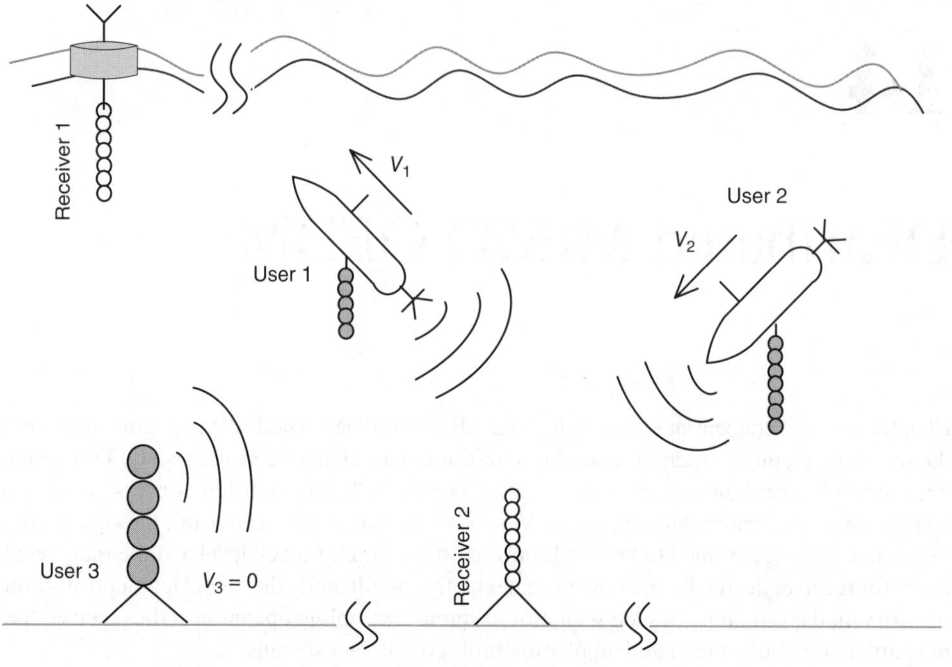

Figure 14.1 Illustration of a distributed underwater MIMO-OFDM system.

14.1 System Model

Consider an underwater system with U users, where the ith user has N_i transmitters. The basic signaling format is zero-padded (ZP) OFDM, with N_i parallel data streams transmitted from N_i transmitters. Hence, the total number of data streams is $N_t = \sum_{i=1}^{U} N_i$.

Within each OFDM block for the ith user, N_i independent data streams are separately channel encoded. Let $s_\mu^{(i)}[k]$ denote the encoded information symbol, e.g., quadratic phase-shift-keying (QPSK) or quadratic amplitude modulation (QAM), to be transmitted on the kth subcarrier by the μth transducer of user i, where $i = 1, 2, \ldots, U$ and $\mu = 1, 2, \ldots, N_i$, and denote $\tilde{s}_\mu^{(i)}(t)$ as the passband signal from the μth transducer of user i. Following the notations in Section 2.1, $\tilde{s}_\mu^{(i)}(t)$ is expressed as

$$\tilde{s}_\mu^{(i)}(t) = 2\Re\left\{\sum_{k=-K/2}^{K/2-1} s_\mu^{(i)}[k] e^{j2\pi f_k t} g(t)\right\}, \qquad t \in [0, T_{\rm bl}]. \tag{14.1}$$

where $g(t)$ is the pulse shaping filter with the Fourier transform denoted by $G(f)$, and f_k denotes the kth subcarrier frequency out of a total of K subcarriers,

$$f_k = f_c + \frac{k}{T} \qquad k = -\frac{K}{2}, \cdots, \frac{K}{2} - 1. \tag{14.2}$$

Consider an underwater acoustic (UWA) multipath channel which consists of $N_{{\rm pa},\nu,\mu}^{(i)}$ discrete paths between the μth transmitter of user i and the νth element. Adopting a time-varying channel model with path-specific Doppler scales in (1.14), the channel impulse response can be expressed as

$$h_{\nu,\mu}^{(i)}(\tau,t)=\sum_{p=1}^{N_{\mathrm{pa},\nu,\mu}^{(i)}} A_{\nu,\mu,p}^{(i)}\delta\left(\tau-(\tau_{\nu,\mu,p}^{(i)}-a_{\nu,\mu,p}^{(i)}t)\right), \tag{14.3}$$

where $A_{\nu,\mu,p}^{(i)}$, $\tau_{\nu,\mu,p}^{(i)}$ and $a_{\nu,\mu,p}^{(i)}$ are the amplitude, the delay at the start of the OFDM block and the Doppler scale for the pth path, respectively.

Assume that the users are quasi-synchronous via some coordinations, and the guard interval is larger than the maximum channel delay spread plus the slight asynchronism among users. As such, there is no interblock interference (IBI), and block-by-block processing is viable. For one OFDM block, the passband signal at the νth element is

$$\tilde{y}_\nu(t)=\sum_{i=1}^{U}\sum_{\mu=1}^{N_i}\sum_{p=1}^{N_{\mathrm{pa},\nu,\mu}^{(i)}} A_{\nu,\mu,p}^{(i)}\tilde{s}_\mu^{(i)}\left((1+a_{\nu,\mu,p}^{(i)})t-\tau_{\nu,\mu,p}^{(i)}\right)+\tilde{n}_\nu(t), \tag{14.4}$$

where $\tilde{n}_\nu(t)$ is the additive noise.

14.2 Multiple-Resampling Front-End Processing

To gain insights on why multiple-resampling front-end processing is needed, we first assume perfect channel state information at receiver. Consider a simple scenario that all the paths between the νth receiving element and the μth transmitter of user i have a common Doppler scale factor $a_\nu^{(i)}$, i.e., $a_{\nu,\mu,p}^{(i)}=a_{\nu,\mu}^{(i)}$, $\forall p$ in (14.3). Similar to the receiver design in [318, Chapter 9] and [391], the transmitted data symbols can be obtained using the maximum-likelihood (ML) rule

$$\{\hat{s}_\mu^{(i)}[k]\}=\arg\min\sum_{\nu=1}^{N_{\mathrm{r}}}\int_0^{T_{\mathrm{bl}}}\left|\tilde{y}_\nu(t)-\sum_{i=1}^{U}\sum_{\mu=1}^{N_i}\sum_{p=1}^{N_{\mathrm{pa},\nu,\mu}^{(i)}} A_{\nu,\mu,p}^{(i)}\tilde{s}_\mu^{(i)}\left((1+a_{\nu,\mu}^{(i)})t-\tau_{\nu,\mu,p}^{(i)}\right)\right|^2 dt. \tag{14.5}$$

Define the channel impulse response at the kth subcarrier f_k as

$$\tilde{h}_{\nu,\mu,k}^{(i)}(t)=\sum_{p=1}^{N_{\mathrm{pa},\nu,\mu}^{(i)}} A_{\nu,\mu,p}^{(i)}e^{-j2\pi f_k(t-\tau_{\nu,\mu,p}^{(i)})}g(t-\tau_{\nu,\mu,p}^{(i)}). \tag{14.6}$$

Eq. (14.5) can be rewritten as

$$\{\hat{s}_\mu^{(i)}[k]\}=\arg\ \min\sum_{\nu=1}^{N_{\mathrm{r}}}\int_0^{T_{\mathrm{bl}}}\left|\tilde{y}_\nu(t)-\sum_{i=1}^{U}\sum_{\mu=1}^{N_i}\sum_{k=-K/2}^{K/2-1} s_\mu^{(i)}[k]\tilde{h}_{\nu,\mu,k}^{(i)}\left(\left(1+a_{\nu,\mu}^{(i)}\right)t\right)\right|^2 dt. \tag{14.7}$$

As shown in [318, Chapter 9], the sufficient test statistics for data detection are:

$$r_{\nu,\mu}^{(i)}[k]=\int_0^{T_{\mathrm{bl}}}\tilde{y}_\nu(t)\left(\tilde{h}_{\nu,\mu,k}^{(i)}\left(\left(1+a_{\nu,\mu}^{(i)}\right)t\right)\right)^* dt, \tag{14.8}$$

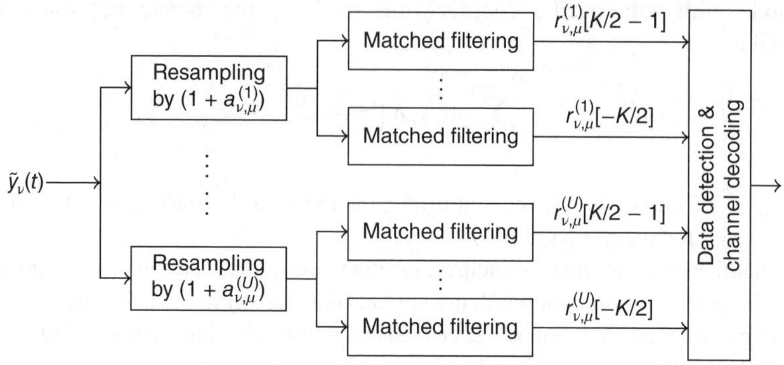

Figure 14.2 Matched-filtering front end with full channel state information at receiver. For the ith user, there are $N_i K$ branches at each receiving element.

where $\nu = 1, \cdots, N_r$, $\mu = 1, \cdots, N_t$, and $k = -K/2, \cdots, K/2-1$. This correlation operation corresponds to matched-filtering of the passband signal with the equivalent channel impulse responses for information symbols at different subcarriers. The correlation operation in (14.8) can be rewritten as

$$r_{\nu,\mu}^{(i)}[k] = \frac{1}{1+a_{\nu,\mu}^{(i)}} \int_0^{T_{\rm bl}} \tilde{y}_\nu \left(\frac{t}{1+a_{\nu,\mu}^{(i)}} \right) \left(\tilde{h}_{\nu,\mu,k}^{(i)}(t) \right)^* dt, \qquad (14.9)$$

where $()^*$ denotes complex conjugate operation. The implementation corresponding to (14.9) is illustrated in Figure 14.2.

The receiver illustrated in Figure 14.2 assumes perfect channel information. We next look into two practical receiver designs in Sections 14.4 and 14.3, respectively, which involves iterative channel estimation, data detection, and channel decoding operations.

14.3 Multiuser Detection (MUD) Based Iterative Receiver

This section presents an iterative receiver based on multiuser channel estimation and data detection, as shown in Figure 14.3. This receiver adopts the frequency-domain oversampling approach discussed in Section 5.2.3 as the receiver front-end, which converts the received continuous-time signal into discrete-frequency samples without information loss. Based on the oversampled frequency-domain samples, a joint channel estimation and multiuser detection will be performed. We next describe the receiver in details.

14.3.1 Pre-processing with Frequency-Domain Oversampling

With an integer frequency-domain oversampling factor $\alpha > 1$, a total of αK frequency-domain samples on each element are obtained for one OFDM block. Define

$$\breve{f}_{\breve{m}} = f_c + \frac{\breve{m}}{\alpha T}, \qquad \breve{m} = -\frac{\alpha K}{2}, \ldots, \frac{\alpha K}{2} - 1, \qquad (14.10)$$

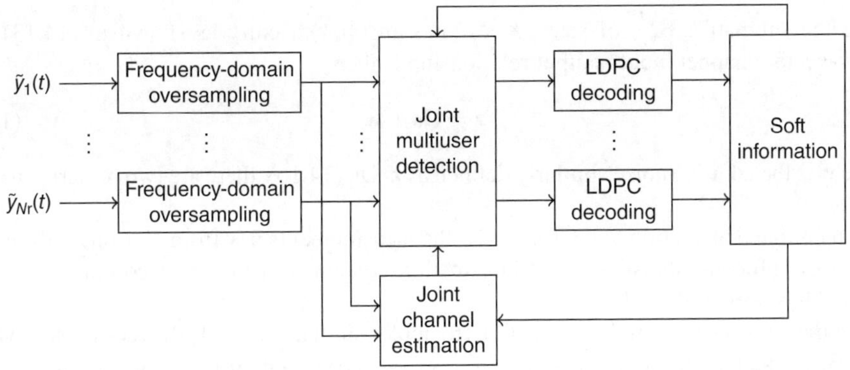

Figure 14.3 MUD-based iterative receiver with joint channel estimation and joint data detection.

where $\breve{m}$ is the index of the oversampled measurements. The measurement $z_\nu[\breve{m}]$ on the frequency $\breve{f}_{\breve{m}}$ is related to $\tilde{y}_\nu(t)$ as

$$z_\nu[\breve{m}] = \int_0^{T_{\rm bl}} \tilde{y}_\nu(t) e^{-j2\pi \breve{f}_{\breve{m}} t} dt, \tag{14.11}$$

which can be implemented by an αK-point FFT operation after padding zeros to the sampled baseband signal.

Due to intercarrier interference (ICI), the measurement on the $\breve{m}$th frequency is potentially affected by all transmitted symbols $s_\mu^{(i)}[k]$, as

$$z_\nu[\breve{m}] = \sum_{i=1}^{U} \sum_{\mu=1}^{N_i} \sum_{k=-K/2}^{K/2-1} \overline{H}_{\nu,\mu}^{(i)}[\breve{m}, k] s_\mu^{(i)}[k] + w_\nu[\breve{m}], \tag{14.12}$$

where $w_\nu[\breve{m}]$ is the additive noise and $\overline{H}_{\nu,\mu}^{(i)}[\breve{m}, k]$ is the coefficient that specifies how the symbol transmitted on the kth subcarrier of the μth transmitter from user i contributes to the output on the $\breve{m}$th subcarrier at the νth element. Following the derivation in Section 5.3, $\overline{H}_{\nu,\mu}^{(i)}[\breve{m}, k]$ can be related to the channel parameters in (14.4) as

$$\overline{H}_{\nu,\mu}^{(i)}[\breve{m}, k] = \sum_{p=1}^{N_{{\rm pa},\nu,\mu}^{(i)}} \frac{A_{\nu,\mu,p}^{(i)}}{1 + a_{\nu,\mu,p}^{(i)}} e^{-j2\pi \breve{f}_{\breve{m}} \frac{\tau_{\nu,\mu,p}^{(i)}}{1+a_{\nu,\mu,p}^{(i)}}} G\left(\frac{\breve{f}_{\breve{m}}}{1 + a_{\nu,\mu,p}^{(i)}} - f_k \right). \tag{14.13}$$

For each element, collect the αK frequency-domain samples into a vector $\mathbf{z}_\nu$, $\nu = 1, \ldots, N_r$. For each transmitter of user i, collect the K transmitted symbols into a vector $\mathbf{s}_\mu^{(i)}$. Now define

$$\mathbf{z} = [\mathbf{z}_1, \mathbf{z}_2, \ldots, \mathbf{z}_{N_r}]^{\rm T}, \qquad \mathbf{s} = [\mathbf{s}_1^{(1)}, \ldots, \mathbf{s}_{N_1}^{(i)}, \ldots, \mathbf{s}_1^{(U)}, \ldots, \mathbf{s}_{N_U}^{(U)}]^{\rm T}. \tag{14.14}$$

Define the channel mixing matrix

$$\overline{\mathbf{H}} = \begin{bmatrix} \overline{\mathbf{H}}_{1,1}^{(1)} & \cdots & \overline{\mathbf{H}}_{1,N_1}^{(1)} & \cdots & \overline{\mathbf{H}}_{1,1}^{(U)} & \cdots & \overline{\mathbf{H}}_{1,N_U}^{(U)} \\ \vdots & \ddots & \vdots & \ddots & \vdots & \ddots & \vdots \\ \overline{\mathbf{H}}_{N_r,1}^{(1)} & \cdots & \overline{\mathbf{H}}_{N_r,N_1}^{(1)} & \cdots & \overline{\mathbf{H}}_{N_r,1}^{(U)} & \cdots & \overline{\mathbf{H}}_{N_r,N_U}^{(U)} \end{bmatrix} \tag{14.15}$$

where the submatrix $\overline{\mathbf{H}}_{\nu,\mu}^{(i)}$ of size $\alpha K \times K$ has an $(\breve{m}, k)$th entry as shown in (14.13). The matrix-vector channel input–output relationship is then

$$\mathbf{z} = \overline{\mathbf{H}}\mathbf{s} + \mathbf{w} \tag{14.16}$$

where $\mathbf{w}$ is the additive noise similarly defined as $\mathbf{z}$. On (14.16), there are two remarks in order.

- As mentioned in Section 5.2.3, for $\alpha > 1$, the measurements in $\mathbf{y}$ from N_r front ends provide a set of sufficient statistics as all information contained in the received continuous-time signal has been collected.
- The matrix-vector formulation in (14.16) looks similar as (13.1) for co-located MIMO OFDM. The key difference is that in co-located MIMO OFDM, the channel matrix $\overline{\mathbf{H}}_{\nu,\mu}^{(i)}$ has its energy mainly concentrated on its main diagonal and several off-diagonals as shown in Figure 5.5, whereas this is not true for the system under consideration. Since different users have significantly different Doppler scales, a single resampling operation does not suffice. For the MUD-based receiver without resampling performed, the main energy of $\overline{\mathbf{H}}_{\nu,\mu}^{(i)}$ could be shifted from the main diagonal and scattered among multiple off-diagonals, and the ICI patterns could be different for different users.

14.3.2 Joint Channel Estimation

The sparse channel estimator described in Chapter 7 can be applied with the following modifications.

- The Doppler search ranges for different data streams are different. For each user, the dominant Doppler scale due to platform motion can be estimated, e.g., by preamble preceding data transmission or training sequences embedded in the transmission.
- During the initialization period of the MUD-based receiver, the data symbols are unknown, and the measurements on pilot subcarriers are severely contaminated by ICI. For the μth data stream of user i, the frequency-domain observation template is generated as

$$\phi_{\nu,\mu}^{(i)}[\breve{m}] = \sum_{k \in \mathcal{S}_{\mathrm{P}}} G\left(\frac{\breve{f}_{\breve{m}}}{1 + \hat{a}_{\nu,\mu}^{(i)}} - f_k\right) s_{\mu}^{(i)}[k], \tag{14.17}$$

  where $\hat{a}_{\nu,\mu}^{(i)}$ is the estimated mean Doppler scale for the μth data stream of user i (note that the Doppler scales can be quite different for different data streams of the same user, which is not applicable to the SUD-based receiver to be developed in Section 14.4). To address the ICI incurred by the unknown data symbols, one strategy is to choose only those measurements with $\phi_{\nu,\mu}^{(i)}[\breve{m}]$ larger than a given threshold for some user i for channel estimation, and exclude other measurements. The other strategy is to use a pre-whitening method to reduce the ICI effect, as discussed in Section 7.3.4. In later iterations, the observations on all subcarriers can be utilized for channel estimation as tentative decisions on all information symbols are available.

14.3.3 Multiuser Data Detection and Channel Decoding

After obtaining the estimated channel mixing matrix $\hat{\bar{\mathbf{H}}}$, joint MIMO detection with *a priori* information fed back from the channel decoder can be applied. Based on the channel input-output model in (14.16), the linear MMSE equalizer developed in Section 8.2.2 can be used. Note that the size of the channel mixing matrix $\hat{\bar{\mathbf{H}}}$ is $(\alpha K N_r \times K N_t)$, so inverting a $(KN_t \times KN_t)$ matrix is computationally demanding. Outputs of the linear MMSE detector are fed into N_t separate channel decoders to recover the information bits from users.

14.4 Single-User Detection (SUD) Based Iterative Receiver

This section considers a practical receiver design for the distributed MIMO-OFDM system. Here, the channel model with path-specific Doppler scales as in (14.3) is adopted; however, it is assumed that data streams from the same user will experience one dominant Doppler scale. A single-user-detection (SUD) based iterative receiver is depicted in Figure 14.4, where a multi-resampling front end in Section 14.2 is adopted, and a MUI cancellation module is added to address the co-channel interference (CCI) among multiple users.

14.4.1 Single-User Decoding

Let $\hat{a}_\nu^{(i)}$ denote an estimated dominant Doppler scale for the N_i received data streams for user i. The receiver for user i will apply the resampling operation on the νth element and then perform OFDM demodulation to generate frequency-domain measurements

$$z_\nu^{(i)}[m] = \int_0^{T_{\rm bl}} \tilde{y}_\nu\left(\frac{t}{1+\hat{a}_\nu^{(i)}}\right) e^{-j2\pi f_m t} dt, \qquad m = -\frac{K}{2}, \cdots, \frac{K}{2}-1 \tag{14.18}$$

where the superscript (i) stresses that the output is associated with user i. The measurement on the mth subcarrier can be expressed as

$$z_\nu^{(i)}[m] = \sum_{\mu=1}^{N_i} \sum_{k=-K/2}^{K/2-1} H_{\nu,\mu}^{(i)}[m,k] s_\mu^{(i)}[k] + \sum_{\ell=1,\ell\neq i}^{U} \chi_\nu^{(\ell\to i)}[m] + w_\nu^{(i)}[m] \tag{14.19}$$

where $\chi_\nu^{(\ell\to i)}[m]$ is the interference from user ℓ to user i on the mth subcarrier, and the channel coefficient can be expressed as

$$H_{\nu,\mu}^{(i)}[m,k] = \sum_{p=1}^{N_{{\rm pa},\nu,\mu}^{(i)}} \xi_{\nu,\mu,p}^{(i)} e^{-j2\pi \frac{m}{T}\bar{\tau}_{\nu,\mu,p}^{(i)}} G\left(\frac{f_m}{1+b_{\nu,\mu,p}^{(i)}} - f_k\right) \tag{14.20}$$

$$b_{\nu,\mu,p}^{(i)} = \frac{a_{\nu,\mu,p}^{(i)} - \hat{a}_\nu^{(i)}}{1+\hat{a}_\nu^{(i)}}, \qquad \xi_{\nu,\mu,p}^{(i)} = \frac{A_{\nu,\mu,p}^{(i)}}{1+b_{\nu,\mu,p}^{(i)}} e^{-j2\pi f_c \bar{\tau}_{\nu,\mu,p}^{(i)}}, \qquad \bar{\tau}_{\nu,\mu,p}^{(i)} = \frac{\tau_{\nu,\mu,p}^{(i)}}{1+b_{\nu,\mu,p}^{(i)}}.$$

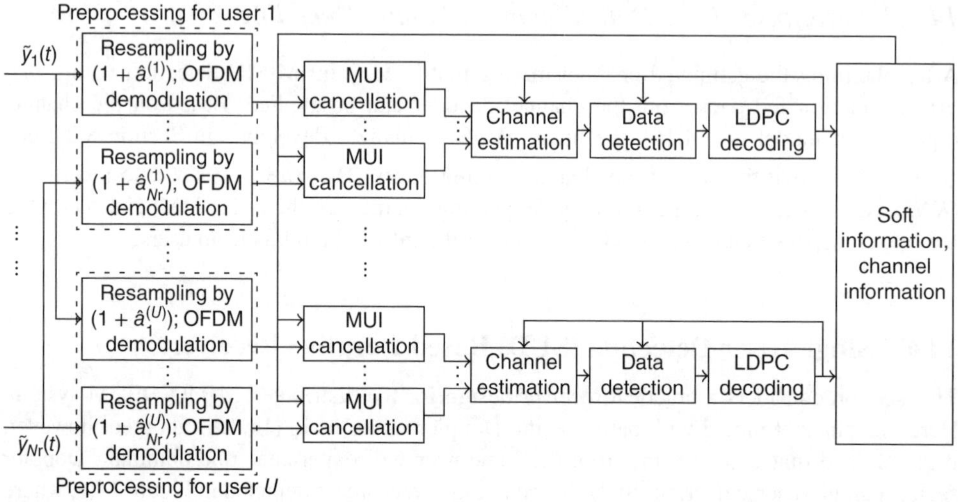

Figure 14.4 SUD-based iterative receiver with MUI cancellation, with UN_r resampling branches.

Suppose that an estimate of $\chi_\nu^{(\ell\to i)}[m]$ is available, the receiver will obtain

$$\bar{z}_\nu^{(i)}[m] = z_\nu^{(i)}[m] - \sum_{\ell=1,\ell\neq i}^{U} \hat{\chi}_\nu^{(\ell\to i)}[m] \tag{14.21}$$

$$= \sum_{\mu=1}^{N_i} \sum_{k=-K/2}^{K/2-1} H_{\nu,\mu}^{(i)}[m,k] s_\mu^{(i)}[k] + \eta_\nu^{(i)}[m] \tag{14.22}$$

where $\eta_\nu^{(i)}[m]$ is the equivalent noise containing both the ambient noise and the residual interference. After MUI cancellation, channel estimation and data detection for co-located MIMO OFDM in Chapter 13 can be directly applied based on $\{\bar{z}_\nu^{(i)}[m]\}_{m=-K/2}^{K/2-1}$ from all $N_{\rm r}$ elements.

14.4.2 MUI Construction

In the SUD-based receiver, the key issue is how to reconstruct the MUI knowing that different users have carried out channel estimation and data detection based on the measurements from different front-ends. Let $\hat{a}_\nu^{(i)}$ and $\hat{a}_\nu^{(\ell)}$ denote the resampling factors used in the front ends of user i and ℓ, respectively.

For the μth data stream from user ℓ, assume that $(\hat{\xi}_{\nu,\mu,p}^{(\ell)}, \hat{\tau}_{\nu,\mu,p}^{(\ell)}, \hat{b}_{\nu,\mu,p}^{(\ell)})$ and $\hat{\mathbf{s}}_\mu^{(\ell)}$ have been estimated. Then, one can construct a virtual signal as

$$\hat{z}_{\nu,\mu}^{(\ell)}(t) = \sum_p (1+\hat{b}_{\nu,\mu,p}^{(\ell)})\hat{\xi}_{\nu,\mu,p}^{(\ell)} e^{j2\pi f_c \hat{\tau}_{\nu,\mu,p}^{(\ell)}} \hat{\tilde{x}}_\mu^{(\ell)}\left((1+\hat{b}_{\nu,\mu,p}^{(\ell)})(t-\hat{\tau}_{\nu,\mu,p}^{(\ell)})\right) \tag{14.23}$$

whose Fourier transform satisfies

$$\hat{Z}_{\nu,\mu}^{(\ell)}(f)|_{f=f_m} = \sum_p \hat{\xi}_{\nu,\mu,p}^{(\ell)} e^{-j2\pi \frac{m}{T} \hat{\tau}_{\nu,\mu,p}^{(\ell)}} \sum_{k=-K/2}^{K/2-1} G\left(\frac{f_m}{1+\hat{b}_{\nu,\mu,p}^{(\ell)}} - f_k\right) \hat{s}_\mu^{(\ell)}[k], \tag{14.24}$$

which is compatible with the channel and symbol estimates. With N_ℓ data streams, one can construct

$$\hat{z}_\nu^{(\ell)}(t) = \sum_{\mu=1}^{N_\ell} \hat{z}_{\nu,\mu}^{(\ell)}(t). \tag{14.25}$$

One can view $\hat{z}_\nu^{(\ell)}(t)$ as the reconstructed signal after the ℓth user's front-end processing, and hence the corresponding version before the resampling operation is $\hat{y}_\nu^{(\ell)}(t) = \hat{z}_\nu^{(\ell)}((1+\hat{a}_\nu^{(\ell)})t)$. Letting $\hat{y}_\nu^{(\ell)}(t)$ pass through the ith user's front-end, the MUI from the user ℓ to user i can be expressed as

$$\hat{\chi}_\nu^{(\ell\to i)}[m] = \int_0^{T_{\rm bl}} \hat{z}_\nu^{(\ell)}\left(\frac{1+\hat{a}_\nu^{(\ell)}}{1+\hat{a}_\nu^{(i)}} t\right) e^{-j2\pi f_m t} dt \tag{14.26}$$

Straightforward manipulation leads to

$$\hat{\chi}_\nu^{(\ell\to i)}[m] = \sum_{\mu=1}^{N_\ell} \sum_{k=-K/2}^{K/2-1} \hat{H}_{\nu,\mu}^{(\ell\to i)}[m,k] \hat{s}_\mu^{(\ell)}[k] \tag{14.27}$$

where $H_{\nu,\mu}^{(\ell\to i)}[m,k]$ can be computed as

$$\hat{H}_{\nu,\mu}^{(\ell\to i)}[m,k] = \sum_{p=1}^{N_{\mathrm{pa},\nu,\mu}^{(\ell)}} \hat{\xi}_{\nu,\mu,p}^{(\ell\to i)} e^{-j2\pi \frac{m}{T} \hat{\tau}_{\nu,\mu,p}^{(\ell\to i)}} G\left(\frac{f_m}{1+\hat{b}_{\nu,\mu,p}^{(\ell\to i)}} - f_k\right) \tag{14.28}$$

with

$$1+\hat{b}_{\nu,\mu,p}^{(\ell\to i)} = \frac{1+\hat{a}_\nu^{(\ell)}}{1+\hat{a}_\nu^{(i)}}(1+\hat{b}_{\nu,\mu,p}^{(\ell)}), \quad \hat{\tau}_{\nu,\mu,p}^{(\ell\to i)} = \frac{1+\hat{a}_\nu^{(i)}}{1+\hat{a}_\nu^{(\ell)}} \hat{\tau}_{\nu,\mu,p}^{(\ell)} \tag{14.29}$$

$$\hat{\xi}_{\nu,\mu,p}^{(\ell\to i)} = \frac{1+\hat{a}_\nu^{(i)}}{1+\hat{a}_\nu^{(\ell)}} \hat{\xi}_{\nu,\mu,p}^{(\ell)} e^{-j2\pi f_c(\hat{\tau}_{\nu,\mu,p}^{(\ell\to i)} - \hat{\tau}_{\nu,\mu,p}^{(\ell)})} \tag{14.30}$$

Hence, the amplitudes, delays, and Doppler scales need to be properly modified when reconstructing the MUI from one user to another user.

14.5 An Emulated Two-User System Using MACE10 Data

The MACE10 experimental setup and OFDM parameter setting can be found in Appendix B. Here we use the single-transmitter signals in MACE10 to emulate a distributed system with two users. The Doppler scales estimated from the single user case will be used to decode the distributed MIMO-OFDM systems.

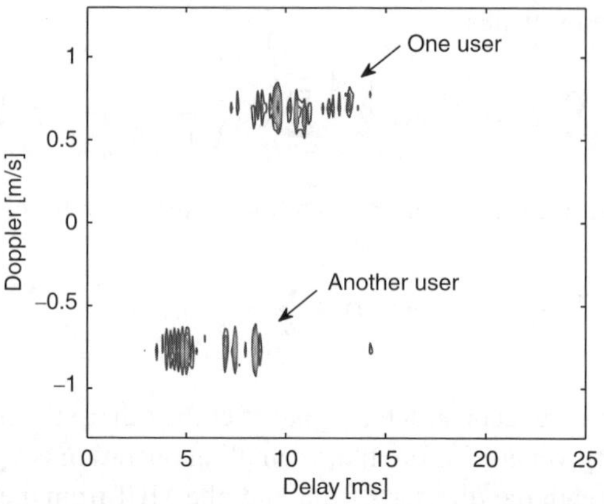

Figure 14.5 Example channel scatting functions for the distributed MIMO setting with two single-transmitter users from MACE10.

Tow 1 has $n = 31$ transmissions, with 20 OFDM blocks in each transmission. The estimated speed of source array is depicted in Figure B.6. To emulate a distributed MIMO-OFDM system with different mobilities, we add the received passband signals of the 1st and the nth transmissions, the 2nd and $(n-1)$th transmissions together, and so on. A total of 15 emulated data sets for a two-user system are obtained while the 16th transmission is excluded. Note that the OFDM blocks in one transmission are reversed in order so that the overlapping blocks have different pilot and data symbols. From Figure B.6, the absolute value of projected velocity was around 1 m/s, which will lead to a frequency shift as about $\pm 7 \sim \pm 10$ Hz while the subcarrier spacing in MACE10 was 4.8 Hz. A measured channel response is plotted in Figure 14.5, where the channel energy for two users are well separated in the Doppler plane. The Doppler spreads are around 0.1 m/s and the delay spread is around 10 ms.

14.5.1 MUD-Based Receiver with and without Frequency-Domain Oversampling

Figure 14.6 shows the performance results by using the MUD-based receiver for both the conventional sampling ($\alpha = 1$) and the frequency-domain oversampling with $\alpha = 2$. The ICI is not limited to only near neighbours, and we assume $\overline{H}^{(i)}_{\nu,\mu}(\breve{m}, k) \neq 0, \forall |\breve{f}_m - f_k| \leq \frac{37}{T}$ for all the channel matrices $\overline{\mathbf{H}}^{(i)}_{\nu,\mu}$ in (14.15). The frequency-domain oversampling method outperforms the conventional sampling uniformly at early iterations. As the number of iterations gets large, the performance gap decreases to a negligible level. In the following, we use the conventional sampling with $\alpha = 1$ for the MUD-based receiver.

14.5.2 Performance of SUD- and MUD-Based Receivers

Figure 14.7 shows the performance for distributed MIMO-OFDM systems with two users, where 8 iterations are used. In the SUD-based receiver, the channel matrix in (14.20) is

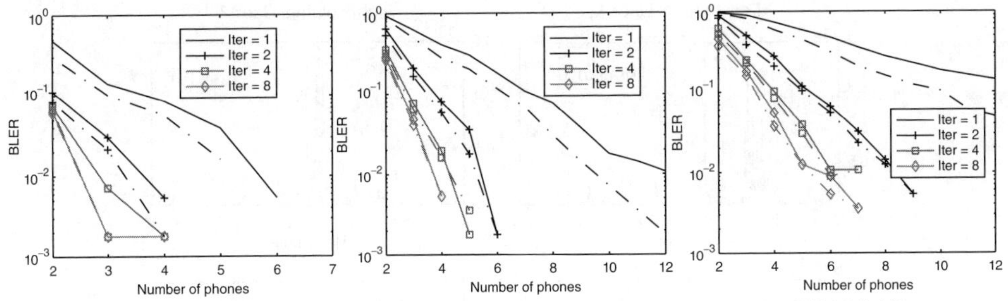

(a) QPSK; 0.55 bits/s/Hz per user (b) 8-QAM; 0.83 bits/s/Hz per user (c) 16-QAM; 1.10 bits/s/Hz per user

Figure 14.6 Distributed MIMO OFDM with two single-transmitter users from MACE10. Solid lines: conventional sampling with $\alpha = 1$; Dash-dotted lines: frequency-domain oversampling with $\alpha = 2$.

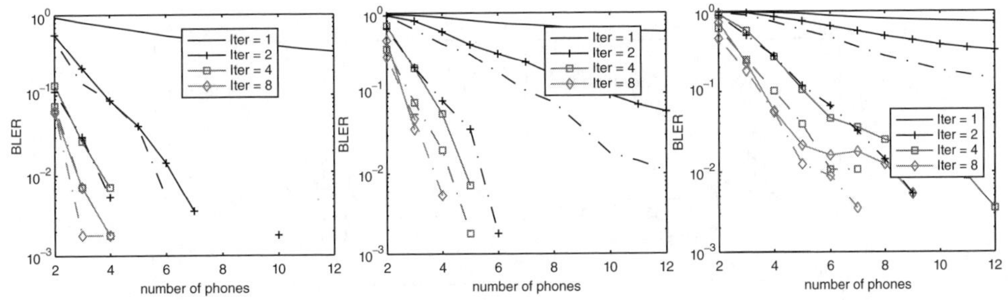

(a) QPSK (b) 8-QAM (c) 16-QAM

Figure 14.7 Distributed MIMO OFDM with two single-transmitter users from MACE10. Dash-dotted lines: MUD-based receiver; Solid lines: SUD-based receiver.

diagonal, thus ignoring the ICI from the same user. When constructing the MUI $\hat{\chi}^{(\ell \to i)}[m]$ in (14.27), the contributions from all the transmitted symbols of user l on each measurement are considered. Both the SUD-based and MUD-based receivers work very well for QPSK, 8-QAM and 16-QAM. The following observations are in order.

- At the first several iterations, the SUD-based receiver is much worse than the MUD-based receiver. This is due to the severe residual MUI at early iterations.
- With continuing iterations, the performance of the SUD-based receiver catches up that of the MUD-based receiver as more and more MUI is cancelled out. For the 8-QAM in this data set, the SUD-based receiver even slightly outperforms the MUD-based counterpart, while on the other hand the MUD-based receiver work slightly better than the SUD-based counterparts for QPSK and 16-QAM.

Figure 14.8 shows an example of the estimated channel impulse responses (CIRs) for the distributed two-user system. It is obvious from Figure 14.8 that as the iteration goes on, the CIRs look like MUI-free ones, which verifies the effectiveness of the MUI cancellation in the SUD-based receiver.

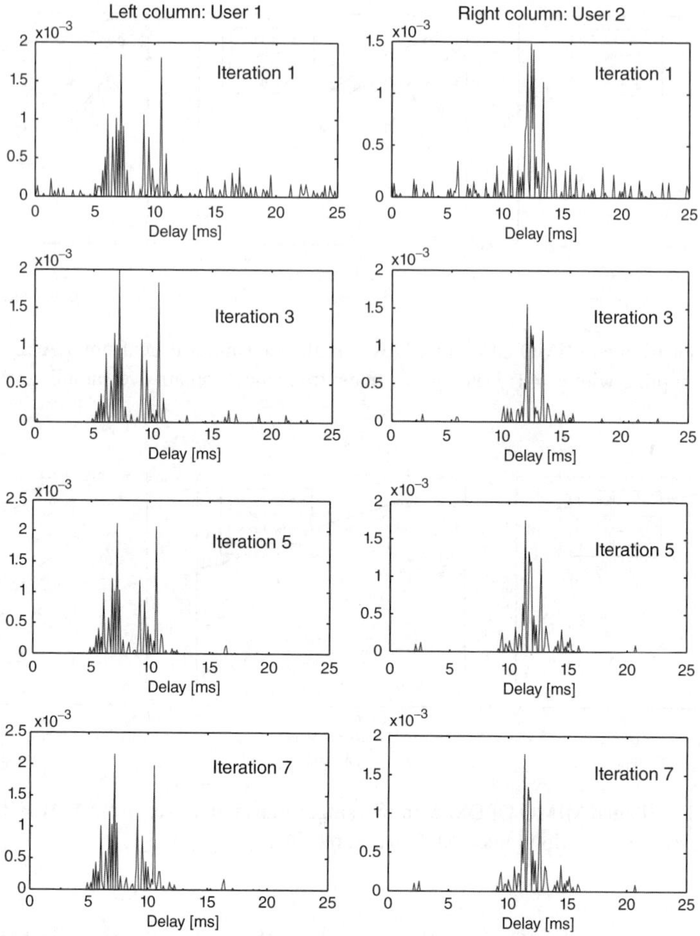

Figure 14.8 Estimated channel impulse responses with the SUD-based receiver, 8-QAM.

The constellation scattering plots in Figure 14.9 show the soft-decision symbols at the output of the MMSE detector for users 1 and 2. The improvement over iterations is clearly observed by comparing subfigures in each column in Figure 14.9. Also, from Figures 14.8 and 14.9, one can see that during the start up stage, the CIRs and scatter plots for both users look very noisy due to the severe MUI, as expected.

14.6 Emulated MIMO OFDM with MACE10 and SPACE08 Data

In this section, the MACE10 and SPACE08 data are used to emulate various distributed multiuser settings. The SPACE08 experimental setup can be found in Appendix B. In particular, the data set collected by receiver S3 (see Figure B.1) in Julian dates 296–297 is used. Note that the user from SPACE08 data has multiple co-located transducers, as specified later.

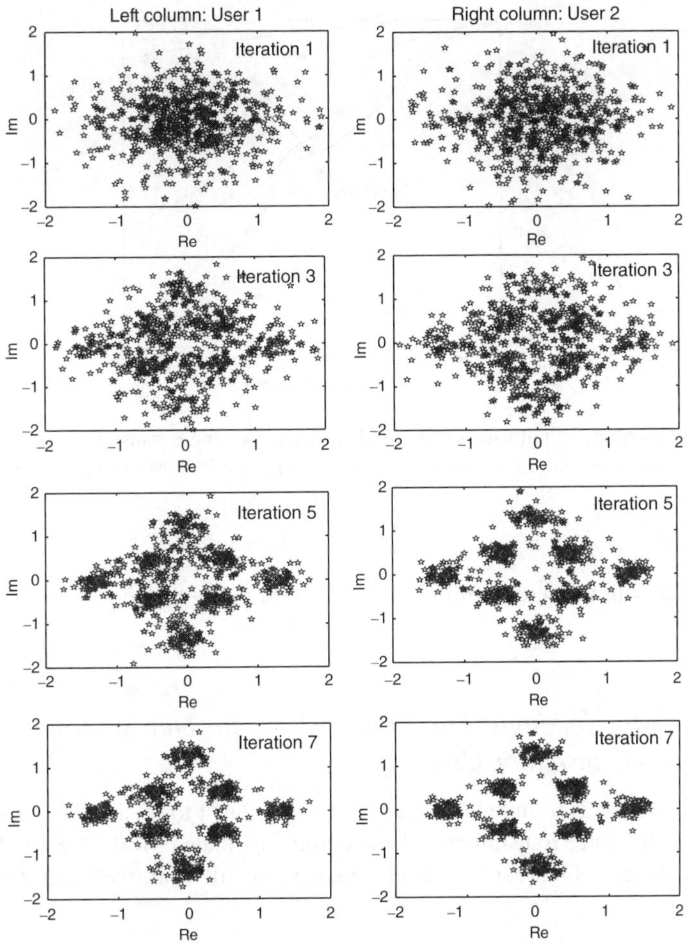

Figure 14.9 Constellation scattering plots at the output of the MMSE detector for users 1 and 2 with the SUD-based receiver, 3 phones are combined. ⋄: 8-QAM constellation points.

14.6.1 One Mobile Single-Transmitter User plus One Stationary Two-Transmitter User

In this setting, one user is from MACE10 with transmissions 1 to 15 (a negative Doppler scale), as shown in Figure B.6. The second user owns the data from SPACE08, with two transmitters. This distributed MIMO OFDM hence has two users and three data streams. As the carrier frequencies and the guard intervals used in MACE10 and SPACE08 were different, we add the baseband signals together on the OFDM block level. Figure 14.10 shows the decoding results for this setting with the SUD-based receiver, using QPSK constellation. Due to the high computation complexity, the MUD-based receiver performance is not reported.

Figure 14.10 shows that the performance at the first iteration is pretty bad due to the severe MUI. The performance improvement from the first iteration to second iteration is impressive,

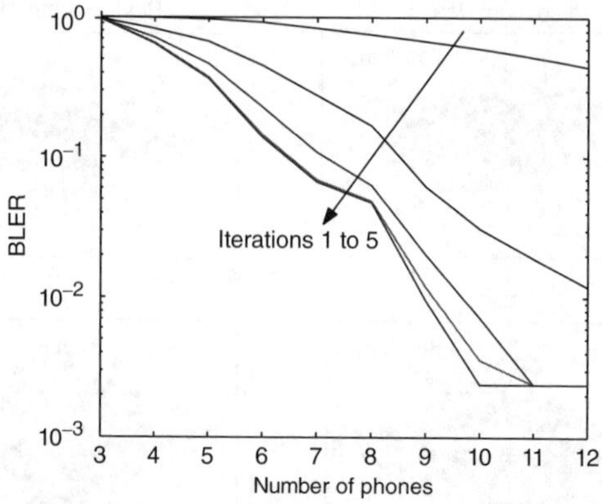

Figure 14.10 Distributed multiuser setting with one mobile single-transmitter user from MACE10 (0.55 bits/s/Hz) plus one stationary two-transmitter user from SPACE08 (1.10 bits/s/Hz). SUD-based receiver, QPSK.

which is similar with the setting in Figure 14.7. Good performance is achieved after four to five iterations.

14.6.2 One Mobile Single-Transmitter User plus One Stationary Three-Transmitter User

In this setting, one user is from MACE10 with transmissions 1 to 15, and the other user is using SPACE08 data with three transmitters. The resulting distributed MIMO OFDM has two users and four data streams. Figure 14.11 shows the system still work well with the SUD-based

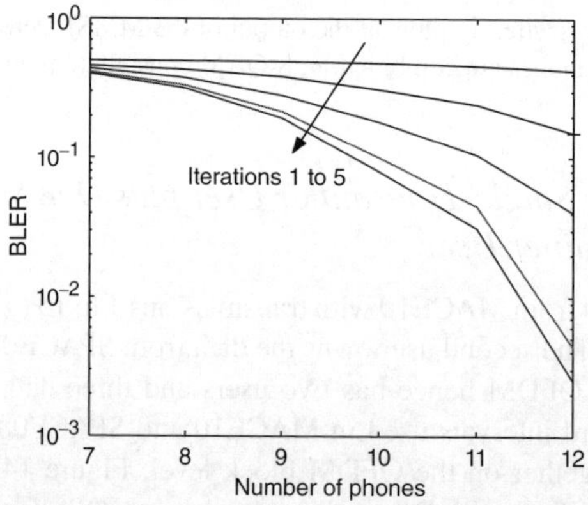

Figure 14.11 Distributed MIMO OFDM with one mobile single-transmitter user from MACE10 (0.55 bits/s/Hz) plus one stationary three-transmitter user from SPACE08 (1.65 bits/s/Hz). SUD-based receiver, QPSK.

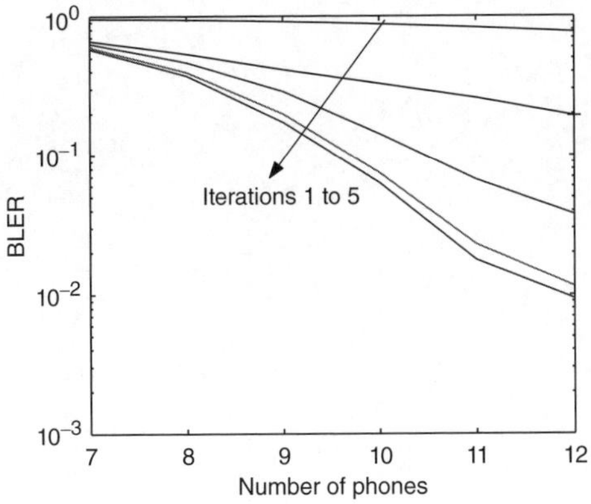

Figure 14.12 Distributed MIMO OFDM with two mobile single-transmitter users (0.55 bits/s/Hz per user) from MACE10 plus one stationary two-transmitter user from SPACE08 (1.10 bits/s/Hz). SUD-based receiver, QPSK.

receiver. Due to the larger number of data streams, the improvement from first iteration to the second iteration is shrunk compared with the settings in Figure 14.7 and Figure 14.10. However, satisfactory performance can still be achieved and the system performance saturates after five iterations.

14.6.3 Two Mobile Single-Transmitter Users plus One Stationary Two-Transmitter User

In this setting, two users are from MACE10 as described in Section 14.5. A third user has data from SPACE08 with two transmitters. Hence, the distributed MIMO OFDM has three users and four data streams. Figure 14.12 depicts the overall coded BLER with the SUD-based receiver. With 12 elements, the BLER after the first iteration is around 0.8, but reduces to below 10^{-2} after 6 iterations.

14.7 Bibliographical Notes

This chapter studies receiver designs for distributed multiuser OFDM systems, where different users have significantly different Doppler scales. This is a problem unique to underwater acoustic communications, due to the slow sound propagation speed, which creates large Doppler deviations among users. Such a problem was first studied in [390, 391], where the multiple-resampling front-end was proposed. The two iterative receivers presented in this chapter are based on the following-on work in [183]. The multiple-resampling concept is also applicable to the single user case when different clusters have different Doppler scales [391]. A receiver design for single carrier transmissions with users having different Doppler scales is provided in [84].

15

Asynchronous Multiuser OFDM

This chapter continues the discussion on the application of OFDM in multiuser underwater acoustic communications. Should signals from multiple users be synchronized at the receiver on the OFDM block level, multiuser communication can be treated as a co-located or distributed multi-input multi-output (MIMO) problem, and techniques developed in Chapters 13 and 14 directly apply. However, due to large propagation delays and lack of a well-defined network infrastructure, synchronization is a challenging task in distributed underwater acoustic systems. Figure 15.1 shows one example system where multiple autonomous underwater vehicles (AUVs) communicate to a fixed cabled network on the sea bottom [159, 421], and another example system where multiple sensor nodes communicate data back to buoys on the water surface. Without a well-defined coordination mechanism, the receiver needs to handle the asynchronous nature of multiuser transmissions explicitly.

This chapter presents a receiver design for asynchronous multiuser transmissions, where all users adopt zero-padded OFDM modulation. To simplify the discussion, this chapter only considers the scenario where all the users have similar mobility patterns, so that the main Doppler effect can be compensated via a resampling operation at the receiver. For an asynchronous distributed multiuser system where different users experience different Doppler distortions, techniques developed in Chapter 14 and this chapter could be combined, although the receiver complexity would increase considerably.

- Section 15.1 describes the system model for asynchronous multiuser OFDM in underwater acoustic channels.
- Section 15.2 introduces the concepts of overlapped truncation and interference aggregation, built on which an asynchronous multiuser reception problem is converted to a quasi-synchronous multiuser reception problem with interference contamination.
- Section 15.3 presents an iterative burst-to-burst OFDM receiver for asynchronous multiuser reception, leveraging the techniques developed for co-located MIMO-OFDM reception in Chapter 13 and for external interference cancellation in Chapter 12.
- Section 15.4 provides an example analysis on the statistic distribution of the maximal misalignment of multiuser transmissions in a data collection network using a classic handshaking protocol.

OFDM for Underwater Acoustic Communications, First Edition. Shengli Zhou and Zhaohui Wang.
© 2014 John Wiley & Sons, Ltd. Published 2014 by John Wiley & Sons, Ltd.

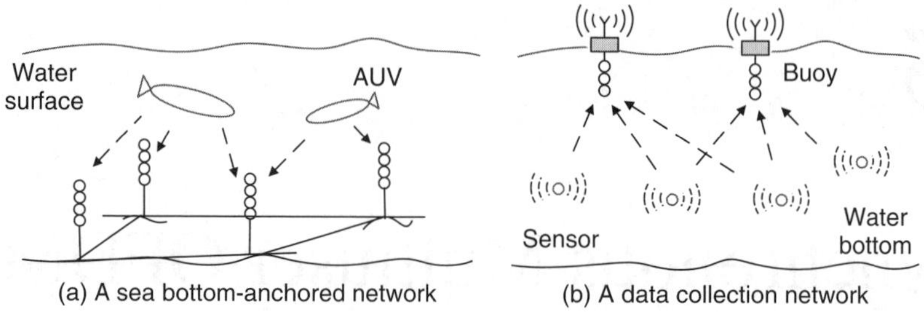

Figure 15.1 Two examples of underwater acoustic networks. The nodes anchored at the water bottom in the first network are connected to a control center via cables. The gateways in the second network can communicate with satellites or ships using radio waves.

- Sections 15.5 and 15.6 contain performance results using both simulation and emulated data sets. The numerical results highlight that the decoding performance degrades as the maximum relative delay among users increases.

The developed asynchronous multiuser reception approach allows the transmissions of multiple users without performing coordination, which could simplify the multi-access control protocol design for certain underwater acoustic networks. On the other hand, the receiver needs a high computational capability to decode the asynchronous transmissions from multiple users.

15.1 System Model for Asynchronous Multiuser OFDM

Consider an underwater system consisting of $N_{\rm u}$ asynchronous users and a receiver equipped with $N_{\rm r}$ co-located receiving elements, and $N_{\rm r} > N_{\rm u}$. Assume that all users use the OFDM block transmission with an identical parameter set. Denote f_k as the kth subcarrier frequency out of a total of K subcarriers,

$$f_k = f_{\rm c} + \frac{k}{T} \qquad k = -\frac{K}{2}, \cdots, \frac{K}{2} - 1 \tag{15.1}$$

where $f_{\rm c}$ is the system center frequency, and T is the OFDM symbol duration. With a guard interval of $T_{\rm g}$ between consecutive OFDM symbols, the total time duration of each OFDM block is thus $T_{\rm bl} := T + T_{\rm g}$. Define $s_\mu[k;n]$ as the symbol transmitted at the kth subcarrier of the nth block from the μth user. The nth transmitted block is expressed as

$$\tilde{s}_\mu(t;n) = 2\Re\left(\sum_{k=-K/2}^{K/2-1} s_\mu[k;n]e^{j2\pi f_k t}g(t)\right), \qquad t \in [0,\ T_{\rm bl}] \tag{15.2}$$

where $g(t)$ is a rectangular window which has nonzero support within $[0, T]$. Assume $N_{\rm bl}$ blocks in each transmission burst. The transmitted signal from the μth user is

$$\tilde{x}_\mu(t) = \sum_{n=1}^{N_{\rm bl}} \tilde{s}_\mu(t - nT_{\rm bl}; n), \qquad t \in [0,\ N_{\rm bl}T_{\rm bl}]. \tag{15.3}$$

Assume a channel model with path-specific Doppler scaling factors. The channel impulse response between the μth user and the νth receiving element during the transmission of the nth block is expressed as

$$h_{\nu,\mu}(\tau, t; n) = \sum_{p=1}^{N_{\mathrm{pa},\nu,\mu}^{(n)}} A_{\nu,\mu,p}^{(n)} \delta(\tau - (\overline{\tau}_{\nu,\mu,p}^{(n)} - a_{\nu,\mu,p}^{(n)} t)) \tag{15.4}$$

where $N_{\mathrm{pa},\nu,\mu}^{(n)}$ is the number of discrete channel paths, $A_{\nu,\mu,p}^{(n)}, \overline{\tau}_{\nu,\mu,p}^{(n)}$ and $a_{\nu,\mu,p}^{(n)}$ are the amplitude, initial delay, and Doppler scaling factor of the pth path, respectively. The triplets $(A_{\nu,\mu,p}^{(n)}, \overline{\tau}_{\nu,\mu,p}^{(n)}, a_{\nu,\mu,p}^{(n)})$ could vary across blocks. The signal from the μth user at the νth receiving element can be expressed as

$$\tilde{y}_{\nu,\mu}(t) = \sum_{n=1}^{N_{\mathrm{bl}}} \tilde{y}_{\nu,\mu}(t - nT_{\mathrm{bl}}; n), \qquad t \in [0, N_{\mathrm{bl}} T_{\mathrm{bl}}] \tag{15.5}$$

with

$$\tilde{y}_{\nu,\mu}(t; n) = \sum_{p=1}^{N_{\mathrm{pa},\nu,\mu}^{(n)}} A_{\nu,\mu,p}^{(n)} \tilde{s}_{\mu}((1 + a_{\nu,\mu,p}^{(n)})t - \overline{\tau}_{\nu,\mu,p}^{(n)}). \tag{15.6}$$

Let ϵ_μ denote the time-of-arrival the μth user, which can be obtained at the receiver by detecting the preamble of this user. On the block level, one can assume that $\epsilon_\mu \le T_{\mathrm{bl}}/2$, as integer block delays can be incorporated by reindexing the blocks. Without loss of generality, assume that $0 = \epsilon_1 \le \epsilon_2 \le \cdots \epsilon_{N_{\mathrm{u}}} \le T_{\mathrm{bl}}/2$, and define $\epsilon_{\max} := \epsilon_{N_{\mathrm{u}}}$. The passband signal at the νth receiving element is the superposition of N_{u} waveforms,

$$\tilde{y}_\nu(t) = \sum_{\mu=1}^{N_{\mathrm{u}}} \tilde{y}_{\nu,\mu}(t - \epsilon_\mu) + \tilde{n}_\nu(t), \tag{15.7}$$

where $\tilde{y}_{\nu,\mu}(t)$ is defined in (15.5), and $\tilde{n}_\nu(t)$ denotes the ambient noise.

15.2 Overlapped Truncation and Interference Aggregation

15.2.1 Overlapped Truncation

To facilitate the decoding operation at the receiver, the received signal is usually truncated into individual processing units. For (quasi-)synchronous multiuser transmissions, the truncation can be easily carried out according to the block structure of the transmitted signal, as described in Chapters 13 and 14. However, for asynchronous transmissions, the block structure of the transmitted signal at the receiver is destroyed. As shown in Figure 15.2, a block from one user can collide with multiple blocks from other users. Different truncation methods could lead to different decoding schemes [156]. One existing method is to synchronize the truncation to the time-of-arrival of one desired user [198, 199, 238, 383], with each truncation having a block length T_{bl}, including one complete block from the desired user and partial blocks from other users. However, this method is not effective when the overlap length of the desired user and others is large.

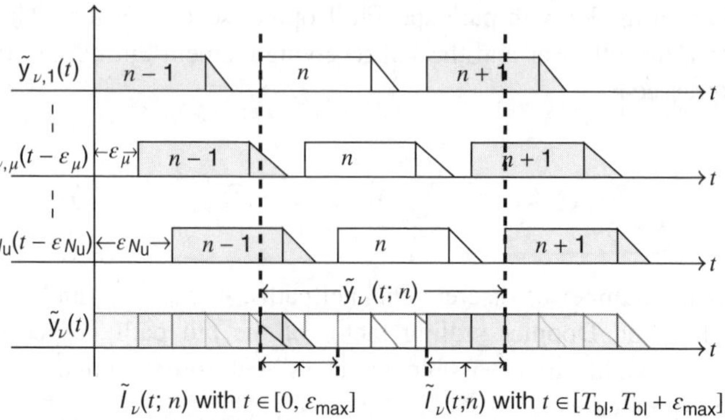

Figure 15.2 Illustration of the overlapped partition of the received signal and the aggregated interference in an asynchronous N_u-user system.

Figure 15.2 shows the overlapped truncation method proposed in [423], where each truncated block length is $\check{T}_{bl} := T_{bl} + \varepsilon_{max}$. The nth truncated block consists of the information from $(3N_u - 2)$ transmitted blocks, including:

(i) part of the $(n-1)$th blocks from users $2 \sim N_u$ at the beginning of this truncation;
(ii) complete information of the nth blocks from the N_u users;
(iii) part of the $(n+1)$th blocks from users $1 \sim (N_u - 1)$ at the end of this truncation.

The received signal within the nth truncation can be expressed as

$$\tilde{y}_\nu(t;n) = \underbrace{\sum_{\mu=1}^{N_u-1} \tilde{y}_{\nu,\mu}(t-\varepsilon_\mu - T_{bl}; n+1)}_{\text{interference from preceding processing units}} + \underbrace{\sum_{\mu=1}^{N_u} \tilde{y}_{\nu,\mu}(t-\varepsilon_\mu; n)}_{\text{desired OFDM signal}}$$
$$+ \underbrace{\sum_{\mu=2}^{N_u} \tilde{y}_{\nu,\mu}(t-\varepsilon_\mu + T_{bl}; n-1)}_{\text{interference from succeeding processing units}} + \tilde{n}_\nu(t;n), \qquad t \in [0, \check{T}_{bl}]. \tag{15.8}$$

With (15.8), an optimal receiver can be designed by treating the asynchronous N_u-user problem as a synchronous $(3N_u - 2)$-user problem [406]. However, solving the problem therein usually requires more efforts than solving a typical synchronous $(3N_u - 2)$-user problem, since the orthogonality of subcarriers of the $(2N_u - 2)$ misaligned users in (15.8) is destroyed.

15.2.2 Interference Aggregation

Rather than modeling the partial block interferences from $(2N_u - 2)$ users individually with the transmitted signal, an *interference aggregation* concept is used to treat the aggregated IBI as one interfering waveform as shown in Figure 15.2, which is formulated as

$$\tilde{I}_\nu(t;n) = \begin{cases} \sum_{\mu=2}^{N_u} \tilde{y}_{\nu,\mu}(t-\epsilon_\mu + T_{bl}; n-1), & t \in [0,\ \epsilon_{max}] \\ 0, & t \in [\epsilon_{max},\ T_{bl}] \\ \sum_{\mu=1}^{N_u-1} \tilde{y}_{\nu,\mu}(t-\epsilon_\mu - T_{bl}; n+1), & t \in [T_{bl},\ \breve{T}_{bl}]. \end{cases} \tag{15.9}$$

The received signal in (15.8) is reformulated as

$$\tilde{y}_\nu(t;n) = \sum_{\mu=1}^{N_u} \tilde{y}_{\nu,\mu}(t-\epsilon_\mu; n) + \tilde{I}_\nu(t;n) + \tilde{n}_\nu(t;n), \qquad t \in [0,\ \breve{T}_{bl}]. \tag{15.10}$$

Hence, the asynchronous N_u-user problem can be regarded as a synchronous N_u-user problem in the presence of an external interference.

Note that the bandwidth of the aggregated interference is taken as identical to that of the useful signal denoted by B. Given the maximum delay ϵ_{max}, one can see that the number of the degrees-of-freedom (DoF) of the interfering waveform, i.e., the time-bandwidth product of the aggregated interference $\lceil 2B\epsilon_{max} \rceil$, does not change as the number of asynchronous users increases.

15.3 An Asynchronous Multiuser OFDM Receiver

15.3.1 The Overall Receiver Structure

Leveraging the overlapped truncation method and the interference aggregation concept, an asynchronous multiuser reception approach was developed in [423] by performing a burst-by-burst decoding with interference cancellation.

Different from the external interference considered in Chapter 12, the time-domain input-output relationship in (15.8) shows that the interference term in (15.10) actually consists of part of useful signals corresponding to the $(n-1)$th and the $(n+1)$th transmitted blocks from the N_u users. If these blocks have been successfully decoded or estimates of transmitted symbols within these blocks are available, one can get initial estimates of the interferences that spill over from these blocks to the nth block, and thus take the estimates as the *a priori* knowledge of the interferences. After subtracting initial estimates of the interferences from the received signal, the joint multiuser decoding and cancellation of the aggregated residual interference can be performed in the nth block.

Prior to the interference subtraction, the passband-to-baseband downshifting and baseband lowpass filtering are performed. Based on the baseband samples, the burst-by-burst asynchronous multiuser receiver for each block consists of the following three steps.

(1) *Interference subtraction*: With the estimated interference passed from the preceding and the succeeding processing units, interference subtraction can be carried out prior to the multiuser decoding;
(2) *Joint multiuser processing with residual interference cancellation*: Techniques for synchronous multiuser decoding and approaches for external interference cancellation developed in Chapter 12 can be used;
(3) *Interference reconstruction*: Based on the multiuser decoding results, the interference of the current block to the preceding and the succeeding blocks will be reconstructed.

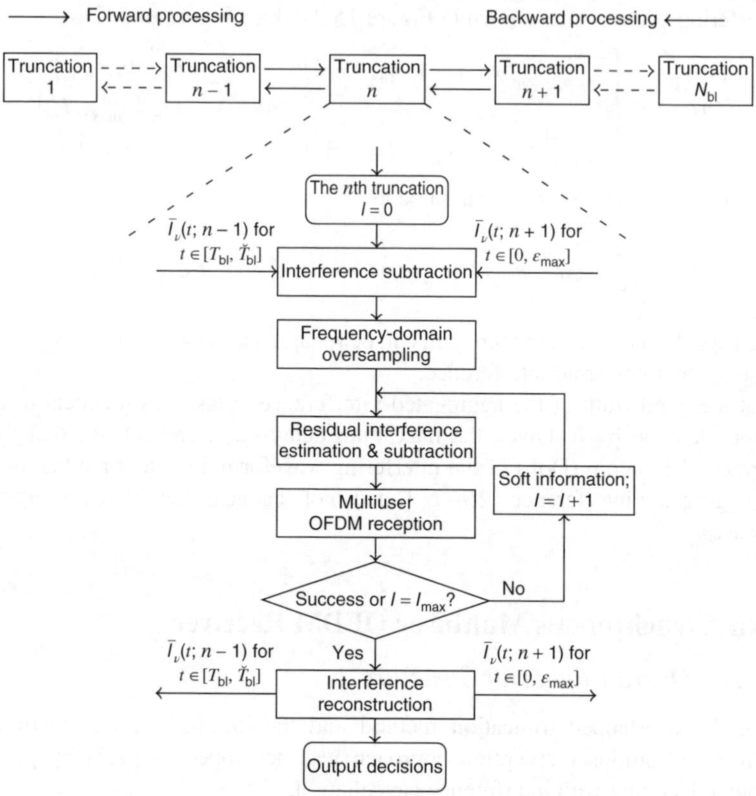

Figure 15.3 Illustration of the burst-by-burst asynchronous multiuser receiver with iterative forward/backward processing, with N_{bl} blocks in each burst.

To improve the interference cancellation performance, an iterative forward/backward processing of blocks within one burst can be performed, as shown in Figure 15.3. As the iteration goes on, the accuracy of interference estimation improves gradually, thus leading to a better multiuser decoding performance, which in turn boosts the performance of interference estimation. Compared with a synchronous multiuser receiver, the burst-by-burst asynchronous multiuser receiver can only perform multiuser decoding after receiving the whole burst from all users, hence incurring a processing latency. Similar to the Viterbi algorithm for channel equalization, by using the *sliding block* techniques proposed in, e.g. [256, 408], the latency can be reduced with a batch-by-batch processing, and the batch size depends on the maximal tolerant latency and the storage capability of the receiver. When the batch size is one, the burst-by-burst iterative receiver degrades to a *block-by-block* receiver.

The detailed descriptions on the receiver modules are provided next.

15.3.2 Interblock Interference Subtraction

Define $y_\nu(t;n)$ and $I_\nu(t;n)$ as the baseband waveforms corresponding to the passband signals $\tilde{y}_\nu(t;n)$ and $\tilde{I}_\nu(t;n)$, respectively. Denote $\bar{I}_\nu(t;n)$ as the reconstructed time-domain interference waveform in baseband (c.f. Section 15.3.5 for details on interference reconstruction). Prior to the receiver processing, subtraction of the initial interference estimate from $y_\nu(t;n)$ in (15.10) leads to

$$\bar{y}_\nu(t;n) = \sum_{\mu=1}^{N_u} y_{\nu,\mu}(t-\varepsilon_\mu;n) + \underbrace{(I_\nu(t;n) - \hat{I}_\nu(t;n))}_{:=\Upsilon_\nu(t;n)} + n_\nu(t;n) \quad (15.11)$$

$$= \sum_{\mu=1}^{N_u} y_{\nu,\mu}(t-\varepsilon_\mu;n) + \Upsilon_\nu(t;n) + n_\nu(t;n), \qquad t \in [0,\ \breve{T}_{\rm bl}] \quad (15.12)$$

where $\Upsilon_\nu(t;n)$ denotes the residual interference. As the residual interference has an identical time duration and bandwidth as the interference $I_\nu(t;n)$, the number of DoF of $\Upsilon_\nu(t;n)$ is also $\lceil 2B\varepsilon_{\max}\rceil$. Note that interference subtraction is performed on the baseband discrete samples, although continuous-time expressions are used in (15.11).

To facilitate the residual interference estimation, an interference parameterization method in Chapter 12 is used. Define the time-bandwidth product $N_{\rm I} := \lceil B\varepsilon_{\max}\rceil$. The residual interference in baseband can be approximated by the Fourier series expansion

$$\Upsilon_\nu(t;n) \approx \begin{cases} \sum_{\ell=-N_{\rm I}/2}^{N_{\rm I}/2-1} c_{\ell,\nu,{\rm he}} e^{j2\pi\frac{\ell}{\varepsilon_{\max}}t}, & t \in [0,\ \varepsilon_{\max}] \\ 0, & t \in [\varepsilon_{\max},\ T_{\rm bl}] \\ \sum_{\ell=-N_{\rm I}/2}^{N_{\rm I}/2-1} c_{\ell,\nu,{\rm ta}} e^{j2\pi\frac{\ell}{\varepsilon_{\max}}t}, & t \in [T_{\rm bl},\ \breve{T}_{\rm bl}] \end{cases} \quad (15.13)$$

where $\{c_{\ell,\nu,{\rm he}}\}$ and $\{c_{\ell,\nu,{\rm ta}}\}$ represent the Fourier series coefficients of the front and end portions of the interference, respectively.

In the initial forward *block-to-block* decoding of the burst-by-burst receiver, the *a priori* knowledge of the interference from the subsequent block is not available. Hence, we set it as zero at the beginning. Once all the blocks have been processed, the initial estimates of the interference spilled over from both neighboring blocks are available. During the following processing, the latest estimates of the decoded blocks are used for interference cancellation.

15.3.3 Time-to-Frequency-Domain Conversion

Note that the interference subtraction is performed in the time domain, while multiuser decoding operates in the frequency domain. An information-lossless transformation of signals between the two domains is thus necessary for iterative processing. To this end, the frequency-domain oversampling approach developed in Section 5.2.3 is used.

Define α as the frequency-domain oversampling factor. The mth frequency component of the time-domain signal $\bar{y}_\nu(t;n)$ can be obtained via

$$z_\nu[m;n] = \int_0^{\breve{T}_{\rm bl}} \bar{y}_\nu(t;n) e^{-j2\pi\frac{m}{\alpha T}t} dt, \quad (15.14)$$

for $m = -\alpha K/2, \cdots, \alpha K/2 - 1$. This can be expanded as

$$z_\nu[m;n] = \sum_{\mu=1}^{N_t} \sum_{k=-K/2}^{K/2-1} H_{\nu,\mu}[m,k;n] s_\mu[k] + \Upsilon_\nu[m;n] + w_\nu[m;n] \tag{15.15}$$

where $H_{\nu,\mu}[m,k;n]$ is the channel coefficient which specifies the contribution of the symbol on the kth subcarrier of the μth transmitter to the mth subcarrier of the νth receiver, $w_\nu[m;n]$ is the ambient noise in the frequency domain, and $\Upsilon_\nu[m;n]$ is the frequency component of the residual interference $\tilde{\Upsilon}_\nu(t;n)$. The mth frequency component of the residual interference $\Upsilon_\nu(t;n)$ can be formulated as

$$\Upsilon_\nu[m;n] = \int_0^{\check{T}_{\rm bl}} \Upsilon_\nu(t;n) e^{-j2\pi \frac{m}{\alpha T} t} dt. \tag{15.16}$$

Substituting (15.13) into (15.16) yields

$$\Upsilon_\nu[m;n] = (u_{\ell,\nu,{\rm he}} + u_{\ell,\nu,{\rm ta}} e^{-j2\pi \frac{m}{\alpha T} T_{\rm bl}}) \rho_{m,\ell} \tag{15.17}$$

with

$$u_{\ell,\nu,{\rm he}} := \varepsilon_{\max} c_{\nu,\ell,{\rm he}}, \qquad u_{\ell,\nu,{\rm ta}} := \varepsilon_{\max} c_{\nu,\ell,{\rm ta}}$$

and

$$\rho_{m,\ell} := \frac{\sin\left(\pi \left(\frac{m}{\alpha T} - \frac{\ell}{\varepsilon_{\max}}\right) \varepsilon_{\max}\right)}{\pi \left(\frac{m}{\alpha T} - \frac{\ell}{\varepsilon_{\max}}\right) \varepsilon_{\max}} e^{-\pi\left(\frac{m}{\alpha T} - \frac{\ell}{\varepsilon_{\max}}\right)\varepsilon_{\max}}.$$

Stack the frequency components $z_\nu[m;n]$, $\boldsymbol{\Upsilon}_\nu[m;n]$ and $w_\nu[m;n]$ into vectors $\mathbf{z}_\nu[n]$, $\Upsilon_\nu[n]$ and $\mathbf{w}_\nu[n]$ of size $\alpha K \times 1$, respectively. Stack $H_{\nu,\mu}[m,k;n]$ into a matrix of size $\alpha K \times K$. Denote $\mathbf{s}_\mu[n]$ as the transmitted symbol vector of size $K \times 1$ from the μth user. Stack $u_{\ell,\nu,{\rm he}}$ and $u_{\ell,\nu,{\rm ta}}$ into vectors $\mathbf{u}_{\nu,{\rm he}}$ and $\mathbf{u}_{\nu,{\rm ta}}$ of size $N_{\rm I} \times 1$, respectively, and stack $\rho_{m,\ell}$ into a matrix $\boldsymbol{\Gamma}$ of size $\alpha K \times N_{\rm I}$, and define

$$\boldsymbol{\Psi} := \left[\boldsymbol{\Gamma},\ \boldsymbol{\Lambda}(T_{\rm bl})\boldsymbol{\Gamma}\right], \qquad [\boldsymbol{\Gamma}]_{m,\ell} := \rho_{m,\ell}, \qquad \mathbf{u}_\nu[n] := \left[\mathbf{u}_{{\rm he},\nu}^{\rm T},\ \mathbf{u}_{{\rm ta},\nu}^{\rm T}\right]^{\rm T}. \tag{15.18}$$

Define a generic diagonal matrix,

$$[\boldsymbol{\Lambda}(\tau)]_{m,m} = e^{-j2\pi \frac{m}{\alpha T} \tau}. \tag{15.19}$$

The input–output relationship in (15.15) can be compactly expressed as

$$\mathbf{z}_\nu[n] = \sum_{\mu=1}^{N_{\rm u}} \boldsymbol{\Lambda}(\varepsilon_\mu) \mathbf{H}_{\nu,\mu}[n] \mathbf{s}_\mu[n] + \boldsymbol{\Upsilon}_\nu[n] + \mathbf{w}_\nu[n] \tag{15.20}$$

where the residual interference vector is compactly expressed as

$$\boldsymbol{\Upsilon}_\nu[n] = \boldsymbol{\Psi} \mathbf{u}_\nu[n]. \tag{15.21}$$

Note that using the baseband sampling rate $B = K/T$, there are $\lceil \check{T}_{\rm bl} B \rceil = \lceil \check{T}_{\rm bl} K/T \rceil$ samples in each processing unit. To avoid information loss during the Fourier transform, the frequency-domain oversampling factor α should satisfy $\alpha K \geq \check{T}_{\rm bl} K/T$, i.e., $\alpha T \geq \check{T}_{\rm bl}$. Based on $\varepsilon_{\max} < T_{\rm bl}/2$, it requires that $\alpha \geq 3(1 + T_g/T)/2$. Taking $\alpha = 2$ for example, the guard interval should satisfy $T_{\rm g} \leq T/3$.

15.3.4 Iterative Multiuser Reception and Residual Interference Cancellation

Based on the frequency-domain input–output relationship in (15.20), an iterative receiver as in Chapter 12 can be used for joint multiuser decoding and residual interference cancellation. The iterative receiver structure is shown in Figure 15.3. The iterative receiver has two steps as follows.

Step 1: Residual interference estimation/subtraction
With the estimated channel matrices $\{\hat{\mathbf{H}}_{\nu,\mu}[n]\}$ and transmitted symbols $\{\hat{\mathbf{s}}_{\mu}[n]\}$ from the last iteration, the frequency-domain interference measurements after subtracting the OFDM components are

$$\boldsymbol{\xi}_{\nu}[n] = \mathbf{z}_{\nu}[n] - \sum_{\mu=1}^{N_{\rm u}} \boldsymbol{\Lambda}(\varepsilon_{\mu})\hat{\mathbf{H}}_{\nu,\mu}[n]\hat{\mathbf{s}}_{\mu}[n] \tag{15.22}$$

$$= \boldsymbol{\Upsilon}_{\nu}[n] + \tilde{\mathbf{w}}_{\nu}[n] \tag{15.23}$$

where the noise term $\tilde{\mathbf{w}}_{\nu}[n]$ is formed by the ambient noise, and the channel and information symbol estimation errors

$$\tilde{\mathbf{w}}_{\nu}[n] := \mathbf{w}_{\nu}[n] + \sum_{\mu=1}^{N_{\rm u}} \boldsymbol{\Lambda}(\varepsilon_{\mu})(\hat{\mathbf{H}}_{\nu,\mu}[n]\hat{\mathbf{s}}_{\mu}[n] - \mathbf{H}_{\nu,\mu}[n]\mathbf{s}_{\mu}[n]). \tag{15.24}$$

Based on the measurements in (15.22) and the interference parameterization in (15.21), the least-squares estimate of the interference vector is

$$\hat{\mathbf{u}}_{\nu}[n] = (\boldsymbol{\Psi}^{\rm H}\boldsymbol{\Psi})^{-1}\boldsymbol{\Psi}^{\rm H}\boldsymbol{\xi}_{\nu}[n]. \tag{15.25}$$

Note that $(\boldsymbol{\Psi}^{\rm H}\boldsymbol{\Psi})^{-1}\boldsymbol{\Psi}^{\rm H}$ only depends on $\varepsilon_{\max}$, hence can be pre-computed before receiver processing.

Step 2: Multiuser channel estimation and data decoding
The desired OFDM component is obtained by subtracting residual interference from the frequency measurements,

$$\bar{\mathbf{z}}_{\nu}[n] = \mathbf{z}_{\nu}[n] - \boldsymbol{\Psi}\hat{\mathbf{u}}_{\nu}[n] = \sum_{\mu=1}^{N_{\rm u}} \boldsymbol{\Lambda}(\varepsilon_{\mu})\mathbf{H}_{\nu,\mu}[n]\mathbf{s}_{\mu}[n] + \check{\mathbf{w}}_{\nu}[n], \tag{15.26}$$

where the equivalent noise term consists of both the ambient noise and the interference estimation error,

$$\breve{\mathbf{w}}_\nu[n] := \boldsymbol{\Psi}(\mathbf{u}_\nu[n] - \hat{\mathbf{u}}_\nu[n]) + \mathbf{w}_\nu[n]. \tag{15.27}$$

With $\{\bar{\mathbf{z}}_\nu[n]\}_{\nu=1}^{N_r}$, the receiver processing is then carried out similar to that for co-located MIMO OFDM in Chapter 13. For example, to reduce the receiver computational complexity, a band-limited intercarrier interference (ICI)-leakage assumption is adopted, i.e.,

$$H_{\nu,\mu}[m,k;n] \approx \begin{cases} H_{\nu,\mu}[m,k;n], & \left|\frac{m}{\alpha} - k\right| \leq D \\ 0, & \text{otherwise.} \end{cases} \tag{15.28}$$

where D is defined as the ICI depth. The sparse channel estimation in Section 7.5 and the factor-graph based MMSE equalization in Section 8.4.3 for a given D can be used with minor modifications. Both hard and soft decisions on the information symbols can be obtained at the decoder output, these being fed back for the residual interference estimation, channel estimation and symbol detection in the next iteration.

Once the parity check conditions of the channel decoders of all users are satisfied, or the number of iterations reaches a predetermined threshold $I_{\max}$, the iteration stops. Similar to the iterative interference cancellation receiver in Chapter 12, the iterative receiver is initialized by first treating the residual interference as the ambient noise when getting the initial estimates of channel matrices and information symbols, which are then used to initialize the iterative operation.

15.3.5 Interference Reconstruction

Once the processing of the nth block stops, with the estimated channel coefficients $\{\hat{H}_{\nu,\mu}[m,k;n]\}$ and the information symbols $\hat{s}_\mu[k]$, the time-domain OFDM waveform in baseband can be reconstructed via the inverse discrete-time Fourier transform

$$\hat{y}_{\nu,\mu}(t;n) = \sum_{m=-\alpha K/2}^{\alpha K/2-1} \sum_{k=-K/2}^{K/2-1} \hat{H}_{\nu,\mu}[m,k;n]\hat{s}_\mu[k]e^{j2\pi \frac{m}{\alpha T}t}, \tag{15.29}$$

for $t \in [0, T_{\rm bl}]$. Based on the aggregated interference representation in (15.9), estimates of the interference components in $I_\nu(t;n-1)$ and $I_\nu(t;n+1)$, which are spilled over from the nth block to the $(n-1)$th and the $(n+1)$th blocks, respectively, can be obtained by replacing $y_{\nu,\mu}(t;n)$ by $\hat{y}_{\nu,\mu}(t;n)$ in (15.9),

$$\hat{I}_\nu(t;n-1) = \begin{cases} \sum_{\mu=2}^{N_{\rm u}} \hat{y}_{\nu,\mu}(t-\varepsilon_\mu + T_{\rm bl}; n-2), & t \in [0, \varepsilon_{\max}] \\ 0, & t \in [\varepsilon_{\max}, T_{\rm bl}] \\ \sum_{\mu=1}^{N_{\rm u}-1} \hat{y}_{\nu,\mu}(t-\varepsilon_\mu - T_{\rm bl}; n), & t \in [T_{\rm bl}, \breve{T}_{\rm bl}]. \end{cases} \tag{15.30}$$

The estimate $\hat{I}_\nu(t;n+1)$ can be similarly computed. These estimates are then passed to the preceding and the succeeding processing units.

15.4 Investigation on Multiuser Asynchronism in an Example Network

As shown in (15.9) and (15.13), the number of degrees of freedom of the interference is decided by its time-bandwidth product $\lceil 2B\varepsilon_{\max}\rceil$. This section provides the analysis of the time duration of the interference in an example network with one data collection unit and multiple sensors, which operates in a *collision-tolerant* fashion by allowing simultaneous transmissions from N_u sensors. For simplicity, assume that the receiver and sensors are at the same water depth, the sensors are uniformly distributed within a circle of diameter D_N, and the receiver is located at the origin.

Suppose that the network operates according to a multiple-access control (MAC) protocol with handshaking. The data collection unit first broadcasts the *clear-to-send* (CTS) frame to allow simultaneous transmissions of N_u active sensors which requested to send packets. Once receiving the CTS frame, each sensor starts the data transmission.

Let d_i denote the distance between the ith active sensor and the receiver. Based on the uniform distribution of the ith sensor within the circle, the probability density function (pdf) of d_i will satisfy

$$f(d_i) \propto 2\pi d_i, \tag{15.31}$$

which leads to the expression:

$$f(d_i) = \frac{8d_i}{D_N^2}, \qquad \text{for } d_i \in [0,\ D_N/2]. \tag{15.32}$$

Assuming that the acoustic waveform propagates along a straight line, the time-of-arrival of the data burst from the ith sensor is thus $\tilde{\varepsilon}_i = 2d_i/c$, with the pdf,

$$g(\tilde{\varepsilon}_i) = \frac{2c^2\tilde{\varepsilon}_i}{D_N^2}, \qquad \text{for } \tilde{\varepsilon}_i \in [0,\ D_N/c] \tag{15.33}$$

where c is the sound speed in water.

Notice that for the block transmissions, the time-of-arrival of the burst $\tilde{\varepsilon}_i$ and the time-of-arrival of each block within the burst $\breve{\varepsilon}_i$ is related via

$$\breve{\varepsilon}_i = [\tilde{\varepsilon}_i]_{\text{mod } T_{\text{bl}}/2} = [2d_i/c]_{\text{mod } T_{\text{bl}}/2}. \tag{15.34}$$

Take $T_{\text{bl}} = 200$ ms and $c = 1500$ m/s as an example. For $d_i = 15$ m, $d_i = 45$ m, and $d_i = 75$ m, one can verify that $\breve{\varepsilon}_i = 20$ ms, $\breve{\varepsilon}_i = 60$ ms, and $\breve{\varepsilon}_i = 0$ ms, respectively. Therefore, the pdf of $\breve{\varepsilon}_i$ is the folded summation of $g(\tilde{\varepsilon}_i)$, with

$$f(\breve{\varepsilon}_i) = \sum_{\ell=0}^{L} g\left(\breve{\varepsilon}_i + \frac{\ell T_{\text{bl}}}{2}\right) = \sum_{\ell=0}^{L} \frac{8(\breve{\varepsilon}_i + \ell T_{\text{bl}}/2)}{D_N^2}, \qquad \text{for } \breve{\varepsilon}_i \in [0,\ T_{\text{bl}}/2] \tag{15.35}$$

where $L = \lceil \frac{2D_N}{T_{\text{bl}}c} \rceil$, and the cumulative distribution function (cdf) of $\breve{\varepsilon}_i$ follows as

$$F(\breve{\varepsilon}_i) = \sum_{\ell=0}^{L} \frac{4(\breve{\varepsilon}_i^2 + \ell T_{\text{bl}})}{D_N^2}, \qquad \text{for } \breve{\varepsilon}_i \in [0,\ T_{\text{bl}}/2]. \tag{15.36}$$

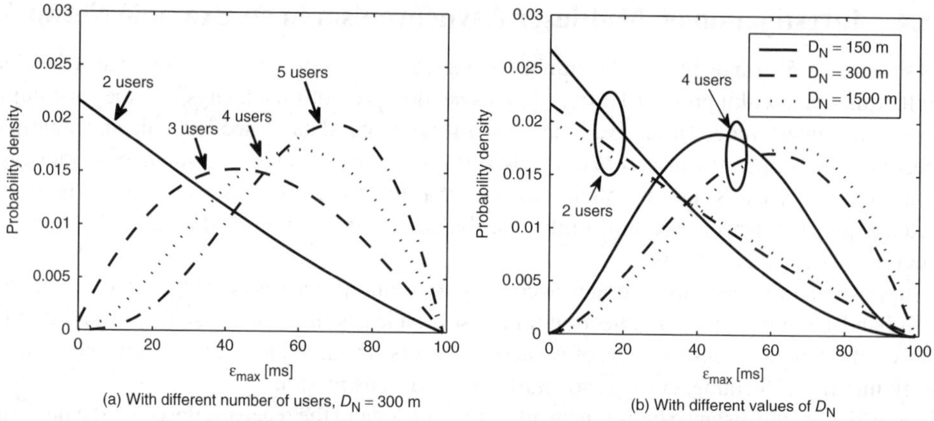

Figure 15.4 Probability density function of the maximum asynchronism on the OFDM block level in an asynchronous multiuser system, where the users are uniformly distributed within a circle of diameter D_N.

Assume that the arrival times of N_u users follow an independent and identical distribution with the pdf $f(\breve{\varepsilon})$ and cdf $F(\breve{\varepsilon})$. The maximum delay ε_{max} is the *range* of the arrival-time sequence [94],

$$\varepsilon_{max} = \max\{\breve{\varepsilon}_1, \breve{\varepsilon}_2, \cdots, \breve{\varepsilon}_{N_u}\} - \min\{\breve{\varepsilon}_1, \breve{\varepsilon}_2, \cdots, \breve{\varepsilon}_{N_u}\}. \tag{15.37}$$

which has the pdf expressed as [94]

$$f_{N_u}(\varepsilon_{max}) = N_u(N_u - 1)\int_{-\infty}^{\infty} f(\breve{\varepsilon})[F(\breve{\varepsilon} + \varepsilon_{max}) - F(\breve{\varepsilon})]^{N_u-2} f(\breve{\varepsilon} + \varepsilon_{max})d\breve{\varepsilon}. \tag{15.38}$$

Substituting (15.35) and (15.36) into (15.38), the distribution of the interference time duration ε_{max} can be obtained.

Using the numerical integration, the pdf of ε_{max} corresponding to different number of users is shown in Figure 15.4, where D_N takes integer multiples of cT_{bl}. One can see that as the number of users increases, the pdf shifts to the large value region of ε_{max} gradually. A similar trend happens as the diameter D_N increases, but the pdf shifts quite slowly.

15.5 Simulation Results

In simulation, the underwater acoustic channel between the transmitter of each user and each receiving element during each block transmission is generated randomly according to the specifications in Section 5.5.1. The channel parameters are $N_{pa} = 10$, $\Delta\tau = 1$ ms, $\Delta P_{pa} = 20$ dB, $T_g = 24.6$ ms, and $v_0 = 0$ m/s. Each transmission burst has $N_{bl} = 10$ blocks.

The ZP-OFDM parameters in the SPACE08 experiment are used for simulation, which are listed in Table B.1. All the users share an identical set of pilot subcarriers. The pilot symbols are drawn randomly from a QPSK constellation, and different users have different pilot symbol sets. The data symbols are encoded with a rate-1/2 nonbinary LDPC code and modulated by a QPSK constellation, which leads to a data rate of each user

$$R = \frac{1}{2} \cdot \frac{|S_D|}{T + T_g} \cdot \log_2 4 = 5.2\,\text{kb/s/user}. \tag{15.39}$$

Throughout this chapter, the block-error rate (BLER) averaged over all users is used as the performance metric. A frequency-domain oversampling factor $\alpha = 2$ is used.

The decoding performance of four receiver processing configurations will be compared.

- *Configuration 1*: A block-by-block multiuser reception: By treating the interference as ambient noise, the iterative multiuser decoding techniques for co-located MIMO in Chapter 13 are used;
- *Configuration 2*: A block-by-block multiuser reception with interference cancellation: By treating the interference as an external interference, the iterative joint multiuser processing and interference cancellation in Section 15.3.4 is performed;
- *Configuration 3*: A block-to-block receiver with forward interference cancellation: After the interference subtraction based on the interference estimate from the preceding block, iterative joint multiuser decoding and residual interference cancellation are performed, as in Section 15.3.4;
- *Configuration 4*: The burst-by-burst receiver with multiple rounds of forward and backward processing.

For fairness of comparison, the frequency-domain oversampling is used for all the configurations, and an identical iteration number threshold of $I_{\max} = 4$ is used in each block processing. Four rounds of forward/backward *block-to-block* processing in configuration 4 are used. In configuration 4, there are 10 blocks within one burst processed in one batch, while for configurations 1 $\sim$ 3, the batch size can be regarded as one.

In terms of the decoding complexity, one can see that configuration 1 has the lowest complexity, the complexities of configurations 2 and 3 are similar, and configuration 4 has about eight times of the complexity of configuration 3 due to the iterative forward and backward processing. Meanwhile, configurations 1 $\sim$ 3 are capable of on-line processing without decoding latency, while configuration 4 suffers a decoding latency of the burst length.

15.5.1 Two-User Systems with Time-Varying Channels

To explore the receiver performance in the time-varying UWA channels, the Doppler rate of each path independently is drawn from a *zero-mean* uniform distribution with standard deviation (std) σ_v m/s according to the path-based channel model in (15.4). To achieve a good decoding performance, the ICI incurred by the channel variation is considered explicitly. For the sake of receiver complexity, a *band-limitedness* assumption of the channel matrix is adopted by assuming the ICI depth $D = 1$.

For ICI estimation with regularly distributed pilots, a progressive decoding procedure described in Section 13.6 is employed. During the iterative processing, the receiver assumes the absence of ICI to get an initial estimate of the transmitted information symbols. Coupled with pilots, the information symbol estimates are then used in the following iterations for channel estimation.

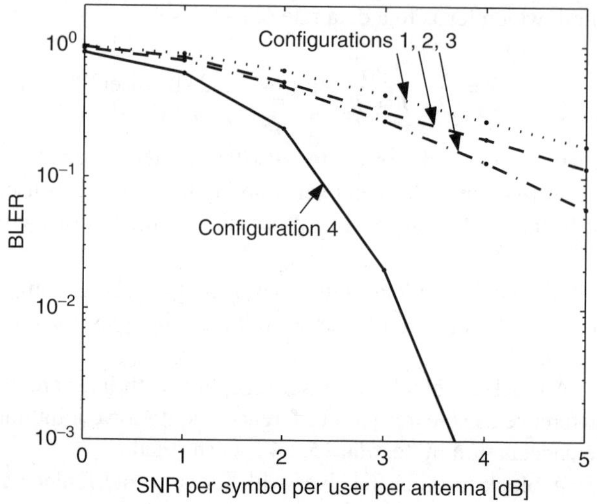

Figure 15.5 Block-error-rate performance of four receiving configurations, $\sigma_\mathrm{v} = 0.1$ m/s.

Assume that three receiving elements at the receiver and that the relative delay of the second user is uniformly distributed within the interval $[0,\ T_{\mathrm{bl}}/2]$. In the channel with a mild Doppler spread $\sigma_\mathrm{v} = 0.1$ m/s, Figure 15.5 shows the BLER performance of the four receiving configurations with different signal-to-noise (SNR) levels. One can see that the conventional receiver without interference cancellation almost cannot work, that the block-by-block interference cancellation brings some performance improvement, and that the burst-by-burst receiver has the best performance. Relative to the one-way message passing in the block-to-block receiver, the two-way message passing in the burst-by-burst receiver improves the decoding performance considerably.

Corresponding to a mild Doppler spread with $\sigma_\mathrm{v} = 0.1$ m/s, Figure 15.6 shows the BLER performance of the burst-by-burst receiver when the relative delay of the second user is uniformly distributed within five consecutive intervals: $[0,\ 0.1]T_{\mathrm{bl}}$, $[0.1,\ 0.2]T_{\mathrm{bl}}$, $[0.2,\ 0.3]T_{\mathrm{bl}}$, $[0.3,\ 0.4]T_{\mathrm{bl}}$ and $[0.4,\ 0.5]T_{\mathrm{bl}}$. One can see that as the relative delay of the second user, i.e., the time duration of the interference increases, the required SNR for successful decoding of the two data streams also increases.

With three receiving elements, the BLER performance of the four receiving configurations in the channel with severe Doppler spreads $\sigma_\mathrm{v} = 0.3$ m/s and $\sigma_\mathrm{v} = 0.5$ m/s are shown in Figure 15.7. Similar to the observation in the channel with a mild Doppler spread, the burst-by-burst receiver with forward and backward message passing outperforms other configurations considerably.

Assuming two receiving elements, the BLER performance of the conventional receiver and the burst-by-burst receiver with and without perfect channel knowledge is shown in Figure 15.8. Relative to the scenario with channel estimation, one can see that the performance gap between the two receiving schemes decreases when perfect channel knowledge is available. One could infer that channel estimation accuracy degrades drastically if the interblock interference is not accounted for.

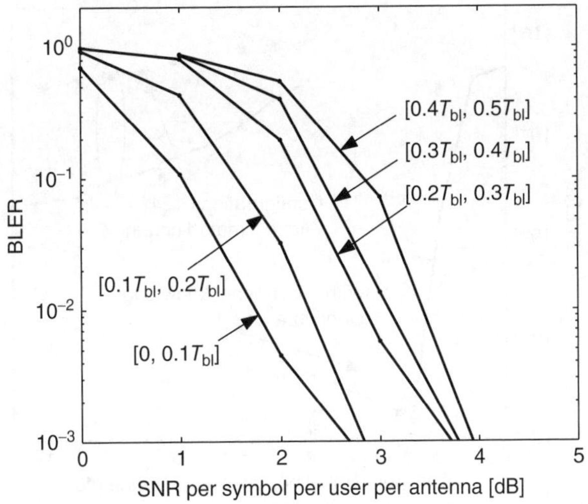

Figure 15.6 Block-error-rate performance of the burst-by-burst iterative receiver in a two-user system with different relative delays, $\sigma_{\text{v}} = 0.1$ m/s; three receiving elements.

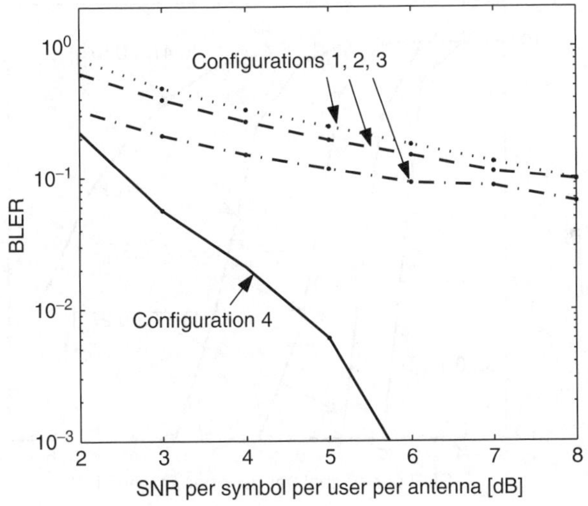

Figure 15.7 Block-error-rate performance of four receiving configurations, solid lines: $\sigma_{\text{v}} = 0.3$ m/s; dashed lines: $\sigma_{\text{v}} = 0.5$ m/s.

15.5.2 Multiuser Systems with Time-Invariant Channels

In the time-invariant channel, the channel matrix $\mathbf{H}_{\nu,\mu}$ satisfies

$$H_{\nu}[m, k; n] \approx \begin{cases} H_{\nu}[m, k; n], & m/\alpha = k \\ 0, & \text{otherwise.} \end{cases} \tag{15.40}$$

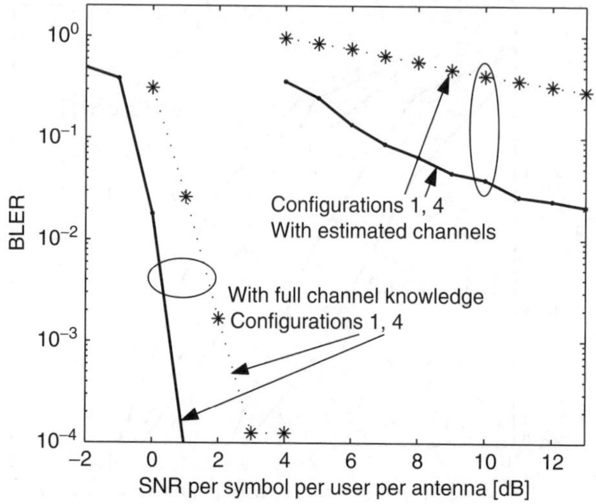

Figure 15.8 Block-error-rate performance of two receiving configurations with and without perfect channel knowledge, two receiving elements are used, and $\sigma_v = 0.3$ m/s.

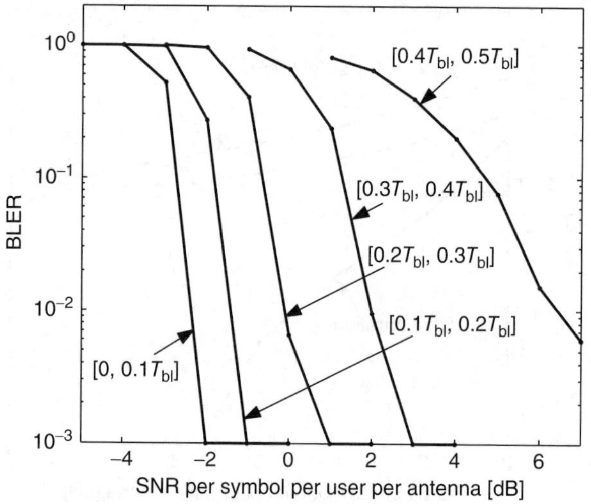

Figure 15.9 Block-error-rate performance of the burst-by-burst iterative receiver in a four-user system with different relative delays, six receiving elements.

The ICI-ignorant receiver processing is adopted. To examine the performance of the burst-by-burst receiver as a function of number of users, we set $\varepsilon_1 = 0$, and assume that the relative delay of the second user is uniformly distributed within a certain interval, and that the delays of users $3 \sim N_u$ are uniformly distributed between zero and the upper bound of this interval. By dividing half of the OFDM block duration $[0, T_{bl}/2]$ into five intervals: $[0, 0.1]T_{bl}$, $[0.1, 0.2]T_{bl}$, $[0.2, 0.3]T_{bl}$, $[0.3, 0.4]T_{bl}$ and $[0.4, 0.5]T_{bl}$, Figures 15.9 and 15.10

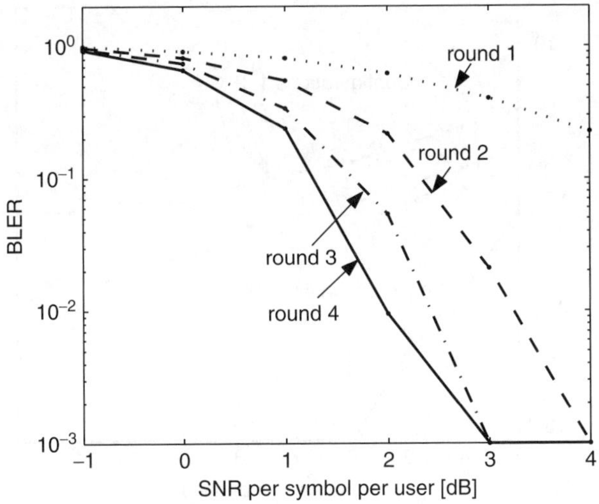

Figure 15.10 Block-error-rate performance of the burst-by-burst iterative receiver with four rounds of forward/backward processing, $\varepsilon_{\max} \sim \mathcal{U}[0.3, 0.4] \times T_{\mathrm{bl}}$, four users, six receiving elements.

show the BLER performance of the burst-by-burst receiver with four asynchronous users and six elements at the receiver. Relative to the BLER performance in the two-user scenario in Figure 15.6, one can see that as the number of users increases, the maximum delay of the users, i.e., the time duration of the interference has more impact on the decoding performance. Meanwhile, one can also observe a considerable performance improvement brought by the iterative forward and backward message passing.

15.6 Emulated Results: MACE10

The data sets collected in the MACE10 experiment are used to test the burst-by-burst receiver for asynchronous N_u-user transmissions; detailed description about the MACE10 experiment can be found in Appendix B. Within the data set, a rate-1/2 nonbinary LDPC code and a QPSK constellation for information bit encoding and mapping are used, leading to a data rate of each user

$$R = \frac{1}{2} \cdot \frac{|S_{\mathrm{D}}|}{T + T_{\mathrm{g}}} \cdot \log_2 4 = 2.7 \quad \text{kb/s/user}. \tag{15.41}$$

The data sets from an asynchronous multiuser system with N_{u} users are emulated as follows. First consecutively divide the received data blocks within each transmission into N_{u} groups and then add the N_{u} groups together, regarding the blocks within each group as the signal from each user. A resampling operation is used to remove the Doppler effect caused by the mobility of the source array before the summation. The emulated system is assumed time-invariant; the channel matrix therefore satisfies (15.40) with ICI depth $D = 0$. Note that due to the existence of ambient noise in the received data blocks, the SNR per user in the emulated data set decreases according to $10\log\,(N_{\mathrm{u}})$ dB as the number of users increases.

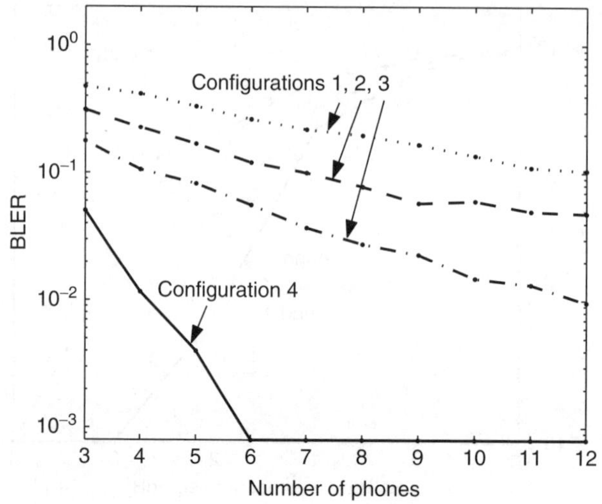

Figure 15.11 Emulated results: Block-error-rate performance of four receiving configurations.

Similar to the simulation setup, the relative delay of each user is uniformly distributed within a certain interval. By setting the distribution interval corresponding to the second user as $[0,\ T_{\mathrm{bl}}/2]$, Figure 15.11 shows the decoding performance of four configurations in simulations, and four rounds of forward/backward *block-to-block* processing in configuration 4 are used. Again, one can see that the conventional multiuser reception approach without interference cancellation almost cannot work, the block-by-block interference cancellation method improves the performance a bit, and the burst-by-burst receiver with the forward and backward message passing is the best.

With different distribution intervals of the relative delay, the BLER performance of the burst-by-burst receiver with two asynchronous users is shown in Figure 15.12. One can see that as the relative delay increases, i.e., the time duration of the interference increases, more receiving elements are required for successful decoding.

Assuming that the relative delays of users are uniformly distributed within $[0,\ T_{\mathrm{bl}}/2]$, Figure 15.13 shows the packet-success rate of the burst-by-burst receiver with different number of users. One can see that as the number of users increases, the decoding performance gets worse gradually. The degradation can be attributed to the increased multiuser interference and the increased ambient noise power due to the generation of the emulated data sets.

To achieve a robust decoding performance of the burst-by-burst receiver, a block-level Reed-Solomon (RS) erasure-correction code over Galois field can be used as an interblock code while the nonbinary LDPC code is used as an intra-block code [247]; the optimal combination between the erasure- and error-correction codes for the layered coding approach has been studied in [41]. With a rate-8/10 block-level *shortened* RS codeword applied for each data subcarrier across a packet consisting of 10 blocks, any two blocks can be received in error (hence erased), while the whole data burst can be recovered. Here, a shortened RS code can be obtained by setting some information symbols as zeros from a RS code of longer length [247]. Figure 15.13 shows the packet-success rate of the burst-by-burst receiver with different number of users. Compared with the packet-success rate without using erasure-correction

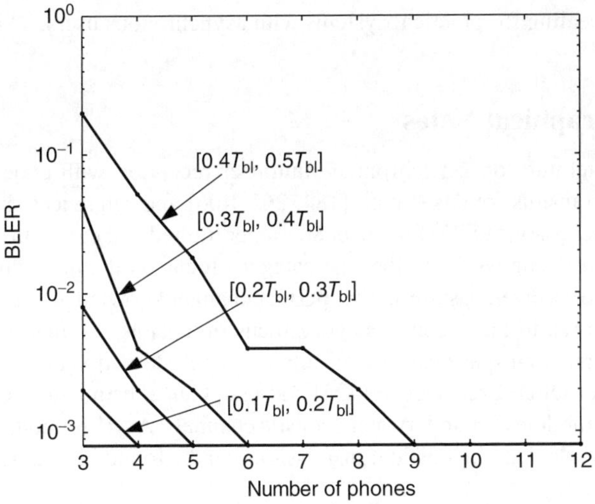

Figure 15.12 Emulated results: Block-error-rate performance of the burst-by-burst receiver with different relative delays, four rounds of forward and backward processing and eight iterations within each block processing.

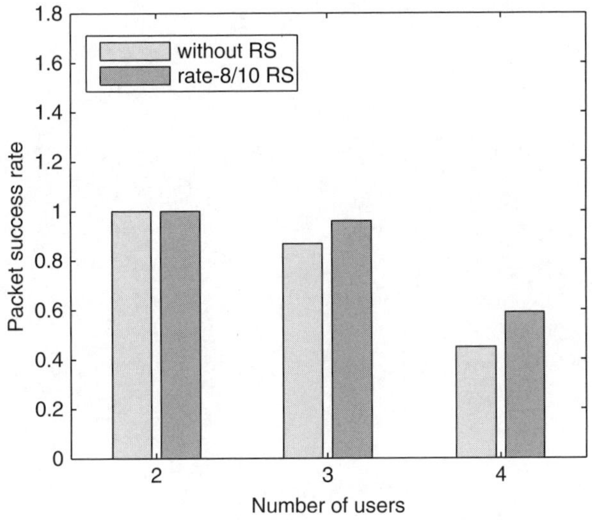

Figure 15.13 Emulated results: Packet-success rate of the burst-by-burst receiver with different number of users with and without a rate-8/10 Reed-Solomon erasure-correction code across 10 blocks, four rounds of forward/backward processing, eight iterations within each block processing.

coding, introducing two redundant blocks leads to a considerable performance improvement. This option is appealing for practical systems with asynchronous users.

15.7 Bibliographical Notes

There is a rich literature on asynchronous multiuser reception with code-division multiple access (CDMA) transmissions; see e.g., [184, 265, 406] and references therein. The related works on the asynchronous OFDM receiver are rather limited and can be broadly grouped into two main categories. The works in the first category focus on the demodulation and decoding modules of the receiver, assuming that perfect channel knowledge is available [198, 238, 383, 386]. The works in the second category focus on channel estimation in asynchronous OFDM systems, e.g., a subspace based semi-blind channel estimation method [199]. The asynchronous multiuser OFDM receiver in [423] address channel estimation, data detection, and channel decoding modules for underwater acoustic channels. Asynchronous multiuser receiver for single carrier underwater acoustic transmissions can be found in [82, 472].

16

OFDM in Relay Channels

Cooperative communication through the use of relay nodes has been extensively studied in recent years for wireless radio systems. Some example relay strategies include:

- *Amplify and forward (AF)*: A relay node simply amplifies the signal received from the source and sends it to the destination [57];
- *Decode and forward (DF)*: A relay node tries to decode the signal received from the source, and sends the re-encoded message to the destination [89];
- *Compress and forward (CF)*: A relay node transmits a quantized and compressed version of the received signal to the destination [7, 89, 237];
- *Quantize, map and forward (QMF)*: A relay quantizes the received signal at the distortion of the noise power, then randomly maps these bits to a transmit Gaussian codeword [13].

This chapter presents two scenarios of relay networks in underwater acoustic channels, where OFDM is used as the underlying modulation. The first scenario considers a dynamic coded cooperation (DCC) scheme in a three-node network, where the half-duplex relay listens until it can decode the message correctly and then switches to the transmission mode. When transmitting, the relay superimposes its transmission on the ongoing transmissions from the source. The second scenario considers a dynamic block cycling (DBC) protocol in a line network with multiple relays. As in DCC, each relay starts the transmission once it successfully decodes the incoming packet. This helps to reduce the end-to-end transmission delay without requiring feedback from the receiver. This chapter is organized as follows.

- Section 16.1 describes the OFDM modulated dynamic coded cooperation.
- Section 16.2 provides one DCC design example based on rate-compatible coding, while Section 16.3 provides another example based on the layered erasure-correction and error-correction coding.
- Section 16.4 presents the DBC protocol in a line network.

16.1 Dynamic Coded Cooperation in a Single-Relay Network

Consider a three-node network as shown in Figure 16.1 consisting of a source, a destination and a relay which helps the transmission from the source to the destination. All the transceivers

OFDM for Underwater Acoustic Communications, First Edition. Shengli Zhou and Zhaohui Wang.
© 2014 John Wiley & Sons, Ltd. Published 2014 by John Wiley & Sons, Ltd.

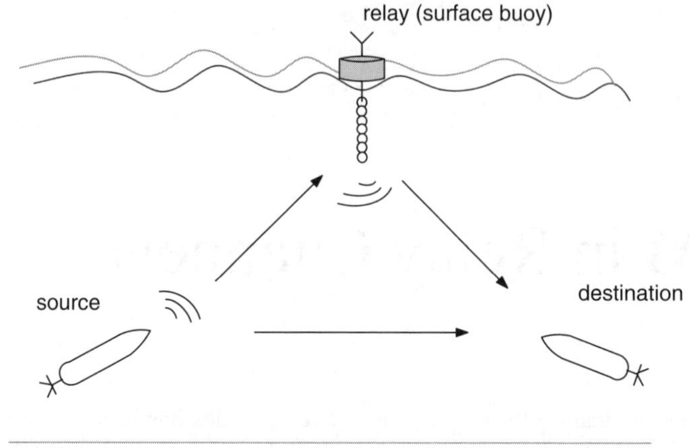

Figure 16.1 One example setup with underwater three-node cooperative communication.

work in a half-duplex fashion, which is the case in underwater acoustic networks. The channels among the source, the relay, and the destination are multipath fading channels with large delay spread. For this reason, OFDM modulation is adopted. Assume that the guard interval between consecutive OFDM symbols is larger than the maximum delay spread plus the offset between the signals from the source and the relay to reach the destination. As such, there is no interblock interference (IBI) between consecutive OFDM blocks at the relay and at the destination.

The source divides a packet into multiple blocks, say N_{bl} blocks, with each block modulated on one OFDM symbol. The source transmits N_{bl} OFDM blocks, which will reach both the relay and the destination. Note that these blocks are related to each other, depending on the coding schemes to be specified later, and the whole packet can be recovered based on a subset of these blocks.

16.1.1 Relay Operations

The relay has two operational phases: the listening phase and the cooperative transmission phase. First, the relay is in the listening phase. For every new OFDM block that it collects, the relay tries to decode the whole packet using the accumulated OFDM blocks. After successfully decoding the transmitted packet before the end of the transmission from the source, the relay switches to the cooperation phase.

16.1.1.1 Cooperation Strategies

Denote N_{li} as the number of OFDM blocks that the relay has used for successful decoding. The relay starts to superimpose its transmission to the ongoing transmission from the source, from the $(N_{\text{li}} + \Delta + 1)$-th block to the N_{bl}-th block, where Δ is an integer to be determined. Two possible cooperation strategies at the relay are the following.

- *Repetition redundancy (RR) cooperation*: The relay regenerates and transmits identical OFDM blocks as the source, from the block index $(N_{\text{li}} + \Delta + 1)$ to N_{bl}.

- *Extra redundancy (ER) cooperation*: The relay generates and transmits different redundant OFDM blocks from the source from the block index $(N_{\text{li}} + \Delta + 1)$ to N_{bl}.

An illustration of RR and ER will be provided in Figure 16.3 for a design example in Section 16.2.

16.1.1.2 Block-level Synchronization

During the cooperative transmission phase, the OFDM blocks from the source and the relay need to be aligned at the block level at the receiver side. This is achieved through a delay control mechanism at the relay.

Define T_{sr}, T_{rd} and T_{sd} as the transmission delays between the source and the relay, the relay and the destination, and the source and the destination, respectively. Denote the start time of the N_{li}-th block at the source as t_0, the relay processing time as T_{proc}, and the relay waiting time as T_{wait}. To synchronize the reception of the $(N_{\text{li}} + \Delta + 1)$-th OFDM block at the destination, the following relationship should be satisfied as illustrated in Figure 16.2:

$$t_0 + T_{\text{sr}} + T_{\text{bl}} + T_{\text{proc}} + T_{\text{wait}} + T_{\text{rd}} \approx t_0 + (\Delta + 1)T_{\text{bl}} + T_{\text{sd}}. \tag{16.1}$$

Hence, the extra waiting time prior to the cooperative transmission at the relay is:

$$T_{\text{wait}} \approx \Delta T_{\text{bl}} - T_{\text{proc}} - (T_{\text{sr}} + T_{\text{rd}} - T_{\text{sd}}). \tag{16.2}$$

The parameter Δ should be taken as a small integer that leads to a nonnegative waiting time T_{wait}. The processing time T_{proc} is known to the relay. The difference $(T_{\text{sr}} + T_{\text{rd}} - T_{\text{sd}})$ depends on the source-relay-destination geometry. In a favorable geometry where $(T_{\text{sr}} + T_{\text{rd}} - T_{sd})$ is small and with a relay having $T_{\text{proc}} < T_{\text{bl}}$, the value of Δ could be as small as one.

The relay needs to have the knowledge of the source-relay distance d_{sr}, the relay-destination distance d_{rd}, and the source-destination distance d_{sd} to determine the waiting time from (16.2). Since the acoustic modems are often equipped with the ranging functionality, the relay needs to probe the source and the destination to obtain d_{sr} and d_{rd}. The source needs to probe the destination to obtain d_{sd}, and conveys it to the relay. For example, the source-destination distance could be put into the packet header, along with the source ID and destination ID.

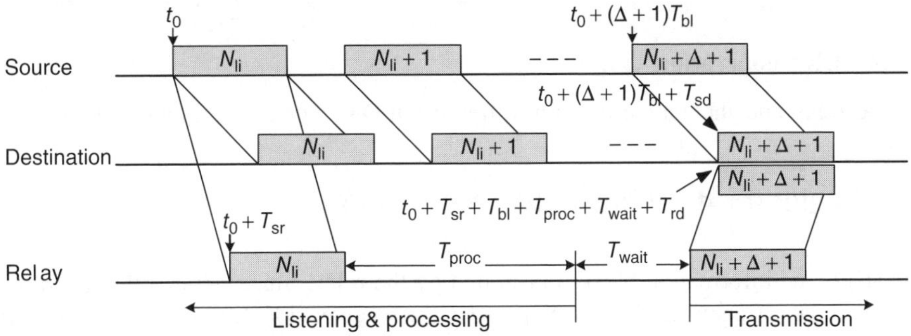

Figure 16.2 Relay introduces a waiting time prior to the cooperative transmission to achieve the block-level synchronization.

The cooperation phase lasts for $N_{\text{bl}} - N_{\text{li}} - \Delta$ OFDM blocks, hence the duration is dynamic depending on the channel quality from the source to the relay.

16.1.2 Receiver Processing at the Destination

Use N_{r} as the number of receive elements at the destination and K as the number of subcarriers in the OFDM modulation. For the first $N_{\text{li}} + \Delta$ blocks, the destination only receives the transmission from the source. After the necessary pre-processing steps which involves Doppler compensation and FFT operation [235], the input–output relationship of the lth received OFDM block at the νth receive element is

$$\mathbf{z}_{\nu}[l] = \mathbf{H}_{\nu,\text{sd}}[l]\mathbf{s}[l] + \mathbf{w}_{\nu}[l], \quad \nu = 1, \dots, N_{\text{r}}, \quad l = 1, \dots, N_{\text{li}} + \Delta, \tag{16.3}$$

where $\mathbf{z}_{\nu}[l]$ is the vector containing the frequency measurements across K subcarriers, $\mathbf{s}[l]$ is the vector of transmitted symbols on K subcarriers, $\mathbf{H}_{\nu,\text{sd}}[l]$ denotes the channel mixing matrix for the channel between the source and the destination, and $\mathbf{w}_{\nu}[l]$ is the ambient noise.

For the last $N_{\text{bl}} - N_{\text{li}} - \Delta$ blocks, the destination receives the superposition of the signals from the source and the relay. The receiver processing depends on the cooperation strategy used.

16.1.2.1 RR Cooperation

Since the relay transmits identical OFDM blocks as the source, the input–output relationship is

$$\begin{aligned}\mathbf{z}_{\nu}[l] &= \underbrace{(\mathbf{H}_{\nu,\text{sd}}[l] + \mathbf{H}_{\nu,\text{rd}}[l])}_{:=\mathbf{H}_{\nu,\text{equ}}[l]}\mathbf{s}[l] + \mathbf{w}_{\nu}[l], \\ &= \mathbf{H}_{\nu,\text{equ}}[l]\mathbf{s}[l] + \mathbf{w}_{\nu}[l], \quad \nu = 1, \dots, N_{\text{r}}, \quad l = N_{\text{li}} + \Delta + 1, \dots, N_{\text{bl}}\end{aligned} \tag{16.4}$$

where $\mathbf{H}_{\nu,\text{rd}}[l]$ denotes the channel mixing matrix for the channel between the relay and the destination. Clearly, an equivalent channel, which consists of multipath arrivals from both the source and the relay, is formed.

16.1.2.2 ER Cooperation

Since the relay and the source transmit different OFDM blocks, the input–output relationship is

$$\mathbf{z}_{\nu}[l] = \mathbf{H}_{\nu,\text{sd}}[l]\mathbf{s}[l] + \mathbf{H}_{\nu,\text{rd}}[l]\tilde{\mathbf{s}}[l] + \mathbf{w}_{\nu}[l], \quad \nu = 1, \dots, N_{\text{r}}, \quad l = N_{\text{li}} + \Delta + 1, \dots, N_{\text{bl}}, \tag{16.5}$$

where $\tilde{\mathbf{s}}[l]$ is the information block transmitted by the relay. Since $\tilde{\mathbf{s}}[l]$ is different from $\mathbf{s}[l]$, two parallel data streams need to be separated at the receiver side, leading to a receiver design problem that has been addressed in the context of multi-input multi-output (MIMO) OFDM.

For the first $N_{\text{li}} + \Delta$ OFDM blocks without cooperation and the next $N_{\text{bl}} - N_{\text{li}} - \Delta$ OFDM blocks during the RR cooperation phase, the destination can adopt a receiver as described in

Chapter 9. For the $N_{\rm bl} - N_{\rm li} - \Delta$ OFDM blocks received during the ER cooperation phase, two data streams in $\mathbf{s}[l]$ and $\tilde{\mathbf{s}}[l]$ need to be separated. The MIMO-OFDM receivers as described in Chapter 13 can be adopted. At the end of the source transmission, the soft information accumulated for all the symbols during the direct transmission phase and the ER cooperation phase are used to decode the whole packet.

16.1.3 Discussion

In the RR cooperation, the relay transmission increases the received power and provides multipath diversity benefits to the last $N_{\rm bl} - N_{\rm li} - \Delta$ OFDM blocks. While in the ER cooperation, the relay could provide both the diversity and the coding benefits through newly added blocks that contain more parity symbols. Hence, ER is expected to have better performance than RR. On the implementation side, however, RR is more convenient than ER.

- The ER cooperation requires the destination to detect the starting point of the cooperation phase. The generalized log-likelihood ratio test (GLRT) was proposed in [224], and a power detector was developed in [190]. For the OFDM signals, one could reserve several subcarriers for the detection of relay cooperation, which however introduces extra overhead.
- For the ER cooperation, the destination needs to demodulate two data streams in the cooperation phase. Hence, the receiver needs to adjust its processing modules between the non-cooperation phase and the ER cooperation phase.

For the RR cooperation, no change is needed at the destination. Note that the OFDM receivers in Chapter 9 perform channel estimation on a block-by-block basis, in order to deal with fast channel variations in underwater environments. Hence, *the receiver does not need to be aware of the existence of a relay*. Further, instead of one relay, multiple relays can be easily added into the OFDM-DCC scheme if using the RR cooperation.

16.2 A Design Example Based on Rate-Compatible Channel Coding

Code design for the dynamic coded cooperation is an important task. The relay deals with truncated codewords with different truncation lengths, while the destination deals with extended codewords in the ER cooperation. Rateless coding has been suggested in [64, 224], which requires a large number of blocks. Multiple turbo codes have been designed in [190] for dynamic coded cooperation. Here, we present a design example leveraging nonbinary rate-compatible (RC) quasi-cyclic LDPC codes [178], which have good performance for short block lengths and low encoding and decoding complexities.

16.2.1 Code Design

This specific design is based on [178], where the rate-compatible code is designed in Galois field (GF) of size 16, with a protograph of 8 information nodes and a set of code rates 8/9, 8/10, …, 8/19. The main difference with [178] is that the *approximate cycle extrinsic* (ACE) algorithm factor is 16 here, resulting in the information length of $8{\times}\log_2(16){\times}2{\times}16 = 1024$ bits.

The 1024 information bits are divided into 8 blocks, with 128 bits per block. The source transmits 8 blocks of information symbols and 6 blocks of parity check symbols, hence the code rate at the source is 8/14. In the ER operation, the relay generates additional redundant blocks of parity check symbols with the lowest code rate to be 8/19. Figure 16.3 illustrates the six possible cases that can occur at the relay, for both RR and ER operations, where $\Delta = 1$. In Case 1 of Figure 16.3, the relay can successfully decode the source message after only receiving the information blocks, while in Case 6 the relay cannot decode the source message in time to perform cooperation.

Figure 16.4 shows the packet error rate (PER) performance of the RC-LDPC codes in AWGN channels with BPSK modulation. All the codes perform about 2~2.5 dB away from the Shannon limit with the BPSK input. Considering the small block length, this is a set of good codes for the dynamic coded cooperation.

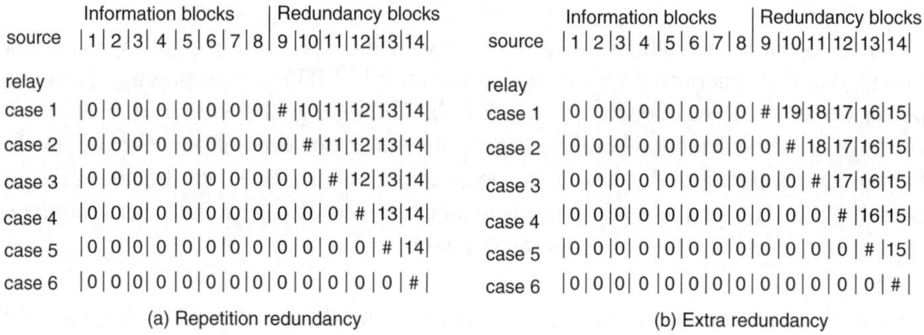
Information blocks | Redundancy blocks
source |1|2|3|4|5|6|7|8|9|10|11|12|13|14|
relay
case 1 |0|0|0|0|0|0|0|0|#|10|11|12|13|14|
case 2 |0|0|0|0|0|0|0|0|0|#|11|12|13|14|
case 3 |0|0|0|0|0|0|0|0|0|0|#|12|13|14|
case 4 |0|0|0|0|0|0|0|0|0|0|0|#|13|14|
case 5 |0|0|0|0|0|0|0|0|0|0|0|0|#|14|
case 6 |0|0|0|0|0|0|0|0|0|0|0|0|0|#|
(a) Repetition redundancy
Information blocks | Redundancy blocks
source |1|2|3|4|5|6|7|8|9|10|11|12|13|14|
relay
case 1 |0|0|0|0|0|0|0|0|#|19|18|17|16|15|
case 2 |0|0|0|0|0|0|0|0|0|#|18|17|16|15|
case 3 |0|0|0|0|0|0|0|0|0|0|#|17|16|15|
case 4 |0|0|0|0|0|0|0|0|0|0|0|#|16|15|
case 5 |0|0|0|0|0|0|0|0|0|0|0|0|#|15|
case 6 |0|0|0|0|0|0|0|0|0|0|0|0|0|#|
(b) Extra redundancy

Figure 16.3 One OFDM-DCC example. The numbers shown are the indexes of the OFDM blocks transmitted. $\Delta = 1$ in this example.

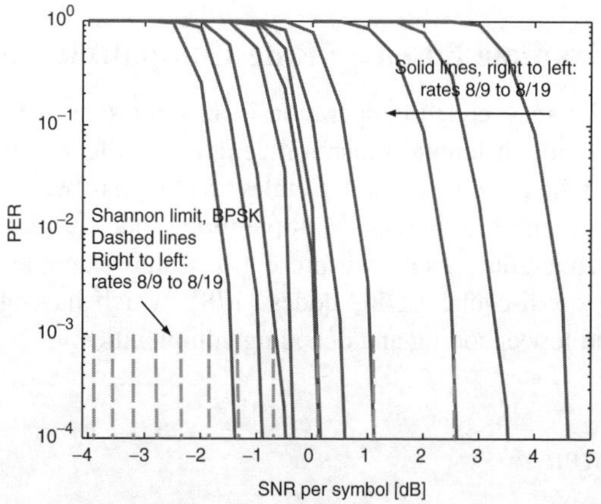

Figure 16.4 Performance of nonbinary RC-LDPC codes over AWGN channel. Dashed lines: Shannon limit with BPSK input; right to left: $8/9, 8/10, \cdots, 8/19$.

16.2.2 Simulation Results

The simulated OFDM system contains 128 data subcarriers, and a BPSK constellation is used for symbol mapping. The channels of source-to-relay, source-to-destination and relay-to-destination are independently generated, each consisting of 16 Rayleigh distributed taps. The channels are quasi-static, meaning that they remain constant for each OFDM block, but change independently from block to block. Perfect channel estimation is assumed at both the relay and the destination.

Test Case 16.2.1 Assume a topology where the relay is much closer to the source than to the destination. Let $\overline{\gamma}_{\rm sd}$ and $\overline{\gamma}_{\rm sr}$ denote the average signal-to-noise-ratio (SNR) of the signal from the source at the destination and the relay, respectively. Let $\overline{\gamma}_{\rm rd}$ denote the average SNR from the relay to the destination. For this topology, set

$$\overline{\gamma}_{\rm sr} = \overline{\gamma}_{\rm sd} + G, \quad \overline{\gamma}_{\rm sd} = \overline{\gamma}_{\rm rd}, \tag{16.6}$$

where G is a constant.

Figure 16.5 depicts the performance of the proposed schemes with one and two receive elements at the destination, respectively, where $G = 20\,$dB. As a comparison, the theoretical outage probability performance bounds assuming perfect channel coding with BPSK input are also included, where the approach in [260] is used to calculate the mutual information for single-input or multi-input OFDM systems with PSK constellations in fading channels. An outage occurs if the total mutual information at the destination after the dynamic coded cooperation is lower than the transmission rate.

As one can see, both RR and ER outperform the noncooperative case by about 1 dB. ER is slightly better than RR. The performance gaps of the RC-LDPC coded system to the outage probabilities vary from 1.5 dB to 3 dB. This is consistent with the performance gap in AWGN channels.

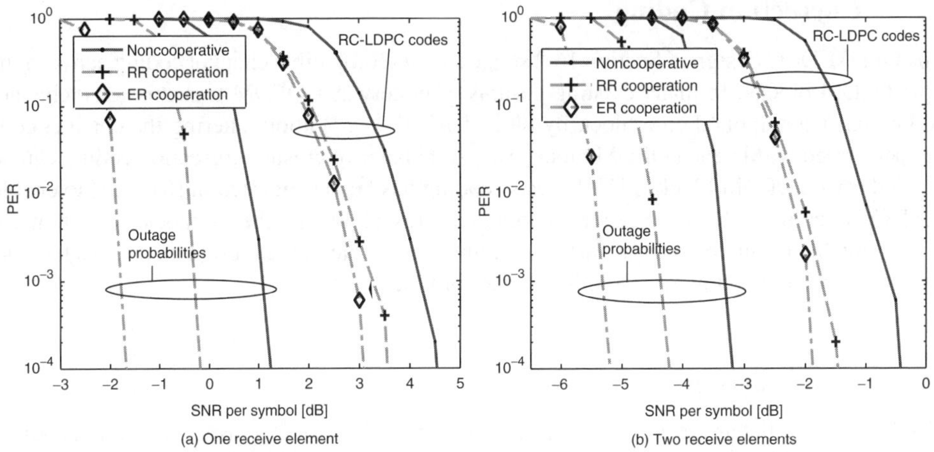

Figure 16.5 Performance comparison with one or two receive elements at the destination.

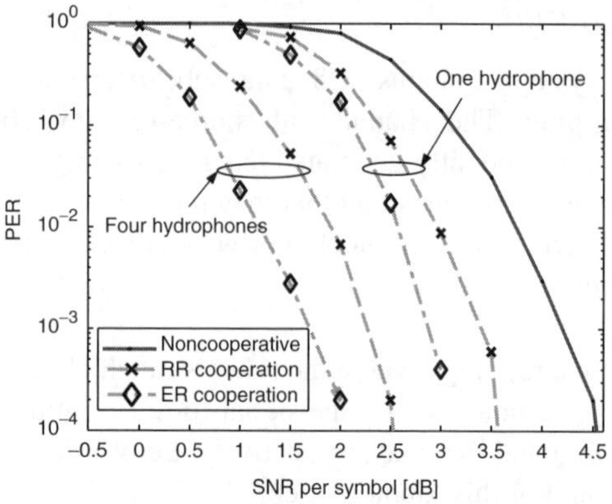

Figure 16.6 Performance with different numbers of hydrophones at the buoy.

Test Case 16.2.2 Consider a UWA network where the water depth is 50 m, the distance d between the source and the destination is 1 km, and the surface buoy is deployed in the middle. A spreading loss of about $\overline{\gamma} \propto d^{-1.5}$ in UWA transmissions is assumed, which leads to $\overline{\gamma}_{\rm sr} \approx \overline{\gamma}_{\rm sd} + 4.5$ dB, and $\overline{\gamma}_{\rm sr} = \overline{\gamma}_{\rm rd}$.

Figure 16.6 depicts performance with different number of hydrophones at the surface buoy, where the destination has one hydrophone. Note that multiple hydrophones at the relay are used for reception purpose only, and the relay has only one transducer for transmission. Clearly, as the number of hydrophones increases at the relay, the performance of dynamic coded relay improves. With four hydrophones, there is 2 dB gain at PER $= 10^{-2}$ for RR and 2.5 dB for ER relative to the noncooperative transmission.

16.3 A Design Example Based on Layered Erasure- and Error-Correction Coding

The OFDM-DCC design in Section 16.2 suggests rate-compatible channel coding across multiple OFDM blocks. In many existing designs of underwater OFDM transmissions, channel coding has been applied on a block-by-block basis [235]. Without altering the channel coding performed within each OFDM block, a separate layer of erasure-correction coding can be applied across OFDM blocks [41]. Rateless coding has been suggested in [64, 224] to enable the DCC operation. However, rateless coding requires a large number of blocks, which might not be suitable for underwater acoustic communications. Next, we present an approach to use erasure-correction coding over a finite number of blocks [74].

16.3.1 Code Design

Here, nonbinary linear precoding is used to perform the interblock erasure-correction coding. Operating over GF(2^8), every eight bits are group into one byte before encoding. First, divide one packet into $I_{\rm bl}$ information blocks, where each block contains P symbols in GF(2^8). Denote the pth symbol of the ith block as $b[i;p]$. An encoder which generates $N_{\rm bl}$ coded symbols from

$I_{\rm bl}$ information symbols is applied as

$$\begin{pmatrix} c[1;p] \\ \vdots \\ c[N_{\rm bl};p] \end{pmatrix} = \begin{pmatrix} 1 & 1 & \cdots & 1 \\ 1 & \alpha & \cdots & \alpha^{I_{\rm bl}-1} \\ \vdots & \ddots & \ddots & \vdots \\ 1 & \alpha^{N_{\rm bl}-1} & \cdots & \alpha^{(N_{\rm bl}-1)(I_{\rm bl}-1)} \end{pmatrix} \begin{pmatrix} b[1;p] \\ \vdots \\ b[I_{\rm bl};p] \end{pmatrix}, \tag{16.7}$$

where α is a primitive element in GF(2^8) [247]. After the erasure-correction coding, each set of P symbols $\{c[l;1], \cdots, c[l;P]\}$ will be forwarded to the error-correction channel encoder to generate the coded symbols to be modulated in the lth OFDM block.

The OFDM blocks which fail in channel decoding will be discarded; (each block has its own CRC flags, as done in e.g., [444]). Thanks to the Vandermonde structure of the code generation matrix in (16.7), any square submatrix drawn from it is guaranteed to be nonsingular and thus invertible in the finite field. Hence, as long as the relay collects $I_{\rm bl}$ correctly decoded blocks, all the information symbols can be recovered, and the whole packet can be regenerated. Note, however, that this layered decoding approach is expected to be suboptimal relative to a joint decoding approach where all accumulated blocks are decoded jointly.

16.3.2 Implementation

The OFDM-DCC scheme with layered coding and RR cooperation was implemented into the modem prototype [444]. For simplicity, the ICI-ignorant receiver from Section 9.1 with the least-squares channel estimator is used. Each OFDM block carries 80 bytes of payload data. Here, we set $I_{\rm bl} = 8$ and $N_{\rm bl} = 18$ for the erasure-correction coding, and hence each packet has 640 bytes of information data.

Two major tasks on implementation have been accomplished.

- *Erasure-correction decoding.* Gaussian elimination is used for matrix inversion over the finite field for erasure-correction decoding. Thanks to the small code length, a very small computational overhead is added to the modem processing.
- *Synchronization.* To achieve the block-level synchronization as described in Section 16.1.1, two changes have been made to the modem: 1) the relay performs a fine synchronization step to locate the starting time of each OFDM block that it has received [39]; and 2) After correctly decoding the packet, the relay needs to hold on its transmission for $T_{\rm wait}$ seconds. A timer is issued, and when it expires, a hardware interrupt is triggered that will get the transmission of the OFDM blocks actually started. This way, the relay can align its transmission to achieve the block-level (quasi-) synchronization for the OFDM blocks received at the destination from both the source and the relay.

16.3.3 An Experiment in Swimming Pool

The experiment was carried out in Aug. 9, 2012, in the Brundage Pool at the University of Connecticut, with the setup shown in Figure 16.7. With the source node and destination node set in two sides of the pool, the relay node was put in three different locations in the middle, as shown in Figure 16.8. The distance from the source node S to the destination node D is about $d = 50$ feet. Relay node was placed between the source node and the destination node.

Figure 16.7 Experimental setup in the swimming pool.

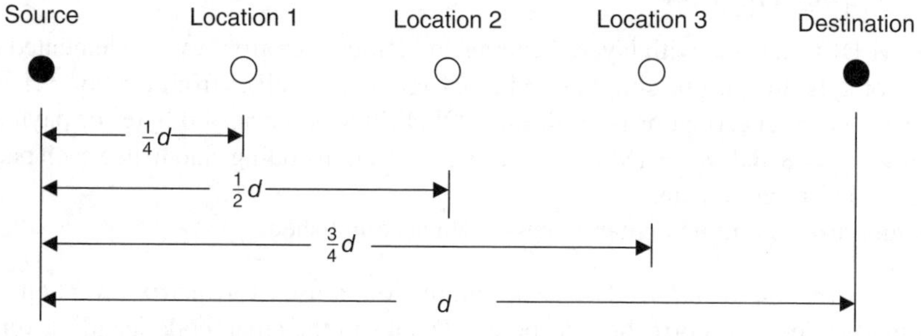

Figure 16.8 There is only one relay used between the source and the destination. The relay can be placed at different locations, as marked.

According to its distance to the source node S, three possible relay locations were included, $d/4, d/2$ and $3d/4$ away from the source node, respectively. In all the three settings, the relay node has the same transmit power as the source node. Since the source, relay, and destination are on a line, $T_{sr} + T_{rd} - T_{sd} = 0$. The value of Δ is set to be one in this experiment as $T_{proc} < T_{bl}$.

A total of 4 scenarios were tested: no relay, and one relay at three different locations. In each scenario, 40 packets of data were recorded at the destination node D, where each packet has $N_{bl} = 18$ OFDM blocks with interblock erasure-correction coding as specified in Section 16.3.2. The input SNRs as measured at the received blocks without relay are high, e.g., about 20 dB. Figure 16.9 shows the channel statistics of one packet from the scenario of relay at $d/4$ away from the source node. Clearly, the RR cooperation leads to an equivalent multipath channel that is stronger than the original multipath channel through signal superposition.

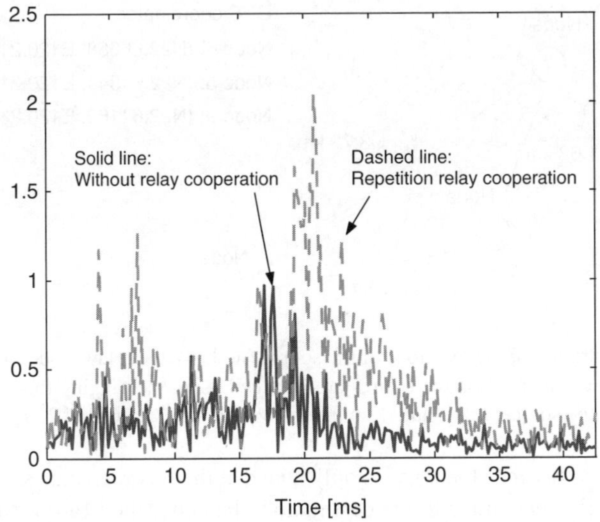

Figure 16.9 Samples of the estimated channels without the relay cooperation and with the relay cooperation.

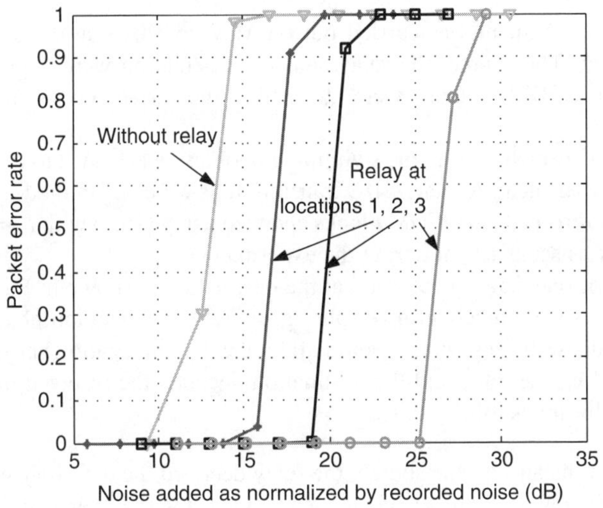

Figure 16.10 The packet error rate is obtained by adding noise to the recorded data at the destination. Note that the relay operation is done online in real time.

Now add white Gaussian noise of different levels to the 40 recorded packets in each test scenario. Figure 16.10 plots the packet error rate (PER) performance in different test scenarios, as a function of the variance of the added noise which is normalized by the variance of the recorded ambient noise in the signal band. Note that during this experiment, the source node has enough transmission power so that the decoding performance at the relay is similar in all

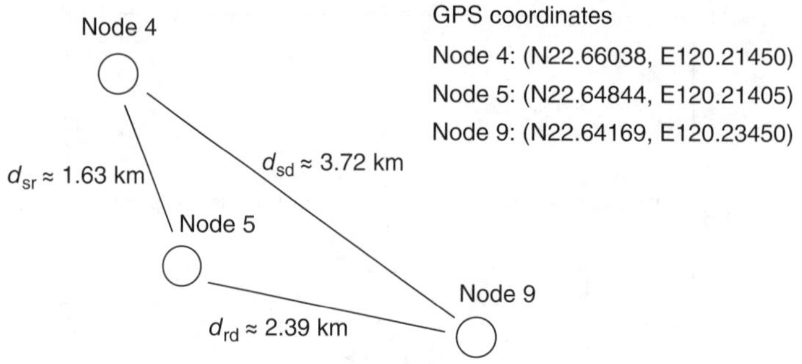

Figure 16.11 Illustration of the locations of the source (node 4), the relay (node 5) and the destination (node 9); The distances marked on the figures are obtained using the ranging function of the modems. The water depths at the source, relay, and destinations were about 27, 26, and 22 meters, respectively.

locations, with $N_{\mathrm{li}} = 8$. Since the noise is only added at the recorded data set at the destination locally, the closer the relay node to the destination, the better the PER performance becomes due to the higher SNRs for the OFDM blocks received at the cooperation phase. This trend is clearly observed in Figure 16.10.

16.3.4 A Sea Experiment

An OFDM-DCC experiment was carried out on May 26, 2013, at the sea near the Kaohsiung City, Taiwan. The source, the relay, and the destination were deployed as shown in Figure 16.11. The OFDM modems were attached to the surface buoys, at a water depth of 6 meters.

The OFDM-DCC firmware from the swimming pool test was loaded to the OFDM modems deployed in this experiment. A total of 189 transmissions were transmitted, and each transmission contained 20 zero-padded OFDM blocks encoded using (16.7) with $I_{\mathrm{bl}} = 8$ and $N_{\mathrm{bl}} = 20$. The block delay was set as $\Delta = 4$ during the experiment.

The waveform of one data set recorded at the destination is shown in Figure 16.12, where the received signals were much stronger during the relay cooperation phase. Note also that there existed impulsive noises, which would affect the communication performance for those affected blocks. Both the relay and the destination decoded the received blocks online. The performance results are as follows.

- Due to the short distance to the source, the relay decoded the data very well. In 188 transmissions, the relay was able to decode the whole packet with the first eight received blocks, and in one transmission, the relay used 9 blocks to decode the packet.
- The destination kept decoding 20 OFDM blocks for each transmission. For each block index from 1 to 20, define the block error rate as the ratio of the number of erroneous blocks to the total number of transmissions. As shown in Figure 16.13, the BLER is around 0.05 before the relay cooperation, and it decreases to around 0.01 after the relay cooperation, when averaged over all 189 transmissions.

  The pilot signal to noise ratio (PSNR), defined as the signal power at the pilot subcarriers to the power at the null subcarriers, is shown in Figure 16.14, averaged over 189 transmissions. A 2.5 dB increase is observed after the relay cooperation.

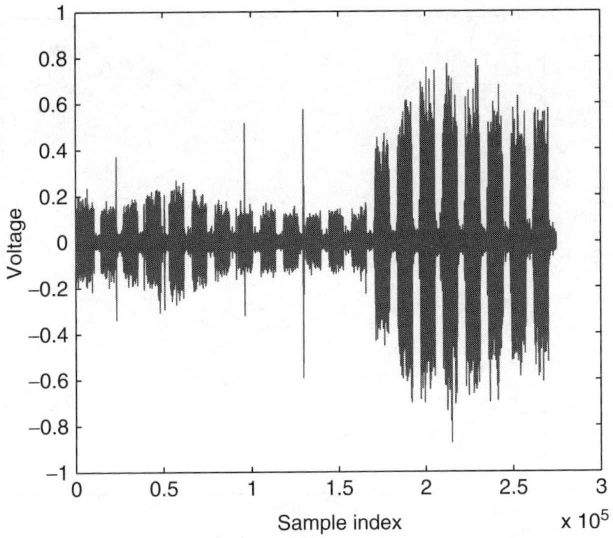

Figure 16.12 One received waveform after bandpass filtering; there are some impulsive noises.

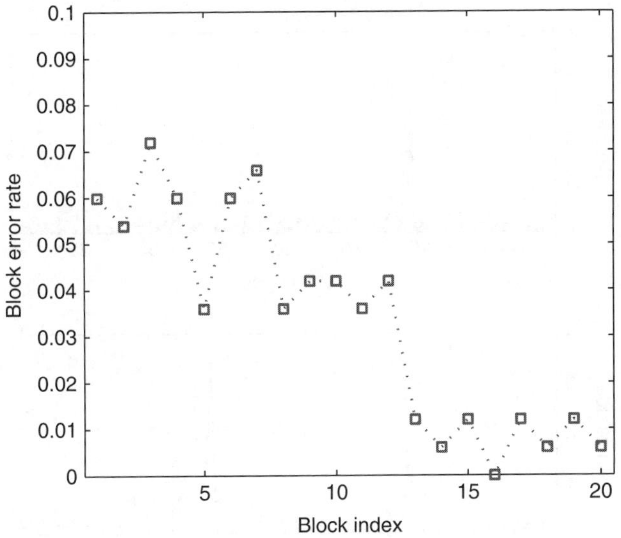

Figure 16.13 The block error rate as a function of the block index.

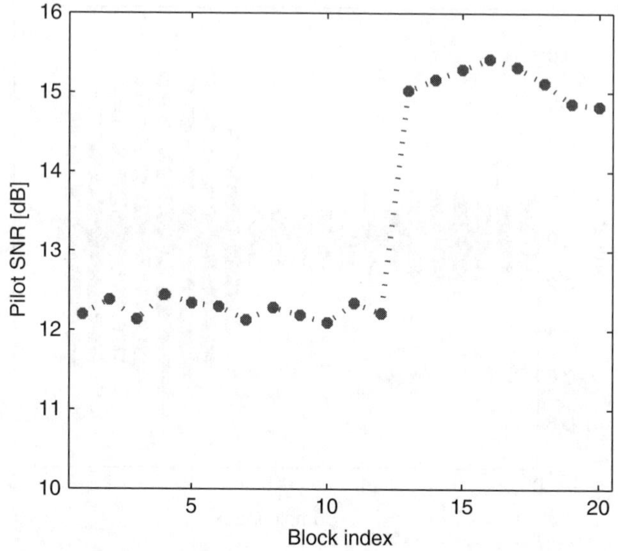

Figure 16.14 The pilot SNR as a function of the block index.

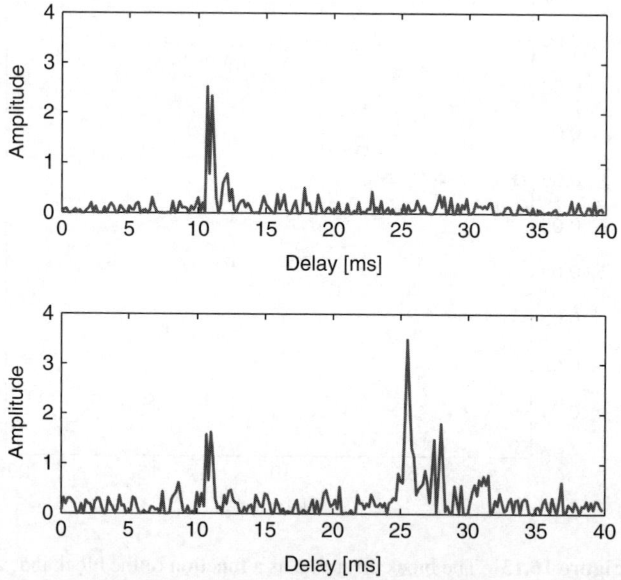

Figure 16.15 The estimated channels before and after the relay cooperation.

Figure 16.15 shows the estimated channels before and after the cooperation. For the composite channel after relay cooperation, the first cluster corresponds to the channel from the source to the destination, and the second cluster corresponds to the channel from the relay to the destination. It can be seen that there is a 15-millisecond gap between the peaks of these two clusters, reflecting the synchronization offset. This sea experiment shows that the relay with the RR strategy improves the performance of the source to destination communication without introducing any changes to the transmission procedure between the source and the relay.

16.4 Dynamic Block Cycling over a Line Network

Consider a linear network with $(M + 1)$ nodes, as illustrated in Figure 16.16(a). The first node indexed by 0 is the source, and the last node indexed by M is the destination. The nodes between the source and destination indexed by $1, 2, \cdots, M - 1$ are relay nodes. Assume that all the nodes operate in the half duplex mode. The linear network topology can be found in many realistic networks. One example is shown in Figure 16.16(b), where nodes in the relay route can be regarded as unequally spaced nodes in a linear network.

16.4.1 Hop-by-Hop Relay and Turbo Relay

Two classic relay protocol over a linear network are shown in Figure 16.17. The first is a hop-by-hop relay protocol, in which signal from source is relayed to destination sequentially by each intermediate node. The second is a turbo relay protocol which was originally proposed for a tree network in [153]. Exploiting the broadcasting nature of wireless transmissions, each relay node and its direct neighbors in this protocol cooperatively relay the message to destination. The relay strategies discussed at the beginning of this chapter, such as AF, DF, and QMF, can be applied to both protocols. Here we consider the extension of the DCC protocol from a single-relay network to a line network with multiple relays.

Compared to the single-relay network, each intermediate node in the line network acts as destination, relay, and source of the single-relay network successively. Moreover, each intermediate node could play dual roles in the single-relay network. For example, one intermediate

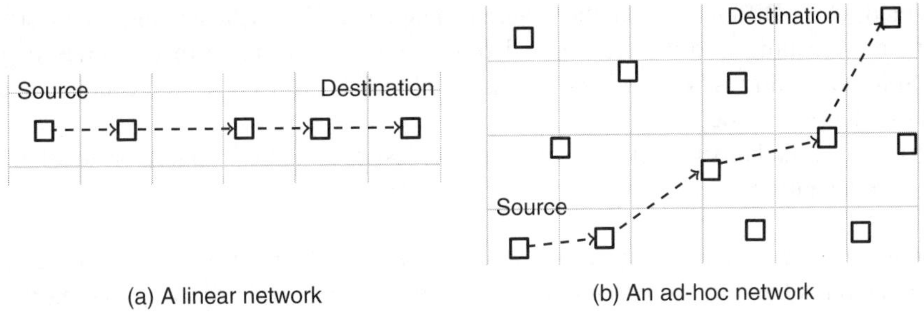

Figure 16.16 Relaying over underwater acoustic networks, with dashed line denoting the relaying route. The source and destination could be autonomous underwater vehicles (AUVs) or bottom-anchored nodes. The nodes in the linear network need not be in a line or equally spaced. (a): an illustrative linear network; (b): a more realistic network showing the context for a linear network.

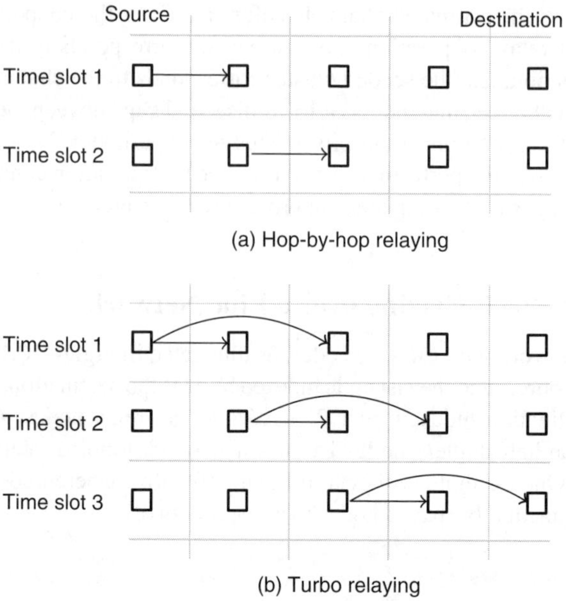

Figure 16.17 Illustration of two relay strategies over a linear network.

node can act as a rely node for its direct downstream neighbor when its upstream neighbor is transmitting, and simultaneously acts as a source node for its downstream nodes which locate beyond the communication range of its upstream neighbor.

In the following sections, we will describe a relay protocol for a linear network, termed *dynamic block cycling (DBC)*, which is easy to operate and resembles the adaptive feature of DCC in single-relay networks [427].

16.4.2 Dynamic Block-Cycling Transmissions

Assume a burst-based transmission, where each burst consists of $N_{\rm bl}$ blocks. Each block can be an OFDM block as explained in Section 16.3.1. A layered channel coding scheme is adopted [41]. Different from the layered channel coding scheme, the $I_{\rm bl}$ information blocks are spreading into $N_{\rm bl}$ transmitted blocks via an erasure-correction code (e.g., the Reed-Solomon coding scheme). Each transmitted block is error-correction coded, and all the nodes use the same codebook.

To relay $N_{\rm bl}$ coded blocks from the source to the destination, the proposed protocol incorporates two strategies.

- *Instantaneous transmission upon successful decoding:* To reduce the latency, each relay node attempts burst decoding and performs parity check when receiving one more block; *immediately after* the information bits within one burst are successfully recovered, it starts relaying the message to downstream nodes.
- *Cyclically synchronized transmissions:* To remove the coordination overhead between the transmitting and receiving nodes, the transmitted block-indices from one node are cyclically

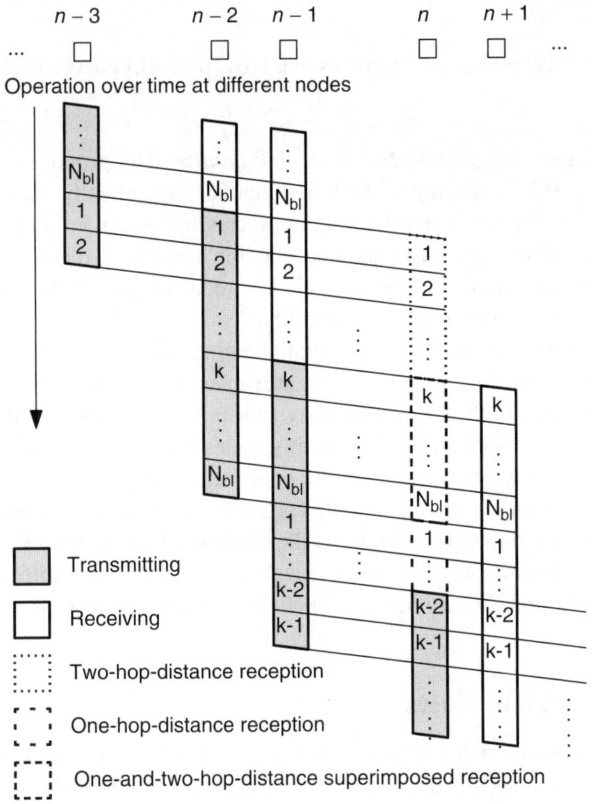

Figure 16.18 Illustration of the cyclically synchronized transmission over a linear network at node n with three temporal progressive listening phases. Two-hop communication distance is assumed as an example.

synchronized with those transmitted by the upstream nodes, as illustrated in Figure 16.18. Specifically, when one node starts transmission, the transmitted block is identical to the block that it will receive from its upstream nodes (denote the transmitted block index as k), so that the signal at the receiving node can be taken as the signal from one source but propagating along multiple paths. Once a node finishes transmission of the block with index N_{bl}, it continues to transmit the blocks with indices $1, \cdots, (k-1)$, thus completing the transmission of total N_{bl} coded blocks.

Two special examples shed light on the dynamics of the proposed protocol.

- In the scenario with a large transmission power such that the destination can directly hear the source, the DBC protocol enables an end-to-end transmission, thus avoids the latency in the hop-by-hop relay;
- In the scenario with a small transmission power such that one node can only hear its direct neighbor, the DBC protocol becomes the classical hop-by-hop relay protocol.

16.4.3 Discussion

Tailored to the UWA channel characteristics, the DBC protocol possesses the following desirable features:

- *Transmission latency adaptation based on link quality:* The proposed protocol adapts its transmission schedule according to the link dynamics. Specifically, for the node with good link quality, it only requires a small number of receiving blocks from upstream transmitters for successful decoding, hence can quickly relay the message to the downstream nodes. For the node with low link quality, a large number of receiving blocks for successful decoding is necessary, so as to maintain a high reliability.
- *Simplicity of implementation:* Given the spatial and temporal variations of UWA channels, the number of transmitting nodes and the transmission schedule during one burst transmission are dynamic. With the cyclically synchronized transmission, the receiving nodes can be oblivious of the number of transmitting nodes and their transmission schedule. This eliminates the coordination overhead of the cooperative relay.
- *Robustness to synchronization offset:* With each node aligning its transmitted block starting point with its own local received blocks on the block level, the proposed cyclically synchronization transmission enables the synchronization of signals from different transmitters at the receiving node, and *this applies to a linear network with non-equally spaced nodes.*

16.5 Bibliographical Notes

Cooperative communication for wireless networking has been extensively explored in the wireless radio systems. Application of relay to underwater systems is very recent, see e.g., [6] for an overview. Optimization of relay-aided underwater acoustic communications was pursued in [57]. OFDM modulated dynamic coded cooperation in a three-node network was studied in [74]. For a regular linear underwater network, a hop-by-hop transmission protocol was investigated in [465]. In [58], the error propagation in both cooperative and multihop transmissions over the regular linear network and the regular plenary network was examined. A dynamic block-cycling based protocol over a linear network was proposed in [427].

17

OFDM-Modulated Physical-Layer Network Coding

This chapter considers a two-way relay network, where two terminals A and B desire to exchange information with the help of a relay R. The conventional approach based on time-division multiplexing requires a total of four time slots for messages exchange, with each terminal having two time slots individually, as illustrated in Figure 17.1(a). Network coding techniques can reduce the number of time slots required.

- *Network-layer network coding (NLNC):* As illustrated in Figure 17.1(b), A and B take turns to send their messages to R in the first two time slots, and R transmits the XOR-ed version of the two messages in the third time slot. Each terminal can recover the message from the other terminal based on the message from the relay and its own message. The scheme applying network coding at the network layer reduces the number of required time slots from four to three by exploiting the broadcasting nature of the message from relay.
- *Physical-layer network coding (PLNC)*: As illustrated in Figure 17.1(c), in the first time slot which is known as the multiple-access (MAC) phase, both A and B transmit their own messages to R simultaneously. In the second time slot which is known as the broadcast (BC) phase, R broadcasts the XOR-ed version of the two messages. This scheme further exploits the superposition nature of wireless signals, and reduces the number of required time slots to two [131, 218, 462, 463].

This chapter considers the application of PLNC in underwater acoustic systems. Examine the scenario in Figure 17.2, where two terminals (e.g., autonomous underwater vehicles or sensors) need to exchange information but cannot reach each other due to the long distance or obstacles in between. By incorporating a relay within the communication ranges of both terminals, PLNC enables the message exchange with two time slots. Moreover, relative to the relay node in radio channels, the relay node (e.g., the buoy as shown in Figure 17.2) in the underwater acoustic setting can be more capable, such as being equipped with multiple receiving elements and strong processing capability.

OFDM for Underwater Acoustic Communications, First Edition. Shengli Zhou and Zhaohui Wang.
© 2014 John Wiley & Sons, Ltd. Published 2014 by John Wiley & Sons, Ltd.

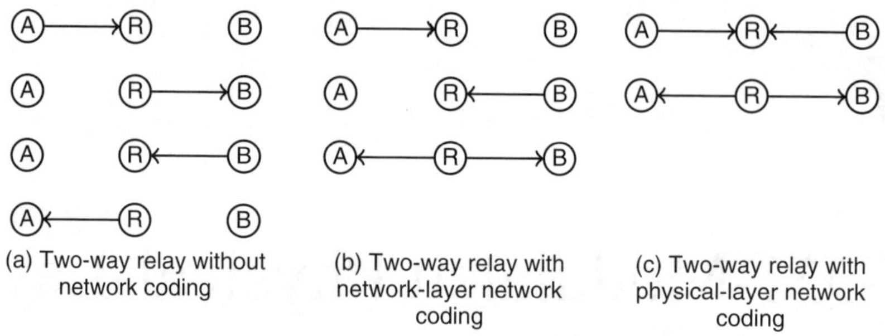

Figure 17.1 Illustration of two-way relay schedules with/without network coding.

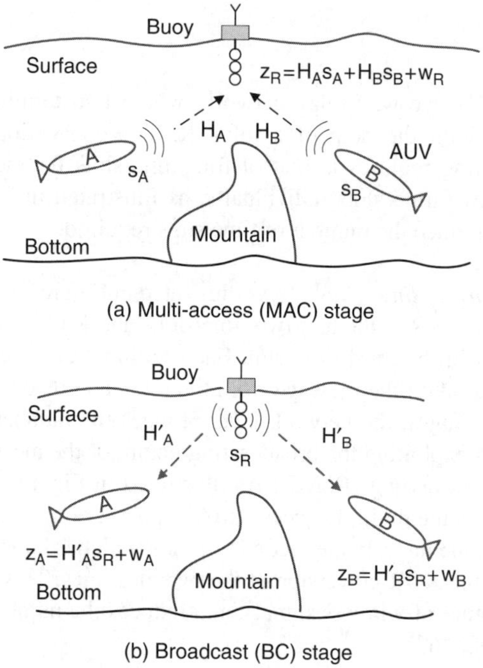

Figure 17.2 Illustration of physical-layer network coding (PLNC): two terminals (AUVs) A and B exchange information in a block manner with each other via the help of the relay (buoy) R which can be equipped with multiple receiving elements. The message exchange consists of a MAC and a BC stage.

With OFDM as the underlying modulation scheme, this chapter presents the receiver design at the relay node in the presence of underwater acoustic time-varying multipath channels.

- Section 17.1 introduces the system model for the OFDM-modulated PLNC at the relay node.
- Section 17.2 presents three iterative receivers to recover the XOR-ed version of the messages.
- Section 17.3 derives outage performance bounds of the relay operation in a time-invariant multipath channel. Sections 17.4 and 17.5 contain performance results based on simulated and emulated data, respectively.

17.1 System Model for the OFDM-Modulated PLNC

Consider a half-duplex two-way relay network, which consists of two single-transmitter terminals A and B, and a relay node R with N_r receiving elements, as shown in Figure 17.2. Assume that the two terminals use an identical parameter set for OFDM modulation. Denote $\mathbf{X}_\mu$ as a channel coded sequence over GF(M) from the μth terminal with μ = A or B. Define T as a modulation mapping operator. The transmitted data symbol vector at the μth terminal is obtained via

$$\mathbf{s}_\mu = \mathrm{T}(\mathbf{X}_\mu), \tag{17.1}$$

where each symbol is drawn from a finite constellation set $\mathcal{M} = \{\alpha_0, \cdots, \alpha_{M-1}\}$. Although not stated explicitly, the symbol vector $\mathbf{s}_\mu$ also contains known symbols as multiplexed pilots and zero symbols at null subcarriers. The transmitted passband waveform $\tilde{s}_\mu(t)$ is then obtained via the OFDM modulation.

The multipath channel between each terminal and receiving element pair follows the model in (1.14) with path-specific Doppler scales. Define $N_{\mathrm{pa},\nu,\mu}$ as the number of paths between the μth terminal and the νth receiving element. The channel impulse response can be written as

$$h_{\nu,\mu}(t;\tau) = \sum_{p=1}^{N_{\mathrm{pa},\nu,\mu}} A_{p,\nu,\mu}\delta(\tau - (\tau_{p,\nu,\mu} - a_{p,\nu,\mu}t)) \tag{17.2}$$

where μ = A or B, $A_{p,\nu,\mu}$, $\tau_{p,\nu,\mu}$, and $a_{p,\nu,\mu}$ are the amplitude, initial delay and Doppler rate of the pth path, respectively.

Assume that both terminals are aware of their distances to the relay node. By adjusting the transmission time at one terminal, signals from the two terminals can be quasi-synchronized at the relay node. The passband signal received at the νth element can be expressed as

$$\begin{aligned}\tilde{y}_\nu(t) = &\sum_{p=1}^{N_{\mathrm{pa},\nu,\mathrm{A}}} A_{p,\nu,\mathrm{A}}\tilde{s}_\mathrm{A}((1+a_{p,\nu,\mathrm{A}})t - \tau_{p,\nu,\mathrm{A}}) \\ &+ \sum_{p=1}^{N_{\mathrm{pa},\nu,\mathrm{B}}} A_{p,\nu,\mathrm{B}}\tilde{s}_\mathrm{B}((1+a_{p,\nu,\mathrm{B}})t - \tau_{p,\nu,\mathrm{B}}) + \tilde{n}(t),\end{aligned} \tag{17.3}$$

where $\tilde{n}(t)$ is the ambient noise.

After the preprocessing operation specified in Chapter 5, the channel impulse response in (17.2) translates into a $K \times K$ channel matrix $\mathbf{H}_{\nu,\mu}$ in the frequency domain, where K is the total number of subcarriers. The (m,k)th element of the channel matrix $\mathbf{H}_{\nu,\mu}$ is related to channel path parameters via (5.25). Define $\mathbf{z}_\nu$ as the frequency measurement vector of the νth receiving element at the relay node, and $\mathbf{w}_\nu$ as the ambient noise vector in the frequency domain. The input–output relationship at the νth receiving element is expressed as

$$\mathbf{z}_\nu = \mathbf{H}_{\nu,A}\mathbf{s}_\mathrm{A} + \mathbf{H}_{\nu,B}\mathbf{s}_\mathrm{B} + \mathbf{w}_\nu. \tag{17.4}$$

For the sake of computational efficiency, a *band-limited ICI leakage* assumption is adopted by approximating

$$H_{\nu,\mu}[m,k] \simeq 0, \qquad \forall |m-k| > D \tag{17.5}$$

where D is the ICI depth. The input–output relationship can be recast as

$$\mathbf{z}_\nu = \mathbf{H}_{\nu,A}\mathbf{s}_\mathrm{A} + \mathbf{H}_{\nu,B}\mathbf{s}_\mathrm{B} + \boldsymbol{\eta}_\nu \tag{17.6}$$

where $\boldsymbol{\eta}_\nu$ is the equivalent noise vector, with each element consisting of the residual ICI and ambient noise,

$$\eta_\nu[m] = \sum_{|m-k|>D} H_{\nu,A}[m,k]s_\mathrm{A}[k] + \sum_{|m-k|>D} H_{\nu,B}[m,k]s_\mathrm{B}[k] + w_\nu[m]. \tag{17.7}$$

Putting $\{\mathbf{z}_\nu\}$ at all receiving elements into a long vector yields

$$\underbrace{\begin{bmatrix}\mathbf{z}_1\\ \vdots \\ \mathbf{z}_{N_\mathrm{r}}\end{bmatrix}}_{:=\mathbf{z}_\mathrm{R}} = \underbrace{\begin{bmatrix}\mathbf{H}_{1,\mathrm{A}}\\ \vdots \\ \mathbf{H}_{N_\mathrm{r},\mathrm{A}}\end{bmatrix}}_{:=\mathbf{H}_\mathrm{A}} \mathbf{s}_\mathrm{A} + \underbrace{\begin{bmatrix}\mathbf{H}_{1,\mathrm{B}}\\ \vdots \\ \mathbf{H}_{N_\mathrm{r},\mathrm{B}}\end{bmatrix}}_{:=\mathbf{H}_\mathrm{B}} \mathbf{s}_\mathrm{B} + \underbrace{\begin{bmatrix}\boldsymbol{\eta}_1\\ \vdots \\ \boldsymbol{\eta}_{N_\mathrm{r}}\end{bmatrix}}_{:=\boldsymbol{\eta}_\mathrm{R}} \tag{17.8}$$

which can be rewritten as

$$\mathbf{z}_\mathrm{R} = \mathbf{H}_\mathrm{A}\mathbf{s}_\mathrm{A} + \mathbf{H}_\mathrm{B}\mathbf{s}_\mathrm{B} + \boldsymbol{\eta}_\mathrm{R}, \tag{17.9}$$

where the size of $\mathbf{z}_\mathrm{R}$ is $N_\mathrm{r}K \times 1$, and the size of $\mathbf{H}_\mathrm{A}$ and $\mathbf{H}_\mathrm{B}$ is $N_\mathrm{r}K \times K$.

The primary task of the relay node at the MAC phase is to recover $\mathbf{X}_\mathrm{R} := \mathbf{X}_\mathrm{A} \boxplus \mathbf{X}_\mathrm{B}$, with the corresponding modulated symbol vector denoted by $\mathbf{s}_\mathrm{R}$. After decoding and successfully recovering $\mathbf{X}_\mathrm{R}$, the relay broadcasts the OFDM modulated symbol $\mathbf{s}_\mathrm{R}$ to both terminals, as shown in Figure 17.2. Both A and B will recover $\mathbf{X}_\mathrm{R}$ from its corresponding received signal. With the estimated $\hat{\mathbf{X}}_\mathrm{R}$, the intended message can be extracted through $\hat{\mathbf{X}}_\mathrm{B} = \hat{\mathbf{X}}_\mathrm{R} \boxplus \mathbf{X}_\mathrm{A}$ at terminal A and $\hat{\mathbf{X}}_\mathrm{A} = \hat{\mathbf{X}}_\mathrm{R} \boxplus \mathbf{X}_\mathrm{B}$ at terminal B.

The receiver design at each terminal during the broadcasting phase is identical to that considered in Chapter 9. This chapter focuses on the MAC phase, where three iterative receivers to recover $\mathbf{X}_\mathrm{R}$ at the relay node are presented next.

17.2 Three Iterative OFDM Receivers

The system model in (17.6) corresponds to a co-located MIMO with two parallel data streams, with the major difference that the goal of the symbol detection and channel decoding modules is to recover $\mathbf{X}_\mathrm{R} = \mathbf{X}_\mathrm{A} \boxplus \mathbf{X}_\mathrm{B}$. The channel estimation module is identical to the one discussed in Chapter 13 with two parallel data streams, and hence the presentation below assumes the availability of the channel state information. In all the iterative receivers, the channel estimation can be included in the iteration loop, so that the channel estimates can be refined by utilizing soft or hard decisions of the transmitted symbols from two terminals, similar to the iterative receivers described in Chapters 13.

17.2.1 *Iterative Separate Detection and Decoding*

In this scheme, the relay node first tries to recover both $\mathbf{X}_A$ and $\mathbf{X}_\mathrm{B}$ individually. The relay encoded symbol $\mathbf{X}_\mathrm{R}$ can then be obtained by doing simple XOR operation. By treating $\mathbf{s}_\mathrm{A}$ and

$\mathbf{s}_B$ as two independent data streams throughout the decoding process, the receiver design is identical to the problem studied in Chapter 13 for the co-located MIMO OFDM. The iterations between data detection and channel decoding are as follows.

(1) The *extrinsic* information $\Pr^{ext}(\mathbf{X}_A)$ and $\Pr^{ext}(\mathbf{X}_B)$ from the two channel decoders is transformed to the *a priori* information $\Pr^{apr}(\mathbf{s}_A)$ and $\Pr^{apr}(\mathbf{s}_B)$ for symbol detection.
(2) Based on the *a priori* information, the symbol detector computes the *extrinsic* information, $\Pr^{ext}(\mathbf{s}_A)$ and $\Pr^{ext}(\mathbf{s}_B)$, which are then transformed into the *a priori* information, $\Pr^{apr}(\mathbf{X}_A)$ and $\Pr^{apr}(\mathbf{X}_B)$, for the decoder.
    For data detection in an ICI-ignorant receiver, calculation of the extrinsic information $\Pr^{ext}(\mathbf{s}_A)$ and $\Pr^{ext}(\mathbf{s}_B)$ has been discussed in Section 8.4.1. In the ICI-aware case, the detectors in Section 8.4.3 can be adopted to compute the extrinsic information $\Pr^{ext}(\mathbf{s}_A)$ and $\Pr^{ext}(\mathbf{s}_B)$.
(3) The channel decoding is applied on $\mathbf{X}_A$ and $\mathbf{X}_B$ separately. The decoder outputs extrinsic information $\Pr^{ext}(\mathbf{X}_A)$ and $\Pr^{ext}(\mathbf{X}_B)$ to the date detection module, and also makes a tentative decision for $\mathbf{X}_A$ and $\mathbf{X}_B$ based on the *a posteriori* information, separately.

The iteration between data detection and channel decoding goes on until two decoders succeed or the number of iterations reaches a predetermined threshold.

17.2.2 Iterative XOR-ed PLNC Detection and Decoding

Since the objective at the relay is to compute $\mathbf{X}_R$, instead of recovering both $\mathbf{X}_A$ and $\mathbf{X}_B$, this scheme directly performs iterative detection and decoding on the XOR-ed codeword $\mathbf{X}_A \boxplus \mathbf{X}_B$; note that $\mathbf{X}_R$ is a valid codeword by itself. The receiver diagram for this scheme is shown in Figure 17.3, where the iterations between data detection and channel decoding are as follows.

(1) The *extrinsic* information $\Pr^{ext}((\mathbf{X}_A \boxplus \mathbf{X}_B))$ from the channel decoder is transformed to the *a priori* information $\Pr^{apr}((\mathbf{s}_A, \mathbf{s}_B))$ for symbol detection. Given that the mapping from $(\mathbf{s}_A, \mathbf{s}_B)$ to $\mathbf{s}_R$, i.e., from $(\mathbf{X}_A, \mathbf{X}_B)$ to $\mathbf{X}_A \boxplus \mathbf{X}_B$ is noninvertible, the transformation is done by equally splitting the probability of each value of $\mathbf{X}_A \boxplus \mathbf{X}_B$ to all corresponding pairs of $(\mathbf{s}_A, \mathbf{s}_B)$.
(2) Based on the *a priori* information, the symbol detector computes the *extrinsic* information, $\Pr^{ext}((\mathbf{s}_A, \mathbf{s}_B))$. The data detectors in Sections 8.4.1 and 8.4.3 need proper modifications, as $(\mathbf{s}_A, \mathbf{s}_B)$ are now correlated, instead of independent as assumed in the I-SDD case. To deliver $\Pr^{apr}((\mathbf{X}_A \boxplus \mathbf{X}_B))$ needed by the decoder, a transformation is done by merging the probability of all pairs of $(\mathbf{s}_A, \mathbf{s}_B)$ corresponding to $\mathbf{s}_R$ as

$$\Pr^{ext}(s_R[k] = \alpha_\ell) = \sum_{(s_A[k], s_B[k]) \in \mathcal{A}(s_R[k]=\alpha_\ell)} \Pr^{ext}(s_A[k] = \alpha_i, s_B[k] = \alpha_j) \tag{17.10}$$

   where $\mathcal{A}(s_R[k] = \alpha_\ell)$ denotes a set formed by all the possible values of $(s_A[k], s_B[k])$, with each value satisfying $X_R[k] = X_A[k] \boxplus X_B[k]$.
(3) Channel decoding is applied on $\mathbf{X}_R = \mathbf{X}_A \boxplus \mathbf{X}_B$ directly. The decoder outputs extrinsic message to the detector and also make a tentative decision on $\mathbf{X}_R$.

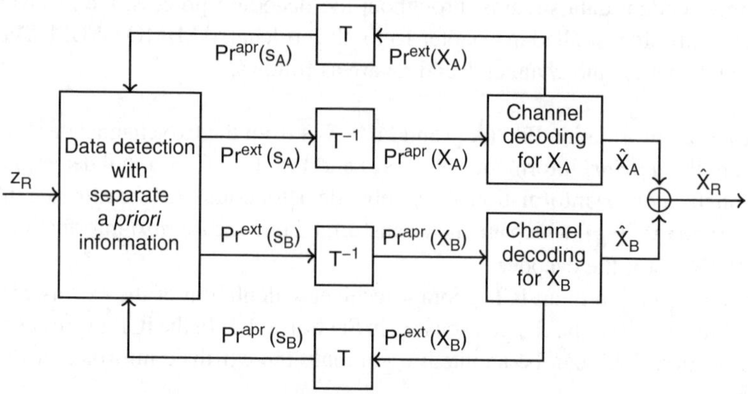

(a) Iterative separate detection and decoding (I-SDD)

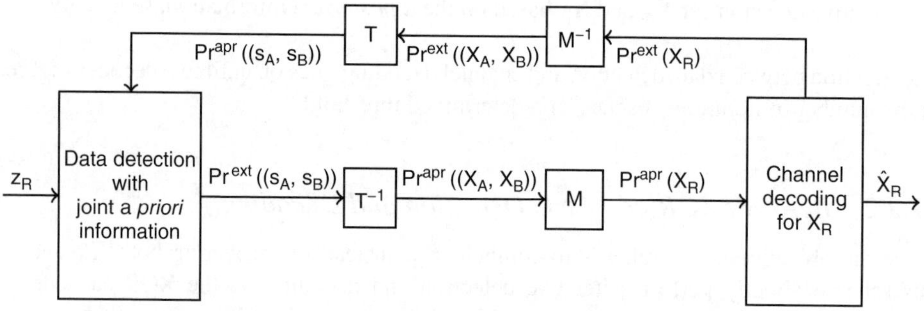

(b) Iterative XOR-ed PLNC detection and decoding (I-XPDD)

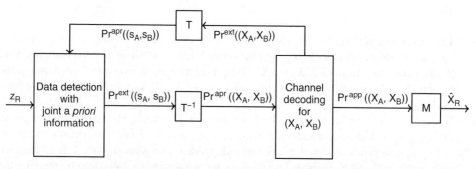

(c) Iterative generalized PLNC detection and decoding (I-GPDD)

Figure 17.3 Three iterative processing schemes; The operator T stands for the mapping between the codeword X and and the symbol vector **s**. The operator M stands for the PLNC mapping between the component codewords $\mathbf{X}_A$, $\mathbf{X}_B$ and the XOR-ed version $\mathbf{X}_R$.

17.2.3 Iterative Generalized PLNC Detection and Decoding

Compared to recovering both $\mathbf{X}_\mathrm{A}$ and $\mathbf{X}_\mathrm{B}$ separately or recovering $\mathbf{X}_\mathrm{A} \boxplus \mathbf{X}_\mathrm{B}$ directly, a better way to exploit all potentials of PLNC is to construct a super-code and perform channel decoding on $(\mathbf{X}_\mathrm{A}, \mathbf{X}_\mathrm{B})$, as proposed in [438]. A super-code over $(\mathbf{X}_\mathrm{A}, \mathbf{X}_\mathrm{B})$ is established by operating over a larger Galois field. For example, with codewords $\mathbf{X}_\mathrm{A}$ and $\mathbf{X}_\mathrm{B}$ over GF(4) together with QPSK modulation, $(\mathbf{X}_\mathrm{A}, \mathbf{X}_\mathrm{B})$ can be viewed a codeword over GF(16). In this scheme, iterative detection and decoding are performed based on the super code over $(\mathbf{X}_\mathrm{A}, \mathbf{X}_\mathrm{B})$. At the end, the PLNC mapping is used to recover $\mathbf{X}_\mathrm{A} \boxplus \mathbf{X}_\mathrm{B}$ on a symbol by symbol basis, based on soft information from the decoder for the super code. The receiver diagram for this scheme is shown in Figure 17.3, with the following iterations between data detection and channel decoding.

(1) The *extrinsic* information $\mathrm{Pr}^{\mathrm{ext}}((\mathbf{X}_A, \mathbf{X}_\mathrm{B}))$ exported from the channel decoder is transformed to the *a priori* information $\mathrm{Pr}^{\mathrm{apr}}((\mathbf{s}_A, \mathbf{s}_\mathrm{B}))$ for symbol detection.
(2) Based on the *a priori* information, the symbol detector computes the *extrinsic* information, $\mathrm{Pr}^{\mathrm{ext}}((\mathbf{s}_\mathrm{A}, \mathbf{s}_\mathrm{B}))$. Same as the I-XPDD case, the data detectors in Sections 8.4.1 and 8.4.3 need proper modifications, as $(\mathbf{s}_\mathrm{A}, \mathbf{s}_\mathrm{B})$ are now correlated. Different from the I-XPDD case, joint probabilities are directly passed to the decoder as the *a priori* information, $\mathrm{Pr}^{\mathrm{apr}}((\mathbf{X}_\mathrm{A}, \mathbf{X}_\mathrm{B}))$.
(3) Channel decoding is applied on the super code $(\mathbf{X}_\mathrm{A}, \mathbf{X}_\mathrm{B})$ with a properly defined parity check matrix. The decoder outputs extrinsic probabilities on the symbol pairs $\{(X_\mathrm{A}[k], X_\mathrm{B}[k])\}$ to the data detector, and also *a posteriori* probabilities for tentative decisions.

The iteration between detection and decoding goes on until the decoder succeeds or the number of iterations reaches a predetermined threshold. In the former case, the XOR-ed codeword $\mathbf{X}_\mathrm{R}$ is obtained via XOR operation based on hard decisions, while in the latter case, the PLNC mapping rule in (17.10) is used to recover $\{X_\mathrm{R}[k]\}$ based on *a posteriori* probabilities of $\{(X_\mathrm{A}[k], X_\mathrm{B}[k])\}$, and thus an estimate of $\mathbf{X}_\mathrm{R}$ is available.

17.3 Outage Probability Bounds in Time-Invariant Channels

Here we present the outage probability analysis with one receive element. In the time-invariant scenario, both $\mathbf{H}_\mathrm{A}$ and $\mathbf{H}_\mathrm{B}$ are diagonal matrices, meaning that there is no ICI. For this simple case, it is possible to derive a tight lower bound and a loose upper bound on the achievable rate for successful decoding at the relay node.

Consider the case where the two nodes are transmitting at the same rate R. Define

$$H_{\min}[k] = \min\left(|H_\mathrm{A}[k]|, |H_\mathrm{B}[k]|\right), \quad k \in \mathcal{S}_\mathrm{D} \tag{17.11}$$

where $\mathcal{S}_\mathrm{D}$ denotes the set of data subcarriers, and $H_\mu[k]$ denotes the kth diagonal element of matrix $\mathbf{H}_\mu$ with $\mu = \mathrm{A}$ or B. To derive the achievable rate for successful relay decoding, it is shown possible, using lattice encoding at both transmitters and lattice decoding at the relay node, to achieve a rate of ([290], Theorem 3)

$$R[k] = \log_2\left[\frac{|H_{\min}[k]|^2}{|H_\mathrm{A}[k]|^2 + |H_\mathrm{B}[k]|^2} + |H_{\min}[k]|^2\bar{\gamma}\right], \quad \forall k \tag{17.12}$$

where $\overline{\gamma}$ is the average SNR per subcarrier. On the other hand, the single-user bound dictates that

$$R[k] \leq \log_2 \left[1 + |H_{\min}[k]|^2 \overline{\gamma}\right], \quad \forall k. \tag{17.13}$$

This bound is valid because if $R[k]$ is larger than the bound, the relay node will not be able to decode the XOR-ed symbol, even given the information of one of the two transmitters (side information). Given such information, the decoding becomes a single user decoding problem.

Assume that there is no power loading across the OFDM subcarriers, as the channel state information is not available at the transmitters. Combining the achievable rate and upper bound, and summing over the data subcarriers S_D, one has

$$R \leq \frac{1}{|S_D|} \sum_{k \in S_D} \log_2 \left[1 + |H_{\min}[k]|^2 \overline{\gamma}\right], \tag{17.14}$$

$$R \geq \frac{1}{|S_D|} \sum_{k \in S_D} \log_2 \left[\frac{|H_{\min}[k]|^2}{|H_A[k]|^2 + |H_B[k]|^2} + |H_{\min}[k]|^2 \overline{\gamma}\right]. \tag{17.15}$$

With the upper and lower bounds on the achievable rate, a loose lower bound and a tight upper bound for the outage probability are obtained

$$P_{out}(R) \geq \Pr\left[\frac{1}{|S_D|} \sum_{k \in S_D} \log_2 \left(1 + |H_{\min}[k]|^2 \overline{\gamma}\right) < R\right], \tag{17.16}$$

$$P_{out}(R) \leq \Pr\left[\frac{1}{|S_D|} \sum_{k \in S_D} \log_2 \left(\frac{|H_{\min}[k]|^2}{|H_A[k]|^2 + |H_B[k]|^2} + |H_{\min}[k]|^2 \overline{\gamma}\right) < R\right]. \tag{17.17}$$

The outage performance bounds are used to benchmark the simulated performance.

17.4 Simulation Results

The ZP-OFDM parameters in the SPACE08 experiment are used for simulation, which are specified in Table B.1. The two terminals share an identical pilot subcarrier sets, but use different pilot symbols which are drawn independently from a QPSK constellation. The data within each OFDM symbol is encoded by a rate-1/2 GF(4) nonbinary near-regular LDPC code of length 672 symbols and modulated using a QPSK constellation, leading to a data rate

$$R = \frac{1}{2} \cdot \frac{|S_D|}{T + T_g} \cdot \log_2 4 = 5.2\,\text{kb/s/user}. \tag{17.18}$$

Based on the channel model depicted in (17.2), the underwater acoustic channel between each terminal and each receive element is generated randomly according to the specifications in Section 5.5.1. Unless specified, The channel parameters are $N_{pa,\nu,\mu} = 10$, $\Delta\tau = 1$ ms, $\Delta P_{pa,\nu,\mu} = 20$ dB, $T_g = 24.6$ ms, and $v_0 = 0$ m/s. In the time-invariant channel, the Doppler rate of each path is set zero. To simulate the time-varying channel, the Doppler rate of each path is drawn independently from a zero-mean uniform distribution according to the standard deviation of path speed σ_v m/s.

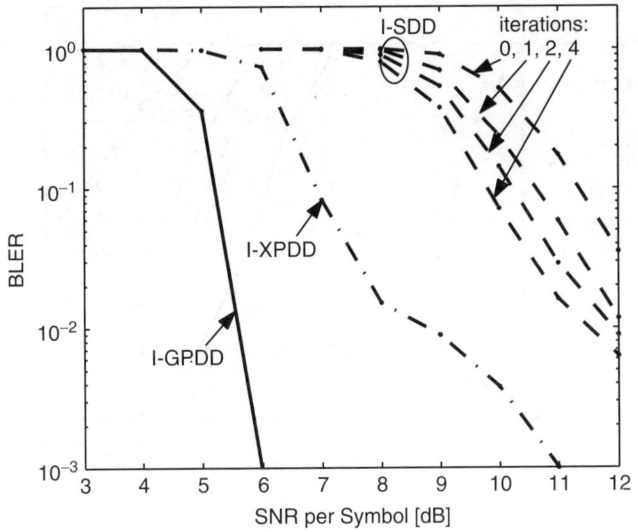

Figure 17.4 Performance comparison of different receivers for OFDM modulated PLNC without ICI in the additive white Gaussian noise channel, one receive element.

To explore the benefit of iterative processing between channel equalization and decoding, the channel estimate is assumed available prior to the symbol detection. As each OFDM symbol is encoded separately, the block-error-rate (BLER) of the XOR-ed symbol is used as the performance merit. We next examine the PLNC decoding algorithms in three channel settings.

17.4.1 The Single-Path Time-Invariant Channel

First consider a single-path time-invariant channel described by (17.2) with $N_{\text{pa},\nu,\mu} = 1$, where the path has a unit amplitude, a random delay and a zero Doppler rate. Figure 17.4 demonstrates the BLER performance of three receivers with a MAP equalizer. One can see that I-GPDD has the best performance, I-XPDD is in the middle, and I-SDD has the worst performance. This is consistent with the observation in [438]. Despite the performance improvement with the iterative processing, there is still a considerable gap between I-SDD and the other two schemes.

17.4.2 The Multipath Time-Invariant Channel

Corresponding to the multipath time-invariant channel described by (17.2) with $N_{\text{pa},\nu,\mu} = 10$ and $a_{p,\nu,\mu} = 0, \forall p$, Figure 17.5 shows performance comparison of three receivers with different numbers of receive elements. Note that in this case the iterative processing for I-XPDD and I-GPDD is not necessary. As a performance benchmark, the lower and upper bounds of the outage probability with single receive element are provided.

One can see from Figure 17.5 that the iterative processing can improve the performance of I-SDD significantly. When the number of receive elements is one, I-SDD performs worse than I-GPDD. This is reasonable because the I-SDD scheme tries to recover both $\mathbf{X}_\text{A}$ and $\mathbf{X}_\text{B}$, which

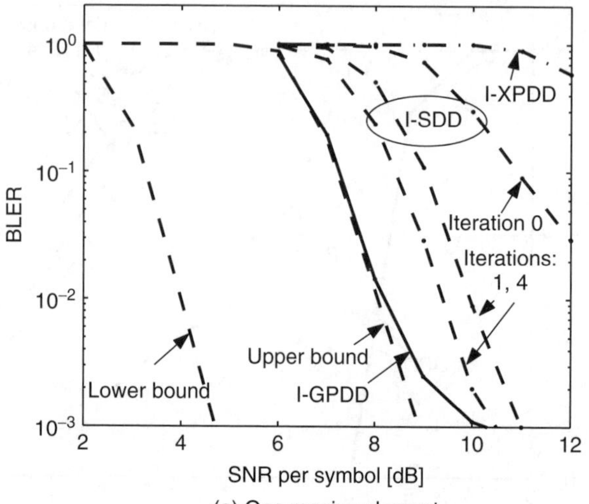

(a) One receive element

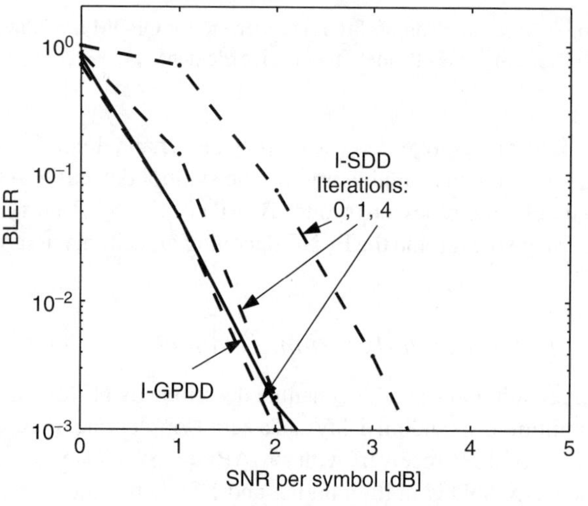

(b) Two receive elements

Figure 17.5 Performance comparison of different receivers for OFDM modulated PLNC without ICI as a function of the number of receive elements.

is under-determined when the relay node has only one receive element. When the number of receive elements is larger than one, the iterative I-SDD receiver can catch up with the I-GPDD scheme with only one iteration. Note that the decoding complexity of I-GPDD is much higher than that of I-SDD.

Meanwhile, Figure 17.5 shows that the performance of I-XPDD degrades significantly in the multipath channel relative to that in the AWGN channel shown in Figure 17.4. Hence, the scheme I-XPDD will not be included in the following simulations.

17.4.3 The Multipath Time-Varying Channel

Now consider doubly selective channels with path-specific Doppler scales, as described by (17.2) with $N_{\mathrm{pa},\nu,\mu} = 10$ and $|a_{p,\nu,\mu}| \geq 0$, $\forall p$. Figure 17.6 shows the performance comparison of I-SDD and I-GPDD with different numbers of receive elements and different standard deviations of path speed. The factor-graph based MMSE equalization is used, and the channel matrix between each terminal and receive element pair is banded with $D = 1$ and $D = 3$ for $\sigma_{\mathrm{v}} = 0.1$ m/s and $\sigma_{\mathrm{v}} = 0.2$ m/s, respectively.

Figure 17.6 shows that the iterative processing improves the performance of I-SDD significantly for both values of σ_{v}. For I-GPDD, iterative processing also boosts its performance when $\sigma_{\mathrm{v}} = 0.2$ m/s as shown in Figure 17.6 whereas not much improvement is seen when $\sigma_{\mathrm{v}} = 0.1$ m/s. With one receive element, I-GPDD outperforms I-SDD considerably. With more than one receive elements, the iterative I-SDD receiver with a lower decoding complexity can catch up with the noniterative I-GPDD scheme by using more iterations.

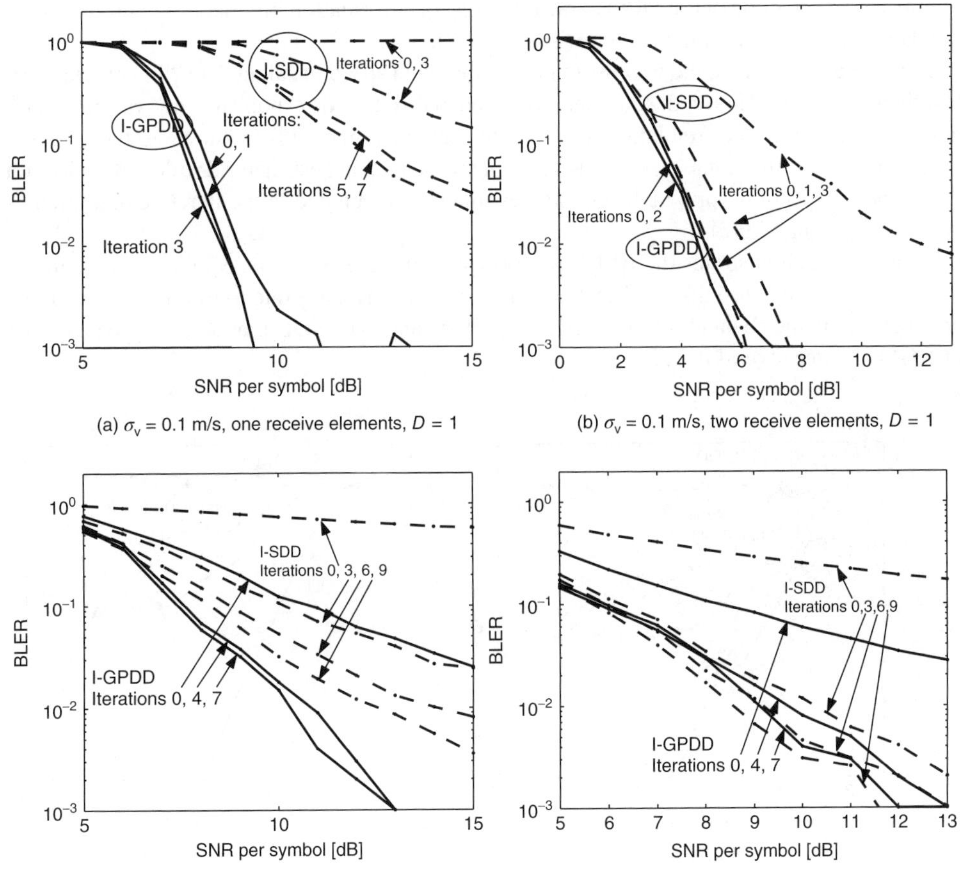

(a) $\sigma_{\mathrm{v}} = 0.1$ m/s, one receive elements, $D = 1$

(b) $\sigma_{\mathrm{v}} = 0.1$ m/s, two receive elements, $D = 1$

(c) $\sigma_{\mathrm{v}} = 0.2$ m/s, two receive elements, $D = 3$

(d) $\sigma_{\mathrm{v}} = 0.2$ m/s, four receive elements, $D = 3$

Figure 17.6 Performance comparison of different receivers for OFDM modulated PLNC with ICI as a function of the number of receive elements and the standard deviation of path speed σ_{v}.

17.5 Experimental Results: SPACE08

The data collected in the SPACE08 experiment are used to test the three iterative receivers. The experimental setup is described in Appendix B. The data sets collected in a two-input multiple-output (TIMO) experimental setting are adopted to emulate a two-way relay system. Specifically, the two co-located transmitters at the source in SPACE08 are taken as two terminals in a two-way relay network, and the receivers labeled as S1 and S3 in SPACE08, which were 60 meters and 200 meters away from the transmitter, respectively, are taken as two individual relay nodes. Each receiver was equipped with twelve elements. Although in a two-way relay network the two terminals do not necessarily have the same distance to the relay node, the TIMO experimental setting in the SPACE08 experiment captures the main feature of UWA channels in the two-way relay network.

The ZP-OFDM parameter setting is listed in Table B.1. Out of the $|\mathcal{S}_P| = 256$ pilot subcarriers, terminal A only transmits nonzero pilot symbols at the even indexed subcarriers, while terminal B transmits the nonzero pilot symbols at the odd indexed subcarriers. With a rate-1/2 nonbinary LDPC code and a QPSK constellation for information bit encoding and mapping, the data rate of each terminal is identical to (17.18).

During the experiment, the waveform height and wind speed in Julian date 298 are relatively low than other days. Hence, the data sets collected in this day correspond to slow-varying channels. Different from the simulation, the channel state information is unknown to the receiver. When processing this data set, the ICI-ignorant sparse channel estimator developed in Section 7.2 is used. Especially for I-SDD, channel estimation is included in the iterative operation as depicted in Figure 9.9(a).

Figure 17.7 demonstrates the BLER performance of three receivers. Similar to the simulation results, one can see that I-XPDD has the worst performance due to multiple paths in the channel, and in the channel setting of S1, the performance of I-SDD gradually approaches that of I-GPDD after several iterations.

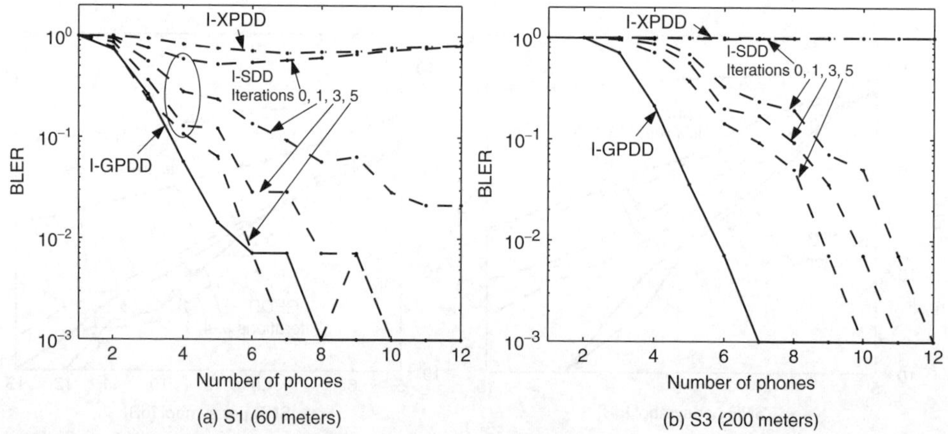

Figure 17.7 SPACE08: Performance comparison of three different receivers in the time-invariant scenario, Julian date 298.

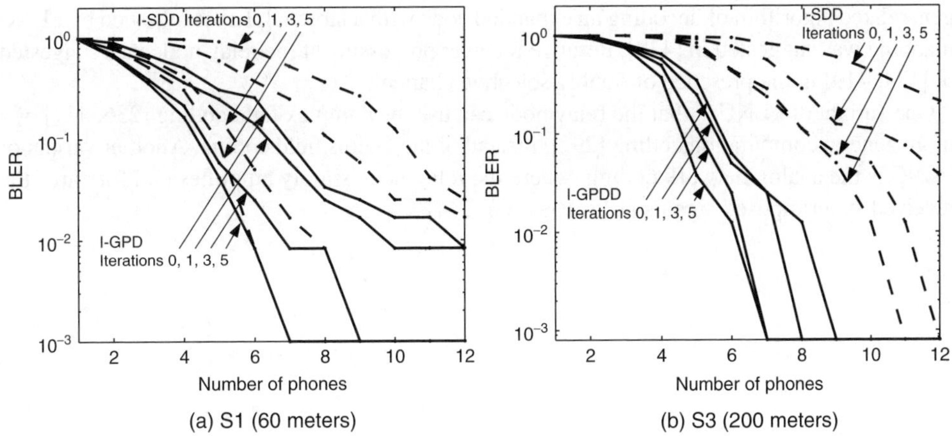

Figure 17.8 SPACE08: Performance comparison of two different receivers in the time-varying scenario, Julian date 300.

The data sets collected on Julian date 300 are used to test the decoding performance of I-SDD and I-GPDD *in the fast-varying channel with large Doppler spreads*. Some distorted files collected in this day are excluded for performance test. To estimate the ICI coefficients with regularly distributed pilots, an ICI-progressive receiver structure developed in Section 9.4 is adopted. With an iterative processing, the receiver in the first iteration assumes the absence of ICI and gets initial estimates of the transmitted information symbols. Then, the ICI-aware receiver processing with ICI depth $D = 1$ is carried out based on both pilot symbols and estimated information symbols.

Figure 17.8 demonstrates the BLER performance of two decoding schemes. One can see that the iterative operation improves the decoding performance of I-SDD and I-GPDD considerably. Although I-GPDD outperforms I-SDD overall, in some scenarios the performance of I-SDD can approach the performance of I-GPDD by using more iterations.

17.6 Bibliographical Notes

The concept of network coding was introduced by Ahlswede, Cai, Li, and Yeung [3] in 2000, where network nodes can combine data streams from multiple sources at into one or several data streams before forwarding. The last decade has witnessed extensive investigation on network coding from various aspects including network code generation [20, 21, 135, 165, 206, 216, 239, 417], decoding algorithm design [244], information theoretical analysis [327], and applications of network coding in a broad range of areas [125, 126, 151, 263], to name a few. Several books are available such as Yeung, 2008 [456], Ho and Lun, 2008 [164], Medard and Sprintson [276].

The so-called physical-layer network coding was introduced in [462, 463] in the context of two-way relay networks. Regarding to the receiver operation at the relay node as covered in this chapter, decoding the XOR-ed codeword directly was suggested in [131, 462], while the generalized algorithm of decoding an expanded code with a larger alphabet followed by PLNC mapping was suggested in [438]. Iterative receiver processing at the relay node was suggested in [174, 419] in the presence of doubly selective channels.

One variant of PLNC is that the relay node can use the complex field coding [236, 417] or a more general computation coding [292], instead of the Galois field coding. Another variant of PLNC is the analog network coding, where the relay node simply amplifies and forwards the received superimposed signals [135, 139, 206, 244].

18

OFDM Modem Development

Chapters 9 to 17 have presented OFDM receiver designs in various system setups. The algorithm development relies on simulations and offline processing of data sets collected from field experiments. To build a real communication system with multiple nodes, acoustic modems with real time processing need to be developed. This chapter describes the modem implementation aspect and presents some example designs of OFDM based acoustic modems.

18.1 Components of an Acoustic Modem

Implementation of an acoustic modem goes much beyond algorithm development. To have a standalone underwater acoustic modem which support point to point communications, the following components are needed.

- *Hardware.* A hardware platform involves various units.
  - A transducer or multiple transducers that can convert the electronic signals into acoustic signals.
  - A hydrophone or an array of hydrophones that can convert the acoustic signals into electronic signals. Note that a transducer can be used for both transmission and reception. If so, transmit-receive (TR) switch mechanism needs to be incorporated.
  - A digital signal processing board that contains DSP chips, and A/D, D/A converters. Communication algorithms are often run inside a DSP chip, and hence an acoustic modem can be viewed as a software defined modem.
  - Power amplifier with matching circuits to the transducer.
  - Preamplifier circuits with interface to the hydrophones.
  - Waterproof housing with the desired depth rating and water-proof connectors that can connect sensor packages to the modem.
  - Battery supply which could be internal or external to the modem.
- *Firmware.* The firmware inside the DSP unit deals with data acquisition and data processing. The functionalities include the following:
  - Data acquisition and buffer management
  - Detection of an incoming packet and synchronization

OFDM for Underwater Acoustic Communications, First Edition. Shengli Zhou and Zhaohui Wang.
© 2014 John Wiley & Sons, Ltd. Published 2014 by John Wiley & Sons, Ltd.

- Demodulation, channel estimation, and data decoding
- Ranging support
- Command interface

The DSP firmware could be frequently updated without changing the hardware platform.

Note that the performance results in published papers, in terms of the data rate and the bit error rate, are often much more impressive than those achieved with existing modems, since the practical modems have limited processing capabilities and less hardware resources.

In addition to point-to-point communications, a modem could also be equipped with networking functionalities. Although networking protocols, such as medium access control, routing, and data transport protocols, are often implemented on a different micro-controller separate from the modem, it is possible to integrate the protocols into the modem itself. In such cases, the modem provides application programming interface (API) for networking applications. Cross layer optimization would be necessary to improve the protocol efficiency, as upper layer protocols can certainly benefit from the channel measurements available at the physical layer.

18.2 OFDM Acoustic Modem in Air

Demonstration of acoustic communication algorithms can be accomplished with low cost in the air, which often serves the educational purposes. A team of undergraduate students at the University of Connecticut has constructed a two-way communication platform based on OFDM modulation [273]. As shown in Figure 18.1, the hardware platform for each node consists of one computer, one speaker, one microphone and one external sound card. The two laptops run Matlab programs for acoustic signal processing. The sampling rate is 44.1 kHz, the bandwidth used is 5.5 kHz, and the center frequency is 12 kHz. A total of 1024 subcarriers are used, and the subcarrier assignment follows that in Table 2.1. With rate 1/2 coding, QPSK modulation and guard interval $T_g = 30$ ms, the achieved data rate was 3.1 kb/s with a single data stream [273].

A MIMO-OFDM implementation has been further explored. As shown in Figure 18.2, the transmitter uses two speakers, and the receiver uses two hydrophones. The data rate increases to 6.2 kb/s due to two parallel data streams, with the same parameters as the single-transmitter case.

18.3 OFDM Lab Modem

The Underwater Sensor Network (UWSN) lab at University of Connecticut has developed an OFDM lab modem prototype based on a DSP development board [444–446]. As shown in Figure 18.3, each OFDM lab modem consists of one development board with the OMAP-L137 chip (having TMS320C6747 DSP inside) from Texas Instruments, one hydrophone as the transmitter, two hydrophones as the receiver, and a small custom made board having one power amplifier and two channels of pre-amplifiers. The lab modem was placed in a gray plastic box with power and RS-232 connectors on its panel. The sampling rate is 48 kHz, the bandwidth is 6 kHz, and the center frequency is 17 kHz. The total number of subcarriers is 1024, and

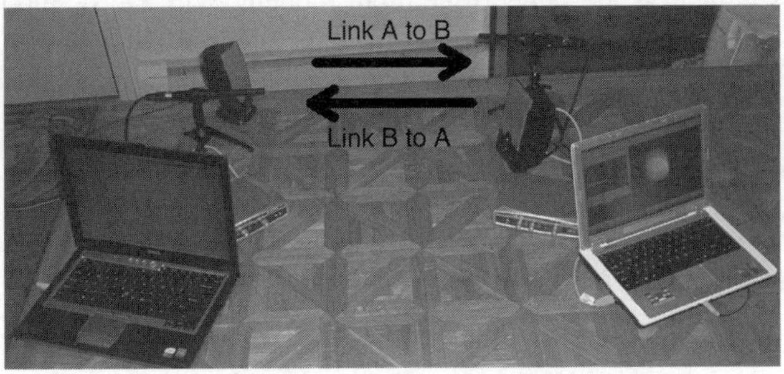

Figure 18.1 Computer-based in-air two-way communication using OFDM modulation.

Figure 18.2 Computer-based in-air MIMO-OFDM demonstration.

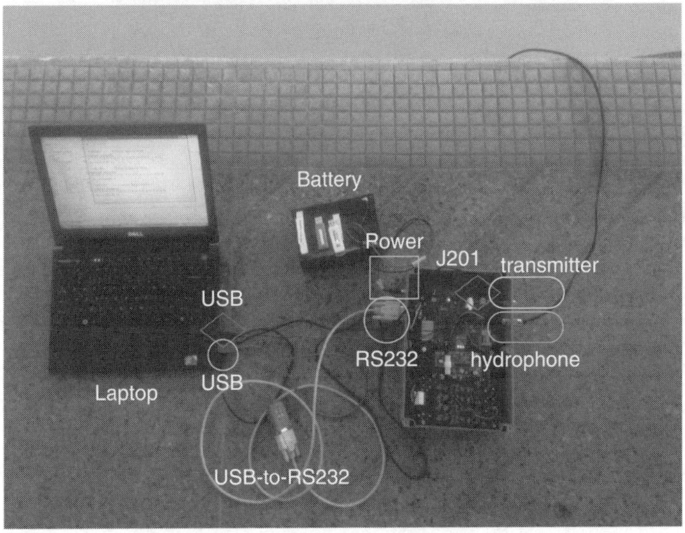

Figure 18.3 The OFDM lab modem as deployed in a swimming pool.

the subcarrier assignment is the same as the in-air modem. With QPSK modulation, rate 1/2 channel coding, and a guard time $T_g = 50$ ms, the data rate is 3.0 kb/s.

These OFDM lab modems are good for network testings in the water lank and in swimming pools [371]. These are good educational tools to have students engaged in hand-on networking experiments in a water environment.

18.4 AquaSeNT OFDM Modem

AquaSeNT, a startup from the University of Connecticut, has licensed the OFDM modem technologies from the University of Connecticut and has launched a commercial modem product in Fall 2012. Figure 18.4 shows a picture of the modem.

Five transmission modes shown in Table 18.1 have been developed. All of them use LDPC channel codes in GF(4), due to the implementation considerations. These 5 modes have different channel code lengths, code rates (r_c) and modulation size (BPSK, QPSK, 16-QAM).

All the modes have 672 data symbols after coding and modulation, which corresponds to $|S_D| = 672$ data subcarriers. The implemented OFDM system has bandwidth $B = 6$ kHz, total number of subcarriers $K = 1024$, which will lead to symbol duration $T = 170.7$ ms. With $T_g = 50$ ms, the payloads in each OFDM data block for the five transmission modes are listed in Table 18.2.

The power level, the transmission mode, and the guard time are controlled by the users. Using these five modes, adaptive modulation and coding can be developed as shown in [413]. The AquaSeNT OFDM modems have been used to test the DCC-OFDM protocol in Chapter 16 and the localization algorithms in Chapter 19. They have also been used to develop networking testbeds as reported in [91, 430].

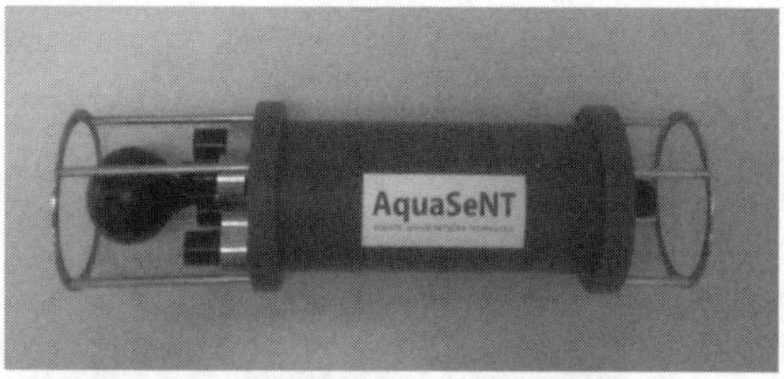

Figure 18.4 The AquaSeNT's OFDM modem (photo courtesy of AquaSeNT).

Table 18.1 Modulation and coding pairs in the OFDM modem

Index	Matrix size	Code rate	Constellation
1	168 × 336	1/2	BPSK
2	336 × 672	1/2	QPSK
3	168 × 672	3/4	QPSK
4	672 × 1344	1/2	16-QAM
5	336 × 1344	3/4	16-QAM

Table 18.2 Payload of the five transmission modes with T_g = 50 ms (note that 4 bytes reserved by the modem physical layer are excluded in the computation)

Transmission mode	Payload per block [bytes]	Data rate [kb/s]
TM 1	38	1.38
TM 2	80	2.90
TM 3	122	4.42
TM 4	164	5.94
TM 5	248	8.99

18.5 Bibliographical Notes

There are several commercial acoustic modem products on the market, e.g., Teledyne Benthos [33], LinkQuest [252], EvoLogics [116], Develogic [95], Kongsberg [219], SonarDyne [345], DSPCOMM [105], and AquaSeNT [10]. A research modem widely used in the research community is the micro-modem from the Woods Hole Oceanographic Institution (WHOI) [127, 138]. A recent research modem, the UNET-2 modem, from National University of Singapore is described in [80]. The implementation of OFDM has been reported in [447] and [444–446]. The prototyping efforts reported in [32, 132, 359, 359, 434] are based on other modulation schemes.

19

Underwater Ranging and Localization

Underwater localization is a topic of great interest and study, and application demands drive the need for better and better solutions. Several current systems feature augmented inertial navigation methods, which use filtering and tracking methods to provide corrections and improvements upon traditional onboard navigational equipments [158, 231]. Aside from these methods, there are several localization techniques based on acoustic signaling.

- Long baseline (LBL) system. Several transponders are installed at the sea floor, and an underwater vehicle interrogates the transponders for round-trip delay estimation followed by triangulation [370]. LBL has good localization accuracy, but it requires long-time calibration.
- Short baseline (SBL) system. A series of closely spaced receivers can be installed on a platform such as a surface ship to monitor the incoming signals from an underwater emitter. The time differences of arrivals (TDoAs) are used for localization.
- Ultra-short-baseline (USBL) system. A small array of hydrophones is used to estimate the angle of arrival (AoA) of the incoming signal from an underwater emitter. The AoA information is combined with a range estimate for much improved localization performance than SBL systems.
- Floating buoy based system. This system acts like a long base line system except that the reference points are surface buoys. There are commercial products – the GPS Intelligent Buoys (GIB) – that route signals from an underwater node to surface buoys [27], and using radio links the surface buoys forward all information to a mother ship, wherein the localization is performed. The floating buoys are easier to deploy and calibrate than LBL systems.

The LBL, SBL, USBL, and GIB systems are commercially available, and one can find technical specifications from the providers. This chapter focuses on two localization schemes that are developed and evaluated using OFDM modem prototypes [62] and [61]. Both approaches are based on range measurements, and are applicable to a network with multiple nodes.

OFDM for Underwater Acoustic Communications, First Edition. Shengli Zhou and Zhaohui Wang.
© 2014 John Wiley & Sons, Ltd. Published 2014 by John Wiley & Sons, Ltd.

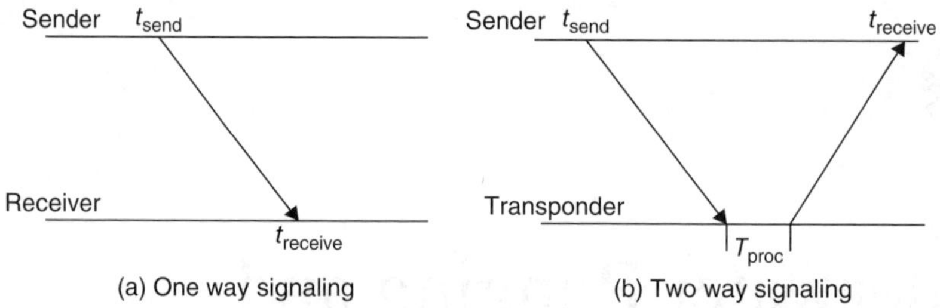

Figure 19.1 Illustration of ranging via measurement of the propagation delay.

19.1 Ranging

Ranging is to determine the distance d between the transmitter and the receiver. Let T_{prop} be the one way signal propagation time between the two nodes, and c be the speed of sound in water. The distance estimate can be found as

$$\hat{d} = \hat{T}_{prop}\hat{c}, \tag{19.1}$$

where both the propagation time and the sound speed need to be estimated in practice.

19.1.1 One-Way Signaling

As shown in Figure 19.1(a), the sender sends a message to the receiver and the receiver records the arrival time of the message. This can be done by cross correlation or other approaches. One way ranging needs clock synchronization between the transmitter and the receiver. The start time can be triggered on a fixed time interval, or the transmission time can be stamped into the message.

$$\hat{T}_{prop} = t_{receive} - t_{send} \tag{19.2}$$

Assuming that there is a clock offset b between the transmitter and the receiver, then the estimate can be modeled as:

$$\hat{T}_{prop} = T_{prop} + b + w \tag{19.3}$$

where w is the measurement noise.

19.1.2 Two-Way Signaling

The transmitter sends a message to a transponder. The transponder receives the message and sends back an immediate acknowledgement, as illustrated in Figure 19.1(b). With a fixed processing delay T_{proc} at the transponder, the one way propagation time is estimated as

$$\hat{T}_{prop} = \frac{1}{2}(t_{receive} - t_{send} - T_{proc}) \tag{19.4}$$

If T_{proc} is accurate, the estimate in (19.4) is unbiased as

$$\hat{T}_{prop} = T_{prop} + w \tag{19.5}$$

Hence, two-way message passing eliminates the need of synchronization of the transmitter and the receiver.

19.1.3 Challenges for High-Precision Ranging

Underwater acoustic channels consist of many propagation paths. The first path is not necessarily the strongest, due to the superposition effect of neighboring paths that may not be resolved individually. Timing accuracy depends on how effectively the first path can be identified, rather than the strongest path. Also, the sound speed might not be constant spatially or temporally. Due to nonuniform sound speed, the propagation path might be slanted rather than a straight line. These effects need to be properly compensated to improve the ranging accuracy.

19.2 Underwater GPS

This section presents the localization approach in [62] based on messages broadcast from multiple surface nodes, coupled with tracking algorithms and implemented on a physical system to provide a complete analysis. With the time-of-arrival measurements, the receiver computes its own localization based on the differences of the travel times from multiple senders to the receiver. The advantage of the proposed localization method is that the broadcast messages can serve an arbitrary number of underwater nodes once they are in range, in contrast to many existing solutions which can only serve a small number of users.

19.2.1 System Overview

Figure 19.2 depicts the system setup considered, with several surface nodes and multiple underwater nodes. The surface nodes are equipped with satellite-based GPS receivers. Relying on the internal pulse provided by the GPS device that is accurate to within 1 microsecond GPS time, the surface nodes are assumed to be well synchronized. At predetermined intervals, the surface nodes sequentially broadcast their current location and time.

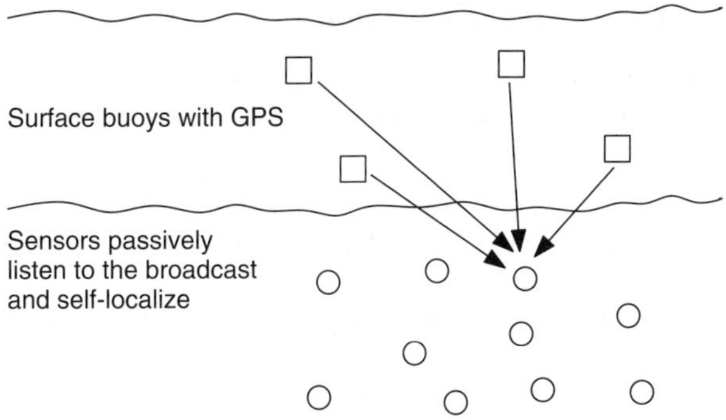

Figure 19.2 An underwater sensor network with multiple surface buoys.

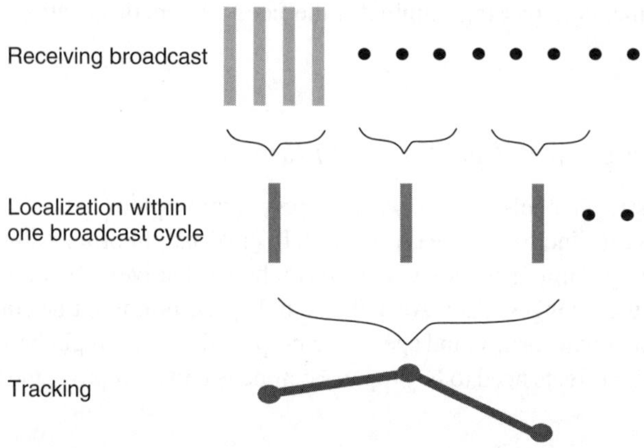

Figure 19.3 One round of surface buoy broadcasting leads to one point estimate, while tracking algorithms improve the localization accuracy when multiple point estimates are available.

The underwater nodes within the broadcast range will detect a series of transmissions and decode those messages. By comparing the reception time with the transmission time encoded in the message, each underwater node can obtain estimates of the time-of-arrival (or time-of-flight) of messages from different surface nodes, based on which it tries to compute its own position. Note that the broadcast from the surface to underwater nodes is a one-way transmission, that localization quality is independent of the number of underwater nodes in the network. Also, it is assumed that there is no additional interference involved among different underwater nodes.

The overall scheme of localization refinement is presented in Figure 19.3. For one round of transmissions, the receiver obtains the travel times from all the surface buoys, based on which a point estimate is available. Once several of these broadcast periods have occurred, individual point estimates may be combined via tracking algorithms to form a more accurate understanding of the current node position. The algorithms at different stages are described in the following sections.

19.2.2 One-Way Travel Time Estimation

Let us focus on one receiver at position (x_r, y_r, z_r). Suppose that there are N surface nodes, at positions (x_n, y_n, z_n), $n = 1, \ldots, N$. Let d_n denote the distance between the receiver node and the nth surface node:

$$d_n = \sqrt{(x_r - x_n)^2 + (y_r - y_n)^2 + (z_r - z_n)^2}. \tag{19.6}$$

Without loss of generality, set the first surface node at the origin, i.e., $x_1 = y_1 = z_1 = 0$, such that

$$d_1^2 = x_r^2 + y_r^2 + z_r^2. \tag{19.7}$$

The actual time of arrival is $t_n = d_n/c$, where c is the sound propagation speed.

The transmission time is encoded at each broadcast message. The receiver needs to estimate the arrival time of each message to provide an estimate on the time of arrival t_n. First, the communication channel is monitored to detect signal arrivals, based on a background noise level monitoring performed by the modem at initialization. When a signal is detected, coarse synchronization can be achieved via a correlation method. After coarse synchronization, the known preamble can be used to estimate the instantaneous underwater channel conditions, and from there, a more refined estimation of the time of arrival is performed via the modified Page test as in [2].

Assume that the synchronization offset is nearly identical across modems with similar hardware and operating software with the GPS synchronization, the localization algorithms will be carried out based on

$$\hat{t}_n = t_n + b + w_n, \quad n = 1, \dots, N. \tag{19.8}$$

19.2.3 Localization

In each round of broadcasting from all surface nodes, a node collects N travel time measurements, and can form a single point estimate of its current position. This is accomplished by way of localization algorithms based on the intersection of spherical surfaces.

Since the bias b is unknown and usually large, time-of-arrival (TOA) based methods are not suitable. The time-difference-of-arrival (TDOA) method cancels the common bias term b by forming

$$\Delta\hat{t}_{n1} = \hat{t}_n - \hat{t}_1, \quad n = 2, \dots, N. \tag{19.9}$$

The distance difference $d_{n1} = d_n - d_1$ is then estimated by

$$\hat{d}_{n1} = c\Delta\hat{t}_{n1}. \tag{19.10}$$

The TDOA method also corrects for clock skew alongside this bias term, due to the nature of the shared GPS clock. Each receiving node will have its own internal clock, which at some update period k will have drifted by an unknown skew factor $\phi(k)$. Each of the surface transmitters, however, will have the same clock skew, and due to the periodic corrections by the GPS clock, this value can be assumed approximately 0 for any period k. Thus, each transmission time can be represented as

$$\hat{t}_n = t_n + b + \phi(k) + w_n, \quad n = 1, \dots, N. \tag{19.11}$$

and again, by taking the difference of the time-of-arrival estimates, this common clock skew is eliminated from the timing estimate.

Next two localization methods are presented, based on the exhaustive search and least-squares formulations.

19.2.3.1 Exhaustive Search

The individual time estimates $\hat{t}_n$ generally have correlated noise in the underwater channel. For simplicity, assume that they are independent and identically distributed (i.i.d.), and hence a maximum likelihood solution can be found through

$$(\hat{x}_r, \hat{y}_r, \hat{z}_r) = \arg\min_{x_r, y_r, z_r} \sum_{n=2}^{N} [c\Delta\hat{t}_{n1} - (d_n - d_1)]^2. \tag{19.12}$$

The solution to (19.12) is found by exhaustive search.

Subtracting a common random variable, $\hat{t}_1$, from all subsequent TOA estimates leads to correlation among measurements by a factor of approximately $1/2$. As such, assuming again i.i.d. measurements, a differencing measurement bias modification can be made as follows:

$$(\hat{x}_r, \hat{y}_r, \hat{z}_r) = \arg\min_{x_r, y_r, z_r} (c\hat{\mathbf{t}}_\Delta - \mathbf{d}_\Delta)^T \boldsymbol{\Sigma}^{-1} (c\hat{\mathbf{t}}_\Delta - \mathbf{d}_\Delta)^T. \tag{19.13}$$

where

$$\hat{\mathbf{t}}_\Delta = \begin{bmatrix} \Delta\hat{t}_{21} \\ \Delta\hat{t}_{31} \\ \vdots \\ \Delta\hat{t}_{n1} \end{bmatrix}, \quad \mathbf{d}_\Delta = \begin{bmatrix} d_2 - d_1 \\ d_3 - d_1 \\ \vdots \\ d_n - d_1 \end{bmatrix}$$

and $\boldsymbol{\Sigma}$ is an $(N-1) \times (N-1)$ normalized covariance matrix

$$\boldsymbol{\Sigma} = \begin{bmatrix} 1 & 1/2 & 1/2 & \dots & 1/2 \\ 1/2 & 1 & 1/2 & \dots & 1/2 \\ \vdots & \vdots & \vdots & \dots & \vdots \\ 1/2 & 1/2 & 1/2 & \dots & 1 \end{bmatrix}. \tag{19.14}$$

19.2.3.2 Least-Squares Solution

Here we present the least-squares solution from [278]. Since $d_n = d_{n1} + d_1$, it follows that

$$(d_{n1} + d_1)^2 = x_n^2 + y_n^2 + z_n^2 - 2x_n x_r - 2y_n y_r - 2z_n z_r + d_1^2, \tag{19.15}$$

which can be simplified as

$$x_n x_r + y_n y_r + z_n z_r = \frac{1}{2}\left([x_n^2 + y_n^2 + z_n^2 - d_{n1}^2]\right) - d_{n1} d_1. \tag{19.16}$$

Define the following matrix and vectors

$$\mathbf{H} = \begin{bmatrix} x_2 & y_2 & z_2 \\ x_3 & y_3 & z_3 \\ \vdots & \vdots & \vdots \\ x_N & y_N & z_N \end{bmatrix}, \quad \mathbf{v} = \begin{bmatrix} -\hat{d}_{21} \\ -\hat{d}_{31} \\ \vdots \\ -\hat{d}_{N1} \end{bmatrix} \tag{19.17}$$

$$\mathbf{u} = \frac{1}{2} \begin{bmatrix} x_2^2 + y_2^2 + z_2^2 - \hat{d}_{21}^2 \\ x_3^2 + y_3^2 + z_3^2 - \hat{d}_{31}^2 \\ \vdots \\ x_N^2 + y_N^2 + z_N^2 - \hat{d}_{N1}^2 \end{bmatrix}, \quad \mathbf{a} = \begin{bmatrix} x_r \\ y_r \\ z_r \end{bmatrix}. \tag{19.18}$$

The least-squares solution can be obtained as

$$\hat{\mathbf{a}} = d_1 \mathbf{H}^\dagger \mathbf{v} + \mathbf{H}^\dagger \mathbf{u}, \tag{19.19}$$

where † stands for pseudo-inverse. Substituting the entries of $\hat{\mathbf{a}}$ into (19.7) yields a quadratic equation for d_1 [278]. Solving for d_1 and substituting the positive root back into (19.19) provides the final solution for the receiver position **a**.

19.2.4 Tracking Algorithms

To further reduce the localization error from a single point measurement, tracking algorithms can be implemented to combine the knowledge of multiple measurements into a more accurate position estimate.

In deciding which tracking approach would be best one needs to consider the scenarios in which the node is being localized. There are two distinct modes in which underwater nodes typically move: either passively, with the water currents as a free-floating node, or actively as an underwater vehicle such as an AUV. Both are characterized primarily by long periods of relatively straight motion at a fairly constant speed. Typically, AUV motion differs in that at certain random intervals, it will change direction according to operator or pre-programmed instruction. Most search patterns for AUVs are defined by spiral paths, or by rectangular search grids. In either case, the vehicle is likely to alter its direction by way of a continuous turn; that is, to make a turn at a fixed angular velocity until the desired heading is achieved (or in the case of a spiral, until the search area is exhausted).

19.2.4.1 Kalman Filter

In the KF, one can model the movement of the node as set of nearly constant velocity ("kinematic") models [22], with a separate model for each possible direction; that is, x, y and z. The state vector at time k is defined as $\boldsymbol{\ell}(k) = [x(k), \dot{x}(k), y(k), \dot{y}(k), z(k), \dot{z}(k)]^T$. The state equation for the Kalman filter at time index $k+1$ based on information from time step k becomes

$$\ell(k+1) = \mathbf{F}(k)\ell(k) + \varsigma(k) \tag{19.20}$$

with measurement

$$\mathbf{p}(k+1) = \mathbf{H}(k+1)\ell(k+1) + \mathbf{w}(k+1) \tag{19.21}$$

where

$$\mathbf{F}(k) = \begin{bmatrix} 1 & \tau & 0 & 0 & 0 & 0 \\ 0 & 1 & 0 & 0 & 0 & 0 \\ 0 & 0 & 1 & \tau & 0 & 0 \\ 0 & 0 & 0 & 1 & 0 & 0 \\ 0 & 0 & 0 & 0 & 1 & \tau \\ 0 & 0 & 0 & 0 & 0 & 1 \end{bmatrix} \tag{19.22}$$

$$\mathbf{H}(k+1) = \begin{bmatrix} 1 & 0 & 0 & 0 & 0 & 0 \\ 0 & 0 & 1 & 0 & 0 & 0 \\ 0 & 0 & 0 & 0 & 1 & 0 \end{bmatrix}, \tag{19.23}$$

$\varsigma(k)$ is process noise, $\mathbf{w}(k)$ is measurement noise and τ is the sampling interval of the discrete model in seconds.

The state covariance is modeled as

$$\mathbf{P}(k+1|k) = \mathbf{F}(k)\mathbf{P}(k|k)\mathbf{F}(k)^{\mathrm{T}} + \mathbf{Q}(k) \tag{19.24}$$

The corresponding process noise has a covariance given as:

$$\mathbf{Q} = \begin{bmatrix} \frac{1}{4}\tau^4 & \frac{1}{2}\tau^3 & 0 & 0 & 0 & 0 \\ \frac{1}{2}\tau^3 & \tau^2 & 0 & 0 & 0 & 0 \\ 0 & 0 & \frac{1}{4}\tau^4 & \frac{1}{2}\tau^3 & 0 & 0 \\ 0 & 0 & \frac{1}{2}\tau^3 & \tau^2 & 0 & 0 \\ 0 & 0 & 0 & 0 & \frac{1}{4}\tau^4 & \frac{1}{2}\tau^3 \\ 0 & 0 & 0 & 0 & \frac{1}{2}\tau^3 & \tau^2 \end{bmatrix} \sigma_\varsigma^2. \tag{19.25}$$

Here, σ_ς is a design parameter that is chosen to match the most likely level of process noise to be experienced by the object in question; which is to say it controls how much the model anticipates the object to maneuver. Given that the object in question is likely to be either stationary or altering its velocity at a slow, steady rate, a process noise level of $\sigma_\varsigma = 0.5\ \mathrm{m/s^2}$ was selected to best emulate this behavior for the test results in this chapter. The filter was initialized with two-point differencing [22].

19.2.4.2 Probabilistic Data Association Filter

Within a tracking window, there might be point estimates which would appear as outliers by a considerable margin, due to timing errors at the physical layer. Due to the assumption implicit to the KF approach that all of our messages are the direct-path propagation, this resulted in a drastic alteration of the point estimate, to the point where it could be classified as a false measurement. In that context, the Probabilistic Data Association Filter (PDAF) offers an improved performance over the standard KF, by allowing outlier estimates such as these to be ignored as false-alarm or clutter detections [23]. The PDAF is very similar to the KF in terms of state equations, presented here for measurement $k+1$:

$$\ell(k+1) = \mathbf{F}(k)\ell(k) + \varsigma(k) \tag{19.26}$$

with measurement

$$\mathbf{p}(k+1) = \mathbf{H}(k+1)\ell(k+1) + \mathbf{w}(k+1) \tag{19.27}$$

where $\mathbf{F}(k)$ and $\mathbf{H}(k+1)$ are given in (19.22) and (19.23), respectively.

The state covariance is modeled similarly as

$$\mathbf{P}(k+1|k) = \mathbf{F}(k)\mathbf{P}(k|k)\mathbf{F}(k)^{\mathrm{T}} + \mathbf{Q}(k). \tag{19.28}$$

The difference is on how to compute $\mathbf{P}(k|k)$. First, let P_{D} denote the probability of detection, which is a design parameter. Operating under the assumption that the measurement is always gated, define

$$b = 2\left(\frac{1 - P_{\mathrm{D}}}{P_{\mathrm{D}}}\right). \tag{19.29}$$

Define the following variables and matrices:

$$\boldsymbol{\nu}(k) = \mathbf{p}(k) - \mathbf{H}(k)\boldsymbol{\ell}(k) \tag{19.30}$$

$$e(k) = e^{-\frac{1}{2}\boldsymbol{\nu}(k)^{\mathrm{T}}\mathbf{S}^{-1}(k)\boldsymbol{\nu}(k)}, \tag{19.31}$$

$$\mathbf{S}(k) = \mathbf{H}(k)\mathbf{P}(k|k-1)\mathbf{H}(k)^{\mathrm{T}} + \mathbf{R}, \tag{19.32}$$

$$\mathbf{W}(k) = \mathbf{P}(k|k-1)\mathbf{H}(k)^{\mathrm{T}}\mathbf{S}^{-1}(k), \tag{19.33}$$

where $\mathbf{R}$ is the observation noise covariance. The probability of no correct measurement available (meaning that the measurement provided is so corrupted as to be "false") is

$$\beta_0(k) = \frac{b}{b + e(k)} \tag{19.34}$$

the probability of a correct measurement is

$$\beta_1(k) = \frac{e(k)}{b + e(k)}. \tag{19.35}$$

Further defining

$$\mathbf{P}_{\mathrm{c}}(k|k) = \mathbf{P}(k|k-1) - \mathbf{W}(k)\mathbf{S}(k)\mathbf{W}(k)^{\mathrm{T}}, \tag{19.36}$$

$$\tilde{\mathbf{P}}(k) = \mathbf{W}(k)(\beta_1(k)\boldsymbol{\nu}(k)\boldsymbol{\nu}(k)^{\mathrm{T}} - \boldsymbol{\nu}(k)\boldsymbol{\nu}(k)^{\mathrm{T}})\mathbf{W}(k)^{\mathrm{T}}, \tag{19.37}$$

the covariance matrix update is as follows:

$$\mathbf{P}(k|k) = \beta_0(k)\mathbf{P}(k|k-1) + \beta_1(k)\mathbf{P}_{\mathrm{c}}(k|k) + \tilde{\mathbf{P}}(k). \tag{19.38}$$

Assume that the measurement is always gated, and that there is only a single target and a single measurement at each time step. Accordingly, P_{D} is the probability that the current measurement is a valid estimate of the node being tracked. Based on experimental data used in this chapter, the number of "false detection" measurements was around 5% of the total samples, and so a value of 0.95 was selected for P_{D}.

19.2.4.3 Interacting Multiple Model (IMM) Filter

For the more complex motion of an active underwater node, an Interacting Multiple Model filter (IMM) was implemented, as the expected maneuvering index of underwater vehicles can easily exceed the threshold for which a single linear filter is likely to have any benefit. To this end, the IMM was a simple two-model filter, with a single, linear, low process noise ($\sigma_{\varsigma} = 0.05$ m/s^2) KF to account for the straight motion travel, and an extended Kalman filter (EKF), configured in a coordinated-turn mode [414]. This validity of the coordinated turn assumption is dependent on the scenario in question, though given the previously described search patterns, it should be sufficiently accurate [119].

The linear KF uses similar system equations as given previously, augmented with an additional column and row of zeros in order to accommodate the use of the EKF's additional state in the IMM. The EKF in this problem uses one of two sets of state equations: the first set is an approximation used when the predicted coordinated turn rate is near 0 ($\hat{\Omega}(k) \approx$), and the second set is used when the predicted coordinated turn rate is greater than some detection threshold ($|\hat{\Omega}(k)| > 0$) [22].

The first set of EKF state equation modifications ($\hat{\Omega}(k) \approx 0$) is as follows:

$$\mathbf{F}_L(k) = \begin{bmatrix} 1 & \tau & 0 & 0 & 0 & 0 & -\frac{1}{2}\tau^2\hat{\dot{\eta}}(k) \\ 0 & 1 & 0 & 0 & 0 & 0 & -\tau\hat{\dot{\eta}}(k) \\ 0 & 0 & 1 & \tau & 0 & 0 & \frac{1}{2}\tau^2\hat{\dot{\xi}}(k) \\ 0 & 0 & 0 & 1 & 0 & 0 & \tau\hat{\dot{\xi}}(k) \\ 0 & 0 & 0 & 0 & 1 & \tau & 0 \\ 0 & 0 & 0 & 0 & 0 & 1 & 0 \\ 0 & 0 & 0 & 0 & 0 & 0 & 1 \end{bmatrix} \tag{19.39}$$

where η and ξ represent the x and y directions, respectively, and $\dot{\eta}$ is the velocity component in the η direction. When $|\hat{\Omega}(k)| > 0$,

$$\mathbf{F}_L(k) = \begin{bmatrix} 1 & \frac{\sin\hat{\Omega}(k)}{\hat{\Omega}(k)}\tau & 0 & -\frac{1-\cos\hat{\Omega}(k)\tau}{\hat{\Omega}(k)} & 0 & 0 & f_{\Omega,1(k)} \\ 0 & \cos\hat{\Omega}(k)\tau & 0 & -\sin\hat{\Omega}(k)\tau & 0 & 0 & f_{\Omega,2(k)} \\ 0 & \frac{1-\cos\hat{\Omega}(k)\tau}{\hat{\Omega}(k)} & 1 & \frac{\sin\hat{\Omega}(k)\tau}{\hat{\Omega}(k)} & 0 & 0 & f_{\Omega,3(k)} \\ 0 & \sin\hat{\Omega}(k)\tau & 0 & \cos\hat{\Omega}(k)\tau & 0 & 0 & f_{\Omega,4(k)} \\ 0 & 0 & 0 & 0 & 1 & \tau & 0 \\ 0 & 0 & 0 & 0 & 0 & 1 & 0 \\ 0 & 0 & 0 & 0 & 0 & 0 & 1 \end{bmatrix} \tag{19.40}$$

where the partial derivatives $f_{\Omega,1(k)},\ldots f_{\Omega,4(k)}$ are found as:

$$\begin{aligned} f_{\Omega,1(k)} =& \frac{(\cos\hat{\Omega}(k)\tau)\tau\hat{\dot{\xi}}(k)}{\hat{\Omega}(k)} - \frac{(\sin\hat{\Omega}(k)\tau)\hat{\dot{\xi}}(k)}{\hat{\Omega}(k)^2} \\ &- \frac{(\sin\hat{\Omega}(k)\tau)\tau\hat{\dot{\eta}}(k)}{\hat{\Omega}(k)} - \frac{(-1+\cos\hat{\Omega}(k)\tau)\hat{\dot{\eta}}(k)}{\hat{\Omega}(k)^2} \end{aligned} \tag{19.41}$$

$$f_{\Omega,2(k)} = -(\sin\hat{\Omega}(k)\tau)\tau\hat{\dot{\xi}}(k) - (\cos\hat{\Omega}(k)\tau)\tau\hat{\dot{\eta}}(k) \tag{19.42}$$

$$\begin{aligned} f_{\Omega,3(k)} =& \frac{(\sin\hat{\Omega}(k)\tau)\tau\hat{\dot{\xi}}(k)}{\hat{\Omega}(k)} - \frac{(1-\cos\hat{\Omega}(k)\tau)\hat{\dot{\xi}}(k)}{\hat{\Omega}(k)^2} \\ &+ \frac{(\cos\hat{\Omega}(k)\tau)\tau\hat{\dot{\eta}}(k)}{\hat{\Omega}(k)} - \frac{(\sin\hat{\Omega}(k)\tau)\hat{\dot{\eta}}(k)}{\hat{\Omega}(k)^2} \end{aligned} \tag{19.43}$$

$$f_{\Omega,4(k)} = (\cos\hat{\Omega}(k)\tau)\tau\hat{\dot{\xi}}(k) - (\sin\hat{\Omega}(k)\tau)\tau\hat{\dot{\eta}}(k) \tag{19.44}$$

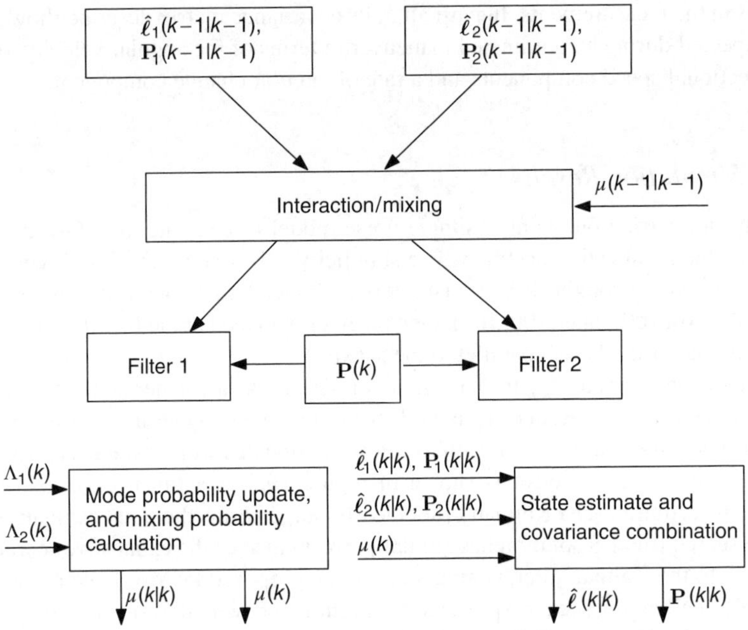

Figure 19.4 Outline of an Interacting Multiple-Model (IMM) filter.

In both cases, the process noise covariance is determined in the following state equations:

$$\mathbf{P}(k+1|k) = \mathbf{F}_L(k)\mathbf{P}(k|k)\mathbf{F}_L(k)^T + \Gamma_{\text{EKF}} Q(k)\Gamma_{\text{EKF}}^T \tag{19.45}$$

where

$$\Gamma_{\text{EKF}} = \begin{bmatrix} \frac{1}{2}\tau^2 & 0 & 0 & 0 \\ \tau & 0 & 0 & 0 \\ 0 & \frac{1}{2}\tau^2 & 0 & 0 \\ 0 & \tau & 0 & 0 \\ 0 & 0 & \frac{1}{2}\tau^2 & 0 \\ 0 & 0 & \tau & 0 \\ 0 & 0 & 0 & \tau \end{bmatrix} \tag{19.46}$$

Based on the assumptions of AUV motion, the value of $\mathbf{Q}(k)$ was selected as:

$$\mathbf{Q}(k) = \begin{bmatrix} r_s & 0 & 0 & 0 \\ 0 & r_s & 0 & 0 \\ 0 & 0 & r_s & 0 \\ 0 & 0 & 0 & r_d \end{bmatrix} \tag{19.47}$$

where $r_s = (1.25 \text{ m/s}^2)^2$ and $r_d = (0.3\pi/180 \text{ rad})^2$ are used in this chapter.

The IMM-CT is outlined in Figure 19.4. It combines a set of filters (in this case a KF and an EKF) and mixes the weighted previous state estimates to determine the current hypothesis of each of the filters. The linear KF is designed as described previously, whereas the nonlinear EKF has a different set of model selection parameters which define how it interprets large differences in the measurements. In particular, its covariance matrix describes how much variation is expected during a coordinated maneuver in terms of the angular velocity, represented as two directional speed components and a rate of angular change component.

19.2.5 Simulation Results

Simulations are carried out using a simple noise model to generate the TOA measurements and evaluate the localization accuracy. For simplicity, z is assumed to be known, and only x and y coordinates are sought. Four transmitters are placed on a square grid with coordinates $(0, 0)$, $(100, 0)$, $(0, 100)$, and $(100, 100)$. One receiver is placed at the $(0, 50)$ point, and moves at a constant rate of 0.125 m/s parallel to the x-axis.

Let $\hat{t}_n$ denote the estimate of t_n from the modem. The TOAs are generated according to (19.8) where b is a fixed large bias, and w_n is i.i.d. zero-mean white Gaussian noise corresponding to a standard deviation of 7.5 m in distance. Position updates were taken every 16 seconds.

The localization position error is shown in Figure 19.5 as a function of total number of measurements acquired. The Kalman Filter clearly outperforms the point estimation based on exhaustive search (the LS solution has similar performance as the exhaustive search).

In addition to the Kalman filter, simulations for the proposed IMM-CT were also run, using the relatively challenging scenario presented in Figure 19.6, with the corresponding RMS position error given by Figure 19.7. The dashed lines depicted in the figure indicate the beginning or end of one of the maneuvers from Figure 19.6. As can be seen in Figure 19.7, the point estimates are drastically improved upon by all three of the trackers, with the KF and and PDAF slightly outperforming the IMM-CT on the straight-path portion after exiting a maneuver. There was no noticeable difference in the performance of the KF and PDAF, which is to be expected, as the scenario did not feature any indirect path propagations.

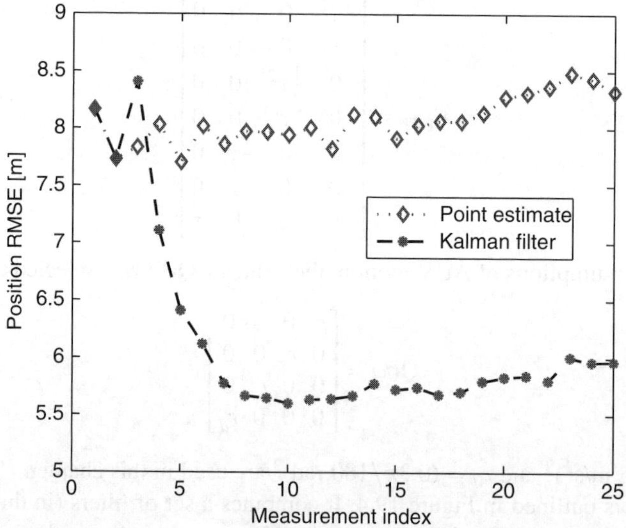

Figure 19.5 Root-mean-squared error (RMSE) as a function of the number of measurements acquired as the receiver moved in a straight line.

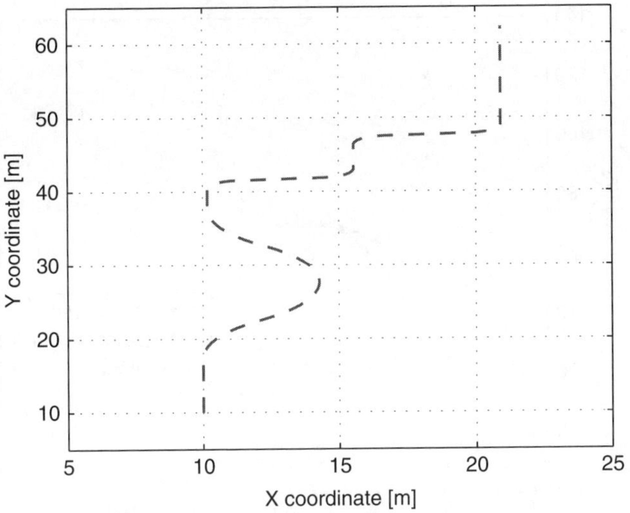

Figure 19.6 Simulation path for the IMM-CT, with distances in meters.

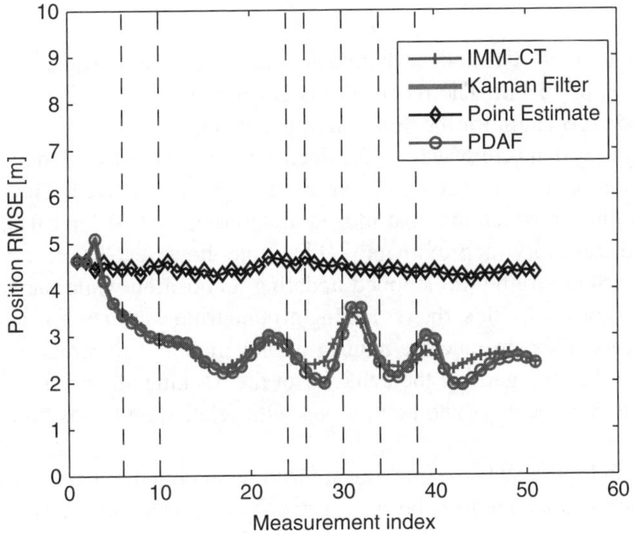

Figure 19.7 Root-mean-squared error (RMSE) as a function of the number of measurements acquired as the receiver moved in the scenario.

19.2.6 Field Test in a Local Lake

One lake test was performed in the Mansfield Hollow lake, located in Mansfield, CT during August 2011. The average depth of the area in which the test occurred is approximately 2.5 m, with minor variations of approximately 0.5 m. During the test, wind speed was minimal,

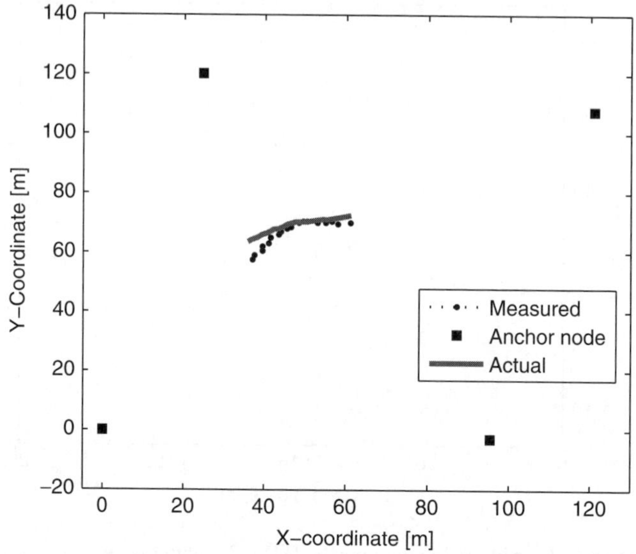

Figure 19.8 Plot of measurements and ground truth of lake test.

typically less than 5 mph. The nodes where positioned in a roughly square formation, with an average separation of 110 m. The receiver was attached to a boat which would freely float inside of this node formation for the duration of the testing.

For this test, the ground-truth was determined via an onboard GPS device which would record the position of the receiver whenever a message was received. The data for a single run is presented, during which the boat moved at approximately 1 knot while moving along a slightly curved trajectory (approximately 10°). Note, however, that at a certain point during the test, the ground-truth GPS stopped updating its position while the boat continued to move. In order to correct for this, the remaining ground truth was extrapolated from the initial GPS measurements. This introduces a nonnegligible amount of uncertainty, but still enables conclusions to be drawn regarding the behavior of the tracking algorithms. The approximate trajectory and measurements of the boat, along with relative node positions are depicted in Figure 19.8.

As can be seen in Figure 19.9, the tracking algorithms smooth the error out over the course of the maneuver and eventually reduce the overall error by a slight amount. Over the whole period, the approximated error never exceeded a combined 5 m. On inspection, the PDAF does not perform much better than the Kalman Filter in most cases. There was one measurement where it clearly offered an improvement, but overall they did not differentiate much in performance. In almost all cases, the IMM-CT was superior to both filters.

19.3 On-Demand Asynchronous Localization

This section presents an on-demand asynchronous localization (ODAL) scheme [61, 271]. The network is established in a manner as shown in Figure 19.10. Several fixed-position anchor

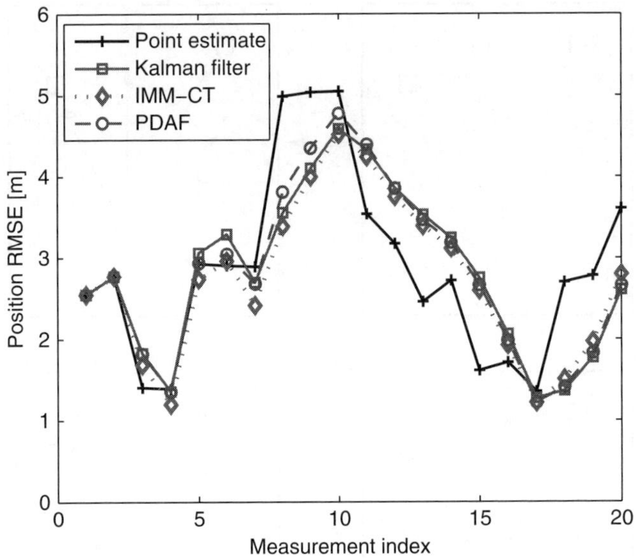

Figure 19.9 Localization error during second lake test.

nodes are deployed throughout the network and are assumed to have perfect knowledge of their own positions. These anchors need only know their own locations, and are not required to be surface nodes or have synchronized clocks, in contrast to the assumptions in Section 19.2.3. All nodes within the network know of the existence of the anchor nodes.

19.3.1 Localization Procedure

The localization process is initiated on demand by one underwater node. Other nodes in the network listen passively and can infer their own positions based on the broadcast messages from the source and the anchor nodes. The localization procedure is as follows.

S1. The source node sends out an initiator message to obtain its position at time $t_{s,s}$. The initiator message contains the sending order for the anchor nodes, indexed by $n = 1, 2, \ldots, N$, and the maximum waiting time δ_n before node n responds to the initiator message. The source node then goes into the listening mode, waiting for the message from the anchors to come in sequentially.

S2. All the anchor nodes operate as follows. Upon receiving the initiator message at $\hat{t}_{s,n}$, node n enters the listening mode and decodes all the messages from nodes $1, \ldots, n-1$, recording the arrival times as $\hat{t}_{k,n}$ where $k = 1, \ldots, n-1$.

After node n finds out that node $n-1$ has transmitted, it can proceed to the transmission mode and sends out its own message at time $t_{n,n}$. In case the message from node $n-1$ is lost, node n will wait the specified maximum time and send out its message at time $t_{n,n} = \hat{t}_{s,n} + \delta_n$. The message from node n contains $\hat{t}_{s,n}$, $\{\hat{t}_{k,n}\}_{\forall k}$, and the transmission time $t_{n,n}$.[1]

[1] This is reasonable for the OFDM modem as each OFDM block contains several tens of bytes upon transmission [444]. A hardware interrupt mechanism is used to control the exact transmission time after a fixed processing delay.

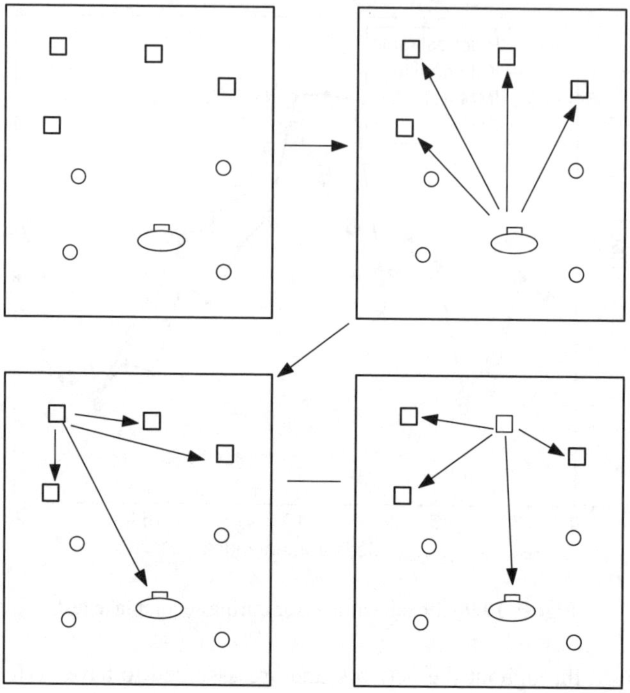

Figure 19.10 A brief overview of the transmission protocol in a sample network. The AUV initiates a transmission to all nodes, subsequently receiving a reply from the first anchor node. As time passes, all anchors reply sequentially until the AUV is finally localized.

S3. The source node receives the reply from node n at time $\hat{t}_{n,\mathrm{s}}$, $n = 1, \ldots, N$. After the reception of the last anchor node, the source node analyzes the collected measurements and computes its own position. Finally the source node sends out a final acknowledgement, which includes its estimated location, to terminate the localization process.

S4. Any passive node in the network can record the arrivals of the messages from the source and the anchor nodes as $\hat{t}_{\mathrm{s,p}}$, $\{\hat{t}_{n,\mathrm{p}}\}_{\forall n}$. Based on these measurements and the measurements from in the received messages, the passive node computes its own position.

For ease of presentation, assume that all the anchor nodes specified in the initiator message have participated in the localization procedure. If there are N_{f} nodes with communication failures, then the effective number of anchor nodes reduces from N to $N - N_{\mathrm{f}}$, while the localization algorithm can still be carried out.

19.3.2 Localization Algorithm for the Initiator

Since all the nodes are asynchronous, the time measurements from each node are subject to an unknown time shift. With the time measurements at the source, and the measurements collected

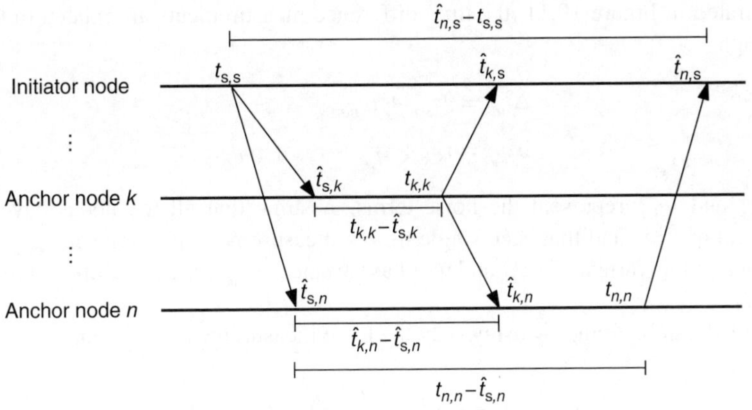

Figure 19.11 Illustration of the timing diagram of the asynchronous localization scheme.

by the anchor nodes, the initiator node computes the differences as:

$$\Delta\hat{t}_{n,s} = (\hat{t}_{n,s} - t_{s,s}) - (t_{n,n} - \hat{t}_{s,n}) \tag{19.48}$$

$$\Delta\hat{t}_{k,n} = (\hat{t}_{k,n} - \hat{t}_{s,n}) - (t_{k,k} - \hat{t}_{s,k}) \tag{19.49}$$

which is illustrated in Figure 19.11. The available measurements at the initiator are

$$\{\Delta\hat{t}_{s,n}\}_{n=1}^{N}, \{\Delta\hat{t}_{k,n}\}_{k=1,n=2}^{n-1,N}. \tag{19.50}$$

Define the distance between the source node and the anchor node n as:

$$d_{s,n} = \sqrt{(x_s - x_n)^2 + (y_s - y_n)^2 + (z_s - z_n)^2}, \tag{19.51}$$

and the distance between anchor nodes k and n as

$$d_{k,n} = \sqrt{(x_k - x_n)^2 + (y_k - y_n)^2 + (z_k - z_n)^2}. \tag{19.52}$$

The one-way propagation delay between the source node and an anchor node is defined as:

$$\tau_{s,n} = \frac{d_{s,n}}{c}, \tag{19.53}$$

while the one-way propagation delay between anchor nodes k and n is defined as:

$$\tau_{k,n} = \frac{d_{k,n}}{c}, \tag{19.54}$$

where c is the propagation speed of sound in water and is approximately 1500 m/s.

As illustrated in Figure 19.11, the time-difference measurements are related to the ground truth through

$$\Delta\hat{t}_{n,\mathrm{s}} = 2\tau_{\mathrm{s},n} + w_{n,\mathrm{s}}, \tag{19.55}$$

$$\Delta\hat{t}_{k,n} = \tau_{\mathrm{s},k} + \tau_{k,n} - \tau_{\mathrm{s},n} + w_{k,n}, \tag{19.56}$$

where $w_{n,\mathrm{s}}$ and $w_{k,n}$ represent the noise terms. Assume that all the nodes have the same measurement quality, and that each single timing measurement has variance σ^2_{mea}. The noise component $w_{n,\mathrm{s}}$ has variance $2\sigma^2_{\mathrm{mea}}$ and $w_{k,n}$ has variance $3\sigma^2_{\mathrm{mea}}$, as can be inferred from (19.48) and (19.49).

The location can be found by using only the local measurements at the initiator node as

$$(\hat{x}_\mathrm{s}, \hat{y}_\mathrm{s}, \hat{z}_\mathrm{s}) = \arg\min_{x_\mathrm{s},y_\mathrm{s},z_\mathrm{s}} \frac{1}{4\sigma^2_{\mathrm{mea}}} \sum_{n=1}^{N} (\Delta\hat{t}_{n,\mathrm{s}} - 2\tau_{\mathrm{s},n})^2. \tag{19.57}$$

This corresponds to a long-base-line (LBL) system.

The ODAL method in [61] uses both the measurements at the initiator and the measurements at all the anchor nodes of the message broadcast sequence. The localization is formulated as

$$(\hat{x}_\mathrm{s}, \hat{y}_\mathrm{s}, \hat{z}_\mathrm{s}) = \arg\min_{x_\mathrm{s},y_\mathrm{s},z_\mathrm{s}} \left\{ \frac{1}{4\sigma^2_{\mathrm{mea}}} \sum_{n=1}^{N} \left(\Delta\hat{t}_{n,\mathrm{s}} - 2\tau_{\mathrm{s},n}\right)^2 \right.$$
$$\left. + \frac{1}{6\sigma^2_{\mathrm{mea}}} \sum_{n=2}^{N} \sum_{k=1}^{n-1} \left[\Delta\hat{t}_{k,n} - (\tau_{\mathrm{s},k} + \tau_{k,n} - \tau_{\mathrm{s},n})\right]^2 \right\}. \tag{19.58}$$

19.3.3 *Localization Algorithm for a Passive Node*

Noting that the system is intended to operate within a network of nodes, and leveraging the broadcast nature of the medium, all the nodes in the network can self-localize by monitoring the traffic resulting from the localization procedure for the initiator. Denote a general passive node p as being located at $(x_p, y_\mathrm{p}, z_\mathrm{p})$ and define the distances as before:

$$d_{\mathrm{s,p}} = \sqrt{(x_\mathrm{s} - x_\mathrm{p})^2 + (y_\mathrm{s} - y_p)^2 + (z_\mathrm{s} - z_\mathrm{p})^2} \tag{19.59}$$

$$d_{n,\mathrm{p}} = \sqrt{(x_n - x_\mathrm{p})^2 + (y_n - y_\mathrm{p})^2 + (z_n - z_\mathrm{p})^2} \tag{19.60}$$

The corresponding propagation delays are

$$\tau_{\mathrm{s,p}} = \frac{d_{\mathrm{s,p}}}{c}, \quad \tau_{n,\mathrm{p}} = \frac{d_{n,\mathrm{p}}}{c} \tag{19.61}$$

The passive node receives the localization message from the initiator, and the responding messages from the anchor nodes. The time differences are taken as:

$$\Delta\hat{t}_{\mathrm{s},n,\mathrm{p}} = (\hat{t}_{n,\mathrm{p}} - \hat{t}_{\mathrm{s,p}}) - (t_{n,n} - \hat{t}_{\mathrm{s},n}), \qquad n = 1, \ldots, N \tag{19.62}$$

$$\Delta\hat{t}_{k,n,\mathrm{p}} = (\hat{t}_{n,\mathrm{p}} - \hat{t}_{k,\mathrm{p}}) - (t_{n,n} - \hat{t}_{k,n}), \quad n = 2, \ldots, N; \quad k = 1, \ldots, n-1; \tag{19.63}$$

The time differences are related to the true propagation delays through

$$\Delta\hat{t}_{s,n,p} = \tau_{s,n} + \tau_{n,p} - \tau_{s,p} + w_{s,n,p}, \quad (19.64)$$

$$\Delta\hat{t}_{k,n,p} = \tau_{k,n} + \tau_{n,p} - \tau_{k,p} + w_{k,n,p}, \quad (19.65)$$

where the equivalent noises $w_{s,n,p}$ and $w_{k,n,p}$ has variance $3\sigma^2_{mea}$.

The passive node initially does not know the position of the initiator node, and localization can be carried out as

$$(\hat{x}_p, \hat{y}_p, \hat{z}_p) = \arg \min_{(x_p,y_p,z_p)} \left\{ \frac{1}{6\sigma^2_{mea}} \sum_{n=2}^{N} \sum_{k=1}^{n-1} \left[\Delta\hat{t}_{k,n,p} - (\tau_{k,n} + \tau_{n,p} - \tau_{k,p})\right]^2 \right\}. \quad (19.66)$$

After hearing the announced position from the initiator node, the estimate $\hat{\tau}_{s,n}$ is available, the passive node can improve its own position through:

$$(\hat{x}_p, \hat{y}_p, \hat{z}_p) = \arg \min_{(x_p,y_p,z_p)} \left\{ \frac{1}{6\sigma^2_{mea}} \sum_{n=2}^{N} \sum_{k=1}^{n-1} \left[\Delta\hat{t}_{k,n,p} - (\tau_{k,n} + \tau_{n,p} - \tau_{k,p})\right]^2 + \frac{1}{6\sigma^2_{mea}} \sum_{n=1}^{N} \left[\Delta\hat{t}_{s,n,p} - (\hat{\tau}_{s,n} + \tau_{n,p} - \tau_{s,p})\right]^2 \right\}. \quad (19.67)$$

Note that an arbitrary number of passive nodes can carry out localization independently, and without any additional broadcasting of their own. The "silent" positioning approach based on the "reactive beaconing" mechanism was first proposed in [76].

19.3.4 Localization Performance Results in a Lake

A test of the ODAL localization system at Mansfield Hollow Lake was carried out on July 24, 2013, as illustrated in Figure 19.12. During the test, 4 anchor nodes were deployed in

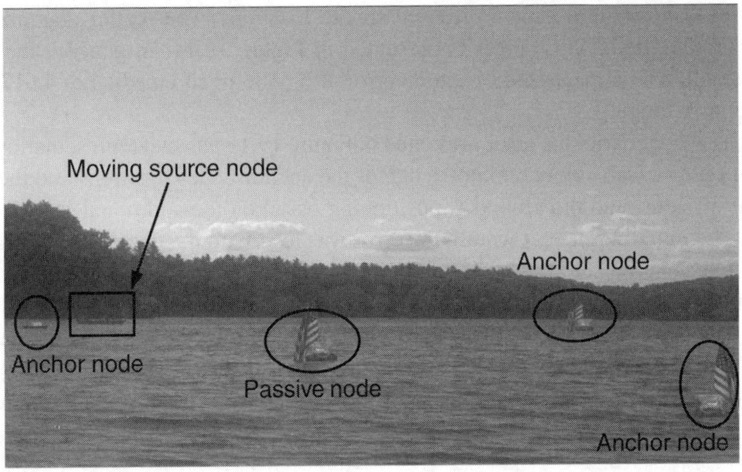

Figure 19.12 A photograph taken during the lake test demonstrating the node locations.

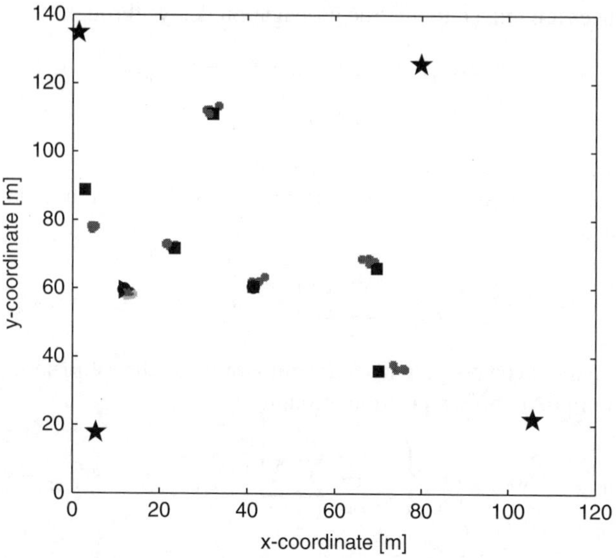

Figure 19.13 Initiator node position estimates during lake test. True positions estimated via laser ranging triangulation (units are meters).

an approximately square composition with sides of approximately 100 m in length. A single passive node was positioned along one edge of this square where it remained for the duration of the test. An initiator node was moved repeatedly through the test and was localized during the periods for which it was stationary. The ground truth of the anchor nodes, the initiator node and the passive nodes are determined using a triangulation with a laser rangefinder and multiple reference points along the shore.

The source node was positioned in five different places in the lake. The ODAL algorithm was run to determine the source position. There were a total of 22 position estimates obtained from the ODAL algorithm at five different source locations. The scatter diagram is shown in Figure 19.13. The location error is presented in Figure 19.14. In general, the errors are relatively small, with a mean RMS location error of 5.6106 m and median of 4.6138 m in the case of the laser ranging.

For the passive scenario, the error presented in Figure 19.15 has two representations: the first is the positioning based solely on knowledge of the anchor nodes positions and the passively heard transmissions, and the second is positioning based on the additional knowledge of the position of the initiating node. Overall, the error was low, with the exception of a rather large outlier, which as in the pool case was determined to be the miss of the line-of-sight path, and was removed. The mean error for the passive position (without initiator data) was 3.2326 m with a median value of 2.2457 m. In the case where initiator information was utilized, the mean value becomes 2.9673 m and the median is 2.0457 m. Thus there is a degree of improvement (approximately 10%) from including the additional initiating node information, as expected.

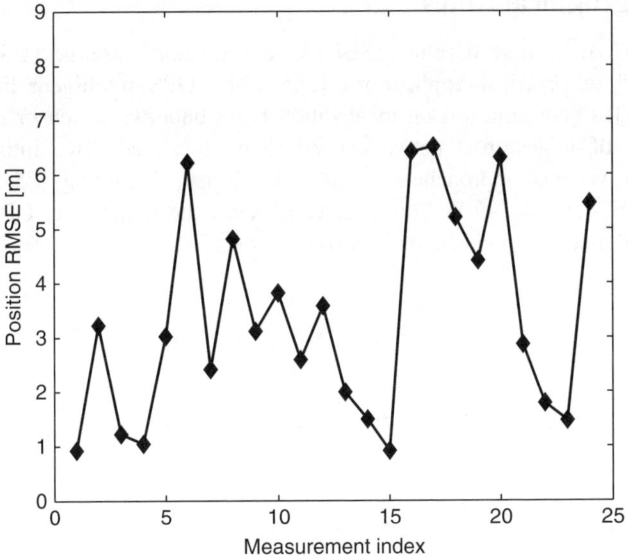

Figure 19.14 Initiator node positioning error during lake test.

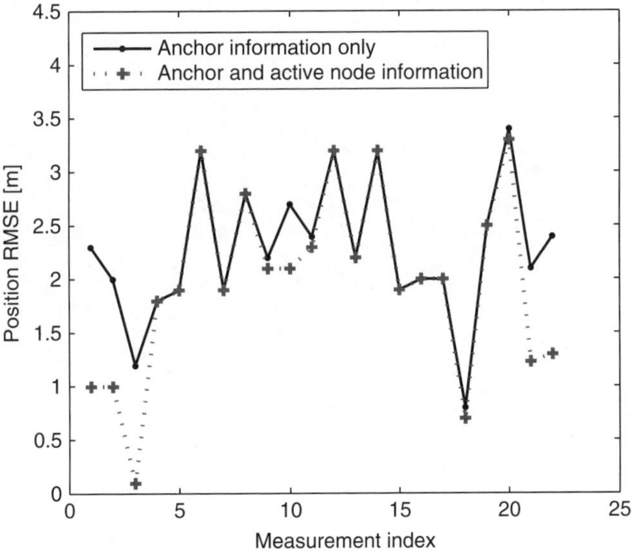

Figure 19.15 Passive node positioning error during lake test.

19.4 Bibliographical Notes

Long baseline (LBL), short baseline (SBL), and ultrashort baseline (USBL) have been extensively used in practical applications [266]. The GPS intelligent buoys (GIB) are described in [27]. Recent interest on localization is on underwater sensor networks, where a large number of nodes in a network need to be localized. Two tutorials [373] and [113] described various approaches recently developed, including [75, 76, 96, 112, 114, 115, 191, 283, 407, 429, 441, 470]. This chapter presents the underwater GPS solution from [62] and the on-demand asynchronous localization from [61, 271].

Appendix A

Compressive Sensing

Since the term compressive sensing was coined a few years ago [53, 100] this subject has been under intensive investigation [24, 56, 70]. It has found broad application in imaging, data compression, radar, and data acquisition to name a few. Overviews on compressive sensing can be found in, e.g., [56, 70]. Extensive references and resources on compressive sensing are also available online [1].

In a nutshell, compressive sensing is a novel paradigm where a signal that is sparse in a known transform domain can be acquired with much fewer samples than usually required by the dimensions of this domain. The only condition is that the sampling process is *incoherent* with the transform that achieves the sparse representation and *sparse* means that most weighting coefficients of the signal representation in the transform domain are zero. While it is obvious that a signal that is sparse in a certain basis can be fully represented by an index specifying the basis vectors corresponding to nonzero weighting coefficients plus the coefficients – determining *which* coefficients are nonzero would usually involve calculating all coefficients, which requires at least as many samples as there are basis functions. The definition of *incoherence* usually states that distances between sparse signals are approximately conserved as distances between their respective measurements generated by the sampling process. In this sense the reconstruction problem has per definition a unique solution.

Research on compressive sensing are mainly centered on two critical issues.

- *Design of sensing matrices* to achieve the *incoherence* property with a feasible sampling scheme;
- *Design of recovery algorithms* to reconstruct the original signal from these samples with tractable computational complexity.

The answers to these questions created the field of compressive sensing, which will be briefly reviewed in this appendix.

OFDM for Underwater Acoustic Communications, First Edition. Shengli Zhou and Zhaohui Wang.
© 2014 John Wiley & Sons, Ltd. Published 2014 by John Wiley & Sons, Ltd.

A.1 Compressive Sensing

A.1.1 Sparse Representation

Consider a signal $\mathbf{y} \in \mathbb{C}^n$ that can be represented in an arbitrary basis, $\{\psi_k\}_{k=1}^n$, with the weighting coefficients x_k. Stacking the coefficients into a vector, $\mathbf{x}$, the relationship with $\mathbf{y}$ is obviously through the transform

$$\mathbf{y} = \left[\psi_1\ \psi_2\ \cdots\ \psi_n\right] \mathbf{x} \tag{A.1}$$

$$:= \boldsymbol{\Psi}\mathbf{x} \tag{A.2}$$

where $\boldsymbol{\Psi}$ is a full rank $n \times n$ matrix. A common example would be a finite length, discrete time signal that one could represent as discrete sinusoids in a limited bandwidth. The matrix $\boldsymbol{\Psi}$ would then be the discrete Fourier transform (DFT) matrix.

In compressive sensing one is particularly interested in any basis that allows a *sparse* representation of $\mathbf{y}$, i.e., a basis $\{\psi_k\}_{k=1}^n$ such that most x_k are zero. Obviously if one knows $\mathbf{y}$, one could always choose some basis for which $\mathbf{y} = \psi_{k_0}$ for some k_0; then all x_k, $k \neq k_0$, would be zero. This trivial case is not of interest, instead one is interested in a predetermined basis that will render a sparse or approximately sparse representation of any $\mathbf{y}$ that belongs to some class of signals.

A.1.2 Exactly and Approximately Sparse Signals

A signal is called s-sparse, if it can be exactly represented by a basis, $\{\psi_k\}_{k=1}^n$, and a set of coefficients x_k, where only s coefficients are nonzero. A signal is called approximately s-sparse, if it can be represented up to a certain accuracy using s nonzero coefficients. Since the desired accuracy depends on the application, signals considered as approximately sparse usually have the property that the reconstruction error decreases super-linearly in s, therefore any required accuracy can be achieved by only sightly increasing s.

As an example of an s-sparse signal, consider the class of signals that are the sum of s sinusoids chosen from the n harmonics of the observed time interval. Now obviously the DFT basis will render an s-sparse representation of any such $\mathbf{y}$, i.e., taking the DFT of any such signal would render only s nonzero values x_k.

An example of approximately sparse signals is when the coefficients x_k, sorted by magnitude, decrease following a power law. This includes smooth signals or signals with bounded variations [56]. In this case the sparse approximation constructed by choosing the s largest coefficients is guaranteed to have an approximation error that decreases with the same power law as the coefficients.

A.1.3 Sensing

So far it was assumed that $\mathbf{y}$ is available, and that one can simply apply the transform into the domain of $\{\psi_k\}_{k=1}^n$ to determine which x_k are relevant (nonzero). Although this case does exist and is important for some forms of data compression, the real application of compressive sensing is the acquisition of the signal from m, possibly noisy, measurements

$$z_\ell = \boldsymbol{\phi}_\ell^{\mathrm{H}}\mathbf{y} + w_\ell, \qquad \ell = 1, \ldots, m \tag{A.3}$$

where here it is assumed that v_ℓ is zero-mean complex Gaussian distributed with variance σ_{w}^2 and the noiseless case is included for $\sigma_{\mathrm{w}}^2 \rightarrow 0$. The signal acquisition process can now be written using a measurement matrix $\mathbf{\Phi}$,

$$\mathbf{z} = \mathbf{\Phi}^{\mathrm{H}}\mathbf{y} + \mathbf{w}, \tag{A.4}$$

where $\mathbf{\Phi} = [\boldsymbol{\phi}_1 \ , \ \boldsymbol{\phi}_2 \ , \ \dots \ , \ \boldsymbol{\phi}_m]$ is an $n \times m$ matrix and $\mathbf{z} = [z_1 \ , \ z_2 \ , \ \dots \ , \ z_m]^{\mathrm{T}}$ is the stacked measurement vector. Substituting (A.1) into (A.4) yields

$$\mathbf{z} = \mathbf{\Phi}^{\mathrm{H}}\mathbf{\Psi}\mathbf{x} + \mathbf{w} := \mathbf{A}\mathbf{x} + \mathbf{w}. \tag{A.5}$$

Since this is a simple linear Gaussian model, it is *well-posed* as long as $\mathbf{A}$ is at least of rank n. By *well posed* we simply mean that there exists some estimator $\hat{\mathbf{x}}$ (or $\hat{\mathbf{y}}$ for that matter), whose estimation error is proportional to the noise variance; therefore as the noise variance approaches zero, the estimation error does as well. This generally requires at least $m \geq n$ measurements if $\mathbf{y}$ is unconstrained in $\mathbb{C}^n$.

A.1.4 Signal Recovery and RIP

The novelty in compressive sensing is that for signals $\mathbf{y}$ that are s-sparse in some $\{\boldsymbol{\psi}_k\}_{k=1}^n$, less measurements are sufficient to make this a *well-posed* problem. The requirement on $\mathbf{A}$ to have at least rank n is replaced by the restricted isometry property (RIP), defined as follows [54]:

For any matrix $\mathbf{A}$ with unit-norm columns one can define the restricted isometry constants δ_s as the smallest number such that,

$$\text{RIP:} \quad 1 - \delta_s \leq \frac{\|\mathbf{A}\mathbf{x}\|_2^2}{\|\mathbf{x}\|_2^2} \leq 1 + \delta_s \tag{A.6}$$

for any $\mathbf{x}$ that is s-sparse. This can be seen as conserving the (approximate) length of s-sparse vectors in the measurement domain and effectively puts bounds on the eigenvalues of any $s \times s$ submatrix of $\mathbf{A}^{\mathrm{H}}\mathbf{A}$.

Now assuming that under all s-sparse vectors, one chooses the estimate $\hat{\mathbf{x}}$ that has the smallest distance to the observations,

$$\hat{\mathbf{x}} = \arg \min_{\{\mathbf{x}:\|\mathbf{x}\|_0 = s, \mathbf{x} \in \mathbb{C}^n\}} \|\mathbf{z} - \mathbf{A}\hat{\mathbf{x}}\|_2^2, \tag{A.7}$$

where the ℓ_0 norm $\|\mathbf{x}\|_0$ is defined as the number of nonzero elements in $\mathbf{x}$. Using the fact that the estimation error $\tilde{\mathbf{x}} := \mathbf{x} - \hat{\mathbf{x}}$ is $(2s)$-sparse, it is easily shown that the estimation error is bounded by

$$\mathbb{E}\{\|\mathbf{x} - \hat{\mathbf{x}}\|_2^2\} \leq \frac{2m\sigma_{\mathrm{w}}^2}{1 - \delta_{2s}}. \tag{A.8}$$

So one can see that the signal recovery problem is *well-posed* as long as $\delta_{2s} < 1$, but since the constant δ_s is monotonic in s, $\delta_s \leq \delta_{s+1}$, and usually increases gradually, it is commonly said that $\mathbf{A}$ obeys RIP if δ_s is not too close to one.

In case of approximately sparse signals, the error caused by noisy observations is additive with the error caused by the approximation as s-sparse. Therefore a good choice of s needs to consider the noise level σ_w^2, since a tradeoff exists between choosing a smaller s that increases the approximation error, but decreases the error caused by the noise due to the monotonic nature of the δ_s and vice versa.

A.1.5 Sensing Matrices

While evaluating the RIP for a particular matrix at hand is (at worst) an NP-hard problem, there are large classes of matrices that obey the RIP with high probability, that is $\delta_s \ll 1$ for any $s \ll m$. Specifically for random matrices like i.i.d. Gaussian or Bernoulli entries, or randomly selected rows of an orthogonal $(n \times n)$ matrix (e.g., the DFT), it can be shown that for $m \geq Cs \log\,(n/s)$ measurements the probability that $\delta_s \geq \delta$ decreases exponentially with m and δ. In other words, as long as one takes *enough* measurements, i.e. increase m, the probability of any such matrix not obeying the RIP for a given threshold δ can be made arbitrarily small. Although the constant C is only loosely specified for the various types of matrices, the fact that the probability decreases exponentially is encouraging as to the number of required measurements. Furthermore it is important to consider that these bounds are on worst cases, so that on the average much fewer measurements, m, will be sufficient.

A.2 Sparse Recovery Algorithms

Section A.1.4 considered the estimator that chooses the solution with minimum distance from the observations between all s-sparse vectors in $\mathbb{C}^n$ to show that the average estimation error is bounded. This is in essence a combinatorial problem, which has exponential complexity. In case s is not known, or for an approximately sparse signal, a natural optimization problem to find a maximally sparse representation of $\mathbf{z}$ is

$$\hat{\mathbf{x}} = \arg \min_{\mathbf{x}} \|\mathbf{x}\|_0 \qquad \text{s.t.} \qquad \|\mathbf{z} - \mathbf{A}\mathbf{x}\|_2^2 \leq \varepsilon. \tag{A.9}$$

A joint cost function can also be used that penalizes less sparse solutions versus a better fit of the observations. This can be achieved using a Lagrangian formulation

$$\hat{\mathbf{x}} = \arg \min_{\mathbf{x}} \frac{1}{2}\|\mathbf{z} - \mathbf{A}\mathbf{x}\|_2^2 + \zeta\|\mathbf{x}\|_0 \tag{A.10}$$

where ζ is a design parameter balancing the sparse approximation error and solution sparsity. Compared to the combinational problem defined in (A.7), the size of problems in (A.9) and (A.10) is increased as all s-sparse vectors for various values of s have to be considered now.

Besides a brute-force approach that searches over all the possible values of $\mathbf{x}$ to find the best fit, algorithms that reconstruct a signal taking advantage of its sparse structure have been used well before the term compressive sensing was coined. The surprising discovery of compressive sensing is that it can be shown that several of these algorithms will – under certain conditions – render the same solution as the combinatorial approach. These conditions largely amount to tighter constraints on the sparsity of $\mathbf{x}$ beyond identifiability.

We next briefly introduce two major types of algorithms: matching pursuit algorithms and the convex ℓ_1-norm relaxation algorithms. Borrowing the message passing and belief propagation principles from the coding community, heuristic algorithms have also been proposed to

Table A.1 Sparse recovery algorithms

Matching pursuit	MP, OMP, StOMP, CoSaMP, IHT
Convex relaxation	IST, TwIST, FISTA, GPSR, SpaRSA, ℓ_1-ℓ_s, ℓ_1-magic, YALL$_1$ [466]
Bayesian framework	Heuristic methods based on belief propagation and message passing; see [25, 102, 197]
Nonconvex relaxation	Relax the ℓ_0-norm to ℓ_p-norm with $p < 1$; see [69]

recover the sparse representation under the Bayesian framework; see, e.g., [25, 102, 197] and references therein. Instead of performing the ℓ_1-norm convex relaxation in the algorithms to be discussed, sparse recovery algorithm based on a nonconvex relaxation by ℓ_p-norm ($p < 1$) has also been investigated [69]. Several example sparse solution solvers under the four categories are listed in Table A.1. A complete summary of sparse solution solvers can be found in e.g., [110].

A.2.1 Matching Pursuits

An alternative approach to the combinatorial problem is based on dynamic programming. In this type of approach the combinatorial problem is circumvented by heuristically choosing which values of $\mathbf{x}$ are nonzero and solving the resulting constrained least-squares problem. The most popular algorithms of this type are greedy algorithms, like Matching Pursuit (MP) or Orthogonal Matching Pursuit (OMP), that identify the nonzero elements of $\mathbf{x}$ in an iterative fashion.

This type of algorithms has been popular mainly because it can be easily implemented and has low computational complexity. Recently it has been shown that this algorithm will also render the optimal solution [387], whereby the constraints are somewhat stronger. This has lead to renewed interest in dynamic programming based solutions, leading to new matching pursuit algorithms, such as the stagewise orthogonal matching pursuit (StOMP), the compressive sampling matching pursuit (CoSaMP) and iterative hard thresholding (IHT); see, e.g., [47, 101, 293] and references therein).

A.2.2 ℓ_1-Norm Minimization

Since the exact formulation using the zero-norm $||\mathbf{x}||_0$ is not amenable to efficient optimization, an immediate choice is its convex relaxation, leading to the following Lagrangian formulation,

$$\text{BP:} \quad \hat{\mathbf{x}} = \arg \min_{\mathbf{x}} ||\mathbf{A}\mathbf{x} - \mathbf{z}||_2^2 + \zeta ||\mathbf{x}||_1, \tag{A.11}$$

where the ℓ_1-norm is defined as $||\mathbf{x}||_1 = \sum_{k=1}^{n} |x_k|$. While the ℓ_1-norm has been used in various applications to promote sparse solutions in the past (see references in [56]), it is now largely popular under the name Basis Pursuit (BP), as introduced in [73]. While originally the term BP was used to designate the case of noiseless measurements and the qualifier Basis Pursuit De-Noising (BPDN) to refer to the case of noisy measurements [73], we will generally refer to both cases simply by BP. Generally the optimal value of ζ in (A.11) is unknown to the solver. Theoretical investigations on the selection of ζ can be found in e.g., [109, 136, 147].

One variant of the convex relaxation in (A.11) is

$$\text{LASSO:} \quad \hat{\mathbf{x}} = \arg \min_{\mathbf{x}} \|\mathbf{z} - \mathbf{A}\mathbf{x}\|_2^2 \qquad \text{s.t.} \qquad \|\mathbf{x}\|_1 \leq \beta \tag{A.12}$$

which is well-known as the least absolute shrinkage and selection operator (LASSO) algorithm in statistics [385]. It can be shown that LASSO and BP are equivalent in that the solution path of LASSO defined by β coincides with that of BP after a reparameterization of $\zeta(\beta)$ [400].

All these algorithms share an aspect that they lead to convex optimization problems, which can be solved efficiently with advanced techniques, such as interior-point methods [73, 212], projected gradient methods [87, 155, 435], or iterative thresholding [28, 92]. A description on these methods can be found in [388].

The discovery that there are conditions under which convex relaxation will render the same result as the combinatorial formulation was the birth of compressive sensing [53, 100]. These conditions usually consider the minimum number of measurements m required to identify an s-sparse signal with high probability, given a certain measurement matrix. For example, in [53] it is shown for m noiseless measurements taken using random rows of the DFT matrix, that if $m > C_M s \log(n)$, any s-sparse signal can be recovered with at least probability $1 - \mathcal{O}(n^{-M})$, where the constant C_M is roughly linear in the parameter M. One immediately notices that this formulation closely resembles the criterion for identifiability, but the constants will take different values.

A.3 Applications of Compressive Sensing

A.3.1 Applications of Compressive Sensing in Communications

So far compressive sensing has been successfully applied in several signal processing fields, specifically in imaging the technology has achieved a certain level of maturity. In communications the range of applications so far has been rather limited, with the exception of channel estimation – although in many variations. To cite a few examples:

- Sparse channel estimation in ultra-wideband, was motivated by the ability to resolve individual arrivals or clusters of arrivals in multipath channels [303].
- Considering mobile radio channels, each path is characterized by a delay and a relative Doppler speed [18, 378].
- Underwater acoustic channels are known to exhibit only few arrivals in a long delay spread with each path having a different Doppler speed [38].

A variation on channel estimation is the combination with active user detection in code division multiple access [9] or spectrum sensing for cognitive radios.

Another proposed application of compressive sensing in communications is coding over the real numbers (versus finite fields as commonly used in coding theory) under a channel model that produces few very large errors (similar to erasures) [54].

A.3.2 Compressive Sensing in Underwater Acoustic Channels

Clearly the motivation to use compressive sensing in channel estimation is the observation that some channels are characterized by sparse multipath, meaning that there are much fewer distinct arrivals as there are baseband channel taps. With this in mind compressive sensing promises to estimate the channel with much less pilot overhead or at higher accuracy with a constant number of pilots. The common assumption is that a sparse multipath channel leads to a baseband channel model where most taps are negligible. Note that in a channel modeled by specular (point) scatterers the number of nonzero baseband taps depends very much on what one defines as negligible. Using instead an oversampled baseband model, the representation of the channel becomes ambiguous, but also more sparse.

In underwater acoustic (UWA) communications, channels are characterized by long delay spread and significant Doppler effects. The long channel delay spread leads to severe inter-symbol interference (ISI) in single carrier transmissions, while in multicarrier approaches like orthogonal frequency division multiplexing (OFDM) the aforementioned Doppler effects destroy the orthogonality of the sub-carriers and lead to intercarrier interference (ICI). On top of high equalization complexity, the ISI or ICI corresponds to a convolution with a time-varying impulse response, leading to a large amount of unknown channel coefficients. While it is well recognized in the community that UWA channels are usually sparse [378], there are major challenges to overcome when applying compressive sensing to exploit channel sparsity.

As an example, Chapter 7 presents a block-by-block OFDM receiver that re-estimates the channel for every OFDM symbol. To apply compressive sensing, one has to consider the following issues: (i) A channel model needs to be established that leads to a sparse representation of the channel coefficients, and is accurate (enough) within the considered time interval. (ii) When placing the pilots, one needs to ensure that ICI from other pilots can be observed. (iii) When estimating the channel based on pilots, ICI from the unknown data symbols has to be accounted for. The channel estimators in Chapter 7 are developed with some strategies to address these issues.

Appendix B

Experiment Description

B.1 SPACE08 Experiment

The Surface Processes and Acoustic Communication Experiment (abbreviated as SPACE08) was conducted at the Air-Sea Interaction Tower (ASIT) from Oct. 14 to Nov. 1, 2008, which is located in around 15 meters of water and 2 miles south of the coast of Martha's Vineyard, MA. The experiment was led by Dr. James Preisig and his team from the Woods Hole Oceanographic Institution (WHOI). Several universities and research institutions have participated in the experiment.

The acoustic source was deployed about 30 meters in a direction of 240° True from the tower, and are formed by five ITC-1007 spherical transducers. One transducer is located approximately 4 meters above the bottom on a stationary tripod, and the remaining four transducers are in a vertical array where the center-to-center spacing of transducers in the array is 50 cm and the top transducer is around 3 meters above the bottom. Any combination of the five transducers can be used for multiple-input and multiple-output (MIMO) transmissions.

As illustrated in Figure B.1, six receiving arrays are located on two paths radiating in orthogonal directions (150° and 240°) from the source at distances of approximately 60 meters, 200 meters and 1000 meters. The arrays are mounted on fixed tripods with the top element 3.25 meters above the sea floor. S1 and S2 have a cross-shaped configuration (vertical and horizontal) with 16 elements on each leg. The spacing between each element is 3.75 cm. S3 and S4 are 24-element vertical arrays with a 5 cm spacing between elements. S5 and S6 are 12-element vertical arrays with a 12 cm spacing between elements. The sampling rate of the transmitter and all the receivers is 1e7/256 Hz. The received signals are recorded *in situ* at the receiving arrays.

During the experiment, a transmission occurs every two hours, leading to 12 recorded files each day. For each transmission, there are 20 OFDM blocks with the parameters specified in Table B.1. A detailed subcarrier allocation is described in Section 2.3.

The significant wave height and average wind speed across all the days are shown in Figure B.2. The significant wave height is calculated as $H = 4\sqrt{m_0}$, where m_0 is the zeroth-moment of the variance spectrum obtained by integration of the variance spectrum.

OFDM for Underwater Acoustic Communications, First Edition. Shengli Zhou and Zhaohui Wang.
© 2014 John Wiley & Sons, Ltd. Published 2014 by John Wiley & Sons, Ltd.

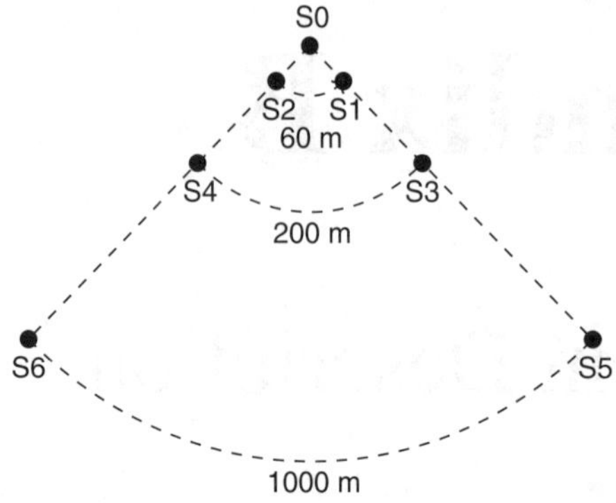

Figure B.1 Setup of the SPACE08 experiment.

Table B.1 OFDM Parameters in SPACE08

Carrier frequency	f_c	13 kHz
Bandwidth	B	9.77 kHz
Symbol duration	T	104.86 ms
Subcarrier spacing	Δf	9.54 Hz
Guard interval	T_g	24.6 ms
# of subcarriers	K	1024
# of data subcarriers	$\vert S_D \vert$	672
# of pilot subcarriers	$\vert S_P \vert$	256
# of null subcarriers	$\vert S_N \vert$	96
Channel coding rate	r_c	0.5

B.2 MACE10 Experiment

The Mobile Acoustic Communication Experiment (abbreviated as MACE10) was carried out off the coast of Martha's Vineyard, Massachusetts, in June 2010. The experiment was led by Mr. Lee Freitag and his team from WHOI. Signals from several universities and research institutions have been transmitted in this experiment.

Source and receiving arrays were deployed in the water with a depth about 95 to 100 meters. The source array has four ITC-1007 transducers. The receivers include two buoy receive arrays (labeled as B1 and B2) and two moored receive arrays (labeled as M1 and M2), as illustrated in Figure B.3. Each buoy receiver has four receive elements, with a 20 cm spacing between elements, and is deployed at a depth of 50 meters. The moored array M1 has 24 elements with a spacing of 5 cm between elements, and is deployed at a depth of 40 meters. The moored array M2 has 12 elements with a spacing of 12 cm between elements, and is deployed at a depth of 40 meters.

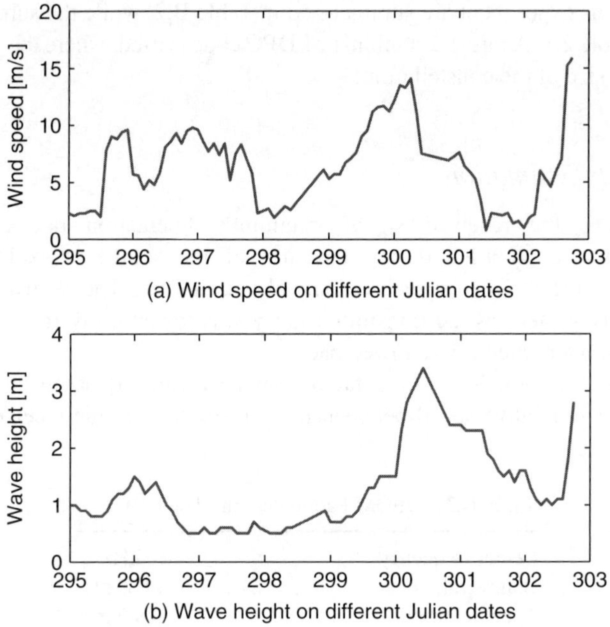

Figure B.2 The average wind speed and significant wave height for selected days in SPACE08.

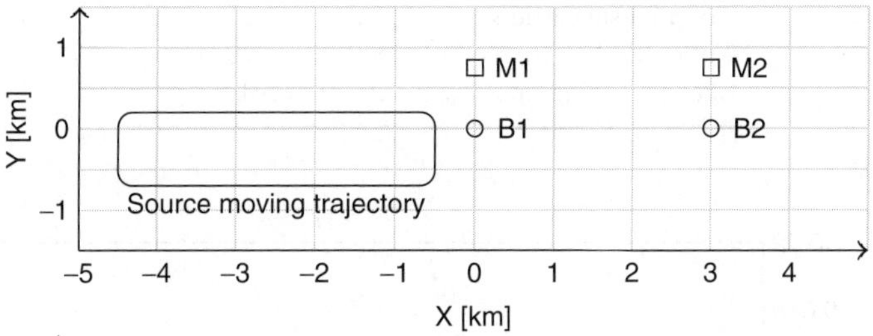

Figure B.3 Experimental setup in MACE10.

B.2.1 Experiment Setup

During the experiment, the receive arrays were stationary, while the source array was towed slowly away from the receivers and then towed back, at a speed around 1 m/s. The relative distance of the transmitter and the receiver M1 changed from 500 m to 4.5 km. The OFDM signals from the University of Connecticut were transmitted in two tows. There are 31 transmissions in each tow, with 20 blocks in each transmission.

Parameters of this experiment are summarized in Table B.2, while the subcarrier allocation is described in Table 2.1. A rate-1/2 nonbinary LDPC code is used, where the size of the Galois field matches the size of the constellation.

B.2.2 Mobility Estimation

Figure B.4 shows the received signal magnitude fluctuation during the first tow. One transmission file recorded during the turn of the source is excluded, where the signal-to-noise-ratio (SNR) of the received signal is quite low. The received signal strength becomes gradually weaker as the transmitter array was towed away from the receiver, and then gradually stronger when it was towed back.

For each block, the Doppler scaling factor can be estimated based on null subcarriers (c.f. Chapter 6). The relative speed between the transmitter and the receiver was estimated

Table B.2 OFDM Parameters in MACE10

Center frequency	f_c	13 kHz
Bandwidth	B	4.883 kHz
Time duration	T	209.7 ms
Frequency spacing	Δ	4.77 Hz
Guard interval	T_g	40.3 ms
# of subcarriers	K	1024
# of data subcarriers	$\|S_D\|$	672
# of pilot subcarriers	$\|S_P\|$	256
# of null subcarriers	$\|S_N\|$	96
Channel coding rate	r_c	0.5

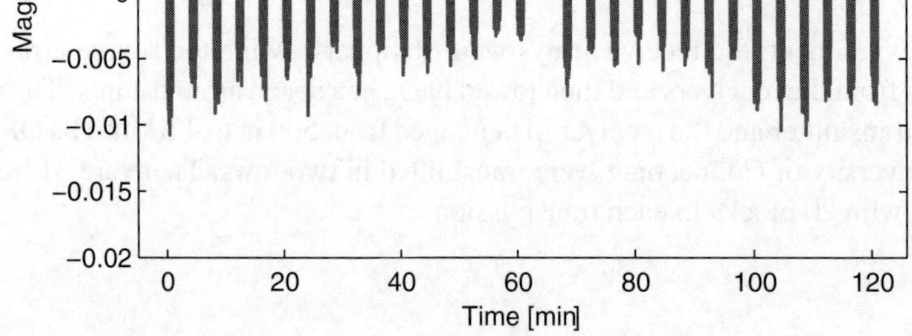

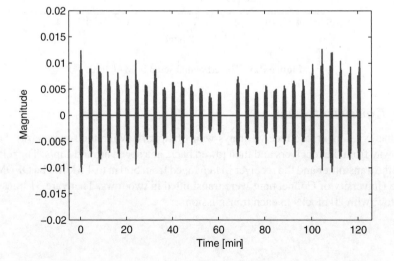

Figure B.4 MACE10: Received signal magnitude fluctuations in tow 1.

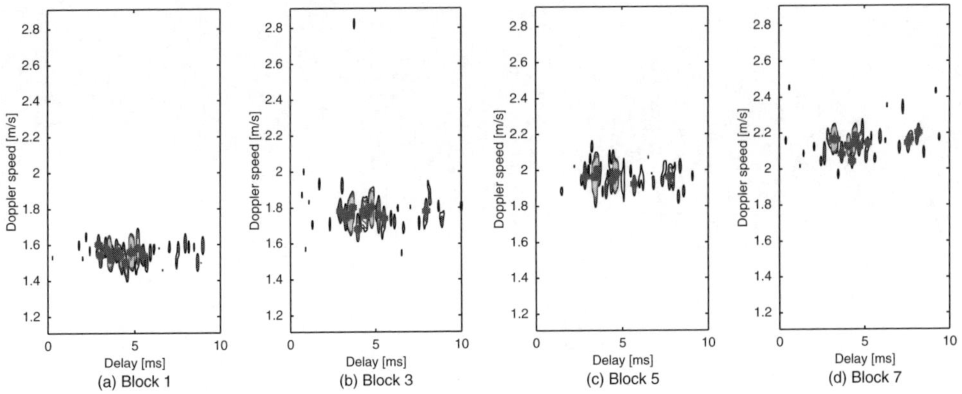

Figure B.5 MACE10: Channel estimates of blocks in one transmission.

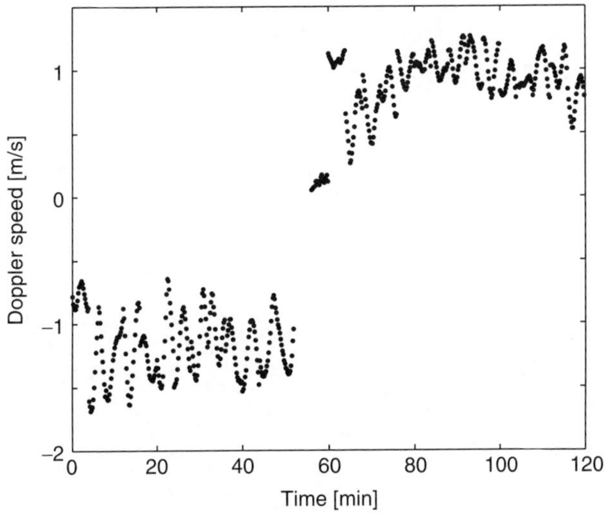

Figure B.6 The estimated relative moving speed in tow 1; MACE10.

as $\hat{v} = \hat{a} \cdot c$, using a nominal sound speed of $c = 1500\,\text{m/s}$. Figure B.5 shows the channel estimates of blocks within one transmission, which reveals a large Doppler speed variation across blocks. Figure B.6 shows the relative speed for the first tow. The relative speed was about 1.2 m/s when the transmitter was moving away from the receiver array, and about 1 m/s when it was towed back. The estimated speed changes from negative to positive after 60 minutes, indicating that the transmitter array began to be towed back from the maximum range at 60th minute.

References

[1] "Compressive sensing resources," http://dsp.rice.edu/cs.
[2] D. A. Abraham and P. K. Willett, "Active Sonar detection in shallow water using the Page Test," *IEEE J. Ocean. Eng.*, vol. 27, no. 1, pp. 35–46, Jan. 2002.
[3] R. Ahlswede, N. Cai, S.-Y. R. Li, and R. W. Yeung, "Network information flow," *IEEE Trans. Inform. Theory*, vol. 46, no. 4, pp. 1204–1216, July 2000.
[4] M. A. Ainslie and J. G. McColm, "A simplified formula for viscous and chemical absorption in sea water," *J. Acoust. Soc. Am.*, vol. 103, no. 3, pp. 1671–1672, Mar. 1998.
[5] I. F. Akyildiz, D. Pompili, and T. Melodia, "Challenges for efficient communication in underwater acoustic sensor networks," *ACM Sigbed Review*, vol. 1, no. 1, pp. 3–8, July 2004.
[6] S. Al-Dharrab, M. Uysal, and T. M. Duman, "Cooperative underwater acoustic communications," *IEEE Communications Magazine*, vol. 51, no. 7, pp. 146–153, July 2013.
[7] G. Al-Habian, A. Ghrayeb, M. Hasna, and A. Abu-Dayya, "Threshold-based relaying in coded cooperative networks," *IEEE Trans. Veh. Technol.*, vol. 60, no. 1, pp. 123–135, Jan. 2011.
[8] E. B. Al-Safadi and T. Y. Al-Naffouri, "Peak reduction and clipping mitigation in OFDM by augmented compressive sensing," *IEEE Trans. Signal Processing*, vol. 60, no. 7, pp. 3834–3839, July 2012.
[9] D. Angelosante, E. Grossi, G. B. Giannakis, and M. Lops, "Sparsity-aware estimation of CDMA system parameters," in *IEEE Workshop on Signal Processing Advances in Wireless Communications*, Perugia, Italy, June 2009.
[10] AquaSeNT, "OFDM Acoustic Modem," http://www.aquasent.com/.
[11] S. Attallah, "Blind estimation of residual carrier offset in OFDM systems," *IEEE Signal Processing Lett.*, vol. 11, no. 2, pp. 216–219, Feb. 2004.
[12] Y. M. Aval and M. Stojanovic, "Fractional FFT demodulation for differentially coherent detection of acoustic OFDM signals," in *Proc. of Asilomar Conf. on Signals, Systems, and Computers*, Nov. 2012.
[13] A. S. Avestimehr, S. N. Diggavi, and D. N. C. Tse, "Wireless network information flow: A deterministic approach," *IEEE Trans. Inform. Theory*, vol. 57, no. 4, pp. 1872–1905, Apr. 2011.
[14] A. Baggeroer, "Acoustic telemetry–an overview," *IEEE Journal of Oceanic Engineering*, vol. 9, no. 4, pp. 229–235, Oct. 1984.
[15] A. B. Baggeroer, "An overview of acoustic communications from 2000–2012," in *Proc. of the Workshop on Underwater Communications: Channel Modelling & Validation*, Italy, Sept. 2012.
[16] A. R. S. Bahai, B. R. Saltzberg, and M. Ergen, *Multi-carrier Digital Communications: Theory and Applications of OFDM*, Kluwer Academic, 2004.

OFDM for Underwater Acoustic Communications, First Edition. Shengli Zhou and Zhaohui Wang.
© 2014 John Wiley & Sons, Ltd. Published 2014 by John Wiley & Sons, Ltd.

[17] L. Bahl, J. Cocke, F. Jelinek, and J. Raviv, "Optimal decoding of linear codes for minimizing symbol error rate (corresp.)," *IEEE Trans. Inform. Theory*, vol. 20, no. 2, pp. 284–287, Mar. 1974.
[18] W. U. Bajwa, J. Haupt, A. Sayeed, and R. Nowak, "Compressed channel sensing: A new approach to estimating sparse multipath channels," *Proc. of the IEEE*, vol. 6, no. 98, pp. 1058–1076, June 2010.
[19] W. U. Bajwa, A. Sayeed, and R. Nowak, "Learning sparse doubly-selective channels," in *Proc. of Allerton Conf. on Communications, Control and Computing*, Sept. 2008, pp. 575–582.
[20] X. Bao and J. T. Li, "Adaptive network coded cooperation (ANCC) for wireless relay networks: matching code-on-graph with network-on-graph," *IEEE Trans. Wireless Commun.*, vol. 7, no. 2, pp. 574–583, Feb. 2008.
[21] X. Bao and J. T. Li, "Generalized adaptive network coded cooperation (GANCC): A unified framework for network coding and channel coding," *IEEE Trans. Commun.*, vol. 59, no. 11, pp. 2934–2938, Nov. 2011.
[22] Y. Bar-Shalom, X. R. Li, and T. Kirubarajan, *Estimation with Applications to Tracking and Navigation: Algorithms and Software for Information Extraction*, Wiley-Interscience, 2001.
[23] Y. Bar-Shalom, P. Willett, and X. Tian, *Tracking and Data Fusion: A Handbook of Algorithms*, YBS Publishing, 2011.
[24] R. Baraniuk, "Compressive sensing," *IEEE Signal Processing Magazine*, vol. 24, no. 4, pp. 118–121, July 2007.
[25] D. Baron, S. Sarvotham, and R. G. Baraniuk, "Bayesian compressive sensing via belief propagation," *IEEE Trans. Signal Processing*, vol. 58, no. 1, pp. 269–280, Jan. 2010.
[26] R. Bäuml, R. Fischer, and J. B. Huber, "Reducing the peak-to-average power ratio of multicarrier modulation by selected mapping," *Electronics Letters*, vol. 32, pp. 2056–2057, 1996.
[27] C. Bechaz and H. Thomas, "GIB system: The underwater GPS solution," in *Proceedings of ECUA*, May 2001.
[28] A. Beck and M. Teboulle, "A fast iterative shrinkage-thresholding algorithm for linear inverse problems," *SIAM Journal on Imaging Sciences*, vol. 2, no. 1, pp. 183–202, 2009.
[29] E. Bejjani and J. C. Belfiore, "Multicarrier coherent communications for the underwater acoustic channel," in *Proc. of OCEANS*, 1996.
[30] S. Benedetto and E. Biglieri, *Principles of Digital Transmission with Wireless Applications*, Kluwer Academic/ Plenum Publishers, 1999.
[31] A. Bennatan and D. Burshtein, "Design and analysis of nonbinary LDPC codes for arbitrary discrete-memoryless channels," *IEEE Trans. Inform. Theory*, vol. 52, no. 2, pp. 549–583, Feb. 2006.
[32] B. Benson, G. Chang, D. Manov, B. Graham, and R. Kastner, "Design of a low-cost acoustic modem for moored oceanographic applications," in *Proc. of WUWNet*, Los Angeles, CA, Sept. 2006.
[33] Benthos, Inc., "Fast and reliable access to undersea data," http://www.benthos.com.
[34] C. R. Berger, W. Chen, S. Zhou, and J. Huang, "A simple and effective noise whitening method for underwater acoustic orthogonal frequency division multiplexing," *Journal of Acoustical Society of America*, vol. 127, no. 4, pp. 2358–2367, Apr. 2010.
[35] C. R. Berger, J. P. Gomes, and J. M. F. Moura, "Sea-trial results for cyclic-prefix OFDM with long symbol duration," in *Proc. of MTS/IEEE OCEANS Conf.*, Santander, Spain, June 2011.
[36] C. R. Berger, J. P. Gomes, and J. M. F. Moura, "Study of pilot designs for cyclic-prefix OFDM on time-varying and sparse underwater acoustic channels," in *Proc. of MTS/IEEE OCEANS Conf.*, Santander, Spain, June 2011.
[37] C. R. Berger, J. Huang, and J. M. F. Moura, "Study of pilot overhead for iterative OFDM receivers on time-varying and sparse underwater acoustic channels," in *Proc. of MTS/IEEE OCEANS Conf.*, Sept. 2011.

[38] C. R. Berger, S. Zhou, J. Preisig, and P. Willett, "Sparse channel estimation for multicarrier underwater acoustic communication: From subspace methods to compressed sensing," *IEEE Trans. Signal Processing*, vol. 58, no. 3, pp. 1708–1721, Mar. 2010.

[39] C. R. Berger, S. Zhou, Z. Tian, and P. Willett, "Precise timing for multiband OFDM in a UWB system,," in *Proc. of ICUWB 2006*, Waltham, Sept. 2006, pp. 269–274.

[40] C. R. Berger, S. Zhou, Z. Tian, and P. Willett, "Performance analysis on an MAP fine timing algorithm in UWB multiband OFDM," *IEEE Trans. Commun.*, vol. 56, no. 10, pp. 1606–1611, Oct. 2008.

[41] C. R. Berger, S. Zhou, Y. Wen, P. Willett, and K. Pattipati, "Optimizing joint erasure- and error-correction coding for wireless packet transmissions," *IEEE Transactions on Wireless Communications*, vol. 7, no. 11, pp. 4586–4595, Nov. 2008.

[42] C. Berrou, A. Glavieux, and P. Thitimajshima, "Near Shannon limit error-correcting coding and decoding: Turbo-codes," in *Proc. of Intl. Conf. on Commun.*, May 1993, vol. 2, pp. 1064–1070.

[43] H. Bessai, *MIMO Signals and Systems*, Springer, 2005.

[44] J. A. C. Bingham, "Multicarrier modulation for data transmission: An idea whose time has come," *IEEE Communications Magazine*, vol. 28, no. 5, pp. 5–14, May 1990.

[45] J. A. C. Bingham, *ADSL, VDSL, and Multicarrier Modulation*, Wiley-Interscience, 2000.

[46] R. E. Blahut, *Theory and Practice of Error Control Codes*, Addison-Wesley, 1983.

[47] T. Blumensath and M. E. Davies, "Iterative hard thresholding for compressed sensing," *Applied and Computational Harmonic Analysis*, vol. 27, no. 3, pp. 265–274, Nov. 2009.

[48] R. Bradbeer, E. Law, and L. F. Yeung, "Using multi-frequency modulation in a modem for the transmission of near-realtime video in an underwater environment," in *Proc. of IEEE International Conference on Consumer Electronics*, June 2003.

[49] D. Brady and J. Catipovic, "Adaptive multiuser detection for underwater acoustical channels," *IEEE J. Ocean. Eng.*, vol. 19, no. 2, pp. 158–165, Apr. 1994.

[50] M. Breiling, S. Muller-Weinfurtner, and J.-B. Huber, "SLM peak-power reduction without explicit side information," *IEEE Commun. Lett.*, vol. 5, no. 6, pp. 239–241, June 2001.

[51] L. M. Brekhovskikh and Yu. P. Lysanov, *Fundamentals of Ocean Acoustics*, Springer, New York, 3 edition, 2003.

[52] G. Caire, G. Taricco, and E. Biglieri, "Bit-interleaved coded modulation," *IEEE Trans. Inform. Theory*, vol. 44, no. 3, pp. 927–946, May 1998.

[53] E. J. Candes, J. Romberg, and T. Tao, "Robust uncertainty principles: Exact signal reconstruction from highly incomplete frequency information," *IEEE Trans. Inform. Theory*, vol. 52, no. 2, pp. 489–509, Feb. 2006.

[54] E. J. Candès and T. Tao, "Decoding by linear programming," *IEEE Trans. Inform. Theory*, vol. 51, no. 12, pp. 4203–4215, Dec. 2005.

[55] E. J. Candès and T. Tao, "Near-optimal signal recovery from random projections: Universal encoding strategies?," *IEEE Trans. Inform. Theory*, vol. 52, no. 12, pp. 5406–5425, Dec. 2006.

[56] E. J. Candes and M. B. Wakin, "An introduction to compressive sampling," *IEEE Signal Processing Magazine*, vol. 25, no. 2, pp. 21–30, Mar. 2008.

[57] R. Cao, L. Yang, and F. Qu, "On the capacity and system design of relay-aided underwater acoustic communications," in *Proc. of IEEE Wireless Communications and Networking Conference*, Apr. 2010.

[58] C. Carbonelli, S.-H. Chen, and U. Mitra, "Error propagation analysis for underwater cooperative multi-hop communications," *Elsevier Ad Hoc Networks*, vol. 7, no. 4, pp. 759–769, June 2009.

[59] C. Carbonelli and U. Mitra, "Clustered channel estimation for UWB multipath antenna systems," *IEEE Trans. Wireless Commun.*, vol. 6, no. 3, pp. 970–981, Mar. 2007.

[60] C. Carbonelli, S. Vedantam, and U. Mitra, "Sparse channel estimation with zero tap detection," *IEEE Trans. Wireless Commun.*, vol. 6, no. 5, pp. 1743–1763, May 2007.

[61] P. Carroll, S. Zhou, K. Mahmood, H. Zhou, X. Xu, and J.-H. Cui, "On-demand collaborative localization for underwater sensor networks," in *Proc. of IEEE/MTS OCEANS Conference*, Hampton Roads, Virginia, Oct. 14-19, 2012.

[62] P. Carroll, S. Zhou, H. Zhou, X. Xu, J.-H. Cui, and P. Willett, "Underwater localization and tracking of physical systems," *Journal of Electrical and Computer Engineering, Special Issue on Underwater Communications and Networks*, 2012, doi:10.1155/2012/683919.

[63] C. G. Carter, "Coherence and time delay estimation," *Proc. of the IEEE*, vol. 75, no. 2, pp. 236–255, Feb. 1987.

[64] J. Castura and Y. Mao, "Rateless coding for wireless relay channels," *IEEE Trans. Wireless Commun.*, vol. 6, no. 5, pp. 1638–1642, May 2007.

[65] J. A. Catipovic, "Performance limitations in underwater acoustic telemetry," *IEEE Journal of Oceanic Engineering*, vol. 15, no. 3, pp. 205–216, July 1990.

[66] CCRMA, "Sampling-rate conversion," https://ccrma.stanford.edu/~jos/resample/.

[67] P. Ceballos and M. Stojanovic, "Adaptive channel estimation and data detection for underwater acoustic MIMO OFDM systems," *IEEE J. Ocean. Eng.*, vol. 35, no. 3, pp. 635–646, July 2010.

[68] R. W. Chang, "High-speed multichannel data transmission with bandlimited orthogonal signals," *Bell Sys. Tech. J.*, vol. 45, pp. 1775–1796, Dec. 1966.

[69] R. Chartrand, "Exact reconstruction of sparse signals via nonconvex minimization," *IEEE Signal Processing Lett.*, vol. 14, no. 10, pp. 707–710, Oct. 2007.

[70] R. Chartrand, R. Baraniuk, Y. C. Eldar, M. A. T. Figueiredo, and J. Tanner, "Introduction to the issue on compressive sensing," *IEEE J. Select. Topics Signal Proc.*, vol. 4, no. 2, pp. 241–243, Feb. 2010.

[71] B. Chen and H. Wang, "Blind estimation of OFDM carrier frequency offset via oversampling," *IEEE Trans. Signal Processing*, vol. 52, no. 7, pp. 2047–2057, July 2004.

[72] R.-R. Chen, R. Peng, A. Ashikhmin, and B. Farhang-Boroujeny, "Approaching MIMO capacity using bitwise Markov chain Monte Carlo detection," *IEEE Trans. Commun.*, vol. 58, no. 2, pp. 423–428, Feb. 2010.

[73] S. S. Chen, D. L. Donoho, and M. A. Saunders, "Atomic decomposition by basis pursuit," *SIAM J. Scientific Computing*, vol. 43, no. 1, pp. 129–159, 2001.

[74] Y. Chen, Z.-H. Wang, L. Wan, H. Zhou, S. Zhou, and X. Xu, "OFDM modulated dynamic coded cooperation in underwater acoustic channels," *IEEE Journal of Oceanic Engineering*, 2014 (to appear).

[75] W. Cheng, A. Y. Teymorian, L. Ma, X. Cheng, X. Lu, and Z. Lu, "Underwater localization in sparse 3D acoustic sensor networks," in *IEEE INFOCOM 2008*, Apr. 2008, pp. 798–806.

[76] X. Cheng, H. Shu, Q. Liang, and H. C. Du, "Silent positioning in underwater acoustic sensor networks," *IEEE Trans. on Vehicular Tech.*, vol. 57, no. 3, pp. 1756–1766, May 2008.

[77] M. Chitre, S. H. Ong, and J. Potter, "Performance of coded OFDM in very shallow water channels and snapping shrimp noise," in *Proc. of MTS/IEEE OCEANS Conf.*, 2005, vol. 2, pp. 996–1001.

[78] M. Chitre, J. R. Potter, and S.-H. Ong, "Optimal and near-optimal signal detection in snapping shrimp dominated ambient noise," *IEEE J. Ocean. Eng.*, vol. 31, no. 2, pp. 497–503, Apr. 2006.

[79] M. Chitre, S. Shahabudeen, and M. Stojanovic, "Underwater acoustic communications and networking: Recent advances and future challenges," *Marine Technology Society Journal*, vol. 42, no. 1, pp. 103–116, 2008.

[80] M. Chitre, I. Topor, and T.-B. Koay, "The UNET-2 modem – an extensible tool for underwater networking research," in *Proceedings of IEEE OCEANS*, Yeosu, May 2012.

[81] T.-D. Chiueh and P.-Y. Tsai, *OFDM Baseband Receiver Design for Wireless Communications*, Wiley, 1 edition, 2007.

[82] S. Cho, H. C. Song, and W. S. Hodgkiss, "Asynchronous multiuser underwater acoustic communications (l)," *J. Acoust. Soc. Am.*, vol. 132, no. 1, pp. 5–8, July 2012.

[83] S. Cho, H. C. Song, and W. S. Hodgkiss, "Multiuser interference cancellation in time-varying channels," *J. Acoust. Soc. Am.*, vol. 131, no. 4, pp. EL163–EL169, Feb. 2012.

[84] S. Cho, H. C. Song, and W. S. Hodgkiss, "Multiuser acoustic communications with mobile users," *J. Acoust. Soc. Am.*, vol. 133, no. 2, pp. 880–890, Feb. 2013.

[85] Y.-S. Choi, P. J. Voltz, and F. A. Cassara, "ML estimation of carrier frequency offset for multicarrier signals in Rayleigh fading channels," *IEEE Trans. Veh. Technol.*, vol. 50, no. 2, pp. 644–655, Mar. 2001.

[86] S. Coatelan and A. Glavieux, "Design and test of a coded OFDM system on the shallow water acoustic channel," in *Proc. of OCEANS*, Sept. 1994.

[87] P. L. Combettes and V. R. Wajs, "Signal recovery by proximal forward-backward splitting," *Multiscale Modeling and Simulation*, vol. 4, no. 4, pp. 1168–1200, 2005.

[88] S. F. Cotter and B. D. Rao, "Sparse channel estimation via matching pursuit with application to equalization," *IEEE Trans. Commun.*, vol. 50, no. 3, pp. 374–377, Mar. 2002.

[89] T. Cover and A. El Gamal, "Capacity theorems for the relay channel," *IEEE Trans. Inform. Theory*, vol. 25, no. 5, pp. 572–584, Sept. 1979.

[90] J.-H. Cui, J. Kong, M. Gerla, and S. Zhou, "The challenges of building mobile underwater wireless networks for aquatic applications," *IEEE Network, Special Issue on Wireless Sensor Networking*, vol. 20, no. 3, pp. 12–18, May 2006.

[91] J.-H. Cui, S. Zhou, Z. Shi, J. O'Donnell, Z. Peng, S. Roy, P. Arabshahi, M. Gerla, B. Baschek, and X. Zhang, "Ocean-TUNE: A community ocean testbed for underwater wireless networks," in *Proc. of the 7th ACM International Conference on UnderWater Networks and Systems (WUWNet)*, Los Angeles, CA, Nov. 5-6, 2012.

[92] I. Daubechies, M. Defrise, and C. De Mol, "An iterative thresholding algorithm for linear inverse problems with a sparsity constraint," *Communications on Pure and Applied Mathematics*, vol. 57, no. 11, pp. 1413–1457, Nov. 2004.

[93] M. C. Davey and D. Mackay, "Monte Carlo simulations of infinite low density parity check codes over GF(q)," in *Proc. of Int. Workshop on Optimal Codes and Related Topics*, Bulgaria, June 9–15 1998.

[94] H. A. David and H. N. Nagaraja, *Order Statistics*, chapter 2, pp. 11–30, John Wiley & Sons, Ltd, 3rd edition, 2003.

[95] Develogic, "Underwater Communication Systems," http://www.develogic.de/.

[96] R. Diamant and L. Lampe, "Underwater localization with time-synchronization and propagation speed uncertainties," in *Proc. of the 8th Workshop on Positioning Navigation and Communication (WPNC)*, Apr. 2011.

[97] R. Diamant, H. Tan, and L. Lampe, "LOS and NLOS classification for underwater acoustic localization," *IEEE Trans. Mobile Computing*, vol. 13, no. 2, pp. 311–323, Feb. 2014.

[98] R. Dinis and A. Gusmao, "On the performance evaluation of OFDM transmission using clipping techniques," in *Proc. of IEEE Vehicular Technology Conf.*, 1999, pp. 2923–2928.

[99] R. Dinis, P. Montezuma, and A. Gusmao, "Performance trade-offs with quasi-linearly amplified OFDM through a two-branch combining technique," in *Proc. of IEEE Vehicular Technology Conf.*, Apr. 1996, pp. 899–903.

[100] D. Donoho, "Compressed sensing," *IEEE Trans. Inform. Theory*, vol. 52, no. 4, pp. 1289–1306, Apr. 2006.

[101] D. L. Donoho, I. Drori, Y. Tsaig, and J. L. Starck, "Sparse solution of underdetermined linear equations by stagewise orthogonal matching pursuit (StOMP)," Stat. dept. tech. rep., Stanford Univ., Palo Alto, CA, Mar. 2006.

[102] D. L. Donoho, A. Maleki, and A. Montanari, "Message-passing algorithms for compressed sensing," *Proc. Nat. Acad. Sci.*, vol. 106, no. 45, pp. 18914–18919, 2009.

[103] C. Douillard, M. Jézéquel, C. Berrou, A. Picart, P. Didier, and A. Glavieux, "Iterative correction of intersymbol interference turbo-equalization," *Euro. Trans. Telecommun.*, vol. 6, no. 5, pp. 507–511, Sep.-Oct. 1995.

[104] R. J. Drost and A. C. Singer, "Factor-graph algorithms for equalization," *IEEE Trans. Signal Processing*, vol. 55, no. 5, pp. 2052–2065, May 2007.

[105] DSPCOMM, "Underwater wireless modem," http://www.dspcomm.com.

[106] T. M. Duman and A. Ghrayeb, *Coding for MIMO Communication Systems*, John Wiley & Sons, Ltd, 2007.

[107] G. F. Edelmann, T. Akal, W. S. Hodgkiss, S. Kim, W. A. Kuperman, and H. C. Song, "An initial demonstration of underwater acoustic communications using time reversal," *IEEE J. Oceanic Eng.*, vol. 31, no. 3, pp. 602–609, Jul. 2002.

[108] G. F. Edelmann, H. C. Song, S. Kim, W. S. Hodgkiss, W. A. Kuperman, and T. Akal, "Underwater acoustic communications using time reversal," *IEEE J. Oceanic Eng.*, vol. 30, no. 4, pp. 852–864, Oct. 2005.

[109] Y. C. Eldar, "Generalized SURE for exponential families: Applications to regularization," *IEEE Trans. Signal Processing*, vol. 57, no. 2, pp. 471–481, Feb. 2009.

[110] Y. C. Eldar and G. Kutyniok, *Compressed Sensing: Theory and Applications*, Cambridge University Press, 1st edition, 2012.

[111] J. Erfanian, S. Pasupathy, and P. G. Gulak, "Reduced complexity symbol detectors with parallel structure for ISI channels," *IEEE Trans. Commun.*, vol. 42, no. 234, pp. 1661–1671, 1994.

[112] M. Erol, H. Mouftah, and S. Oktug, "Localization techniques for underwater acoustic sensor networks," *IEEE Communications Magazine*, vol. 48, no. 12, pp. 152–158, June 2010.

[113] M. Erol-Kantarci, H. T. Mouftah, and S. Oktug, "A survey of architectures and localization techniques for underwater acoustic sensor networks," *IEEE Communications Surveys & Tutorials*, vol. 13, no. 3, pp. 487–502, third quarter 2011.

[114] M. Erol-Kantarci, S. Oktug, L. Filipe, M. Vieira, and M. Gerla, "Performance evaluation of distributed localization techniques for mobile underwater acoustic sensor," *Ad Hoc Networks*, vol. 9, no. 1, pp. 61–72, Jan. 2011.

[115] R. M. Eustice, L. L. Whitcomb, H. Singh, and M. Grund, "Experimental results in synchronous-clock one-way-travel-time acoustic navigation for autonomous underwater vehicles," in *IEEE IROS 2009*, Oct. 2009.

[116] EvoLogics, "Underwater Acoustic Modems," http://www.evologics.de/.

[117] B. Farhang-Boroujeny, H. Zhu, and Z. Shi, "Markov Chain Monte Carlo algorithms for CDMA and MIMO communication systems," *IEEE Trans. Signal Processing*, vol. 54, no. 5, pp. 1896–1909, May 2006.

[118] K. Fazel and S. Kaiser, *Multi-Carrier and Spread Spectrum Systems: From OFDM and MC-CDMA to LTE and WiMAX*, John Wiley & Sons, Ltd. 2nd edition, 2008.

[119] B. Ferreira, A. Matos, and N. Cruz, "Estimation approach for AUV navigation using a single acoustic beacon," *Sea Technology*, vol. 51, no. 12, pp. 54–58, Dec. 2010.

[120] B. L. Floch, M. Alard, and C. Berrou, "Coded orthogonal frequency division multiplex," *Proceedings of the IEEE*, vol. 83, no. 6, pp. 982–996, June 1995.

[121] J. A. Flynn, J. A. Ritcey, D. Rouseff, and W. L. J. Fox, "Multichannel equalization by decision-directed passive phase conjugation: Experimental results," *IEEE J. Oceanic Eng.*, vol. 29, no. 3, pp. 824–836, Jul. 2004.

[122] G. D. Forney, "Maximum-likelihood sequence estimation of digital sequences in the presence of intersymbol interference," *IEEE Transactions on Information Theory*, vol. 18, no. 3, pp. 363–378, May 1972.

[123] G. J. Foschini and M. J. Gans, "On limits of wireless communication in a fading environment when using multiple antennas," *Wireless Personal Communications*, vol. 6, no. 3, pp. 311–335, Mar. 1998.

[124] M. Fossorier, "Quasi-cyclic low-density parity-check codes from circulant permutation matrix," *IEEE Transactions on Information Theory*, vol. 50, no. 8, pp. 1788–1793, 2004.

[125] C. Fragouli, J.-Y. Le Boudec, and J. Widmer, "Network coding: An instant primer," *ACM SIGCOMM Computer Communication Review*, vol. 36, no. 1, Jan. 2006.

[126] C. Fragouli and E. Soljanin, "An algebraic approach to network coding," *Foundations and Trends ® in Networking*, vol. 2, no. 2, pp. 135–269, Jan. 2007.

[127] L. Freitag, M. Grund, S. Singh, J. Partan, P. Koski, and K. Ball, "The WHOI Micro-Modem: An acoustic communications and navigation system for multiple platforms," in *Proceeding of OCEANS*, Washington DC, 2005.

[128] L. Freitag, M. Stojanovic, S. Singh, and M. Johnson, "Analysis of channel effects on direct-sequence and frequency-hopped spread-spectrum acoustic communications," *IEEE Journal of Oceanic Engineering*, vol. 26, no. 4, pp. 586–593, Oct. 2001.

[129] B. Friedlander, "On the Cramer- Rao bound for time delay and Doppler estimation (corresp.)," *IEEE Trans. Inform. Theory*, vol. 30, no. 3, pp. 575–580, May 1984.

[130] B. Friedlander, "Random projections for sparse channel estimation and equalization," in *Proc. of Asilomar Conf. on Signals, Systems, and Computers*, Pacific Grove, CA, Oct. 2006.

[131] S. Fu, K. Lu, T. Zhang, Y. Qian, and H.-H. Chen, "Cooperative wireless networks based on physical layer network coding," *IEEE Wireless Communications*, vol. 17, no. 6, pp. 86–95, Dec. 2010.

[132] T. Fu, D. Doonan, C. Utley, R. Iltis, R. Kastner, and H. Lee, "Design and development of a software-defined underwater acoustic modem for sensor networks for environmental and ecological research," in *Proc. of OCEANS*, Sept. 2006.

[133] T. Fusco and M. Tanda, "Blind synchronization for OFDM systems in multipath channels," *IEEE Trans. Commun.*, vol. 8, no. 3, pp. 1340–1348, Mar. 2009.

[134] G. Davidoff, P. Sarnak, and A. Valette, *Elementary Number Theory, Group Theory, and Ramanujan Graphs*, Cambridge University Press, 2003.

[135] H. Gacanin and F. Adachi, "Broadband analog network coding," *IEEE Trans. Wireless Commun.*, vol. 9, no. 5, pp. 1577–1583, May 2009.

[136] N. P. Galatsanos and A. K. Katsaggelos, "Methods for choosing the regularization parameter and estimating the noise variance in image restoration and their relation," *IEEE Trans. Image Processing*, vol. 1, no. 3, pp. 322–336, July 1992.

[137] R. G. Gallager, *Low Density Parity Check Codes*, Cambridge, MA: MIT Press, 1963.

[138] E. Gallimore, J. Partan, I. Vaughn, S. Singh, J. Shusta, and L. Freitag, "The WHOI Micromodem-2: A scalable system for acoustic communications and networking," in *Proc. of MTS/IEEE OCEANS Conference*, Sept. 2010.

[139] F. Gao, R. Zhang, and Y.-C. Liang, "Channel estimation for OFDM modulated two-way relay networks," *IEEE Trans. Signal Processing*, vol. 57, no. 11, pp. 4443–4455, Nov. 2009.

[140] P. J. Gendron, "Orthogonal frequency division multiplexing with on-off-keying: Noncoherent performance bounds, receiver design and experimental results," *U.S. Navy Journal of Underwater Acoustics*, vol. 56, no. 2, pp. 267–300, Apr. 2006.

[141] X. Geng and A. Zielinski, "An eigenpath underwater acoustic communication channel model," in *Proc. of MTS/IEEE OCEANS Conf.*, San Diego, CA , USA, Oct. 1995, pp. 1189–1196.

[142] A. Gershman and N. Sidiropoulos (eds), *Space-Time Processing for MIMO Communications*, John Wiley & Sons, Ltd, 2005.

[143] M. Ghosh, "Analysis of the effect of impulse noise on multicarrier and single carrier QAM systems," *IEEE Trans. Commun.*, vol. 44, no. 2, pp. 145–147, Feb. 1996.

[144] G. B. Giannakis, Z. Liu, X. Ma, and S. Zhou, *Space Time Coding for Broadband Wireless Communications*, John Wiley & Sons Inc., Dec. 2006.

[145] A. Goldsmith, *Wireless Communications*, Cambridge University Press, 2005.

[146] E. Golovins, *Pilot-Assisted OFDM Systems: Low-Complexity Algorithms for Optimised Channel Estimation and Detection*, VDM Verlag Dr. Müller, 1st edition, 2010.

[147] G. H. Golub, M. Heath, and G. Wahba, "Generalized cross-validation as a method for choosing a good ridge parameter," *IEEE Transactions on Image Processing*, vol. 21, no. 2, pp. 215–223, May 1979.

[148] J. Gomes, A. Silva, and S. Jesus, "Adaptive spatial combining for passive time-reversed communications," *J. Acoust. Soc. Am.*, vol. 124, no. 2, pp. 1038–1053, Aug. 2008.

[149] J. Gomes, A. Silva, and S. Jesus, "Experimental – assessment of time-reversed OFDM underwater communications," *J. Acoust. Soc. Am.*, vol. 123, no. 5, pp. 3891–3891, 2008.

[150] Q. Guo and P. Li, "LMMSE turbo equlization based on factor graph," *IEEE J. Select. Areas Commun.*, vol. 26, no. 2, pp. 311–319, Feb. 2008.

[151] Z. Guo, B. Wang, P. Xie, W. Zeng, and J.-H. Cui, "Efficient error recovery using network coding in underwater sensor networks," *Elsevier Ad Hoc Networks, Special Issue on Underwater Networks*, vol. 7, pp. 791–802, June 2009.

[152] A. S. Gupta and J. Preisig, "A geometric mixed norm approach to shallow water acoustic channel estimation and tracking," *Elsevier J. of Physical Commun.*, vol. 5, no. 2, pp. 119–128, June 2012.

[153] O. Gurewitz, A. de Baynast, and E. W. Knightly, "Cooperative strategies and achievable rate for tree networks with optimal spatial reuse," *IEEE Trans. Inform. Theory*, vol. 53, no. 10, pp. 3596–3614, Oct. 2007.

[154] C. El Hajjar, *Synchronization Algorithms for OFDM Systems (IEEE802. 11a, DVB-T): Analysis, Simulation, Optimization and Implementation Aspects*, Fraunhofer IRB Verlag, 1st edition, 2009.

[155] E. T. Hale, W. Yin, and Y. Zhang, "A fixed-point continuation method for l_1-regularized minimization with applications to compressed sensing," Caam tech. rep. tr07-07, Rice Univ., Houston, TX, July 2007.

[156] K. A. Hamdi, "Precise interference analysis of OFDMA time-asynchronous wireless ad-hoc networks," *IEEE Trans. Wireless Commun.*, vol. 9, no. 1, pp. 134–144, Jan. 2010.

[157] L. L. Hanzo, M. Munster, B. J. Choi, and T. Keller, *OFDM and MC-CDMA for Broadband Multi-User Communications, WLANs and Broadcasting*, Wiley-IEEE Press, 2003.

[158] R. Hartman, W. Hawkinson, and K. Sweeney, "Tactical underwater navigation system (TUNS)," in *IEEE/ION Position, Location and Navigation Symposium*, Fairfax, Virginia, USA, May 2008, pp. 898–911.

[159] R. Hayford, D. Nagle, and J. Catipovic, "AUTEC undersea cellular network," in *Proc. of the ACM Intl. Workshop on Underwater Networks (WUWNet)*, Berkeley, California, Nov. 2009.

[160] R. Headrick and L. Freitag, "Growth of underwater communication technology in the U.S. Navy," *IEEE Communications Magazine*, vol. 47, no. 1, pp. 80–82, Jan. 2009.

[161] Heat, Light & Sound Research, Inc., "ONR Ocean Acoustics library," http://oalib.hlsresearch.com/.

[162] W. Henkel and B. Wagner, "Trellis shaping for reducing the peak-to-average ratio of multitone signals," in *Proc. of IEEE Intl. Symp. on Information Theory*, June 1997.

[163] G. R. Hill, M. Faulkner, and J. Singh, "Reducing the peak-to-average power ratio in OFDM by cyclically shifting partial transmit sequences," *Electronics Letters*, vol. 36, no. 6, pp. 560–561, Mar. 2000.

[164] T. Ho and D. Lun, *Network Coding: An Introduction*, Cambridge University Press, 1st edition, 2008.

[165] T. Ho, M. Medard, R. Koetter, D. R. Karger, M. Effros, J. Shi, and B. Leong, "A random linear network coding approach to multicast," *IEEE Trans. Inform. Theory*, vol. 52, no. 10, pp. 4413–4430, Oct. 2006.

[166] M. L. Honig, *Advances in Multiuser Detection*, Wiley-IEEE Press, 1 edition, 2009.

[167] I. Honkala and A. Klapper, "Bounds for the multicovering radii of Reed-Muller codes with applications to stream ciphers," *Designs, Codes and Cryptography*, vol. 23, no. 2, pp. 131–146, July 2001.

[168] A. Hottinen, O. Tirkkonen, and R. Wichman, *Multi-antenna Transceiver Techniques for 3G and Beyond*, John Wiley & Sons, 2003.

[169] X.-Y. Hu and E. Eleftheriou, "Binary representation of cycle tanner-graph GF(2^b) codes," *Proc. of Intl. Conf. on Commun.*, vol. 27, no. 1, pp. 528–532, June 2004.

[170] X.-Y. Hu, E. Eleftheriou, and D.-M. Arnold, "Regular and irregular progressive edge-growth tanner graphs," *IEEE Transactions on Information Theory*, vol. 51, no. 1, pp. 386–398, Jan. 2005.

[171] J. Z. Huang, C. R. Berger, S. Zhou, and J. Huang, "Performance and complexity comparisons of basis pursuit alglrithms for sparse channel estimation in underwater OFDM," in *Proc. of MTS/IEEE OCEANS Conf.*, Sydney, Australia, June 2010.

[172] J. Huang, J.-Z. Huang, C. R. Berger, S. Zhou, and P. Willett, "Iterative sparse channel estimation and decoding for underwater MIMO-OFDM," *EURASIP Journal on Advances in Signal Processing*, vol. 2010, Article ID 460379, 11 pages, 2010. doi:10.1155/2010/460379.
[173] J. Huang, L. Liu, W. Zhou, and S. Zhou, "Large-girth nonbinary QC-LDPC codes of various lengths," *IEEE Transactions on Communications*, vol. 58, no. 12, pp. 3436–3447, Dec. 2010.
[174] J. Huang, Z.-H. Wang, S. Zhou, and Z. Wang, "Turbo equalization for OFDM modulated physical layer network coding," in *Proc. of IEEE Workshop on Signal Processing Advances in Wireless Communications*, San Francisco, USA, June 2011.
[175] J. Huang, S. Zhou, and P. Willett, "Nonbinary LDPC coding for multicarrier underwater acoustic communication," *IEEE Journal on Selected Areas in Communications*, vol. 26, no. 9, pp. 1684–1696, Dec. 2008.
[176] J. Huang, S. Zhou, and P. Willett, "Structure, property, and design of nonbinary regular cycle codes," *IEEE Transactions on Communications*, vol. 58, no. 4, Apr. 2010.
[177] J. Huang, S. Zhou, J. Zhu, and P. Willett, "Group-theoretic analysis of Cayley-graph-based cycle GF(2^p) codes," *IEEE Transactions on Communications*, vol. 57, no. 6, pp. 1560–1565, June 2009.
[178] J. Huang, W. Zhou, and S. Zhou, "Structured nonbinary rate-compatible low-density parity-check codes," *IEEE Commun. Lett.*, vol. 15, no. 9, pp. 998–1000, Sept. 2011.
[179] J. Huang and J.-K. Zhu, "Linear time encoding of cycle GF(2^p) codes through graph analysis," *IEEE Communications Letters*, vol. 10, pp. 369–371, May 2006.
[180] J.-Z. Huang, S. Zhou, J. Huang, C. R. Berger, and P. Willett, "Progressive inter-carrier interference equalization for OFDM transmission over time-varying underwater acoustic channels," *IEEE J. Select. Topics Signal Proc.*, vol. 5, no. 8, pp. 1524–1536, Dec. 2011.
[181] J.-Z. Huang, S. Zhou, J. Huang, J. Preisig, L. Freitag, and P. Willett, "Progressive intercarrier and co-channel interference mitigation for underwater acoustic MIMO-OFDM," *Wireless Communications and Mobile Computing*, Feb. 2012, doi:10.1002/wcm.1251.
[182] J.-Z. Huang, S. Zhou, and Z.-H. Wang, "Robust initialization with reduced pilot overhead for progressive underwater acoustic OFDM receivers," in *Proc. of MILCOM Conf.*, Nov. 2011, pp. 406–411.
[183] J.-Z. Huang, S. Zhou, and Z.-H. Wang, "Performance results of two iterative receivers for distributed MIMO-OFDM with large Doppler deviations," *IEEE J. Ocean. Eng.*, vol. 38, no. 2, pp. 347–357, 2013.
[184] A. L. Hui and K. B. Letaief, "Successive interference cancellation for multiuser asynchronous DS/CDMA detectors in multipath fading links," *IEEE Trans. Commun.*, vol. 46, no. 3, pp. 384–391, Mar. 1998.
[185] S.-J. Hwang and P. Schniter, "Efficient multicarrier communication for highly spread underwater acoustic channels," *IEEE J. Select. Areas Commun.*, vol. 26, no. 9, pp. 1674–1683, Dec. 2008.
[186] S. Ibrahim, J.-H. Cui, and R. Ammar, "Surface-level gateway deployment for underwater sensor networks," in *Proc. of MILCOM Conf.*, Oct. 2007.
[187] S. Ibrahim, J.-H. Cui, and R. Ammar, "Efficient surface gateway deployment for underwater sensor networks," in *Proc. of IEEE Symposium on Computers and Communications*, July 2008.
[188] R. A. Iltis, "Iterative joint decoding and sparse channel estimation for single-carrier modulation," in *Proc. of Intl. Conf. on ASSP*, Las Vegas, NV, Apr. 2008, pp. 2689–2692.
[189] H. Imai and S. Hirakawa, "A new multilevel coding method using error-correcting codes," *IEEE Trans. Inform. Theory*, vol. 23, no. 3, pp. 371–377, May 1977.
[190] K. Ishibashi, K. Ishii, and H. Ochiai, "Dynamic coded cooperation using multiple turbo codes in wireless relay networks," *IEEE J. Select. Topics Signal Proc.*, vol. 5, no. 1, pp. 197–207, Feb. 2011.
[191] M. T. Isik and O. B. Akan, "A three dimensional localization algorithm for underwater acoustic sensor networks," *IEEE Transactions on Wireless Communications*, vol. 8, no. 9, Sept. 2009.
[192] E. Jacobsen and P. Kootsookos, "Fast, accurate frequency estimators [DSP Tips & Tricks]," *IEEE Signal Processing Magazine*, vol. 24, no. 3, pp. 123–125, May 2007.

[193] H. Jafarkhani, *Space-Time Coding: Theory and Practice*, Cambridge University Press, 2005.

[194] M. Jankiraman, *Space-Time Codes and MIMO Systems*, Artech House Publishers, 2004.

[195] F. B. Jensen, W. A. Kuperman, M. B. Portor, and H. Schmidt, *Computational Ocean Acoustics*, Springer New York, New York, 2nd edition, 2011.

[196] W.-G. Jeon and K.-H. Chang, "An equalization technique for orthogonal frequency-division multiplexing systems in time-variant multipath channels," *IEEE Trans. Commun.*, vol. 47, no. 1, pp. 27–32, Jan. 1999.

[197] S. Ji, Y. Xue, and L. Carin, "Bayesian compressive sensing," *IEEE Trans. Signal Processing*, vol. 56, no. 6, pp. 2346–2356, June 2008.

[198] H. Jung and M. D. Zoltowski, "On the equalization of asynchronous multiuser OFDM signals in fading channels," in *Proc. of Intl. Conf. on ASSP*, May 2004, vol. 4, pp. 765–768.

[199] H. Jung and M. D. Zoltowski, "Semiblind multichannel identification in asynchronous multiuser OFDM systems," in *Proc. of Asilomar Conf. on Signals, Systems, and Computers*, Pacific Grove, CA, Oct. 2004.

[200] D. Jungnickel and S. A. Vanstone, "Graphical codes revisited," *IEEE Transactions on Information Theory*, vol. 43, pp. 136–146, Jan. 1997.

[201] T. Kang and R. A. Iltis, "Iterative carrier frequency offset and channel estimation for underwater acoustic OFDM systems," *IEEE J. Select. Areas Commun.*, vol. 26, no. 9, pp. 1650–1661, Dec. 2008.

[202] T. Kang, H. C. Song, and W. S. Hodgkiss, "Long-range multi-carrier acoustic communication in deep water using a towed horizontal array," *J. Acoust. Soc. Am.*, vol. 131, no. 6, pp. 4665–4671, June 2012.

[203] T. Kang, H. C. Song, W. S. Hodgkiss, and J. S. Kim, "Long-range multi-carrier acoustic communications in shallow water based on iterative sparse channel estimation," *J. Acoust. Soc. Am.*, vol. 128, no. 6, Dec. 2010.

[204] G. Z. Karabulut and A. Yongacoglu, "Sparse channel estimation using orthogonal matching pursuit algorithm," in *Proc. of IEEE Vehicular Technology Conf.*, Los Angeles, CA, Sept. 2004.

[205] S. A. Kassam, *Signal Detection in Non-Gaussian Noise*, Springer, 1st edition, 1987.

[206] S. Katti, S. Gollakota, and D. Katabi, "Embracing wireless interference: analog network coding," in *Proc. of Conf. on Applications, Technologies, Architectures, and Protocols for Computer Communications*, 2007, pp. 397–408.

[207] K. Kebkal and R. Bannasch, "Sweep-spread carrier for underwater communication over acoustic channels with strong multipath propagation," *J. Acoust. Soc. Am.*, vol. 112, no. 5, pp. 2043–2052, Nov. 2002.

[208] D. B. Kilfoyle and A. B. Baggeroer, "The state of the art in underwater acoustic telemetry," *IEEE Journal of Oceanic Engineering*, vol. 25, no. 1, pp. 4–27, Jan. 2000.

[209] D. B. Kilfoyle, J. C. Preisig, and A. B. Baggeroer, "Spatial modulation over partially coherent multiple-input/multiple-output channels," *IEEE Transactions on Signal Processing*, vol. 51, no. 3, pp. 794–804, Mar. 2003.

[210] D. B. Kilfoyle, J. C. Preisig, and A. B. Baggeroer, "Spatial modulation experiments in the underwater acoustic channel," *IEEE J. Ocean. Eng.*, vol. 30, no. 2, pp. 406–415, Apr. 2005.

[211] B. C. Kim and I. T. Lu, "Parameter studies of OFDM underwater communications systems," in *Proc. of MTS/IEEE Oceans Conference*, Providence, Rhode Island, September 11-14, 2000.

[212] S.-J. Kim, K. Koh, M. Lustig, S. Boyd, and D. Gorinevsky, "An interior-point method for large-scale l_1-regularized least squares," *IEEE J. Select. Topics Signal Proc.*, vol. 1, no. 4, pp. 606–617, Dec. 2007.

[213] D. Kimura, R. Pyndiah, and F. Guilloud, "Construction of parity-check matrices for non-binary LDPC codes," in *Proc. of 4th Intl. Symp. on Turbo Codes and Related Topics*, Munich, Germany, April 3-7, 2006.

[214] C. H. Knapp and C. G. Carter, "The generalized correlation method for estimation of time delay," *IEEE Trans. Acoustics, Speech and Signal Processing*, vol. 24, no. 4, pp. 320–327, Aug. 1976.

[215] W. Koch and A. Baier, "Optimum and sub-optimum detection of coded data disturbed by time-varying intersymbol interference," in *Proc. of Global Telecommunications Conference*, Dec. 1990, vol. 3, pp. 1679–1684.

[216] R. Koetter and M. Médard, "An algebraic approach to network coding," *IEEE/ACM Transactions on Networking*, vol. 11, no. 5, pp. 782–795, Oct. 2003.

[217] R. Koetter, A. C. Singer, and M. Tuchler, "Turbo equalization," *IEEE Signal Processing Magazine*, vol. 21, no. 1, pp. 67–80, Jan. 2004.

[218] T. Koike-Akino, P. Popovski, and V. Tarokh, "Optimized constellations for two-way wireless relaying with physical network coding," *IEEE J. Select. Areas Commun.*, vol. 27, no. 5, pp. 773–787, June 2009.

[219] Kongsberg Maritime, "Hydroacoustic telemetry link (HTL)," http://www.km.kongsberg.com.

[220] Y. Kou, W.-S. Lu, and A. Antoniou, "New peak-to-average power-ratio reduction algorithms for multicarrier communications," *IEEE Transactions on Circuits and Systems*, vol. 51, no. 9, pp. 1790–1800, Sept. 2004.

[221] B. S. Krongold and D. L. Jones, "PAR reduction in OFDM via active constellation extension," *IEEE Transactions on Broadcasting*, vol. 49, no. 3, pp. 258–268, Sept. 2003.

[222] J. J. Kroszczyński, "Pulse compression by means of linear-period modulation," *Proc. of the IEEE*, vol. 57, no. 7, pp. 1260–1266, July 1969.

[223] F. R. Kschischang, B. J. Frey, and H.-A. Loeliger, "Factor graphs and the sum-product algorithm," *IEEE Trans. Inform. Theory*, vol. 42, no. 2, pp. 498–519, Feb. 2001.

[224] K. Kumar and G. Caire, "Coding and decoding for the dynamic decode and foward relay protocol," *IEEE Trans. Inform. Theory*, vol. 55, no. 7, pp. 3186–3205, July 2009.

[225] W. K. Lam and R. F. Ormondroyd, "A coherent COFDM modulation system for a time-varying frequency-selective underwater acoustic channel," in *Proc. of the 7th International Conference on Electronic Engineering in Oceanography*, June 1997, pp. 198–203.

[226] E. G. Larsson, G. Liu, J. Li, and G. B. Giannakis, "Joint symbol timing and channel estimation for OFDM based WLANs," *IEEE Commun. Lett.*, vol. 5, no. 8, pp. 325–327, Aug. 2001.

[227] E. G. Larsson and P. Stoica, *Space-Time Block Coding for Wireless Communications*, Cambridge University Press, 2003.

[228] N. Lashkarian and S. Kiaei, "Class of cyclic-based estimators for frequency-offset estimation of OFDM systems," *IEEE Trans. Commun.*, pp. 2139–2149, Dec. 2000.

[229] J. D. Laster and J. H. Reed, "Interference rejection in digital wireless communications," *IEEE Signal Processing Magazine*, vol. 14, no. 3, pp. 37–62, May 1997.

[230] E. Lawrey and C. J. Kikkert, "Peak to average power ratio reduction of OFDM signals using peak reduction carriers," in *Proc. of the Fifth International Symposium on Signal Processing and Its Applications*, 1999, pp. 737–740.

[231] C. Lee, P. Lee, S. Hong, and S. Kim, "Underwater navigation system based on inertial sensor and Doppler velocity log using indirect feedback kalman filter," *Offshore and Polar Engineering*, vol. 15, no. 2, pp. 88–95, June 2005.

[232] G. Leus and P. A. Van Walree, "Multiband OFDM for covert acoustic communications," *IEEE J. Select. Areas Commun.*, vol. 26, no. 9, pp. 1662–1673, Dec. 2008.

[233] B. Li, J. Huang, S. Zhou, K. Ball, M. Stojanovic, L. Freitag, and P. Willett, "MIMO-OFDM for high rate underwater acoustic communications," *IEEE J. Ocean. Eng.*, vol. 34, no. 4, Oct. 2009.

[234] B. Li and M. Stojanovic, "Alamouti space time coded OFDM for underwater acoustic communications," in *Proc. of MTS/IEEE OCEANS Conference*, Sydney, Australia, May 24–27, 2010.

[235] B. Li, S. Zhou, M. Stojanovic, L. Freitag, and P. Willett, "Multicarrier communication over underwater acoustic channels with nonuniform Doppler shifts," *IEEE J. Ocean. Eng.*, vol. 33, no. 2, pp. 198–209, Apr. 2008.

[236] G. Li, A. Cano, J. Gomez-Vilardebo, G. B. Giannakis, and A. I. Perez-Neira, "High-throughput multi-source cooperation via complex-field network coding," *IEEE Trans. Wireless Commun.*, vol. 10, no. 5, pp. 1606–1617, May 2011.

[237] Q. Li, S. Ting, and C.-K. Ho, "A joint network and channel coding strategy for wireless decode-and-forward relay networks," *IEEE Trans. Commun.*, vol. 59, no. 1, pp. 181–193, Jan. 2011.

[238] Q. Li, J. Zhu, X. Guo, and C. N. Georghiades, "Asynchronous co-channel interference suppression in MIMO OFDM systems," in *Proc. of Intl. Conf. on Commun.*, June 2007, pp. 5744–5750.

[239] S.-Y. Li, R. W. Yeung, and N. Cai, "Linear Network Coding," *IEEE Transactions on Information Theory*, vol. 49, no. 2, pp. 371–381, Feb. 2003.

[240] W. Li and J. C. Preisig, "Estimation of rapidly time-varying sparse channels," *IEEE J. Ocean. Eng.*, vol. 32, no. 4, pp. 927–939, Oct. 2007.

[241] X. Li and L. J. Cimini Jr., "Effects of clipping and filtering on the performance of OFDM," *IEEE Commun. Lett.*, vol. 2, no. 5, pp. 131–133, May 1998.

[242] X. Li and J. A Ritcey, "Bit-interleaved coded modulation with iterative decoding," *IEEE Communications Letters*, vol. 1, no. 6, pp. 169–171, Nov. 1997.

[243] Y. (G.) Li and G. Stüber, *OFDM for Wireless Communications*, Springer, 2006.

[244] Z. Li, X.-G. Xia, and B. Li, "Achieving full diversity and fast ML decoding via simple analog network coding for asynchronous two-way relay networks," *IEEE Trans. Commun.*, vol. 57, no. 12, pp. 3672–3681, Dec. 2009.

[245] A. Ligeti, *Single Frequency Network Planning*, Royal Institute of Technology, Stockholm, Sweden, 1999, Ph.D. Disertation.

[246] W. C. Lim, B. Kannan, and T. T. Tjhung, "Joint channel estimation and OFDM synchronisation in multipath fading," in *Proc. of Intl. Conf. on Commun.*, 2004, pp. 983–987.

[247] S. Lin and D. J. Costello, *Error Control Coding*, Prentice Hall, 2nd edition, 2004.

[248] S. Lin, S. Song, L. Lan, L. Zeng, and Y.-Y. Tai, "Constructions of nonbinary quasi-cyclic LDPC codes: a finite field approach," in *Proc. of Inform. Theory and Applications (ITA) Workshop 2006*, UCSD, 2006.

[249] J. Ling and J. Li, "Gibbs-sampler-based semiblind equalizer in underwater acoustic communications," *IEEE Journal of Oceanic Engineering*, vol. 37, no. 1, pp. 1–13, Jan. 2012.

[250] J. Ling, X. Tan, T. Yardibi, J. Li, M. L. Nordenvaad, H. He, and K. Zhao, "On bayesian channel estimation and FFT-based symbol detection in MIMO underwater acoustic communications," *IEEE Journal of Oceanic Engineering*, vol. 39, no. 1, pp. 59–73, Jan. 2014.

[251] J. Ling, T. Yardibi, X. Su, H. He, and J. Lia, "Enhanced channel estimation and symbol detection for high speed multi-input multi-output underwater acoustic communications," *J. Acoust. Soc. Am.*, vol. 125, no. 5, pp. 3067–3078, May 2009.

[252] LinkQuest, Inc., "Underwater acoustic modems," http://www.link-quest.com/html/uwm_hr.pdf.

[253] S. Litsyn, *Peak Power Control in Multicarrier Communications*, Cambridge, 1st edition, 2007.

[254] H. Liu and U. Tureli, "A high-efficiency carrier estimator for OFDM communications," *IEEE Commun. Lett.*, vol. 2, no. 4, pp. 104–106, Apr. 1998.

[255] L. Liu, S. Zhou, and J. Cui, "Prospects and problems of wireless communications for underwater sensor networks," *Wireless Communications and Mobile Computing*, vol. 8, no. 8, pp. 977–994, Oct. 2008.

[256] R. Liu and K. K. Parhi, "Low-latency low-complexity architectures for Viterbi decoders," *IEEE Trans. Circuits and Systems*, vol. 56, no. 10, pp. 2315–2324, Oct. 2009.

[257] Z. Liu and T. C. Yang, "On overhead reduction in time reversed OFDM underwater acoustic communications," *IEEE Journal of Oceanic Engineering*, 2014 (to appear).

[258] H.-A. Loeliger, "An introduction to factor graphs," *IEEE Signal Processing Magazine*, vol. 21, no. 1, pp. 28–41, Jan. 2004.

[259] H.-A. Loeliger, J. Dauwels, J. Hu, S. Korl, L. Ping, and F. Kschischang, "The factor graph approach to model-based signal processing," *Proc. of the IEEE*, vol. 95, no. 6, pp. 1295–1322, June 2007.

[260] B. Lu, G. Yue, and X. Wang, "Performance analysis and design optimization of LDPC coded MIMO OFDM systems," *IEEE Trans. Signal Processing*, vol. 52, no. 2, pp. 348–361, Feb. 2004.

[261] M. Luby, M. Mitzenmacher, A. Shokrollahi, and D. A. Spielman, "Analysis of low density codes and improved designs using irregular graphs," in *Proc. of the 30th Annual ACM Symposium on Theory of Computing*, 1998, pp. 249–258.

[262] M. G. Luby, M. Mitzenmacher, M. A. Shokrollahi, and D. A. Spielman, "Efficient erasure correcting codes," *IEEE Trans. Inform. Theory*, vol. 47, no. 2, pp. 569–584, Feb. 2001.

[263] D. E. Lucani, M. Médard, and M. Stojanovic, "Network coding schemes for underwater networks: The benefits of implicit acknowledgement," in *Proc. of the ACM Intl. Workshop on Underwater Networks (WUWNet)*, 2007, doi:10.1145/1287812.1287819.

[264] M. Luise, M. Marselli, and R. Reggiannini, "Low-complexity blind carrier frequency recovery for OFDM signals over frequency-selective radio channels," *IEEE Trans. Commun.*, vol. 50, no. 7, pp. 1182–1188, 2002.

[265] J. Luo, K. R. Pattipati, and P. K. Willett, "A sliding window PDA for asynchronous CDMA, and a proposal for deliberate asynchronicity," *IEEE Trans. Commun.*, vol. 57, no. 12, pp. 1970–1974, Dec. 2003.

[266] X. Lurton, *An Introduction to Underwater Acoustics: Principles and Applications*, Springer, 2 edition, 2010.

[267] X. Ma, C. Tepedelenlioglu, G. B. Giannakis, and S. Barbarossa, "Non-data-aided carrier offset estimations for OFDM with null subcarriers: Identifiability, algorithms, and performance," *IEEE Journal on Selected Areas in Communications*, vol. 19, no. 12, pp. 2504–2515, Dec. 2001.

[268] E. MacCurdy, *The Notebooks of Leonardo Da Vinci*, Konecky & Konecky, 1st edition, 2002.

[269] D. J. C. Mackay, "Good error-correcting codes based on very sparse matrices," *IEEE Transactions on Information Theory*, vol. 45, no. 2, pp. 399–431, Mar. 1999.

[270] D. J. C. MacKay, *Information Theory, Inference, and Learning Algorithms*, Cambridge University Press, Cambridge, UK, 1st edition, 2003.

[271] K. Mahmood, K. Domrese, P. Carroll, H. Zhou, X. Xu, and S. Zhou, "Implementation and field testing of on-demand asynchronous underwater localization," in *Proc. of Asilomar Conf. on Signals, Systems, and Computers*, Nov. 2013.

[272] Y. Mao and A. H. Banihashemi, "A heuristic search for good LDPC codes at short block lenghts," in *Proc. of Intl. Conf. on Commun.*, Helsinki, Finland, June 2001, vol. 1, pp. 41–44.

[273] S. Mason, R. Anstett, N. Anicette, and S. Zhou, "A broadband underwater acoustic modem implementation using coherent OFDM," in *Proc. of National Conference for Undergraduate Research (NCUR)*, San Rafael, California, Apr. 2007.

[274] S. Mason, C. R. Berger, S. Zhou, and P. Willett, "Detection, synchronization, and Doppler scale estimation with multicarrier waveforms in underwater acoustic communication," *IEEE Journal on Selected Areas in Communications*, vol. 26, no. 9, pp. 1638–1649, Dec. 2008.

[275] J. L. Massey, "Coding and modulation in digital communication," in *Proc. Intl. Zurich Seminar Digital Commun.*, Zurich, Switzerland, Mar. 1974.

[276] M. Medard and A. Sprintson, *Network Coding: Fundamentals and Applications*, Academic Press, 1st edition, 2011.

[277] H. Medwin and C. S. Clay, *Fundamentals of Acoustical Oceanography*, Academic Press, 1st edition, 1997.

[278] G. Mellen, M. Patcher, and J. Raquet, "Closed-form solution for determining emitter location using time difference of arrival measurements," *IEEE Trans. Aerospace and Electronic Systems*, vol. 39, no. 3, pp. 1056–1058, 2003.

[279] D. Mestdagh and P. Spruyt, "A method to reduce the probability of clipping in DMT-based transceivers," *IEEE Transactions on Communications*, vol. 44, no. 10, pp. 1234–1238, Oct. 1996.
[280] H. Meyr, M. Moeneclaey, and S. A. Fechtel, *Digital Communication Receivers, Synchronization, Channel Estimation, and Signal Processing*, Wiley-Interscience, 2nd edition, 1997.
[281] H. Minn, V. K. Bhargava, and K. B. Letaief, "A robust timing and frequency synchronization for OFDM systems," *IEEE Trans. Wireless Commun.*, vol. 2, no. 4, pp. 822–839, July 2003.
[282] H. Minn, V. K. Bhargava, and K. B. Letaief, "A combined timing and frequency synchronization and channel estimation for OFDM," *IEEE Trans. Commun.*, vol. 54, no. 3, pp. 1081–1096, June 2006.
[283] D. Mirza and C. Schurgers, "Motion-aware self-localization for underwater networks," in *Proc. of ACM International Workshop on Underwater Networks (WUWNet)*, San Francisco, California, USA, Sept. 2008, pp. 51–58.
[284] A. F. Molisch, *Wireless Communications*, John Wiley & Sons, Ltd, 2005.
[285] P. H. Moose, "A technique for orthogonal frequency division multiplexing frequency offset correction," *IEEE Trans. Commun.*, vol. 42, no. 10, pp. 2908–2914, Oct. 1994.
[286] M. Morelli and U. Mengali, "An improved frequency offset estimator for OFDM applications," *IEEE Commun. Lett.*, pp. 75–77, Mar. 1999.
[287] A. K. Morozov and J. C. Preisig, "Underwater acoustic communications with multi-carrier modulation," in *Proc. of MTS/IEEE OCEANS Conference*, Boston, MA, Sept. 18–21, 2006.
[288] S. H. Muller and J. B. Huber, "OFDM with reduced peak-to-average power ratio by optimum combination of partial transmit sequences," vol. 33, no. 5, pp. 368–369, Feb. 1997.
[289] S. Myung, K. Yang, and J. Kim, "Quasi-cyclic LDPC codes for fast encoding," *IEEE Transactions on Information Theory*, vol. 51, no. 8, pp. 2894–2901, 2005.
[290] W. Nam, S.-Y. Chung, and Y. H. Lee, "Capacity of the Gaussian two-way relay channel to within 1/2 bit," *IEEE Trans. Inform. Theory*, vol. 56, no. 11, pp. 5488–5494, Nov. 2010.
[291] A. B. Narasimhamurthy, M. K. Banavarand, and C. Tepedelenliogu, *OFDM Systems for Wireless Communications*, Morgan & Claypool Publishers, 2010.
[292] B. Nazer and M. Gastpar, "Computation over multiple-access channels," *IEEE Trans. Inform. Theory*, vol. 53, no. 10, pp. 3498–3516, Oct. 2007.
[293] D. Needell and J. A. Tropp, "CoSaMP: Iterative signal recovery from incomplete and inaccurate samples," *Appl. Comp. Harmonic Anal.*, vol. 26, pp. 301–321, 2009.
[294] R. Negi and J. Cioffi, "Pilot tone selection for channel estimation in a mobile OFDM system," *IEEE Transactions on Consumer Electronics*, vol. 44, no. 3, pp. 1122–1128, Aug. 1998.
[295] P. Nicopolitidis, G. I. Papadimitriou, and A. S. Pomportsis, "Adaptive data broadcasting in underwater wireless networks," *IEEE J. Ocean. Eng.*, vol. 35, no. 3, pp. 623–634, July 2010.
[296] H. Niu, M. Shen, J. A. Ritcey, and H. Liu, "A factor graph approach to iterative channel estimation and LDPC decoding over fading channels," *IEEE Trans. Wireless Commun.*, vol. 4, no. 4, pp. 1345–1350, July 2005.
[297] X. X. Niu, P. C. Ching, and Y. T. Chan, "Wavelet based approach for joint time delay and Doppler stretch measurements," *IEEE Trans. Aerosp. Electron. Syst.*, vol. 35, no. 3, pp. 1111–1119, July 1997.
[298] H. Ochiai, "A novel trellis-shaping design with both peak and average power reduction for OFDM systems," *IEEE Trans. Commun.*, vol. 52, no. 11, pp. 1916–1926, Nov. 2004.
[299] S. Ohno and G. B. Giannakis, "Capacity maximizing MMSE-optimal pilots for wireless OFDM over frequency-selective block Rayleigh-fading channels," *IEEE Trans. Inform. Theory*, vol. 50, no. 9, pp. 2138–2145, Sept. 2004.
[300] R. OŃeill and L. B. Lopes, "Performance of amplitude limited multitone signals," in *Proc. of IEEE Vehicular Technology Conf.*, June 1994, pp. 1675–1679.
[301] E. Panayirci, H. Dogan, and H. V. Poor, "A Gibbs sampling based MAP detection algorithm for OFDM over rapidly varying mobile radio channels," in *Proc. of Global Telecommunications Conference*, Nov. 2009.

[302] A. Papoulis, *Probability, Random Variables and Stochastic processes*, McGraw Hill, 1st edition, 1965.

[303] J. L. Paredes, G. R. Arce, and Z. Wang, "Ultra-wideband compressed sensing: Channel estimation," *IEEE J. Select. Topics Signal Proc.*, vol. 1, no. 3, pp. 383–395, Oct. 2007.

[304] A. Paulraj, R. Nabar, and D. Gore, *Introduction to Space-Time Wireless Communications*, Cambridge University Press, 2003.

[305] C. M. Payne, *Principles of Naval Weapons Systems*, Naval Institute Press, 2nd edition, 2010.

[306] R. Peled and A. Ruiz, "Frequency domain data transmission using reduced computational complexity algorithms," in *Proc. of Intl. Conf. on ASSP*, Denver, CO, 1980, pp. 964–967.

[307] K. Pelekanakis, H. Liu, and M. Chitre, "An algorithm for sparse underwater acoustic channel identification under symmetric α-stable noise," in *Proc. of MTS/IEEE OCEANS Conf.*, Santander, Spain, June 2011.

[308] R.-H. Peng and R.-R. Chen, "Design of nonbinary quasi-cyclic LDPC cycle codes," in *Proc. of ITW'07*, Tahoe City, CA, Sept. 2-6 2007, pp. 13–18.

[309] R.-H. Peng, R.-R. Chen, and B. Farhang-Boroujeny, "Markov Chain Monte Carlo detectors for channels with intersymbol interference," *IEEE Trans. Signal Processing*, vol. 58, no. 4, pp. 2206–2217, Apr. 2010.

[310] D. Pompili and I. Akyildiz, "Overview of networking protocols for underwater wireless communications," *IEEE Communications Magazine*, vol. 47, no. 1, pp. 97–102, Jan. 2009.

[311] K. Popat and R. W. Picard, "Cluster-based probability model and its application to image and texture processing," *IEEE Trans. Image Processing*, vol. 6, no. 2, pp. 268–284, Feb. 1997.

[312] M. B. Porter, "The KRAKEN normal mode program," Technical report, Heat, Light & Sound Research, Inc., La Jolla, CA, 2001.

[313] M. B. Porter and Y.-C. Liu, "Finite-element ray tracing," *Theoretical and Computational Acoustics*, vol. 2, pp. 947–966, 1994.

[314] C. Poulliat, M. P. Fossorier, and D. Declercq, "Using binary images of nonbinary LDPC codes to improve overall performance," in *Proc. of 4th Intl. Symp. on Turbo Codes and Related Topics*, Munich, Germany, April 3–7, 2006.

[315] R. Prasad, *OFDM for Wireless Communication Systems*, Artech House Publishers, 2004.

[316] J. Preisig, "Acoustic propagation considerations for underwater acoustic communications network development," in *Proc. First ACM International Workshop on Underwater Networks (WUWNet)*, Los Angeles, CA, Sept. 2006.

[317] J. G. Proakis and D. K. Manolakis, *Digital Signal Processing*, Prentice Hall, 4th edition, 2006.

[318] J. G. Proakis and M. Salehi, *Digital Communications*, McGraw-Hill, 5th edition, 2008.

[319] F. Qu and L. Yang, "Orthogonal space-time block-differential modulation over underwater acoustic channels," in *Proc. of MTS/IEEE OCEANS Conference*, Vancouver, BC, Canada, Sept. 29–Oct. 4 2007.

[320] F. Qu and L. Yang, "Basis expansion model for underwater acoustic channels?," in *Proc. of MTS/IEEE OCEANS Conf.*, Quèbec City, Quèbec, Sept. 2008.

[321] A. Quazi and W. Konrad, "Underwater acoustic communications," *IEEE Communications Magazine*, vol. 20, no. 2, pp. 24–30, Mar. 1982.

[322] O. Rabaste and T. Chonavel, "Estimation of multipath channels with long impulse response at low SNR via an MCMC method," *IEEE Trans. Signal Processing*, vol. 55, no. 4, pp. 1312–1325, Apr. 2007.

[323] L. R. Rabiner, R. W. Schafer, and C. M. Rader, "The chirp z-transform algorithm," *IEEE Trans. Audio and Eletronics*, vol. 17, no. 2, pp. 86–92, June 1969.

[324] A. Radosevic, R. Ahmed, T. M. Duman, J. G. Proakis, and M. Stojanovic, "Adaptive OFDM modulation for underwater acoustic communications: Design considerations and experimental results," *IEEE Journal of Oceanic Engineering*, 2014, to appear.

[325] A. Rafati, H. Lou, and C. Xiao, "Soft-decision feedback turbo equalization for LDPC-coded MIMO underwater acoustic communications," *IEEE Journal of Oceanic Engineering*, vol. 39, no. 1, pp. 90–99, Jan. 2014.

[326] M. R. Raghavendra and K. Giridhar, "Improving channel estimation in OFDM systems for sparse multipath channels," *IEEE Signal Processing Lett.*, vol. 12, no. 1, pp. 52–55, Jan. 2005.

[327] A. Ramamoorthy, J. Shi, and R. D. Wesel, "On the capacity of network coding for random networks," *IEEE Trans. Inform. Theory*, vol. 51, no. 8, pp. 2878–2885, Aug. 2005.

[328] T. Richardson and R. Urbanke, *Modern Coding Theory*, Cambridge University Press, 2008.

[329] G. Rojo and M. Stojanovic, "Peak-to-average power ratio (PAR) reduction for acoustic OFDM systems," *Marine Technology Society Journal*, vol. 44, no. 4, pp. 30–41, July/August 2010.

[330] D. Rouseff, D. R. Jackson, W. L. J. Fox, C. D. Jones, J. A. Ritcey, and D. R. Dowling, "Underwater acoustic communication by passive-phase conjugation: Theory and experimental results," *IEEE J. Oceanic Eng.*, vol. 26, no. 4, pp. 821–831, Oct. 2001.

[331] S. Roy, T. M. Duman, V. McDonald, and J. G. Proakis, "High rate communication for underwater acoustic channels using multiple transmitters and space-time coding: Receiver structures and experimental results," *IEEE J. Ocean. Eng.*, vol. 32, no. 3, pp. 663–688, July 2007.

[332] L. Rugini, P. Banelli, and G. Leus, "Simple equalization of time-varying channels for OFDM," *IEEE Commun. Lett.*, vol. 9, no. 7, pp. 619–621, July 2005.

[333] H. Sari, G. Karam, and I. Jeanclaude, "Transmission techniques for digital terrestrial TV broadcasting," *IEEE Communications Magazine*, vol. 33, no. 2, pp. 100–109, Feb. 1995.

[334] T. M. Schmidl and D. C. Cox, "Robust frequency and timing synchronization for OFDM," *IEEE Trans. Commun.*, vol. 45, no. 12, pp. 1613–1621, Dec. 1997.

[335] H. Schmidt, "OASES user guide and reference manual," Technical report, Department of Ocean Engineering, Massachusetts Institute of Technology, Mar. 2011.

[336] J. Senne, A. Song, M. Badiey, and K. B. Smith, "Parabolic equation modeling of high frequency acoustic transmission with an evolving sea surface," *J. Acoust. Soc. Am.*, vol. 132, no. 3, pp. 1311–1318, Sept. 2012.

[337] C. E. Shannon, "A mathmatical theory of communication," *Bell System Technical Journal*, vol. 27, pp. 623–656, Oct. 1948.

[338] B. S. Sharif, J. Neasham, O. R. Hinton, and A. E. Adams, "A computationally efficient Doppler compensation system for underwater acoustic communications," *IEEE J. Ocean. Eng.*, vol. 25, no. 1, pp. 52–61, Jan. 2000.

[339] M. Sharp and A. Scaglione, "Application of sparse signal revocery to pilot-assisted channel estimation," in *Proc. of Intl. Conf. on ASSP*, Las Vegas, NV, Apr. 2008.

[340] W. Shieh and I. Djordjevic, *OFDM for optical communications*, Academic Press, 1st edition, 2009.

[341] D. Siegmund, *Sequential Analysis: Tests and Confidence Intervals*, Springer, 1985.

[342] M. K. Simon and M.-S. Alouini, *Digital Communication over Generalized Fading Channels: A Unified Approach to the Performance Analysis*, John Wiley & Sons, Inc., 2000.

[343] A. Singer, J. Nelson, and S. Kozat, "Signal processing for underwater acoustic communications," *IEEE Communications Magazine*, vol. 47, no. 1, pp. 90–96, Jan. 2009.

[344] K. B. Smith, "Convergence, stability, and variability of shallow water acoustic predictions using a split-step fourier parabolic equation model," *J. Comp. Acoust.*, vol. 9, no. 1, pp. 243–285, 2001.

[345] Sonardyne International Ltd., "Wideband acoustic data logger," http://www.sonardyne.com.

[346] A. Song, A. Abdi, M. Badiey, and P. Hursky, "Experimental demonstration of underwater acoustic communication by vector sensors," *IEEE Journal of Oceanic Engineering*, vol. 36, no. 3, pp. 454-461, July 2011.

[347] A. Song and M. Badiey, "Time reversal multiple-input/multiple-output acoustic communication enhanced by parallel interference cancellation," *Journal of Acoustical Society of America*, vol. 131, no. 1, pp. 281–291, Jan. 2012.

[348] A. Song, M. Badiey, V. McDonald, and T. C. Yang, "Time reversal receivers for high rate multiple-input/multiple-output communication," *IEEE J. Ocean. Eng.*, vol. 34, no. 4, pp. 525–538, Oct. 2011.

[349] A. Song, M. Badiey, A. E. Newhall, J. F. Lynch, H. A. DeFerrari, and B. G. Katsnelson, "Passive time reversal acoustic communications through shallow-water internal waves," *IEEE J. Ocean. Eng.*, vol. 35, no. 4, pp. 756–765, Oct. 2010.

[350] A. Song, M. Badiey, H.-C. Song, W. S. Hodgkiss, M. B. Porter, and the KauaiEx Group, "Impact of ocean variability on coherent underwater acoustic communications during the Kauai experiment (KauaiEx)," *J. Acoust. Soc. Am.*, vol. 123, no. 2, pp. 856–865, 2008.

[351] H. Song and J. R. Cruz, "Reduced-complexity decoding of Q-ary LDPC codes for magnetic recording," *IEEE Trans. Magn.*, vol. 39, pp. 1081–1087, Mar. 2003.

[352] H. C. Song, W. S. Hodgkiss, W. A. Kuperman, T. Akal, and M. Stevenson, "Multiuser communications using passive time reversal," *IEEE J. Oceanic Eng.*, vol. 32, no. 4, pp. 915–926, Oct. 2007.

[353] H. C. Song, W. S. Hodgkiss, W. A. Kuperman, T. Akal, and M. Stevenson, "High-frequency acoustic communications achieving high bandwidth efficiency," *J.Acoust. Soc. Am.*, vol. 126, no. 2, pp. 561–563, Aug. 2009.

[354] H. C. Song, W. S. Hodgkiss, W. A. Kuperman, W. J. Higley, K. Raghukumar, and T. Akal, "Spatial diversity in passive time reversal communications," *J. Acoust. Soc. Am.*, vol. 120, no. 4, pp. 2067–2076, Apr. 2006.

[355] H. C. Song, W. S. Hodgkiss, W. A. Kuperman, M. Stevenson, and T. Akal, "Improvement of time-reversal communications using adaptive channel equalizers," *IEEE J. Ocean. Eng.*, vol. 31, no. 2, pp. 487–496, Apr. 2006.

[356] H. C. Song, W. A. Kuperman, and W. S. Hodgkiss, "Basin-scale time reversal communications," *J.Acoust. Soc. Am.*, vol. 125, pp. 212–217, 2009.

[357] H. C. Song, P. Roux, W. S. Hodgkiss, W. A. Kuperman, T. Akal, and M. Stevenson, "Multiple-input/multiple-output coherent time reversal communications in a shallow water acoustic channel," *IEEE J. Ocean. Eng.*, vol. 31, no. 1, pp. 170–178, Jan. 2006.

[358] H. W. Sorenson, *Kalman Filtering: Theory and Application*, IEEE Press, 1st edition, 1985.

[359] E. Sozer and M. Stojanovic, "Reconfigurable acoustic modem for underwater sensor networks," in *Proc. First ACM International Workshop on Underwater Networks*, Los Angeles, CA, Sept. 2006.

[360] M. Stojanovic, "Recent advances in high-speed underwater acoustic communications," *IEEE Journal of Oceanic Engineering*, vol. 121, no. 2, pp. 125–136, Apr. 1996.

[361] M. Stojanovic, "Low complexity OFDM detector for underwater channels," in *Proc. of MTS/IEEE OCEANS Conference*, Boston, MA, Sept. 18–21, 2006.

[362] M. Stojanovic, "On the relationship between capacity and distance in an underwater acoustic communication channel," in *Proc. First ACM International Workshop on Underwater Networks (WUWNet)*, Los Angeles, CA, Sept. 2006.

[363] M. Stojanovic, "OFDM for underwater acoustic communications: Adaptive synchronization and sparse channel estimation," in *Proc. of Intl. Conf. on ASSP*, Las Vegas, NV, Apr. 2008.

[364] M. Stojanovic, "MIMO OFDM over underwater acoustic channels," in *Proc. of Asilomar Conf. on Signals, Systems, and Computers*, Nov. 2009.

[365] M. Stojanovic, "A method for differentially coherent detection of OFDM signals on doppler-distorted channels," in *IEEE Sensor Array and Multichannel Signal Processing Workshop (SAM)*, Oct. 2010.

[366] M. Stojanovic, J. Catipovic, and J. G. Proakis, "Adaptive multichannel combining and equalization for underwater acoustic communications," *J. Acoust. Soc. Am.*, vol. 94, no. 3, pp. 1621–1631, Sept. 1993.

[367] M. Stojanovic, J. Catipovic, and J. G. Proakis, "Phase-coherent digitial communications for underwater acoustic channels," *IEEE J. Ocean. Eng.*, vol. 19, no. 1, pp. 100–111, Jan. 1994.

[368] M. Stojanovic and J. Preisig, "Underwater acoustic communication channels: Propagation models and statistical characterization," *IEEE Communications Magazine*, vol. 47, no. 1, pp. 84–89, Jan. 2009.

[369] M. Stojanovic and Z. Zvonar, "Multichannel processing of broad-band multiuser communication signals in shallow water acoustic channels," *IEEE J. Oceanic Eng.*, vol. 21, no. 2, pp. 156–166, Apr. 1996.

[370] R. Stuart, "Acoustic digital spread spectrum: An enabling technology," *Sea Technology*, vol. 46, no. 10, Oct. 2005.

[371] Y. Su, Y. Zhang, S. Le, H. Mo, L. Wei, Y. Huang, Z. Peng, and J.-H. Cui, "A versatile lab testbed for underwater sensor networks," in *Proc. of IEEE/MTS OCEANS Conference*, San Diego, CA, Sept. 23–26, 2013.

[372] H. Sun, W. Shen, Z.-H. Wang, S. Zhou, X. Xu, and Y. Chen, "Joint carrier frequency offset and impulsive noise estimation for underwater acoustic OFDM with null subcarriers," in *Proc. of IEEE/MTS Oceans Conference*, Hampton Roads, Virginia, Oct. 14–19, 2012.

[373] H. P. Tan, R. Diamant, W. K. G. Seah, and M. Waldmeyer, "A survey of techniques and challenges in underwater localization," *Ocean Engineering*, vol. 38, no. 14–15, pp. 1663–1676, Oct. 2011.

[374] J. Tao, J. Wu, Y. R. Zheng, and C. Xiao, "Enhanced MIMO LMMSE turbo equalization: algorithm, simulations and undersea experimental results," *IEEE Transactions on Signal Processing*, vol. 59, no. 8, pp. 3813–3823, Aug. 2011.

[375] J. Tao and Y. R. Zheng, "Turbo detection for MIMO-OFDM underwater acoustic communications," *International Journal of Wireless Information Networks*, vol. 20, no. 1, pp. 27–38, Mar. 2013.

[376] J. Tao, Y. R. Zheng, C. Xiao, T. C. Yang, and W.-B. Yang, "Channel equalization for single carrier MIMO underwater acoustic communications," *EURASIP J. on Advanced Signal Proc.*, 2010, doi:10.1155/2010/281769.

[377] J. Tao, Y. R. Zheng, C. Xiao, and T.C. Yang, "Robust MIMO underwater acoustic communications using Turbo block decision-feedback equalization," *IEEE Journal of Oceanic Engineering*, vol. 35, no. 4, pp. 948–960, Oct. 2010.

[378] G. Tauböck, F. Hlawatsch, D. Eiwen, and H. Rauhut, "Compressive estimation of doubly selective channels in multicarrier systems: Leakage effects and sparsity-enhancing processing," *IEEE J. Select. Topics Signal Proc.*, vol. 4, no. 2, pp. 255–271, Apr. 2010.

[379] I. E. Telatar, "Capacity of multi-antenna Gaussian channels," *Bell Laboratories Technical Memorandum*, 1995.

[380] J. Tellado, *Multicarrier Modulation with Low PAR: Applications to DSL and Wireless*, Springer, 1st edition, 2000.

[381] J. Tellado and J. M. Cioffi, "Efficient algorithms for reducing PAR in multicarrier systems," in *Proc. of IEEE Intl. Symp. on Information Theory*, Aug. 1998.

[382] C. Tellambura, "Improved phase factor computation for the PAR reduction of an OFDM signal using PTS," *IEEE Commun. Lett.*, vol. 5, no. 4, pp. 135–137, Apr. 2001.

[383] T. A. Thomas and F. W. Vook, "Asynchronous interference suppression in broadband cyclic-prefix communications," *Proc. of IEEE Wireless Communications and Networking Conference*, vol. 1, pp. 568–572, Mar. 2003.

[384] T. Tian, C. Jones, and J. Villasenor, "Rate-compatible low-density parity check codes," in *Proc. of ISIT*, Chicago, USA, 2004.

[385] R. Tibshirani, "Regression shrinkage and selection via the lasso," *J. Roy. Statist. Soc., ser. B*, vol. 58, no. 1, pp. 267–288, 1996.

[386] A. M. Tonello, "Asynchronous multicarrier multiple access: Optimal and sub-optimal detection and decoding," *Bell Labs Tech. J.*, vol. 7, no. 3, pp. 191–217, 2003.

[387] J. A. Tropp and A. C. Gilbert, "Signal recovery from random measurements via orthogonal matching pursuit," *IEEE Trans. Inform. Theory*, vol. 53, no. 12, pp. 4655–4666, Dec. 2007.

[388] J. A. Tropp and S. J. Wright, "Computational methods for sparse solution of linear inverse problems," *Proc. of the IEEE*, vol. 98, no. 6, pp. 948–958, June 2010.

[389] D. Tse and P. Viswanath, *Fundamentals of Wireless Communication*, Cambridge University Press, 2005.

[390] K. Tu, T. Duman, J. Proakis, and M. Stojanovic, "Cooperative MIMO-OFDM communications: Receiver design for Doppler-distorted underwater acoustic channels," in *Proc. of 44th Asilomar Conf. Signals, Systems, and Computers*, Pacific Grove, CA, Nov. 2010.

[391] K. Tu, T. Duman, M. Stojanovic, and J. Proakis, "Multiple-resampling receiver design for OFDM over Doppler-distorted underwater acoustic channels," *IEEE J. Ocean. Eng.*, vol. 38, no. 2, pp. 333–346, Apr. 2013.

[392] K. Tu, D. Fertonani, T. M. Duman, M. Stojanovic, J. G. Proakis, and P. Hursky, "Mitigation of intercarrier interference for OFDM over time-varying underwater acoustic channels," *IEEE Journal of Oceanic Engineering*, vol. 36, no. 2, pp. 156–171, Apr. 2011.

[393] M. Tüchler, A. C. Singer, and R. Koetter, "Minimum mean squared error equalization using a priori information," *IEEE Trans. Signal Processing*, vol. 50, no. 3, pp. 673–683, Mar. 2002.

[394] U. Tureli and H. Liu, "A high-efficiency carrier estimator for OFDM communications," *IEEE Communications Letters*, vol. 2, no. 4, pp. 104–106, Apr. 1998.

[395] U. Tureli, H. Liu, and M. D. Zoltowski, "OFDM blind carrier offset estimation: ESPRIT," *IEEE Transactions on Communications*, vol. 48, no. 9, pp. 1459–1461, Sept. 2000.

[396] G. Ungerboeck, "Channel coding with multilevel/phase signals," *IEEE Trans. Inform. Theory*, vol. 28, no. 1, pp. 55–67, Jan. 1982.

[397] R. Urich, *Principles of Underwater Sound*, 3rd ed. New York: McGraw-Hill, 1983.

[398] M. Vajapeyam, S. Vedantam, U. Mitra, J. C. Preisig, and M. Stojanovic, "Distributed spacetime cooperative schemes for underwater acoustic communications," *IEEE J. Ocean. Eng.*, vol. 33, no. 4, pp. 489–501, Oct. 2008.

[399] J.-J. van de Beek, M. Sandell, and P. O. Borjesson, "ML estimation of time and frequency offset in OFDM systems," *IEEE Transactions on Signal Processing*, vol. 45, no. 7, pp. 1800–1805, July 1997.

[400] E. van den Berg and M. Friedlander, "Probing the Pareto frontier for basis pursuit solutions," *SIAM J. Sci. Comput.*, vol. 31, no. 2, pp. 890–912, 2008.

[401] P. Van Eetvelt, G. Wade, and M. Tomlinson, "Peak to average power reduction for OFDM schemes by selective scrambling," *Electronics Letters*, vol. 32, no. 21, pp. 1963–1964, Oct. 1996.

[402] H. Van Trees, *Detection, Estimation, and Modulation Theory*, John Wiley & Sons, Inc., New York, 1st edition, 1968.

[403] H. Van Trees, *Optimum Array Processing,* Detection, Estimation, and Modulation Theory (Part IV). John Wiley & Sons, Inc., New York, 1st edition, 2002.

[404] P. van Walree and R. Otnes, "Wideband properties of underwater acoustic communication channels," in *Proc. of the Workshop on Underwater Communications: Channel Modelling & Validation*, Italy, Sept. 2012.

[405] P. A. van Walree and R. Otnes, "Ultrawideband underwater acoustic communication channels," *IEEE Journal of Oceanic Engineering*, vol. 38, no. 4, pp. 678–688, Oct. 2013.

[406] S. Verdu, *Multiuser Detection*, Cambridge University Press, New York, 1998.

[407] L. F. M. Vieira, U. Lee, and M. Gerla, "Phero-trail: a bio-inspired location service for mobile underwater sensor networks," *IEEE Journal on Selected Areas in Communications*, vol. 28, no. 4, pp. 553–563, Feb. 2010.

[408] A. J. Viterbi, "An intuitive justification and a simplified implementation of the MAP decoder for convolutional codes," *IEEE J. Select. Areas Commun.*, vol. 16, no. 2, pp. 260–264, Feb. 1998.

[409] B. Vucetic and J. Yuan, *Space-Time Coding*, John Wiley & Sons, Ltd, 2003.

[410] A. Wald, "Sequential tests of statistical hypotheses," *Annals of Mathematical Statistics*, vol. 16, no. 2, pp. 117–186, June 1945.

[411] H. Wan, R.-R. Chen, J. W. Choi, A. Singer, J. Preisig, and B. Farhang-Boroujeny, "Markov chain Monte Carlo detection for frequency-selective channels using list channel estimates," *IEEE J. Select. Topics Signal Proc.*, vol. 5, no. 8, pp. 1537–1547, Dec. 2011.

[412] L. Wan, Z.-H. Wang, S. Zhou, TC Yang, and Z. Shi, "Performance comparison of Doppler scale estimation methods for underwater acoustic OFDM," *Journal of Electrical and Computer Engineering, Special Issue on Underwater Communications and Networks*, 2012, doi:10.1155/2012/703243.

[413] L. Wan, H. Zhou, X. Xu, Y. Huang, S. Zhou, Z. Shi, and J.-H. Cui, "Field tests of adaptive modulation and coding for underwater acoustic OFDM," in *Proc. of the 8th ACM International Conference on UnderWater Networks and Systems (WUWNet)*, Kaohsiung, Taiwan, Nov. 11–13, 2013.

[414] H. Wang, T. Kirubarajan, and Y. Bar-Shalom, "Precision large scale air traffic surveillance using IMM estimator with assignment," *IEEE Transactions on Aerospace and Electronic Systems*, pp. 255–266, Jan. 1999.

[415] L. Wang, J. Tao, C. Xiao, and T. C. Yang, "Low-complexity turbo detection for single-carrier LDPC coded MIMO underwater acoustic communications," *Wireless Communications and Mobile Computing*, vol. 13, no. 4, pp. 439–450, Mar. 2013.

[416] L. Wang, J. Tao, and Y. R. Zheng, "Single-carrier frequency-domain turbo equalization without cyclic prefix or zero padding for underwater acoustic communications," *J. Acoust. Soc. Am.*, vol. 132, no. 6, pp. 3809–3817, Dec. 2012.

[417] T. Wang and G. B. Giannakis, "Mutual information jammer-relay games," *IEEE Trans. Information Forensics and Security*, vol. 3, no. 2, pp. 290–303, June 2008.

[418] Z. Wang and G. B. Giannakis, "Wireless multicarrier communications: Where Fourier meets Shannon," *IEEE Signal Processing Magazine*, vol. 17, no. 3, pp. 29–48, May 2000.

[419] Z.-H. Wang, J. Huang, S. Zhou, and Z. Wang, "Iterative receiver processing for OFDM modulated physical-layer network coding in underwater acoustic channels," *IEEE Trans. Commun.*, vol. 61, no. 2, pp. 541–553, Feb. 2013.

[420] Z.-H. Wang, S. Zhou, J. Catipovic, and J. Huang, "OFDM in deep water acoustic channels with extremely long delay spread," in *Proc. of the ACM Intl. Workshop on Underwater Networks (WUWNet)*, Woods Hole, MA, Sept. 2010.

[421] Z.-H. Wang, S. Zhou, J. Catipovic, and J. Huang, "Factor-graph based joint IBI/ICI mitigation for OFDM in underwater acoustic multipath channels with long-separated clusters," *IEEE J. Ocean. Eng.*, vol. 37, no. 4, pp. 680–694, Oct. 2012.

[422] Z.-H. Wang, S. Zhou, J. Catipovic, and P. Willett, "Parameterized cancellation of partial-band partial-block-duration interference for underwater acoustic OFDM," *IEEE Trans. Signal Processing*, vol. 60, no. 4, pp. 1782–1795, Apr. 2012.

[423] Z.-H. Wang, S. Zhou, J. Catipovic, and P. Willett, "Asynchronous multiuser reception for OFDM in underwater acoustic communications," *IEEE Trans. Wireless Commun.*, vol. 12, no. 3, pp. 1050–1061, Mar. 2013.

[424] Z.-H. Wang, S. Zhou, G. B. Giannakis, C. R. Berger, and J. Huang, "Frequency-domain oversampling for zero-padded OFDM in underwater acoustic communications," *IEEE J. Ocean. Eng.*, vol. 37, no. 1, pp. 14–24, Jan. 2012.

[425] Z.-H. Wang, S. Zhou, J. C. Preisig, K. R. Pattipati, and P. Willett, "Per-cluster-prediction based sparse channel estimation for multicarrier underwater acoustic communications," in *Proc. of IEEE Intl. Conf. on Signal Processing, Communications and Computing*, Xi'an, China, Sept. 2011.

[426] Z.-H. Wang, S. Zhou, J.C. Preisig, K. R. Pattipati, and P. Willett, "Clustered adaptation for estimation of time-varying underwater acoustic channels," *IEEE Trans. Signal Processing*, vol. 60, no. 6, pp. 3079–3091, June 2012.

[427] Z.-H. Wang, S. Zhou, Z. Wang, B. Wang, and P. Willett, "Dynamic block-cycling over a linear network in underwater acoustic channels," in *Proc. of the 7th ACM International Conference on UnderWater Networks and Systems (WUWNet)*, Los Angeles, CA, Nov. 5–6, 2012.

[428] J. Ward, M. Fitzpatrick, N. DiMarzio, D. Moretti, and R. Morrissey, "New algorithms for open ocean marine mammal monitoring," in *Proc. of MTS/IEEE OCEANS Conf.*, Providence, RI, USA, Sept. 2000, pp. 1749–1752.

[429] S. E. Webster, R. M. Eustice, H. Singh, and L. L. Whitcomb, "Preliminary deep water results in single-beacon one-way-travel-time acoustic navigation for underwater vehicles," in *IEEE International Conference on Robotics and Automation*, Apr. 2007.

[430] L. Wei, Z. Peng, H. Zhou, J.-H. Cui, S. Zhou, Z. Shi, and J. O'Donnell, "Long Island Sound testbed and experiments," in *Proc. of IEEE/MTS OCEANS conference*, San Diego, CA, Sept. 23–26, 2013.

[431] S. Weinstein, "The history of orthogonal frequency-division multiplexing," *IEEE Communications Magazine*, vol. 47, no. 11, pp. 26–35, Nov. 2009.

[432] S. B. Weinstein and P. M. Ebert, "Data transmission by frequency-division multiplexing using the discrete Fourier transform," *IEEE Transactions on Communication Technology*, vol. 19, no. 5, pp. 628–634, Oct. 1971.

[433] S. B. Wicker, *Error Control Systems for Digital Communication and Storage*, Prentice-Hall, Inc., 1995.

[434] J. Wills, W. Ye, and J. Heidemann, "Low-power acoustic modem for dense underwater sensor networks," in *Proc. of WUWNet*, Los Angeles, CA, Sept. 2006.

[435] S. J. Wright, R. D. Nowak, and M. A. Figueiredo, "Sparse reconstruction by separable approximation," *IEEE Trans. Signal Processing*, vol. 57, no. 7, pp. 2479–2493, July 2009.

[436] C.-J. Wu and D. W. Lin, "Sparse channel estimation for OFDM transmission based on representative subspace fitting," in *Proc. of IEEE Vehicular Technology Conf.*, Stockholm, Sweden, May 2005.

[437] C.-J. Wu and D. W. Lin, "A group matching pursuit algorithm for sparse channel estimation for OFDM transmission," in *Proc. of Intl. Conf. on ASSP*, Toulouse, France, May 2006.

[438] D. Wubben and Y. Lang, "Generalized sum-product algorithm for joint channel decoding and physical-layer network coding in two-way relay systems," in *Proc. of Global Telecommunications Conference*, Miami, FL, Dec. 6–10, 2010.

[439] H. Wymeersch, *Iterative Receiver Design*, Cambridge University Press, 1st edition, 2007.

[440] H. Wymeersch, H. Steendam, and M. Moeneclaey, "Log-domain decoding of LDPC codes over GF(q)," in *Proc. of Intl. Conf. on Commun.*, Paris, France, June 2004, pp. 772–776.

[441] T. Xing and J. Li, "Cooperative positioning in underwater sensor networks," *IEEE Transactions on Signal Processing*, vol. 58, no. 11, pp. 5860–5871, Nov. 2010.

[442] X. Xu, Z.-H. Wang, S. Zhou, and L. Wan, "Parameterizing both path amplitude and delay variations of underwater acoustic channels for block decoding of orthogonal frequency division multiplexing," *J. Acoust. Soc. Am.*, vol. 31, no. 6, pp. 4672–4679, 2012.

[443] X. Xu, S. Zhou, A. Morozov, and J. Preisig, "Per-survivor processing for underwater acoustic direct-sequence spread spectrum communications," *J. Acoust. Soc. Am.*, vol. 133, no. 5, May 2013.

[444] H. Yan, L. Wan, S. Zhou, Z. Shi, J.-H. Cui, J. Huang, and H. Zhou, "DSP based receiver implementation for OFDM acoustic modems," *Elsevier J. of Physical Commun.*, 2011, doi:10.1016/j.phycom.2011.09.001.

[445] H. Yan, S. Zhou, Z. Shi, J.-H. Cui, L. Wan, J. Huang, and H. Zhou, "DSP implementation of SISO and MIMO OFDM acoustic modems," in *Proc. of MTS/IEEE OCEANS Conference*, Sydney, Australia, May 24–27, 2010.

[446] H. Yan, S. Zhou, Z. Shi, and B. Li, "A DSP implementation of OFDM acoustic modem," in *Proc. of the ACM Intl. Workshop on Underwater Networks (WUWNet)*, Montréal, Québec, Canada, Sept. 2007.

[447] Z. Yan, J. Huang, and C. He, "Implementation of an OFDM underwater acoustic communication system on an underwater vehicle with multiprocessor structure," *Frontiers of Electrical and Electronic Engineering in China*, vol. 2, no. 2, pp. 151–155, 2007.

[448] T. C. Yang, "Phase-coherent underwater acoustic communications: Building a high-data-rate wireless communication network in the ocean," *NRL Review*, pp. 53–63, 2001.

[449] T. C. Yang, "Temporal resolutions of time-reversal and passive-phase conjugation for underwater acoustic communications," *IEEE Journal of Oceanic Engineering*, vol. 28, no. 2, pp. 229–245, Apr. 2003.

[450] T. C. Yang, "Differences between passive-phase conjugation and decision-feedback equalizer for underwater acoustic communications," *IEEE Journal of Oceanic Engineering*, vol. 29, no. 2, pp. 472–487, Apr. 2004.

[451] T. C. Yang, "Correlation-based decision-feedback equalizer for underwater acoustic communications," *IEEE Journal of Oceanic Engineering*, vol. 30, no. 4, pp. 865–880, Oct. 2005.

[452] T. C. Yang, "Underwater telemetry method using doppler compensation," U.S. Patent 6512720, Jan. 2003.

[453] K. Yao, F. Lorenzelli, and C.-E. Chen, *Detection and Estimation for Communication and Radar Systems*, Cambridge University Press, 1st edition, 2013.

[454] Y. Yao and G. B. Giannakis, "Blind carrier frequency offset estimation in SISO, MIMO, and multiuser OFDM systems," *IEEE Trans. Commun.*, pp. 173–183, Jan. 2005.

[455] S. Yerramalli and U. Mitra, "Optimal resampling of OFDM signals for multiscale-multilag underwater acoustic channels," *IEEE Journal of Oceanic Engineering*, vol. 36, no. 1, pp. 126–138, Jan. 2011.

[456] R. W. Yeung, *Information Theory and Network Coding*, Springer, 1st edition, 2008.

[457] A. G. Zajic and G. F. Edelmann, "Feasibility study of underwater acoustic communications between buried and bottom-mounted sensor network nodes," *IEEE Journal of Oceanic Engineering*, vol. 38, no. 1, pp. 109–116, Jan. 2013.

[458] Y. V. Zakharov and V. P. Kodanev, "Multipath-Doppler diversity of OFDM signals in an underwater acoustic channel," in *IEEE International Conference on Acoustics, Speech, and Signal Processing*, June 2000, vol. 5, pp. 2941–2944.

[459] E. Zehavi, "8-PSK trellis codes for a Rayleigh channel," *IEEE Transactions on Communications*, vol. 40, no. 5, pp. 873–884, May 1992.

[460] J. Zhang and Y. R. Zheng, "Bandwidth-efficient frequency-domain equalization for single carrier multiple-input multiple-output underwater acoustic communications," *J. Acoust. Soc. Am.*, vol. 128, no. 5, pp. 2910–2919, Nov. 2010.

[461] J. Zhang and Y. R. Zheng, "Frequency-domain turbo equalization with soft successive interference cancellation for single carrier MIMO underwater acoustic communications," *IEEE Trans. Wireless Commun.*, vol. 10, no. 9, pp. 2872–2882, Sep. 2011.

[462] S. Zhang and S.-C. Liew, "Channel coding and decoding in a relay system operated with physical-layer network coding," *IEEE J. Select. Areas Commun.*, vol. 27, no. 5, pp. 788–796, June 2009.

[463] S. Zhang, S. C. Liew, and P. P. Lam, "Hot topic: Physical layer network coding," in *Proc. of 12th MobiCom*, New York, NY, 2006, pp. 358–365.

[464] W. Zhang, X. Ma, B. Gestner, and D. Anderson, "Designing low-complexity equalizers for wireless systems," *IEEE Communications Magazine*, vol. 47, no. 1, pp. 56–62, Jan. 2009.

[465] W. Zhang, M. Stojanovic, and U. Mitra, "Analysis of a linear multihop underwater acoustic network," *IEEE J. Ocean. Eng.*, vol. 35, no. 4, pp. 961–970, Oct. 2010.

[466] Y. Zhang, "User's guide for YALL1: Your algorithms for l1 optimization," Caam tech. rep. tr09-17, Rice Univ., Houston, TX, June 2009.

[467] L. Zheng and D. Tse, "Diversity and multiplexing: A fundamental tradeoff in multiple-antenna channels," *IEEE Transactions on Information Theory*, vol. 49, no. 5, pp. 1073–1096, May 2003.

[468] Y. R. Zheng, C. Xiao, T. C. Yang, and W.-B. Yang, "Frequency-domain channel estimation and equalization for shallow-water acoustic communications," *Elsevier J. of Physical Commun.*, vol. 3, pp. 48–63, Mar. 2010.

[469] W. Zhou, Z.-H. Wang, J. Huang, and S. Zhou, "Blind CFO estimation for Zero-Padded OFDM over underwater acoustic channels," in *Proc. of MTS/IEEE Oceans Conf.*, KONA, Hawaii, Sept. 2011.

[470] Z. Zhou, J.-H. Cui, and S. Zhou, "Efficient localization for large-scale underwater sensor networks," *Ad Hoc Networks*, vol. 8, pp. 267–279, 2010.

[471] Y. Zhu, D. Guo, and M. L. Honig, "A message-passing approach for joint channel estimation, interference mitigation, and decoding," *IEEE Trans. Wireless Commun.*, vol. 8, no. 12, pp. 6008–6018, Dec. 2009.

[472] Z. Zvonar, D. Brady, and J. Catipovic, "An adaptive decentralized multiuser receiver for deep-water acoustic telemetry," *J. Acoust. Soc. Am.*, vol. 101, no. 4, pp. 2384–2387, Apr. 1997.

Index

OFDM for Underwater Acoustic Communications, First Edition. Shengli Zhou and Zhaohui Wang.
© 2014 John Wiley & Sons, Ltd. Published 2014 by John Wiley & Sons, Ltd.